NANOPARTICLES:
NEW RESEARCH

NANOPARTICLES: NEW RESEARCH

SIMONE LUCA LOMBARDI
EDITOR

Nova Science Publishers, Inc.
New York

For permission to use material from this book please contact us:
Telephone 631-231-7269; Fax 631-231-8175
Web Site: http://www.novapublishers.com

LIBRARY OF CONGRESS CATALOGING-IN-PUBLICATION DATA

Lombardiá, Simone Luca.
Nanoparticles : new research / Simone Luca Lombardiá.
p. cm.
ISBN 978-1-60456-704-5 (hardcover)
1. Nanostructures. 2. Nanostructured materials. 3. Nanoparticles. I. Title.
QC176.8.N35L66 2008
620'.5--dc22 2008016990

Published by Nova Science Publishers, Inc. ✦ New York

CONTENTS

PREFACE

This book provides leading edge research from around the globe in the field of nanoparticles which is a small particle with at least one dimension less than 100 nm. A nanoparticle is an amorphous or semicrystalline zero dimensional (0D) nano structure with at least one dimension between 10 and 100nm and a relatively large (¡Ý 15%) size dispersion.Nanoparticle research is currently an area of intense scientific research, due to a wide variety of potential applications in biomedical, optical, and electronic fields.

The process of nanoparticle formation under laser ablation of solids in liquids is described in Chapter 1. Critical parameters are discussed that govern the properties of nanoparticles ejected into the surrounding liquid. These parameters are laser wavelength, pulse duration, interaction of individual nanoparticles with laser beam inside the liquid. A review of previous results is presented on the properties of nanoparticles of noble metals. Recent data on laser-assisted generation of other metals is given, including the formation of alloyed nanoparticles. Micro- and nanostructuring of the target upon its laser ablation in liquid environment is discussed. Examples are given of the influence of the surrounding liquid on the chemical composition of generated nanoparticles. The laser control over the size distribution of nanoparticles in liquids is demonstrated either by spatial profiling of laser beam intensity or proper tuning of laser wavelength into plasmon resonance of nanoparticles. Recent results are given on excitation of high energy levels of media under laser exposure to laser pulses of picosecond range of duration.

Carbon nanoparticles, such as fullerenes, nanotubes and nanodiamonds, have been considered as promising building blocks in nanotechnology for broadscale nanostuctured devices and materials, e.g., microchips, nanorobots and biosensors, carriers for controlled drug and gene delivery or tracers for novel imaging technologies. Soon after macroscopic production was established, fullerenes, as donors and acceptors of electrons, started to be targeted for use in photodynamic therapy against tumor and microbial cells and also for quenching oxygen radicals. It has been discovered that nanodiamonds can also act as antioxidant and anti-inflammatory agents.

The interaction of carbon nanoparticles with cells and tissues has been investigated mainly using suspensions of these particles in cell culture media or other fluids. Relatively little is known about the influence of layers consisting of carbon nanoparticles on cell-substrate adhesion. Carbon nanoparticles can be advantageously used for surface modification of various artificial materials developed for the construction of tissue replacements and other body implants. In the form of films deposited on the surface of implants, carbon nanoparticles can not only improve the mechanical and other physical properties of a body implant, but can also enhance the attractiveness of the implant for cell colonization. The latter effect is probably mediated by the surface nanostructure of the films, which, at least to a certain degree, mimics the architecture of physiological cell adhesion substrates, such as extracellular matrix molecules. On nanostructured substrates, the cell adhesion-mediating molecules are adsorbed in the appropriate spectrum, amount and spatial conformation that make specific sites on these molecules (e.g. amino acid sequences like RGD) accessible to cell adhesion receptors. In addition, the nanostructured surfaces have been reported to enhance the adsorption of vitronectin, preferred by osteoblasts for their adhesion. The electrical conductivity of carbon nanotube films has also had beneficial effects on the growth and maturation of osteoblasts. Thus, the nanoparticles deposited on the bone-anchoring parts of bone, joint or dental replacements can improve the integration of these devices with the surrounding bone tissue. In addition, carbon nanoparticles admixed into polymers, designed for the fabrication of three-dimensional scaffolds for bone tissue engineering, could decorate the walls of the pores in these materials, and thus promote the ingrowth of bone cells. The interaction of cells with substrates modified with carbon nanoparticles could be further intensified by functionalizing these particles with various chemical functional groups or biomolecules, including KRSR-containing adhesion oligopeptides, recognized preferentially by osteoblasts.

This interdisciplinary review, based mainly on their own results, deals with the interaction of bone-derived cells with glass, polymeric and silicon substrates modified with fullerenes C_{60}, nanotubes, nanocrystalline diamond as well as other carbon allotropes, such as amorphous hydrogenated carbon or pyrolytic graphite. In general, these modifications have resulted in improved adhesion, growth, viability and osteogenic differentiation of these cells. The use of carbon and/or metal-carbon composite nanoparticles for the construction of patterned surfaces for regionally-selective adhesion and directed growth of cells is also discussed in Chapter 2.

Chapter 3 reviews the novel synthesis of inorganic nanoparticles stabilized by an organic shell layer with investigation into their specific characteristics. How to design organic-inorganic hybrid nanoparticles in chemical solution process effectively is indispensable to uncover their unprecedented natures due to the fusion of merits of both organic and inorganic substances. Expressly, the two synthetic parameters are a focus of this chapter: the physical construction of the inorganic core and the chemical constituents introduced. Concretely, the catalytic activities of nanoparticles are controllable by modifying the crystal faces and the surface area of their inorganic cores in connection with their shape and size. In another case, when functional units, as an organic shell, are attached to the surface of nanoparticles, external stimuli such as electric field or light are utilized to transform the characteristics of nanoparticles themselves. It is also interesting to bring metal coordination polymers into an inorganic core, leading to the first isolation of alkyl chain stabilized-metal coordination nano-polymers with ferromagnetism. The nanomaterials presented here are listed as Pt nano-cube, size-selected Au nanoparticles, metal hexacyanoferrate coordination nano-polymers, and metal nanoparticles functionalized by bifferocene, anthraquinone, porphyrin and triphenylene derivatives. The synthetic procedure and their remarkable characteristics with some application examples are demonstrated.

In Chapter 4, the authors used PEG/polyamine block copolymers and highly stabilized PEGylated GNPs (PEG-GNPs) was facilely synthesized by autoreduction of $HAuCl_4$ using poly(ethylene glycol)-*b*-poly(2-(*N,N*-dimethylamino)ethyl methacrylate) (PEG-*b*-PAMA) without addition of the further reduction reagents. The condition of GNP synthesis, such as pH and the block copolymer concentration, was investigated. Furthermore, a function of target molecule recognition was added to the PEG-GNPs and the ability of the target molecule recognition was investigated. In order to investigate the interaction between the GNPs surface and PEG-*b*-PAMA, the technique of the GNPs stabilization was applied for the surface modification of GNPs using PEG-*b*-PAMA and commercial GNPs. The GNP modification conditions, such as pH, block copolymer concentration and PAMA chain length, were investigated, and the dispersion stability of the PEG-GNPs under various conditions was evaluated in detail.

The aim of Chapter 5 is to discuss the advantages of use of nanoparticles and quantum dots as tags for electrochemical bioassays.

Nanoscale materials bring new opportunities for electrochemical biosensing. Quantum dots are very stable (comparing to enzyme labels), they offer high sensitivity (thousands of atoms can be released from one nanoparticle) and wide variety of nanoparticles brings possibility for multiplex detection of several biomolecules.

Inorganic nanoparticles have received great interests due to their attractive electronic, optical, and thermal properties as well as catalytic properties and potential applications in the fields of physics, chemistry, biology, medicine, and material science and their different interdisciplinary fields. Great deals of synthetic methods such as aqueous- and organic-phase approaches have been reported to synthesize high-quality inorganic nanoparticles, but the stability of colloid nanoparticles obtained is still a great problem. In addition, most of these methods for preparing nanoparticles were based on the use of organic ligand, surfactant or polymeric templates. However, the addition of organic ligand, surfactant or polymeric templates may introduce contamination in the final product due to the formation of undesired byproduct in the reaction medium, which makes them unsuitable for further application, especially in biology. On the other hand, fluorohores such as dyes has been employed as optical probe by many groups for carrying out bioassay such as detecting target DNA, protein, etc. However, because one DNA or other probes can only be labeled with one or a few fluorophores, the fluorescence signal is too weak to be detected when the target concentration is low. Furthermore, most organic dyes suffer serious photobleaching, often resulting in irreproducible signals for ultratrace bioanalysis. Silica coating of inorganic nanomaterial or fluorohores appears as an attractive alternative for providing enhanced colloidal or photostable stability and suitable functional groups for bioconjugation. Hybrid nanoparticles based on silica has been pursued vigorously in the last few years and in this review chapter the authors provide a perspective on the present status of the subject. Described approaches include encapsulating inorganic nanomaterials using SiO_2 sphere, fluorophore-doped silica nanoparticles, constructing muti-functional hybrid materials using SiO_2 sphere as supporting material, and some important applications of hybrid nanoparticle based on silica. Finally, summary and outlook of these hybrid nanoparticles are also presented in Chapter 6.

Nanofluids, a dispersion of nanoparticles in conventional heat transfer fluids, promises to offer enhanced heat transfer performance. Nanofluids are a new, innovative class of heat transfer fluids and represent a rapidly emerging field where nanoscale science and thermal engineering meet. In recent years, nanofluids have attracted great interest from researchers of multi-disciplines because of their superior thermal properties and potential applications in important fields such as microelectronics, microfluidics, transportation, and biomedical. Published research works have shown that nanofluids possess higher effective thermal conductivity and effective thermal diffusivity compared to their base fluids and the magnitudes of these properties increase remarkably with increasing nanoparticle volume fraction. Particle size and shape as well as fluid temperature are also found to influence the enhancement of thermal conductivity of the nanofluids. Despite numerous theoretical studies on model development for nanofluids, there is no widely accepted model available due to inconclusive heat transfer mechanisms of nanofluids. There are also many inconsistencies in reported experimental results and controversies in the proposed mechanisms for enhanced thermophysical properties of nanofluids. In Chapter 7, the current state-of-research in nanofluids including synthesis, potential applications, experimental and analytical studies on the effective thermal conductivity, thermal diffusivity, and viscosity of nanofluids are critically reviewed. Results from the authors' extensive theoretical and experimental studies on thermophysical and electrokinetic properties of nanofluids are also summarized.

Since the first report in 1995 that indicated that the addition of nanometer copper metal particles to traditional heat transfer liquids, such as water and Ethylene Glycol, could result in an enhanced effective thermal conductivity which was considerably higher than that predicted by traditional mean-field theory, tremendous experimental studies have been conducted. Most of these support this conclusion, but there is some variation in the degree of the level of the enhancement. These investigations include experiments on metal nanoparticle/liquid suspensions, metal oxide nanoparticle/liquid suspensions, liquid nanoparticle/liquid suspensions, and carbon nanotube/liquid suspensions, and cover a wide range of nanoparticle shapes and volume fractions. These investigations include variations in the size of the nanoparticle, the type of nanoparticle material, the type of base fluid, the effect of surfactants, the effect of pH value of the suspension, the temperature of the suspension, and the ultrasonic vibrating time. Moreover, a number of experimental techniques and methods have been adopted to measure the effective thermal conductivity of different types of suspensions, including steady-state cut-bar methods, transient hot-wire methods and other less well known methods.

The theoretical exploration of the fundamental mechanisms for enhanced effective thermal conductivity of nanoparticle suspensions has resulted in a number of theoretical models. There are basically two different mechanisms proposed for these variations, the Brownian motion effect and the agglomeration effect. In addition to these, several other alternatives, such as the effect of the adsorption layer of the fluid molecules on the surface of the nanoparticles and the thermophoresis of the nanoparticles inside the suspension have been proposed.

The Brownian motion mechanism in nanoparticle suspensions is a relatively new concept, which has grown in acceptance with the addition of additional experimental evidence. While the agglomeration mechanism is not new, it assumes that the agglomeration of the nanoparticles replaces the micron size and/or millimeter size particles in the mean-field theory. Both major mechanisms are supported by the comparisons between predictions of models and reported experimental results. Moreover, there is almost equal support for both from numerical simulation studies and experimental observations.

An extensive review of the available experimental reports and theoretical developments indicates that a majority of the experiments have confirmed the enhanced effective thermal conductivities of nanoparticle suspensions and indicated that these are greater than the values predicted from mean-field theory models. In addition, the mean-field theory can not explain why the effective thermal conductivity of nanoparticle suspensions could have this unusually high enhancement, or why it increases with increases in the bulk temperature, and with decreases in the nanoparticle size. Finally, the evolution process of newly developed Brownian motion models and agglomeration models is discussed in detail.

As a result, what is clear from the available research is that it is necessary to illuminate why effective thermal conductivities of nanoparticle suspensions are higher than the predictions of mean-field theory models. Extensive study on the contributions of the mechanisms of Brownian motion, agglomeration and correlation of both is critically needed. Chapter 8 will help the development of a more robust understanding of the new phenomenon and the further application of effective thermal conductivity of nanoparticle suspensions in the future.

One of the most important findings for nanofluids is the discovery of a significant enhancement and a strong temperature effect in their thermal conductivity. Research confirms that the standard rules for mixtures is not capable of describing the nanoparticle's contribution to the thermal conductivity, and some researchers believe that the Brownian motion is the key mechanism affecting the variation of thermal conductivity with temperature. In Chapter 9, the authors review existing experimental results and explore different approaches to estimating the effective thermal conductivity of nanofluids from recent theoretical models. Focus is primarily on an explanation of the temperature-dependence on the thermal conductivity of nanofluids. It is found that the mobility of nanoparticles is theoretically conjectured as having a key role in determining the temperature effect on the thermal conductivity of a nanofluid. The limitations of existing models and future research directions are discussed.

Chapter 10 aims to review the immense influence that nanotechnology has having along the last years in all areas of knowledge, as well as, the method of obtaining of nanoparticles, trough of clays modification and Sol-gel synthesis and the used of the nanoparticles obtained in the formation of nanocomposites with polyolefin, also is discussed the interaction nanoparticle-polymer, and the used of a new compatibilizer agent in the properties of nanocomposites. Is presented one of its main applications and the main capital investment from the governments is the packaging industry of food and pharmaceutical products. This is due, among many other factors, to the possibility of using nano-sensors that when combined with the package communicate to the consumer/producer, as well an in control situations, the characteristics of package material such as: microbiological quality, storage conditions (humidity, temperature, light, etc), time-temperature treatment and shelf-life, among others. Finally is shows some trends for the application of nanoparticles.

Combination of Nano-sciences & techniques and electrochemistry makes electrochemical techniques are promising in many aspects, such as electroanalysis, bio-analysis, power & energy, etc. At the same time, the introduction of Nano-sciences & techniques in electrochemistry also expands the unknown filed of the interdiscipline.

In Chapter 11, the authors will discuss the applications of nanoparticles (NPs) in electrochemistry including the following threads 1) why scientists introduce the NPs into electrochemistry; in this part, we summarize the develop of Nano-sciences & techniques; mainly discuss some basic theories and concepts about NPs and electrochemistry and how scientists use NPs in electrochemistry; 2) progresses of NPs application in electrochemistry, in this part, they summarize the achieved marvelous achievements in this area; 3) challenge of the NPs application in electrochemistry, in this part, they summarize the challenges that electrochemists are encountering with and bring up their opinions and strategies.

In Chapter 12, the authors report that the thermal conductivity (TC) of heat transfer nanofluids containing Ni coated single wall carbon nanotube (SWNT) can be enhanced by applied magnetic field. A reasonable explanation for these interesting results is that Ni coated nanotubes form aligned chains under applied magnetic field, which improves thermal conductivity via increased contacts. On longer holding in magnetic field, the nanotubes gradually move and form large clumps of nanotubes, which eventually decreases the TC. When they reduce the magnetic field strength and maintain a smaller field right after TC reaches the maximum, the TC value can be kept longer compared to without magnetic field. We attribute gradual magnetic clumping to the gradual cause of the TC decrease in the magnetic field. The authors also found that the time to reach the maximum peak value of TC is increased as the applied magnetic field is reduced. Scanning Electron Microscopy (SEM) images show that the Ni coated nantubes are aligned well under the influence of a magnetic field. Transmission Electron Microscopy (TEM) images indicate that nickel remains attached onto the nanotubes after the magnetic field exposure.

Proteins and peptides as therapeutic agents have been used for clinical applications. While due to their short *in vivo* half life, easily undergoing hydrolysis, low permeability to biological membrane barriers and poor stability in gastrointestinal tract, parenteral administration such as subcutaneous, intravenous, or intramuscular injections is the common way for these drugs delivery. Alternative injection, the oral route presents many advantages such as avoidance injection pain and discomfort, elimination of possible infections caused by the use of needles. Therefore, many strategies to improve their oral delivery efficiency have been studied. One of the most potential strategies is usage of micro- and nanoparticles of functional polymers as protein drug carriers. Chapter 13 summarizes the nanoparticles based on different polymers as an oral administration system for protein and peptide drugs delivery. It primarily describes the pH-sensitive nanoparticles. Depending on the fact that pH of the human gastrointestinal tract increases progressively from the stomach to the intestine, nanoparticles remain insoluble in the stomach and disintegrate at the higher pH. Secondly, mucoadhesive polymer-coated nanoparticles are described, which can intensify the contact between the polymer and the mucus surface either by non covalent bonds or covalent disulfide bonds. Thirdly, bacteria degradable polymer-coated nanoparticles for colon delivery are briefly reported. Fourthly, a pH-sensitive and mucoadhesive polymer based on nano- and microparticles is included. Finally, some success of oral protein drug *in vivo* deliveries are enumerates. All of these nanoparticles can be expected to improve the proteins absorption.

In: Nanoparticles: New Research
Editor: Simone Luca Lombardi, pp. 1-37

ISBN: 978-1-60456-704-5

Chapter 1

FORMATION OF NANOPARTICLES UNDER LASER ABLATION OF SOLIDS IN LIQUIDS

G.A. Shafeev
Wave Research Center, Prokhorov General Physics Institute,
Russian Academy of Sciences, Moscow, Russia

Abstract

The process of nanoparticle formation under laser ablation of solids in liquids is described. Critical parameters are discussed that govern the properties of nanoparticles ejected into the surrounding liquid. These parameters are laser wavelength, pulse duration, interaction of individual nanoparticles with laser beam inside the liquid. A review of previous results is presented on the properties of nanoparticles of noble metals. Recent data on laser-assisted generation of other metals is given, including the formation of alloyed nanoparticles. Micro- and nanostructuring of the target upon its laser ablation in liquid environment is discussed. Examples are given of the influence of the surrounding liquid on the chemical composition of generated nanoparticles. The laser control over the size distribution of nanoparticles in liquids is demonstrated either by spatial profiling of laser beam intensity or proper tuning of laser wavelength into plasmon resonance of nanoparticles. Recent results are given on excitation of high energy levels of media under laser exposure to laser pulses of picosecond range of duration.

Introduction

The advent of lasers opened a new branch of research of interaction of radiation with matter. The primary "eye-visible" effect of laser action on a solid target is removing of some material from the target surface within the laser spot. This process was called "laser ablation" from a Latin word *ablatio*, which means removal.

The process of laser ablation of solids in liquids has attracted much attention of researchers during the last decade. This is due mainly to the simplicity of the experimental setup. Many modern laboratories (and not only physical ones) are equipped with lasers, and synthesis of nanoparticles is a strong temptation. The process proceeds in one step and results in immediate formation of nanoparticles in the liquid in which the target is immersed. The main feature of the process is that ideally the liquid contains only nanoparticles made of the target material and the liquid. There are no counter-ions or residuals of reducing agents left in the solution. For this reason laser ablation of solids in liquids can be considered as a method of nanoparticle synthesis, which is an alternative to chemical methods.

Commercially available laser sources are characterized by a number of parameters, such as peak power, average power, wavelength of emission, pulse repetition rate, etc. If the final purpose of ablation of a target immersed into a liquid is the synthesis of nanoparticles with desired properties, such as their chemical composition, size distribution, concentration, etc., then the above mentioned laser parameters are of different importance to the properties of desired nanoparticles. Also, the nature of the liquid plays a significant role in the final properties of nanoparticles generated under laser ablation.

The objective of this text is to outline the relative importance of various experimental parameters to the properties of nanoparticles synthesized by laser ablation of a solid target immersed into a liquid.

General Setup of Laser Ablation in Liquids

Figure 1 shows a generalized setup of experiments on laser ablation of solids in a liquid environment. This setup may vary from one research group to another, though their common features are the same.

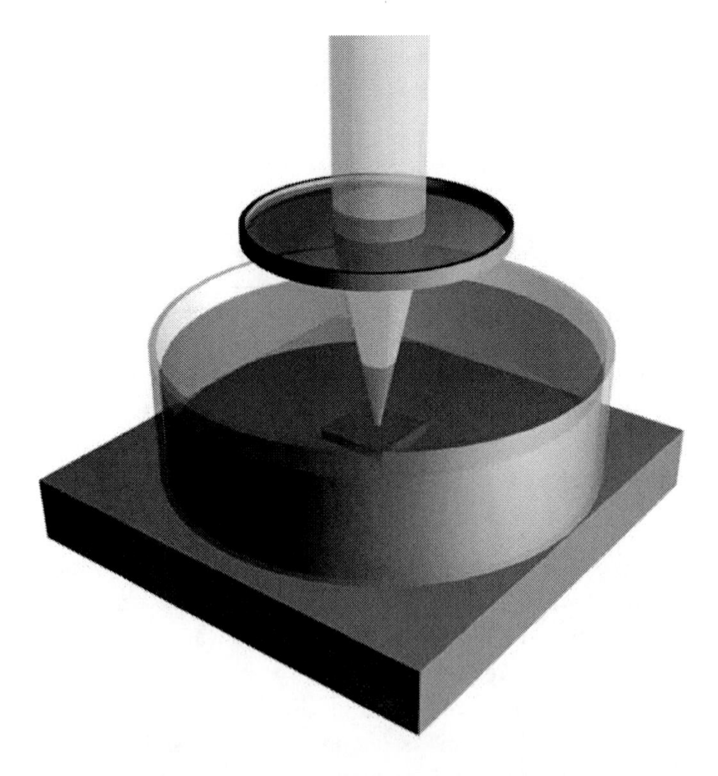

Figure 1. Experimental setup for laser ablation of solids in a liquid environment.

Laser radiation is focused onto a solid target immersed in a liquid. It is assumed that the liquid is transparent at the laser wavelength, otherwise the focalization of the beam would be problematic due to absorption of laser radiation in it. The simplest way is working with the free surface of the liquid, which allows avoiding additional reflection at the interface "covering glass/air". However, the use of volatile liquids, such as acetone, ethanol, etc., requires covering the liquid with a window that is transparent at the laser wavelength.

Experimental Technique

The experimental setup for laser ablation in a liquid environment is perfectly simple. It is usually assumed that the liquid that surrounds the solid target is transparent at laser wavelength while the laser radiation is absorbed by the target. A solid target is placed under a thin (several millimeters) layer of liquid and is exposed to laser radiation through this layer. Different pulsed laser sources can be used, e.g., a Nd:YAG laser at 1.06 μm output and its harmonics, a Cu vapor laser, a Ti:sapphire laser, etc. UV excimer lasers are less common, since most liquids and NP absorb in the UV region. The only necessary requirement is that the laser beam is sufficiently powerful to induce local melting of the target. Usually the laser beam is focused onto the target using an appropriate optics to a certain size of laser spot (see Figure 1). In some experiments the target is rotated under the laser beam to avoid the exposure of the same area. Some research uses a sealed-off cell to avoid oxidation of NP by air oxygen, but basically the setup is the same. It is the density of the laser energy (in J/cm^2), or so-called *fluence*, that determines the temperature of the target and the possibility to produce surface melting and eventual generation of NP. The irradiation of the metal surface results in fast removal of the material that is confined to the laser spot. The ejected nanoparticles remain in the liquid that surrounds the target, resulting in formation of a so-called colloidal solution. Unlike *real* solutions that contain ionic or molecular species, colloidal solutions also contain particles, e.g., NP and clusters. Due to accumulation of NP in the surrounding liquid, their prolonged interaction with laser radiation is possible. Therefore, the thickness of the liquid layer above the target is also an important experimental parameter that may influence the properties of generated NP.

Different laser parameters are of different importance to the efficiency of NP generation.

1. Pulse Duration

Usually, pulsed laser sources are used for generation of NP in liquids with pulse duration from hundreds of femtoseconds to hundreds of nanoseconds. The NP under laser ablation in liquids are formed owing to sputtering of the molten layer by the recoil pressure of the liquid that surrounds the target. Therefore, the necessary condition of NP synthesis is melting of the target material. In general, the temperature distribution under laser exposure of solids can be found solving a heat conduction equation with corresponding boundary conditions. However, in the case of short laser pulses of the duration mentioned above, the complicated problem of temperature calculation can be significantly simplified. This simplification is based on the fact that the heat diffusion length l_d from laser-exposed areas of the target during the duration of the laser pulse t_p is small compared to laser spot size d. Indeed, in a typical experiment on pulsed laser ablation d ~ 10 μm, while the heat diffusion length is much less. Assuming that the laser beam has a flat-top profile (typical of excimer lasers, metal vapor lasers, etc.) one may suggest that the absorbed laser energy is spent for heating of the target layer whose thickness is of order of $(at_p)^{1/2}$, where a stands for the heat diffusion coefficient of the target material, while the area of this layer coincides with the laser spot. Unlike to what one might believe, the presence of liquid around the target does not alter noticeably the temperature inside the laser spot. The reason is that the heat diffusion coefficient for liquids is even smaller than that of solids. Presence of the liquid may affect only the average temperature rise of the target if the repetition rate of laser pulses is elevated, e.g., of order of 1 kHz and more. Simple heat balance equation leads for the following expression of the temperature T within the laser spot:

$$T \approx \frac{Aj}{c\rho h},$$

where A stands for absorptivity of the target at the laser wavelength, A = 1- R, where R is the reflectivity coefficient at laser wavelength, c stands for heat capacity of the target material, □ is density of the target material, and h is the heat diffusion length inside the target. One can see that in a quite natural way the temperature is proportional to the energy density of the laser beam, or so called *fluence* j. The heat diffusion length h depends on the heat diffusivity of the target material:

$$h \propto \sqrt{at_p},$$

where in turn a = k/c□, where k is the heat conduction coefficient of the target, and t_p stands for laser pulse duration. The longer is the laser pulse t_p, the thicker is the layer of the material which is heated by absorbed laser energy. The above-made estimation of the temperature rise T is made assuming superficial absorption of laser radiation. If α is the coefficient of absorption of laser radiation, then this condition can be written as follows:

$$\alpha^{-1} << h$$

Of course some energy is consumed for heating and evaporation of a liquid adjacent to the laser spot, but this energy is small compared to the absorbed one due to low thermal conductivity of liquids.

In case of metal targets laser radiation is absorbed by free electrons that transfer their energy to the metal lattice within 3- 5 picoseconds. Virtually no heat exchange with the bulk of the solid target occurs, and the absorbed laser energy is spent for heating of the layer within the absorption depth α^{-1} of laser radiation.

The absorptivity A of the target surface is a complex parameter. For a smooth metallic surface it can be calculated using reference data on both real and imaginary part of a complex dielectric function of the material. However, as soon as the surface of the target is not flat and is characterized by a certain relief, the absorptivity of the target may largely deviate from its theoretical value. This is due to the dependence of absorptivity on the angle of incidence of radiation. Since the target material is dispersed into surrounding liquid as nanoparticles during laser ablation in liquids, the formation of the relief indeed occurs.

2. Laser Wavelength

As soon as laser ablation of metals is considered, any wavelength is appropriate. However, laser radiation can be absorbed by NP that are generated under ablation of the target. The majority of NP absorb in UV region, which imposes certain drawback on the use of excimer UV lasers for generation of NP.

3. Repetition Rate

Nanoparticles are ejected from the solid target at each pulse provided that the absorbed laser energy is sufficient to melt it. Therefore, the higher is the repetition rate of laser pulses, the higher is the rate of NP generation. However at high repetition rate the target may be screened from the beam by gas bubbles that have remained from previous pulses. This can be avoided using either a flow cell or elevated velocity of scanning (rotation) of the target.

Historical Review

Formation of nanoparticles under laser ablation of solids either in gas or in vacuum has been extensively explored during the last decade. Understanding of the mechanisms of cluster formation is needed to control the process of Pulsed Laser Deposition (PLD) that is widely used now for deposition of a large variety of compounds. Formation of nanoparticles under laser ablation of solids in liquid environment has been studied to a lower extent. A number of wet chemical methods of nanoparticles (NP) preparation are known up to date. Laser ablation of solids in liquid environment emerged as an alternative technique that is capable of producing "pure" NP without counter-ions and surface-active substances. Primary motivation for generation of NP by laser exposure of solid targets in liquids was the hope that pure NP would be more efficient in the process of Surface Enhanced Raman Scattering (SERS), since no other compounds can obscure the surface of a metallic NP [1-5]. Formation of both large variety of NP under laser ablation of corresponding solids has been reported during the last decade [6-17]. Further studies showed that the laser ablation in liquids is a complex process with a number of experimental parameters. In particular, chemical interaction of ejected nano-sized species with hot liquid vapors at high transient pressure results in large diversity of final particles that remain in the liquid. Also, surface tension at the interfaces is dominant force that governs the dynamics of interaction of molten target material with vapors of surrounding liquid on a nanometer scale. Under pulsed laser exposure the liquid also undergoes chemical changes that may be stipulated by ejected NP that act in this case as a catalyst. Initially liquid substance may become supercritical in the vicinity of the target, which alters its reactivity and opens new channels of chemical reactions especially in case of aqueous solutions.

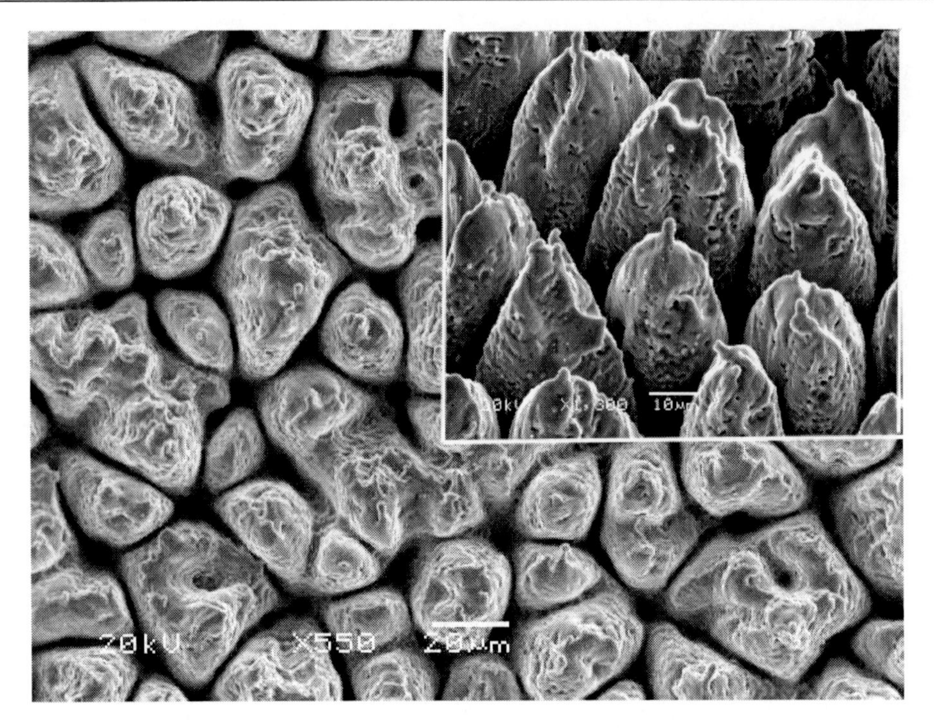

Figure 2. SEM top view of a brass target produced by scanning beam of a Cu vapor laser. Inset shows the enlarged view of the same surface tilted for 35°. Scale bar denotes 20 μm and 10 μm in inset.

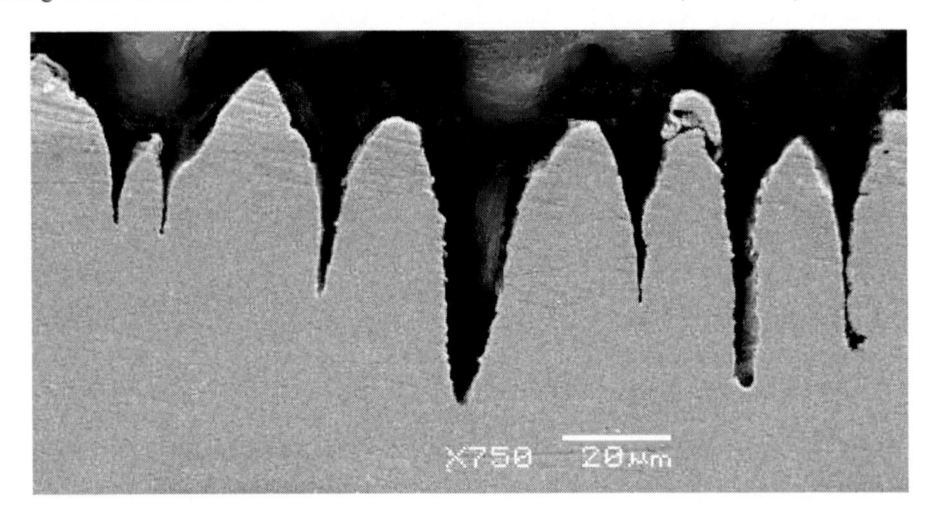

Figure 3. SEM view of a cross section of a brass target subjected to laser beam of a Cu vapor laser under the layer of ethanol. Space bar denotes 20 μm.

Interaction of metal nanoparticles with laser light proceeds via its absorption by free electrons. Free electrons exhibit plasmon resonance, whose position is determined by both their concentration, their effective mass (that depends on the solid) and particle size. NP of noble metals, e.g., Au or Ag, show a strong selective absorption in the visible, so the suspensions of these NP in liquids are colored. Plasmon resonances of other metallic NP are situated in the UV region [18, 19]. Though if the liquid is transparent at the onset of laser exposure of the target, the appearance of NP may change this situation, and the laser radiation is absorbed by the colloidal solution. This may lead to poor control of the laser spot size on the target due to either thermal lensing inside the solution or different non-linear phenomena induced by laser beam in NP [20-22]. So, the thickness of the liquid layer above the target is also one of the experimental parameters that are essential during ablation in liquids.

Depending on the dwell time of the laser beam on the target, a more or less significant amount of the target material is transferred into the surrounding liquid as NP. The target relief is therefore modified, and a crater is formed in the target in case of its exposure to a stationary laser beam. To avoid the formation of the crater the target is either rotated or is scanned under the laser beam. However, a periodic relief arises on the target surface at sufficiently high number of laser shots of order of 10^4. The structures are densely packed micro-cones, and their period linearly varies with the size of laser spot on the target as shown in Figure 2 [23].

These micro-cones are separated by long channels with high aspect ratio that run deep inside the target body. The formation of these micro-structures alters the properties of NP generated by laser ablation, since the conditions of vapor expansion inside the channels are different from those on a flat solid-liquid interface. In other words, the molten areas of the target are no longer on its surface but are situated mostly in the channels. Indeed, the micro-cones are oriented in a way to reflect the incident laser beam. Therefore, the properties of NP generated from a smooth target surface at the onset of laser exposure are different from those at later stages when the micro-relief has been formed.

The cross section of a brass target subjected to radiation of a Cu vapor laser in ethanol is presented in Figure 3. One can see deep channels that separate the microstructures formed in the target upon laser ablation. The melt is formed predominantly on the bottom of channels due to both increased angle of incidence of laser radiation compared to a flat surface and to reflection of laser light from side surface of micro-cones.

The molten layer is pushed along the cone surface and is partially dispersed into surrounding liquid as NP. Other part of melt solidifies on cone surface forming in some cases typical tips on their tops as can be seen in Figure 2.

Formation of periodic micro-relief on the target surface inexorably alters the character of temperature distribution along its surface during the laser pulse. The temperature distribution is smooth on initial flat target surface, and in case of a flat top profile of laser beam intensity the temperature profile is also almost flat except for side effects. As soon as periodic micro-relief is formed, the temperature profile consists of numerous “hot spots” located in the channels between micro-cones. The energy density (fluence) inside these spots exceeds by far the fluence on the initial surface without micro-relief. Since the size of NP depends on the fluence, the size distribution of NP generated via laser ablation of the micro-structured target should be different from that generated at the first stages of laser ablation when the target is still flat.

Unlike laser ablation in vacuum where synthesized NP rapidly leave the laser beam and never come back due to sticking on the chamber walls, NP synthesized by ablation in liquids remain in it and may return into the laser beam during their motion. The efficiency of coupling of radiation to nanoparticles depends on the proximity of laser wavelength λ to plasmon resonance of charge carriers. The energy from electrons to the lattice is transferred within 3-5 ps [24], and the temperature T of a nanoparticle of radius R can be estimated on the basis of conventional heat diffusion equation [25]. For $8\pi kR/l \ll 1$ one obtains $T \sim 2\pi I_0 kR^2/\lambda k_{liq}$, that is the temperature of the nanoparticle in the laser beam is proportional to its geometric cross-section. Here k_{liq} stands for the thermal conductivity of the surrounding liquid which is assumed to be constant during its evaporation for the sake of simplicity. Note that due to the small size of NP their temperature is proportional to the peak power density of the laser beam I_0 (Watts per square centimeter). The extinction coefficient κ under large detuning from the plasmon resonance is close to that of the bulk metal. However, in the vicinity of plasmon resonance $\kappa=\kappa(\lambda)$ shows resonant behavior, and the temperature T of the particle strongly depends on the laser wavelength.

The main difference between laser ablation in vacuum and in liquid environment is a short free path of ablated species in the latter case. Indeed, the formation of NP under laser irradiation of a metal target immersed into a liquid proceeds via local melting of the metal. The adjacent liquid layer is heated to almost the same temperature owing to heat transfer from the metal. The thickness h of this layer can be estimated using the heat diffusion coefficient of a liquid $a = 10^{-3}$ cm^2/s and the diffusion time equal to duration of the laser pulse $\tau_p = 20$ ns. Then the thickness of the liquid layer $h \sim (a\tau_p)^{1/2} = 0.3$ μm. This thin layer has a temperature much higher than boiling point of the liquid at normal pressure and is hence in the vapor phase. The pressure in this vapor layer can be roughly estimated as the vapor pressure of the liquid at temperature of the substrate. This means that in typical experimental conditions all liquids are in vapor phase at pressure of the order of hundreds atmospheres. The thickness of the molten layer on the target is about the length of heat diffusion into the metal, $h_m \sim (a_m\tau_p)^{1/2}$. $a_m = 1.7$ and 0.13 cm^2/s for Ag and Au, respectively, which gives the heat diffusion length of 1.9 μm for Ag and 0.5 μm for Au. This is the maximal thickness of the reservoir for generation of nanoparticles that is gained at fluence much higher than the melting threshold. That is why the size of nanoparticles is almost independent on the pulse width, from hundreds of nanosecond through hundreds of femtosecond. At lower fluence the thickness of the melt is smaller than h_m. Expanding vapors of the liquid splash this reservoir resulting in the removal of the molten layer. Note that formation of nanoparticles via evaporation of the metal is unlikely, since the pressure of metal vapor at a temperature close to melting is too low compared to vapor pressure of the surrounding liquid. Surface tension stabilizes the molten drop of the metal, while the pressure of surrounding vapor of the liquid tends to split it. As a first approximation one may suggest that the size of a stable drop of metal can be found from the following relation: $2\sigma/R \sim p_{liq}$, where σ is the surface tension coefficient of the metal, R is the radius of the drop, and p_{liq} is the pressure of the vapor of the liquid surrounding the target. Substitution of these parameters for Ag ablated in water for R = 30 nm and T = 1000° C gives $p_{liq} \sim 10^8$ Pa, which is a reasonable value. The melting temperature decreases with the size of the nanoparticles, so at sufficiently small R the particle melts under the laser pulse, but the vapor pressure of surrounding liquid is not high enough to cause its further splitting.

Laser Ablation of an Ag Target in Liquid Environment

Historically Ag NP were the first ones synthesized by laser ablation of a silver plate in water. Various laser sources have been successfully used for this synthesis ranging from a nanosecond Nd:YAG through Ti:sapphire femtosecond lasers [21,26,27]. Ablation of a metallic Ag immersed into water by radiation of a Cu vapor laser at wavelength of 510.6 nm and pulse duration of 20 ns results in visible coloration of the liquid; it takes on a yellowish color. The optical spectra of Ag NP generated by laser ablation in different liquids are presented in Figure 3. The maximum of absorption is at about 400 nm, which is typical for so called plasmon absorption of Ag nanoparticles in water [18,19]. The intensity of the plasmon band increases with irradiation time, while its position does not vary significantly. Ag NP are slowly oxidized by air oxygen dissolved in the liquid, and the position of plasmon resonance shifts to the red with time due to formation of oxide layer with high refractive index [22].

The Transmission Electron Microscope (TEM) view of Ag nanoparticles produced by ablation of an Ag target in acetone is shown in Figure 5. The average size of particles is around 15 nm. Moreover, they are flat disks, and their thickness is of order of few nanometers [15]. This is clearly seen in those areas where the TEM image of intersection of two or more Ag particles appears darker than each separate particle.

Smaller Ag NP are synthesized using sodium dodecyl sulphate (SDS) as a surface active substance [26]. Laser pulses of various durations have been used for synthesis of Ag NP in water ranging from ns to fs domain [21, 27].

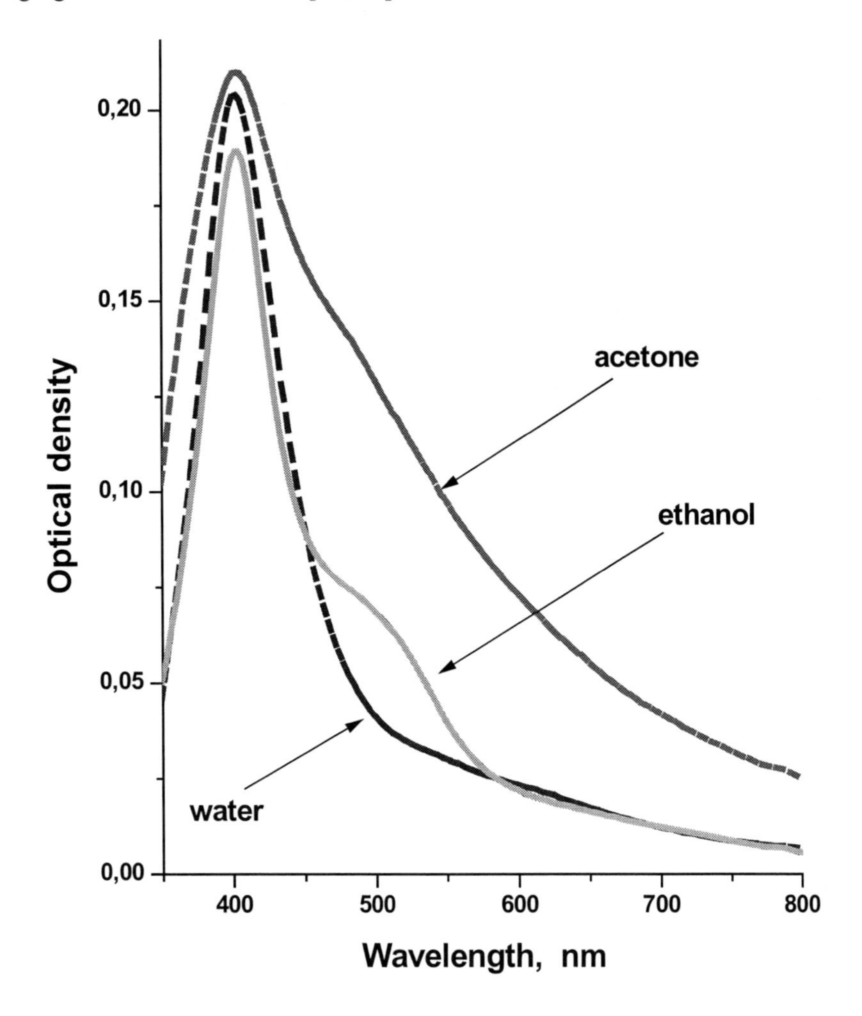

Figure 4. Absorption spectrum of Ag nanoparticles obtained by ablation of Ag in different liquids. The plasmon peak of Ag is well pronounced at ca 400 nm.

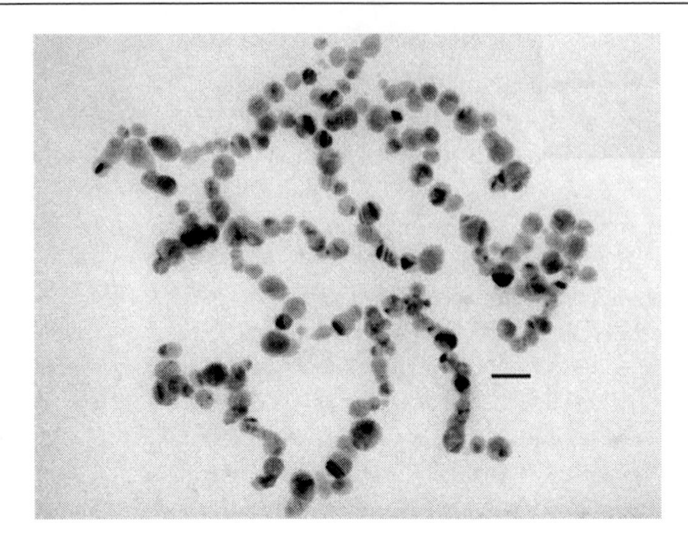

Figure 5. TEM view of Ag nanoparticles obtained by ablation of a bulk Ag in acetone. Scale bar denotes 50 nm.

Laser Ablation of an Au Target in Liquid Environment

Gold NP exhibit a well-pronounced dependence of their absorption spectrum on their shape and size. In general, a metallic ellipsoid is characterized by three different plasmon resonances. In case of a spherical particle all three of them are degenerated, and Au NP show the maximum of absorption around 520 nm independent on the way of their preparation. Elongated Au NP (nanorods) are characterized by two absorption bands that correspond to so called transverse and longitudinal plasmon resonances. The position of the latter varies with aspect ratio of Au nanorods, and for sufficiently high aspect ratio the longitudinal resonance moves from the visible to the near IR range of the spectrum [28].

The optical spectra of water after laser exposure of metallic Au show well-pronounced absorption band at ca 520 nm that is characteristic of transverse plasmon resonance [18, 19]. At higher laser fluence elongated Au NP are also synthesized, which is accompanied by the appearance of a red wing. Au NP prepared in water in absence of surface-active agents are metastable against precipitation, and their significant precipitation is observed within several days after preparation. Precipitation of NP is accompanied by appearance of a wide red wing that indicates the presence of elongated Au NP with various aspect ratios. Synthesis of a stable colloidal solution of Au NP in water has been reported by several groups [29-32].

The coupling of laser radiation to the colloidal solution can lead to the modification of the size distribution of Au NP. This is confirmed by TEM view of Au particles as obtained after the ablation of Au in water. Laser radiation modifies not only the size of the particles, but also their shape. As-obtained particles are elongated, while those exposed to laser radiation are the nano-disks similar to Ag nanoparticles. Average size of Au as-obtained after ablation by a Cu vapor laser in H_2O in absence of surface-active substances is 20 nm, while with further laser exposure of the solution alone their size diminishes to ca 10 nm.

Au NP produced at high laser fluence on the target are characterized by elongated shape independent of the nature of the surrounding liquid. This effect has been observed for laser ablation in alkanes [33] as well as in water. Figure 6 shows the TEM view of Au NP synthesized by laser ablation of a gold target in water. The majority of NP has a spherical shape, while some NP have aspect ratio ranging from 2 to 10. Accordingly, the absorption spectrum of this solution is characterized by two maxima, one at ca 540-560 nm and the second having a wide red wing centered at longer wavelengths.

The rate of generation of Ag and Au NP depends on laser fluence. For instance, it is 0.6 and 1.2 mg/hour, for respective metals, with a laser spot of 30 μm and fluence of 30 J/cm^2.

Nanoparticle size can be drastically reduced by the use of aqueous solutions of surfactants, which cover the particles just after their ablation and thus prevent them from further agglomeration. Sodium dodecyl sulphate (SDS) is one of the most popular agents capable of reducing the mean size of Au nanoparticles down to 5 nm during nanosecond laser ablation of gold [29-31]. Ablation of a gold target with shorter laser pulses, e.g., femtosecond ones, results in synthesis of bimodal distribution function of nanoparticle size at elevated laser fluence and almost mono-dispersed Au NP at lower fluence of 50 J/cm^2 [32]. Another stabilizing agent for Au NP are cyclodextrins [34]. Their addition provides perfect stability of Au NP as small as 2-2.5 nm synthesized by laser ablation with femtosecond pulses.

NP of another noble metal Pt can also be synthesized by laser ablation of a metallic Pt target. In presence of SDS their size can be as small as 5 nm [35].

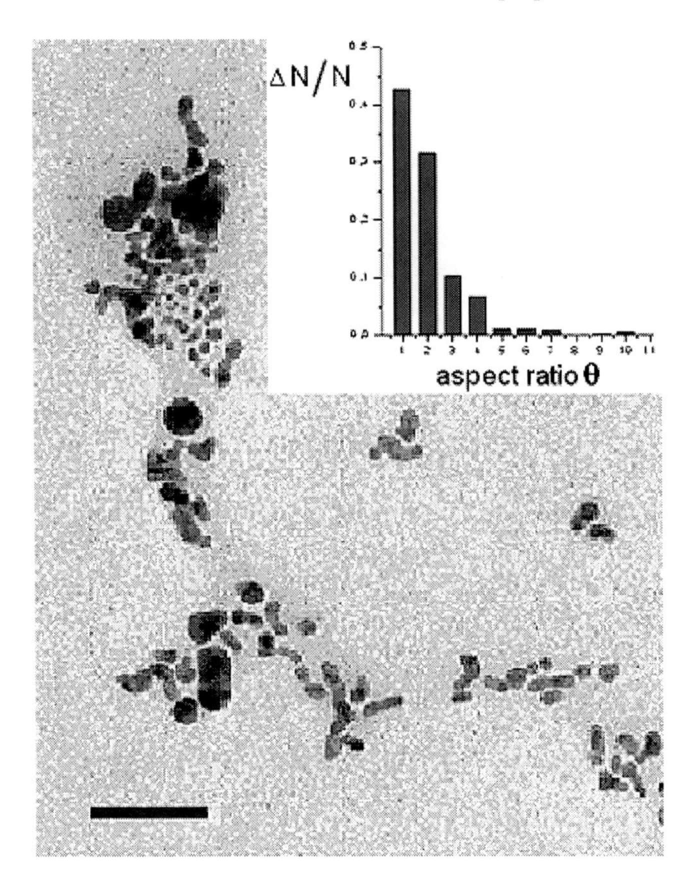

Figure 6. TEM view of Au NP obtained by ablation of a gold target immersed into water at elevated laser fluence of 90 J/cm^2 at 1.06 μm, pulse width of 130 ns.

Interaction of Nanoparticles with Laser Beam

Coupling of laser radiation to isolated NP proceeds via its absorption by free electrons. Then the absorbed energy is thermalized and transferred to NP material. Since these electrons are characterized by some plasmon resonance, the efficiency of this coupling strongly depends on the detuning of laser radiation from maximum of absorption. This interaction is not important in case of laser generation of NP in vacuum, since NP do not come back to the laser beam. In liquid environment NP remain in the liquid and do come back inside the beam due to convective flows. Repeated interaction of NP with laser beam leads to observation of some interesting phenomena described below.

Fragmentation of NP under Laser Exposure in Liquids

Fragmentation of NP under their laser exposure in liquids has been reported as early as the synthesis of NP itself [2, 6, 30]. The process manifests itself as a gradual decrease of the size of NP upon laser irradiation. As a result, the distribution function of particle size shifts to lower dimensions. The dependence of the NP size on the laser wavelength has been reported [36, 37] indicating the spectral dependence of the fragmentation process. Optical constants of metallic targets do not vary largely in the wavelength range of lasers used for NP synthesis, so fragmentation of NP should be ascribed to interaction of synthesized NP themselves with the laser beam inside the liquid.

The physical processes that lead to fragmentation of NP in the laser beam are still under discussion. The most probable reason leading to reduction of NP size are the instabilities that develop at the interface of molten NP with the vapors of surrounding liquid [25]. Indeed, a NP may absorb enough energy from the laser beam to undergo the phase transition into a liquid state. The temperature of NP exceeds at this moment the boiling temperature of surrounding liquid. The latter is therefore vaporized and forms a shell around the NP. The pressure of vapors in this shell is around 10^8 Pa, and its possible asymmetry would result in the break of molten NP inside it into smaller parts. The most probable process is splitting of the molten NP into two equal particles, since, first, this corresponds to lowest value of total surface energy, and second, is the lowest mode of perturbation of the surface of molten NP. This hypothesis is in qualitative agreement with the fact that the average size of NP exposed to laser radiation in various liquids decreases with the decrease of boiling temperature of the liquid. The balance is gained when the pressure inside the NP of radius R is of order of the pressure of the surrounding liquid p_{liq}: $2□/R \sim p_{liq}$, where □ stands for the surface tension of molten NP.

Fragmentation of the NP under their laser exposure introduces *feedbacks* into the system “laser radiation – NP”. Indeed, the average size of NP under their laser exposure is inversely proportional to the peak intensity of the laser beam inside the liquid. The size of NP decreases upon exposure till NP become so small that the energy absorbed from the laser pulse is not sufficient for melting. The system therefore demonstrates a negative feedback and auto-stabilization. On the other hand, the melting temperature of a NP decreases with its size, which is a positive feedback. Finally, small NP formed via fragmentation from bigger ones do not absorb enough energy from the laser beam to be molten, and the ensemble of NP acquires its upper limit of NP size.

Shape-selective Fragmentation

The fragmentation of NP under laser exposure of their suspension in a liquid can be shape-selective, if the spectrum of plasmon resonance depends on their shape. This can be illustrated by shape-selective fragmentation of Au NP having elongated shape (nanorods). Au nanorods are characterized by two absorption peaks, one of them corresponds to so called transverse resonance (TR), while the second is called longitudinal one (LR). These resonances correspond to oscillations of free electrons in the directions across the nanorods axis and along it, respectively [28]. The position of TR near 560-570 nm is common for Au nanorods with any aspect ratio, while the position of their LR is a linear function of the aspect ratio and moves to the red with its increase [38-41]. If laser radiation is tuned into TR, then all nanorods are molten and fragmented into spherical NP at sufficiently high peak power of the laser beam [40]. However, tuning the laser wavelength into LR allows selective fragmentation of Au nanorods with aspect ratios which are in resonance with laser radiation [42].

This fragmentation (splitting) is illustrated in Figure 7 in which the evolution of absorption spectrum of suspension of Au NP with wide distribution of aspect ratio is presented. Au NP are especially convenient for this type of experiments since their absorption spectrum depends on their shape. The radiation of a green line of a Cu vapor laser fits well the transverse plasmon resonance of Au NP, which is common for all Au NP and does not depend on their aspect ratio. On the contrary, the yellow line of a Cu vapor laser is absorbed by non-spherical (elongated) Au NP. Therefore, the evolution of the absorption spectrum is different if various laser outputs are used for NP exposure.

One can see gradual decrease of the optical density of the suspension in the red that corresponds to fragmentation of Au nanorods irrespective to their aspect ratio (Figure 7, 1). On the contrary, exposure of the suspension to radiation that is preferentially absorbed by Au nanorods with the aspect ratio 2 – 3 results in the decrease of the optical density just at the laser wavelength (Figure 7, 2). This decrease indicates the depopulation of the suspension with nanorods that are in resonance with laser radiation. Of course the laser fluence should not be too high otherwise nanorods with "non-resonant" aspect ratio will also be fragmented, since their absorption wings are very wide.

Au nanorods with aspect ratio about 10 absorb in the near IR region. Exposure of their suspension to radiation of a Nd:YAG laser operating at its fundamental output at 1.06 μm results in self-stabilization of fragmentation: no changes of absorption spectrum of the solution occur as soon as long nanorods are fragmented [42].

Similar fragmentation of Au NP is also observed in ethanol and acetone. Their size in these liquids is somewhat smaller than in water and is of 7 nm under the same laser fluence. As a consequence, these colloidal solutions are stable against sedimentation. Finally, the addition of a surface-active substance, e.g., poly(vinyl pyrrolidone) (PVP) with molecular mass of 10^4, to ethanol in which the ablation is carried out further decreases the average size of Au nanoparticles down to ca 4 nm at otherwise equal conditions.

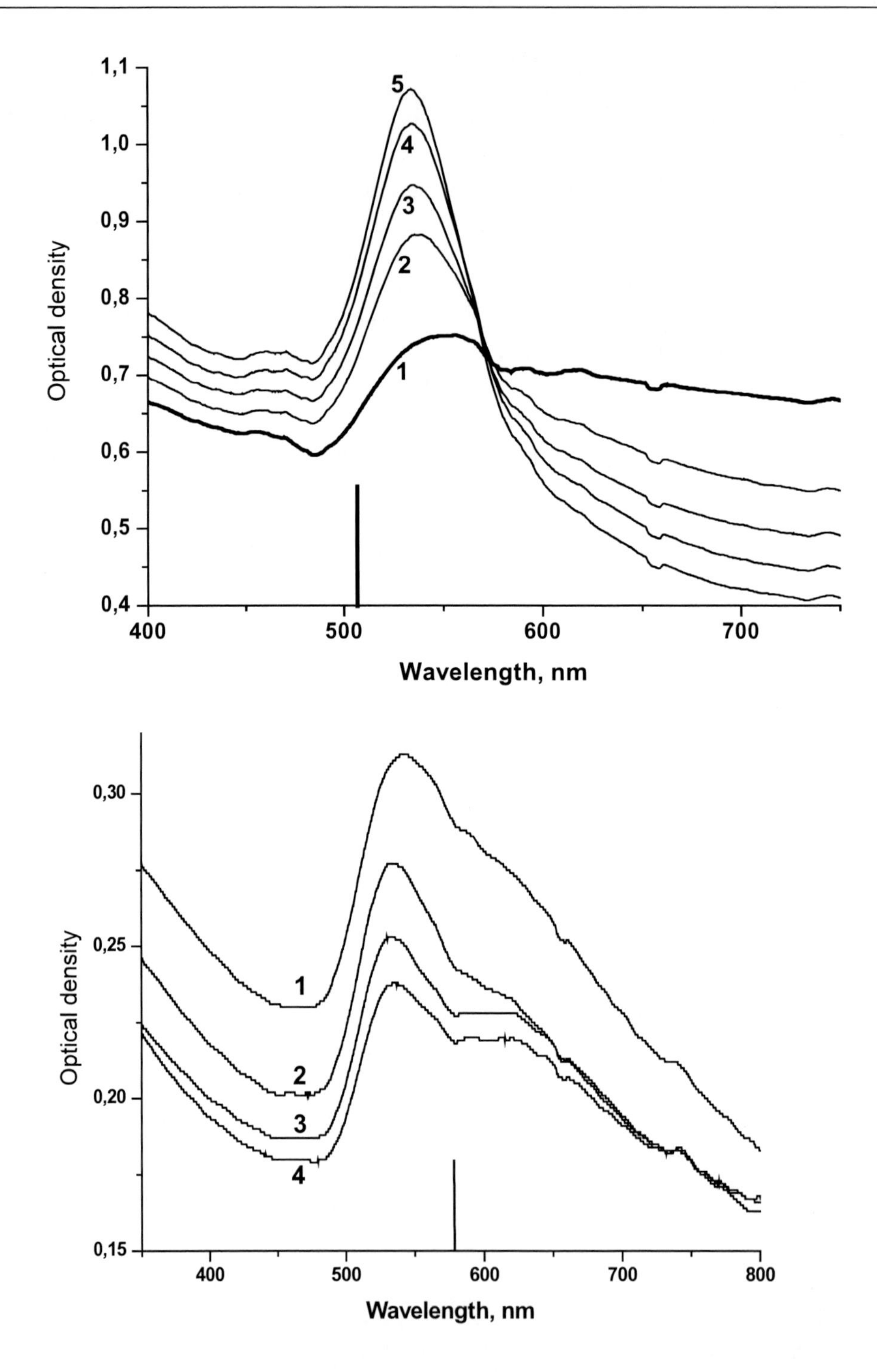

Figure 7. Evolution of optical density of suspension of Au nanorods in water under exposure to radiation of a Cu vapor laser at 510.6 nm (1) and 578.2 nm laser output. Vertical lines indicate the position of the laser wavelength. Curve (1) in both figures corresponds to the spectrum of initial colloidal solution.

Formation of the Au-Ag Alloy under Laser Irradiation of Nanoparticles

Direct interaction of NP with laser beam inside the liquid may lead to their melting during each laser pulse at sufficiently high laser peak intensity. If the colloidal solution contains NP of various materials, their absorption of laser radiation may be different. Laser exposure of a mixture of different NP may lead to formation of alloyed NP, and at least one kind of NP should be molten by laser beam. Recent publications [43-45] have reported the formation of gold-silver alloy nanoparticles in liquid environment under laser exposure of core-shell particles obtained by a chemical method. The spectrum of core shell (non-alloyed) nanoparticles is not a linear combination of the spectra of monometallic colloids [45]. On the contrary, the spectrum of a mixture of individual colloids is composed of spectra of monometallic particles. Upon laser exposure, these peaks disappear, and the single alloy peak arises. Laser exposure of Ag-Au core-shell NP may lead to both Ag shell removal from Au NP and their alloying depending on laser wavelength and its fluence [46]. Individual NP of various materials have to come into contact to form an alloy, and therefore the laser alloying of NP is concentration-dependent process. The position of plasmon resonance of alloyed NP is a linear combination of plasmon frequencies of individual NP. Formation of alloyed AuAg NP is accompanied by peculiar modification of the absorption spectrum indicating an intermediate phase of the Au-Ag alloy formation [47-49]. The temporal evolution of the optical density of the mixture is presented in Figure 8.

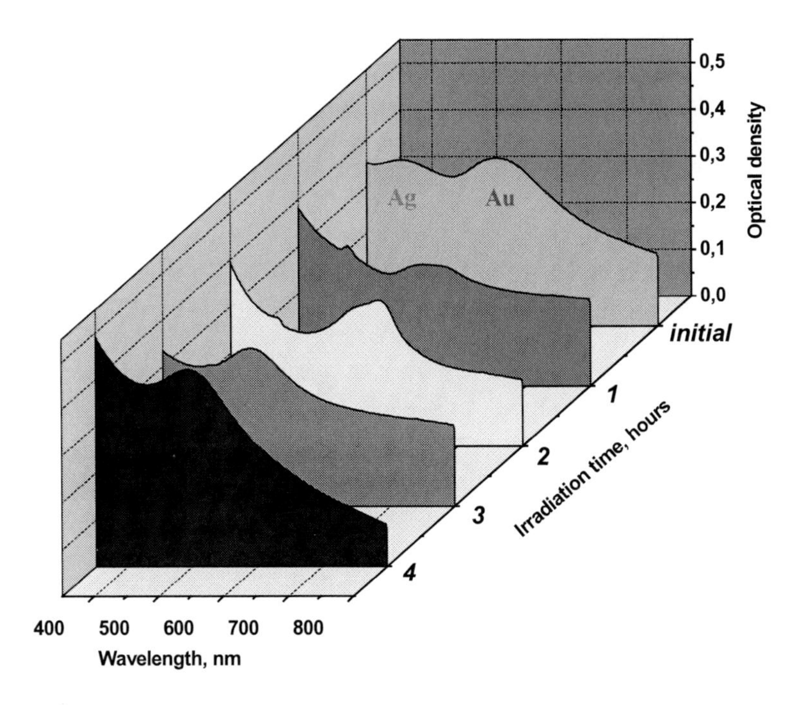

Figure 8. Evolution of the absorption spectrum of the mixture of both Au and Ag NP synthesized by laser ablation of respective metals in ethanol under laser exposure at 55 J/cm^2 of a Cu vapor laser radiation (wavelength of 510.6 nm).

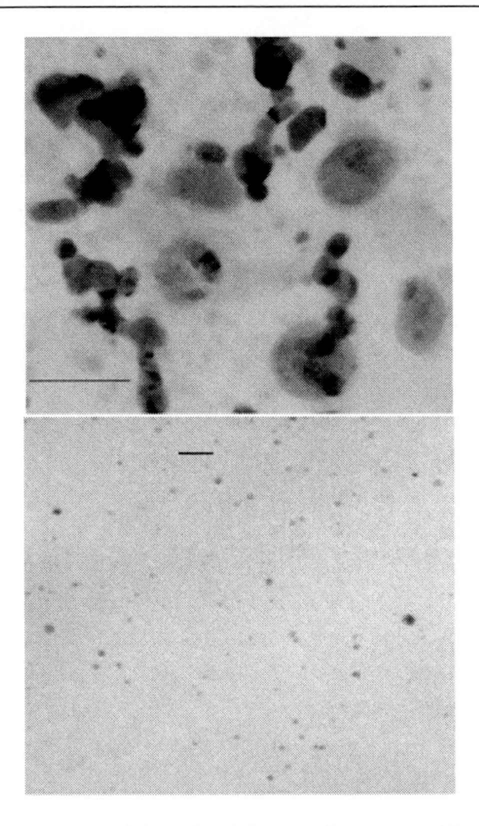

Figure 9. TEM view of Au-Ag nanoparticles after 2 hours of exposure (a). Concentration of PVP is 0.1 g/l, Cu vapor laser, laser fluence of 9 J/cm^2. Brighter particles are Ag while the darker ones are Au. Scale bar denotes 40 nm. Alloyed Au-Ag nanoparticles obtained by exposure of the mixture of individual colloids in ethanol during 4 hours (b). Scale bar denotes 100 nm. Average size of alloyed Au-Ag particles is 7 nm.

The formation of an intermediate phase of the alloy is observed only in case of sufficiently large initial NP of each metal (50 -70 nm). Smaller NP of 10-20 nm in diameter also show the deviation of the maximum of plasmon resonance from its stoichiometric position but this effect is weak.

The position of plasmon frequency of an Au-Ag alloy varies continuously with relative concentration of metals and is always in between the maxima of plasmon resonance at 400 nm and 520 nm of Ag and Au, respectively. In case of individual particles, however, the maximum of absorption falls out of this interval after certain exposure time and is significantly red-shifted after 1-2 hours of exposure. With further exposure the maximum returns to the interval corresponding to an Au-Ag alloy and remains there for even longer exposures indicating the formation of the alloy. The TEM view of NP at the stage of anomalous spectrum (Figure 8, spectrum 2) reveals multiple contacts between Au and Ag nanoparticles that are well observed due to different atomic mass of both metals.

With further exposure these hybrid particles are alloyed, and the final position of the absorption maximum corresponds to the Ag-Au alloy (Figure 8, spectrum 4). The size of alloyed nanoparticles is about 5 nm, which is much smaller than initial ones.

Formation of alloyed nanoparticles is sensitive to the presence of surface-active substances in the liquid. The rate of alloy formation increases under addition of PVP (10^{-5} M).

However, the alloying is inhibited at given laser intensity at elevated concentrations of PVP exceeding 5×10^{-5} M [48]. The rate of alloying decreases with dilution of the mixture, presumably due to the increase of the average distance between the nanoparticles. It is pertinent to note that the surface-active substance, e.g., PVP, apparently remains untouched under laser exposure of NP in spite of their elevated temperature. Raman analysis shows no variation in the glassy carbon content before and after laser exposure of suspensions of NP with PVP, though this should be so if the pyrolysis of the surfactant on hot NP took place. This might be explained by the preferential localization of PVP at the interface of a vapor bubble that is formed around the NP during their laser exposure.

The absorption spectrum of the Au-Ag alloyed NP is closer to the wavelength of a Cu vapor laser (510.6 nm) than the plasmon maxima of individual NP. Therefore, the efficiency of coupling of laser radiation increases upon formation of the alloy. As a result, the average size of alloyed Au/Ag NP is much smaller than that of individual ones [48, 49]. This is illustrated in Figure 9 where TEM images of NP are presented at different stages of Au-Ag alloy formation.

Similar behavior is observed under exposure of the mixture of colloidal solutions of Cu/Ag and Cu/Au. The alloyed NP are formed under sufficiently long laser exposure of the mixture, and the final position of the plasmon resonance depends on ratio of individual colloids in it.

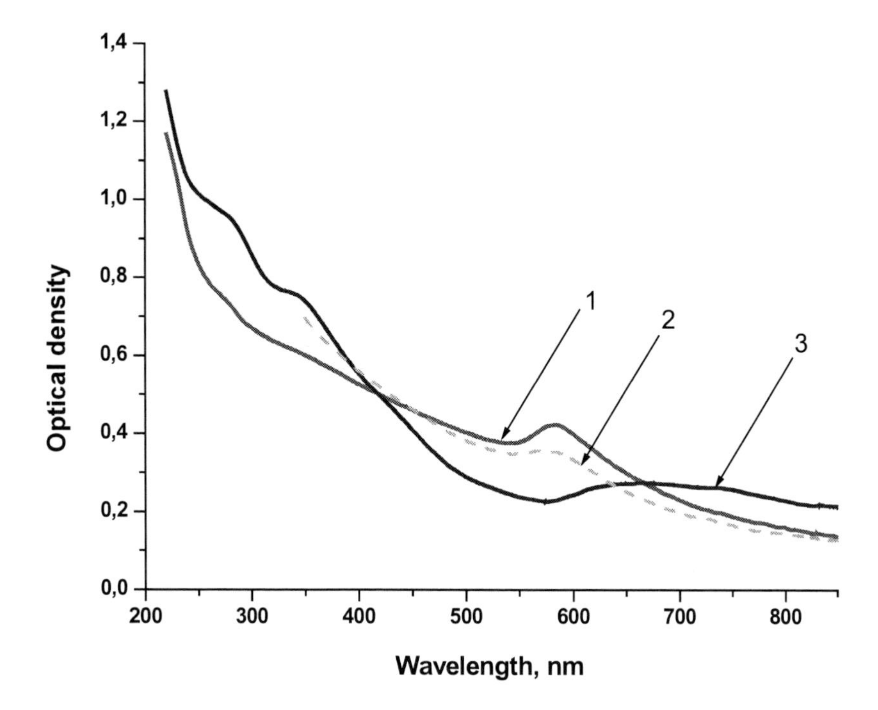

Figure 10. Absorption spectra of the liquids after laser ablation of a bulk Cu target. Ethanol (1), acetone (2), and water (3). Cu vapor laser, fluence of 30 J/cm^2.

Nanoparticles of Cu, Brass, and Bronze

Copper is chemically active and easily reacts with vapors of surrounding liquid. The chemical interaction may take place both during the ablation and after it due to contact with air oxygen dissolved in the colloidal solution. Typically, the plasmon resonance of Cu nanoparticles lies in the visible. Cu NP synthesized by chemical means in anaerobic conditions show the absorption peaks from 570 to 590 nm [49, 50].

Laser ablation of Cu target in various liquids also leads to formation of NP that are characterized by absorption in the visible [51]. The peak position is around 590 nm, which agrees with previously reported data for Cu NP (see Figure 10). Ablation of a Cu target in water leads to formation of compounds containing only ionic species Cu^{+} without any plasmon peak. The stability of generated Cu NP to slow oxidation is different. Namely, Cu NP produced by ablation in ethanol are slowly oxidized. This is manifested by gradual disappearance of the peak of plasmon resonance (Figure 10). Remarkably, the initial plasmon peak can be restored by exposure of aged colloidal solution to the radiation of a Cu vapor laser at sufficiently high fluence of several tens of J/cm^2.

On the contrary, ablation of a Cu target in acetone results in a stable spectrum of the colloidal solution of NP. The difference in behaviour of Cu nanoparticles in these two liquids stems from their images obtained with a transmission electron microscope. Cu NP generated by ablation in ethanol are low-contrast entities with average size of 5 nm. Cu NP in acetone are metallic as one can conclude from their high contrast and well pronounced peak of plasmon resonance. However, in this instance they are embedded into some amorphous cloud [51]. Raman analysis of the dried suspension indicates that this cloud consists of glassy carbon. The layer of glassy carbon protects Cu NP from oxidation with air oxygen and preserves their plasmon resonance in the visible. Note that decomposition of acetone on Cu NP that occurs during laser ablation of a Cu target is highly chemically selective - it takes place on Cu nanoparticles but not on Ag ones in similar experimental conditions.

Cu NP can also be synthesized via exposure of a suspension of micro-particles of CuO to laser radiation [52, 53]. At first stages of the exposure the micro-particles are broken into smaller entities, and then their reduction occurs by surrounding 2-propanol. It is worth mentioning that Ag NP can also be produced by exposure of suspended Ag_2O micro-particles in an appropriate hydrocarbon liquid to laser radiation. In both cases, fragmentation of micro-particles into smaller particles occurs due to absorption of laser energy by inter-band transition of corresponding oxide, while the subsequent formation of metallic NP takes place due to chemical reduction of oxide by surrounding liquid that acts like a reducing agent.

Brass NP can be synthesized in a similar way. Unlike Cu NP they are stable to oxidation for at least several months. The position of the plasma resonance of brass nanoparticles depends slightly on the type of laser source used for their synthesis and lies near 510-515 nm. The presence of brass NP in the liquid is corroborated by the X-ray diffractometry of the evaporated suspension. Bronze NP synthesized by laser ablation of a bronze target do not show any well-defined absorption maximum, instead a wide plateau in the green range of spectrum is observed in the absorption spectrum of their colloidal solution. However the absorption spectrum is merely different from that of Cu NP. The mean size of NP of these Cu alloys is around 20-30 nm.

Internal Segregation of Brass NP

Apart from fragmentation described above brass NP undergo drastic modification under their laser exposure in a liquid. This is an internal segregation of a nanoparticle that is repulsion of one of its components to its periphery [54]. The material of the nanoparticle is subjected to additional pressure due to its small radius. For a liquid nanoparticle, the capillary pressure is given by the expression p=2□/R, where □ is the surface tension coefficient and R is the particle radius, which is equal for Cu NP to 120 MPa for R=20 nm. In addition to this pressure, the nanoparticle during the laser pulse is subjected to the vapor pressure of the surrounding liquid, which reaches a maximum simultaneously with the particle temperature and decreases rapidly upon expansion of the vapor shell. The pressure of ethanol vapors surrounding the particle with a temperature of 1500 K is estimated at about 20 MPa from the equation of state of a real gas. Note that the pressure of nanoparticle-substance vapors at the melting temperature for most metals is negligibly low compared to the above values.

As synthesized Cu NP are not stable towards oxidation by air oxygen. The oxidation is accompanied by disappearance of the peak of plasmon resonance of Cu NP as shown in Fig. 11. It is pertinent to note that the spectra of NP of Cu alloys are more stable with time. This is due to the formation of a shell around them as described below.

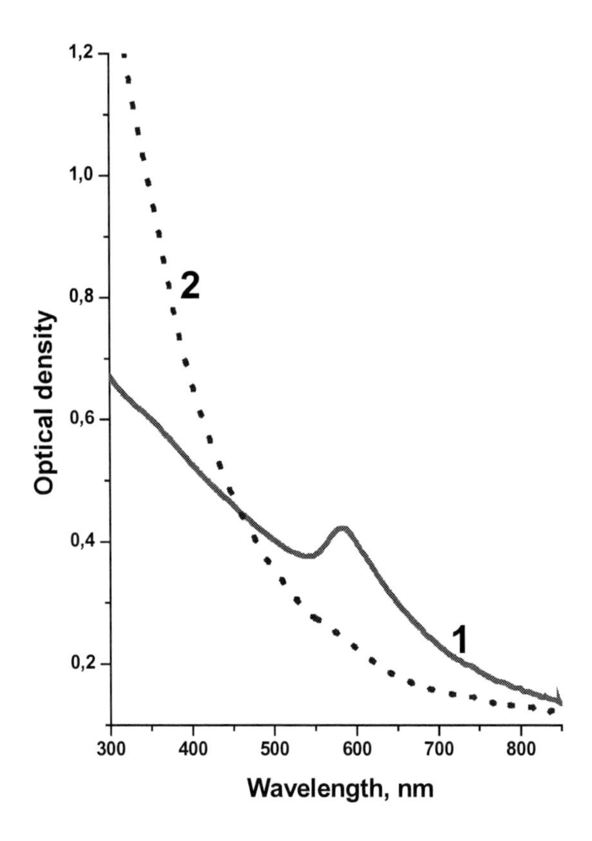

Figure 11. Evolution of absorption spectrum of Cu NP produced by laser ablation of a bulk Cu target in ethanol. As prepared (1), 6 months later (2). Ablation was carried out using a Cu vapor laser.

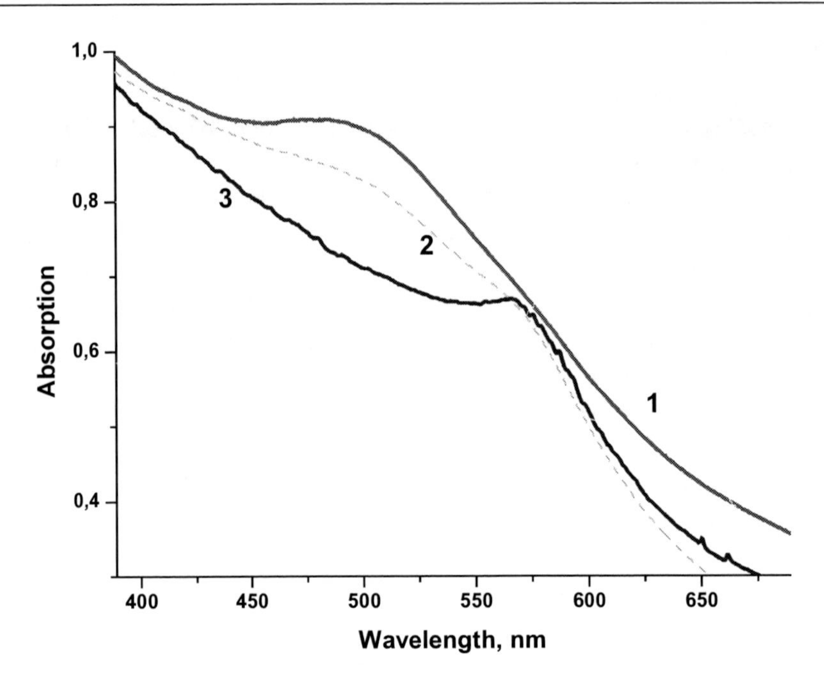

Figure 12. Evolution of the absorption spectrum of brass NP upon exposure of their colloidal solution in ethanol to radiation of a 130 ns Nd:YAG laser. As synthesized brass NP (1), 10 (2), and 90 min of exposure (3).

Laser radiation acting on the ensemble of brass NP in the liquid changes their absorption spectrum. The absorption spectrum of brass NP evolves to that of Cu NP (Figure 12). With laser exposure the peak corresponding to Cu NP becomes noticeable. Upon further action on the colloidal solution, the peak corresponding to brass nanoparticles disappears, and only the plasmon resonance peak of Cu NP remains. Qualitative evidence of the effect is the absence of small brass nanoparticles (with a radius smaller than 5 nm) upon the laser ablation of a brass target in ethanol. Estimates show that the time of diffusion of zinc atoms to a distance of about the diameter of a nanoparticle in the liquid state is of order of several ns, which is comparable with duration of the laser pulses commonly used in experiments [54]. Small nanoparticles rapidly lose zinc upon laser irradiation and are transformed to copper NP oxidized by air oxygen.

A similar shell is observed under laser exposure of bronze NP. In this case the shell thickness is only few nm due to lower content of the component with low melting temperature (Sn) (Figure 13). Plasmon resonance of bronze NP is less pronounced than that of brass NP though is merely different from both Cu and brass NP.

Small radius of NP is responsible for decrease of their melting temperature compared to bulk material. Also, it has been theoretically demonstrated that the melting temperature is different for different shapes of NP composed of the same material [55].

Internal segregation of NP under laser exposure leads to repulsion of the components with lower melting temperature. In this sense one may speak about purification of the core material under laser exposure.

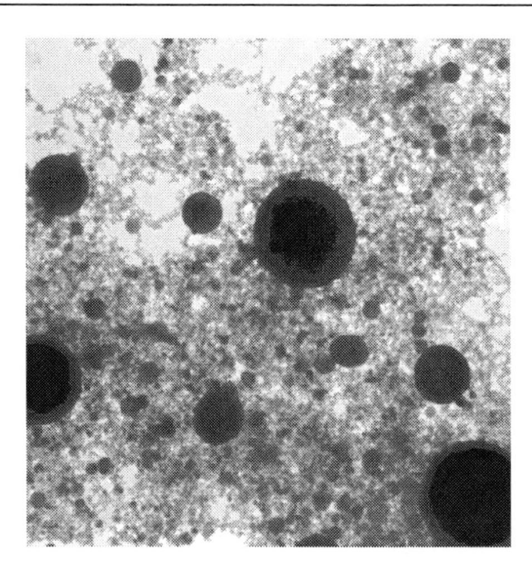

Figure 13. TEM view of core-shell bronze NP synthesized by laser ablation of corresponding targets (8% Sn, 92%Cu) and then subjected to laser radiation in suspension. The diameter of big NP is about 40 nm.

Self-influence of a Femtosecond Laser Beam

Remarkably, the initial average size of nanoparticles produced by ablation of metallic targets is almost independent on the duration of laser pulse from 100 fs to 100 ns and is of order of 10 nm. This implies that the size is determined rather by thermal properties of the target. Indeed, the layer of metal molten by a laser pulse is the only source from which the nanoparticles are formed. The difference in size becomes pronounced under sufficiently long laser exposures when the majority of generated nanoparticles pass through the laser beam in the liquid. In this case the key parameter that determines the efficiency of coupling is detuning of laser wavelength from the plasmon resonance. In certain cases a good coupling of laser radiation to the ensemble of NP is achieved via non-linear transformation of laser radiation by NP themselves. Indeed, the wavelength of a femtosecond Ti:sapphire laser (810 nm) is too far from the plasmon resonances of both Au and Ag NP. However, prolonged laser exposure of the colloidal solution leads to drastic difference in their average size (14 nm vs 4 nm) [21]. Also, there is a significant amount of even smaller Ag nanoparticles whose size is below the measuring limits of the microscope (< 2 nm). The ablation is accompanied by intense generation of the second harmonics of laser radiation at 405 nm. This wavelength exactly fits the plasmon resonance of Ag NP, while for Au nanoparticles it is significantly detuned. The effect of second harmonics generation (SHG) on small metal clusters is well documented in the literature [56, 57]. Despite too small size of metal clusters compared to the laser wavelength, SHG can be very effective due to their high density, so that the conditions of spatial synchronism are easily fulfilled. Several models account for high value of the second order non-linear susceptibility $\chi^{(2)}$ and for efficient generation of the second harmonics. This is assigned to quantum size effect for wave functions of electrons confined to small metal spheres, which gives rise to quadruple oscillations and, consequently, to non-zero $\chi^{(2)}$ even for nanoparticles with a center of symmetry. The observed difference in particle size is attributed to the self-influence of laser radiation upon generation of nanoparticles via laser ablation in liquids. Indeed, appearing nanoparticles double the frequency of laser radiation. Owing to exact matching to plasmon resonance of Ag NP, their temperature can be much higher than under exposure to initial laser radiation. This causes fragmentation of nanoparticles that are in resonance with the second harmonics that is Ag ones.

The third order non-linear susceptibility of NP $\chi^{(3)}$ is responsible for non-linear lensing of laser radiation inside the colloidal solution. As a result, the focusing conditions change during the accumulation of NP in the beam path, and the ablation rate decreases even though laser radiation is largely detuned from plasmon resonance of NP.

In general, the limiting factor of interaction of NP immersed in a liquid with a femtosecond laser beam is generation of a white continuum. This continuum consumes a significant part of laser energy and reduces the efficiency of interaction with NP. This is well pronounced for pulse durations less than 60 fs.

Influence of the Nature of the Liquid

Ablation of a Ti Target

The composition of NP produced by laser ablation of Ti was found to be dependent on the nature of the liquid [16] (see Figure 14). Namely, laser ablation of Ti in ethanol results in formation of Ti nanoparticles having the cubic structure. It should be reminded that the initial polycrystalline Ti plate has tetragonal structure. It is known that the cubic phase of Ti is metastable and exists only at elevated temperature > 600° C. At the same density of laser energy laser ablation of Ti target immersed into dichlorethane leads to the formation of nanoparticles of titanium carbide TiC. Four the most intense peaks of TiC can be distinguished on the X-ray diffractogram of the evaporated suspension. Finally, laser ablation of Ti in water results in the formation of nanoparticles having the composition of non-stoichiometric oxide TiO_x, where x = 1.04. The width of the peaks indicates the small size of the nanoparticles. All the diffractograms contains several unidentified peaks.

Formation of TiC via ablation of Ti in dichlorethane may be due to catalytic action of Ti nanoparticles, since no carbon-containing particles have been observed during ablation of either Si or Au in this liquid under otherwise equal experimental conditions. It is worthwhile to mention that the surface of Ti target after laser ablation in dichlorethane does not contain any detectable by X-ray diffraction amount of TiC. This indicates the formation of TiC nanoparticles via chemical reaction of ejected metal nanoparticles with liquid vapors.

Unlike to what is believed, NP generated by laser ablation in liquids are always crystalline, and the fraction of amorphous component is usually very low.

Oxidation state of NP material is strongly influenced by the nature of surrounding liquid even in case noble metals, e.g., Au. It was confirmed by ablation of Au in water, ethanol, and chloroform [58]. X-ray Photoelectron Spectroscopy (XPS) analysis of dried colloids showed different degree of oxidation of Au. The formation of a gold–chlorine compound is suggested in case of ablation of Au in chloroform with 5 ns laser shots at wavelength of 532 nm.

Another example of the influence of the liquid on the chemical composition of NP generated via laser ablation of a solid target is copper. In this case final NP are metallic only under ablation in acetone, while water and ethanol lead to non-metallic NP.

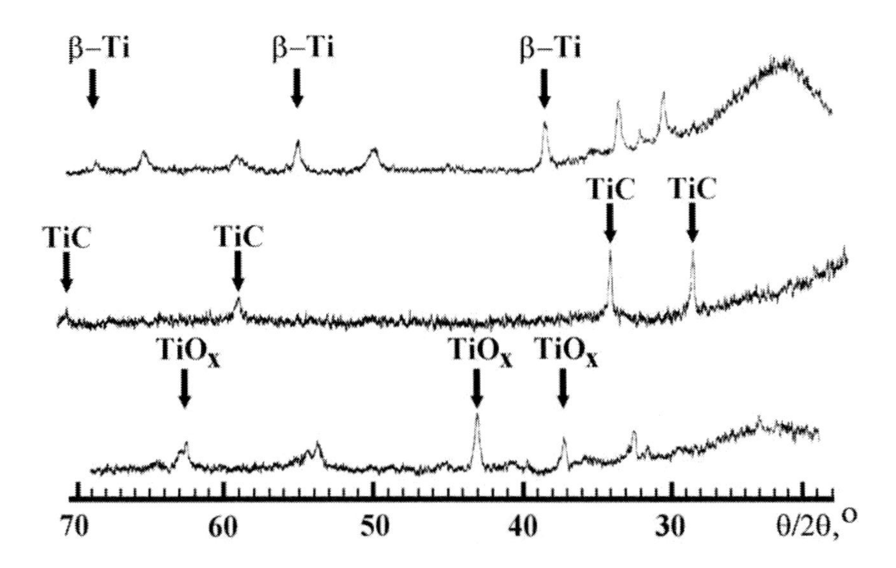

Figure 14. Difractograms of NP synthesized by laser ablation of metallic Ti target in different liquids. Ethanol (upper curve), dichlorethane (middle), and water (bottom curve). The size of nanoparticles calculated from broadening of diffraction peaks is 25, 35, and 25 nm, respectively. Cu vapor laser, fluence of 4 J/cm^2.

Ablation of Sn

Tin is a metal with low melting point, and laser fluence required for its ablation is much lower than for other metals mentioned above. A certain dependence of the particles composition on the concentration of a surface-active substance is found in this case [59].

Non-stoichiometric tin oxide SnO_{2-x} is formed under laser ablation of a tin target in aqueous solutions containing SDS [60]. Ablation of Sn in either water or ethanol at moderate laser fluence (less than 0.2 J/cm^2 at nanosecond pulse duration) leads to formation of Sn NP. This is corroborated both by X-ray diffraction of the dried suspension and its optical absorption spectra.

Since melting temperature of bulk Sn is low, Sn NP are liquid at room temperature. In general, this is due to high fraction of surface atoms in NP compared to their total number. However, evaporation of their colloidal solution in absence of surface-active substances leads to aggregation, and X-ray diffraction shows narrow peaks of Sn NP about 100 nm in diameter. A gold plate immersed into colloidal solution is rapidly covered by numerous layers of Sn. Once fixed on a large substrate, Sn NP are solidified and form a kind of nano-contacts to the substrate. Mixing Sn NP and Au NP in ethanol leads to wetting of Au NP by liquid Sn and therefore to specific core-shell NP. Their TEM view is presented in Figure 15.

The thickness of the shell depends on the volume of the Sn NP. Large Sn NP are solidified just after the contact with Au NP, and no shell is formed. Smaller Sn NP envelop Au NP resulting in a shell with homogeneous thickness. Diffuse entities are tentatively assigned to tin oxide.

At picosecond pulse duration the resulting NP consist mostly of tin oxide, which is confirmed by their characteristic absorption peak at 360 nm.

W and Mo NP

Other metals that might interact with surrounding liquid during ablation remain metallic and are oxidized very slowly with air oxygen that is dissolved in liquid. This concerns NP of either W or Mo upon their ablation in water. These metals with high melting temperature are dispersed into surrounding liquid as metallic NP, which is corroborated by both X-ray diffraction and optical spectra of corresponding colloidal solutions.

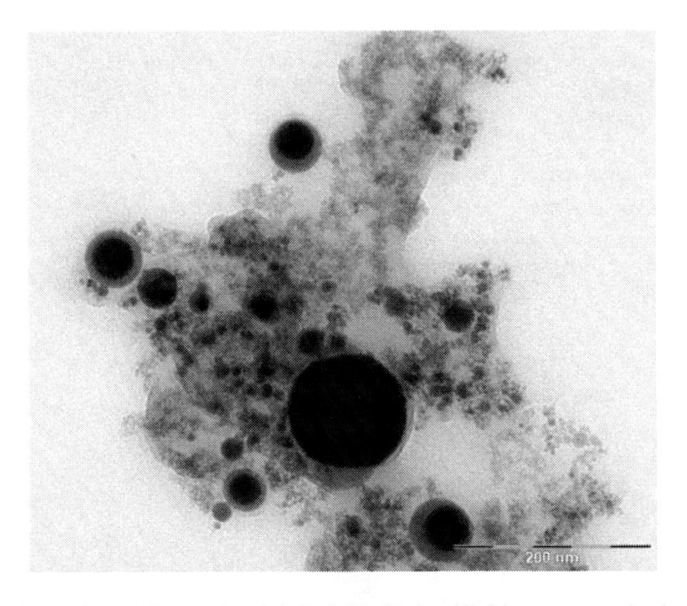

Figure 15. TEM view of Au NP wetting by Sn NP. Both colloids were synthesized by ablation of corresponding metallic targets in ethanol using radiation of a Cu vapor laser. Scale bar denotes 200 nm.

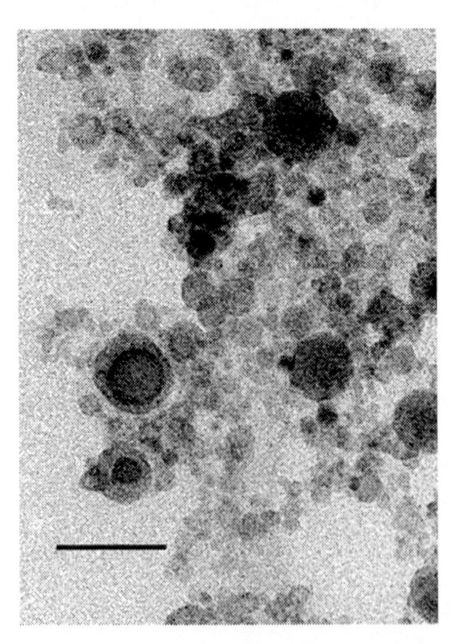

Figure 16. TEM view of W NP synthesized by ablation of a W target in ethanol by radiation of a Cu vapor laser. Scale bar denotes 50 nm.

Their morphology alters during several months of storage at room temperature due to oxidation, which results in the formation of core-shell NP (see Figure 16). Like many other metals, the oxide layer prevents further oxidation of the metallic core. Similar result has been reported for W NP obtained by ablation of a W target in ethanol using high-repetition-rate copper bromide laser radiation [61]. The authors deduce the formation of non-stoichiometric oxide WO_{3-x} around W NP on the basis of absorption spectra of the colloidal solution.

Carbon NP can also be obtained by laser ablation of a graphite target immersed into liquid [62].

A peculiar feature of laser ablation in water is enhanced solubility of some compounds in it under high temperature and high transient pressure that are established during the laser pulse. This leads to so called hydrothermal synthesis of NP of unusual morphology, e.g., platelets or disks of $Pb(Zr,Ti)O_3$ [63]. Aqueous solutions at pressure and temperature that exceed its critical value take on the ability to dissolve solids that are not soluble in normal conditions. Synthesis of compounds via their dissolution in supercritical compounds is called the hydrothermal one. In nature this type of synthesis occurs in the vicinity of underwater volcano, which results in formation of various minerals. Hydrothermal synthesis is typical for non-metallic compounds, such as oxides of more complex ceramics. The material of the target is dispersed and dissolved in a supercritical medium during the laser pulse. Upon cooling the solubility drops, and the formation of NP occurs from oversaturated solution. Apparently, similar process may take place under laser ablation of single crystal rutile in an aqueous environment [64].

The specific surface of NP is very high, for instance, 1 ml of Au colloid prepared by laser ablation of a bulk target at typical density has the surface of 30 m^2. High surface favors catalytic reactions in liquids. Not only NP material but also the liquid itself may undergo chemical changes during laser ablation. In some cases these changes are catalyst-specific that is occur with NP of some specific metals. Laser exposure of most metallic targets (Sn, brass, etc.) in ethanol leads to its noticeable modification towards formation of products with higher molecular mass. Indeed, a wide absorption band centered at 350 nm is observed that correspond to a viscous non-volatile liquid that remains after evaporation of the colloidal solution synthesized by laser ablation. This product might be a low-molecular mass polyethylene, which may also form a kind of shell on the synthesized NP. Formation of this product occurs mostly on the stage of exposure of NP to laser radiation, and catalytic effects of its formation mediated by NP of various compositions are not excluded.

Modeling of Distribution Function

Modeling of distribution function of nanoparticles size can be performed by a numerical solution of kinetic equation on the basis of first principles. The following processes should be taken into account: generation of nanoparticles by laser ablation of a solid target immersed into liquid, aggregation of nanoparticles, their fragmentation and escape of NP adsorbed on the wall of a reactor [25, 65]. In the specific conditions of laser ablation in liquids, the last term can be negligibly small unlike laser ablation in vacuum. On the contrary, fragmentation of the particles in the laser beam is the main channel of modification of their distribution function. The peculiar feature of ablation in liquids is simultaneous action of all 4 processes, since the generated NP remain in the liquid and, in particular, on the optical path of laser beam. Formation of NP due to coagulation of individual metal atoms ejected from the target is unlikely, since the necessary condition of their synthesis is melting of the target. NP are ejected from the target at the very beginning of ablation.

Melting of NP and their subsequent fragmentation may occur only in those areas of the liquid where the peak power of laser radiation is sufficiently high. The model takes into account a factor proportional to the ratio of the volume of laser beam waist to total volume of the liquid. In typical experimental conditions this factor is about 10^{-4}, which explains the need of prolonged laser exposures required to alter the distribution function at measurable level.

Laser exposure of NP in liquid in absence of the metal target results in their fragmentation due to absorption of laser energy by individual particles. Under laser fluence typically used NP are molten during the laser pulse, and their splitting is due to hydrodynamic instabilities of the molten metal embedded into a vapor pocket. According to the model derived above, the size of the particles goes to 0 under long exposure. In reality for each value of laser fluence there is a certain minimal size of particles that are too small to absorb the energy sufficient for their melting from laser beam. The model that takes into account all above-mentioned considerations shows good qualitative agreement with the experimentally measured distribution function [25].

Influence of Intensity Distribution of the Laser Beam on the Shape of Nanoparticles

The influence of the laser beam profile onto the morphology of NP generated via laser ablation of solids in liquids is not studied so far and is often ignored in current scientific literature. The origin of this influence is a certain value of the energy density of the laser beam, which is needed to melt the target material. The size and shape of NP generated via laser ablation are sensitive to the laser fluence. Typically higher fluence leads to higher average size of generated NP. Also, in case of laser ablation of Au at elevated laser fluence the generated NP have elongated shape (see Figure 6). The term "elevated" should be understood in term of laser fluence needed to melt the target surface. One may suggest that this is due to specific way of interaction of the melt bath within the laser spot with surrounding medium. Indeed, sputtering of the melt is due to splashing of the molten layer on the target surface by high pressure vapors of surrounding liquid. However, the ejection of the melt in the central areas of the laser spot is different from that at its periphery. The melt ejected in the spot center has high probability to return to the target surface being pushed by the expanding vapor cloud. On the contrary, ejection of the melt from the periphery of the melt bath very likely results in the formation of NP, since the lateral dimensions of the expanding vapor pocket above the target are quite close to the laser spot size. This means that only the edges of the molten area contribute to NP formation, while the central part of the laser spot provides relatively small fraction of NP.

These considerations suggest the following experiment on NP generation in which the laser beam that causes ablation of a target immersed into liquid has another type of symmetry of its intensity distribution different from that of a flat-top or Gaussian profile. A good possibility is formation of a periodic intensity distribution in the plane of the target. Such periodic distribution may occur spontaneously due to, for example, interference of the laser beam with Surface Electromagnetic Wave excited in the solid by this beam. Another possibility is exposure of the target to the interference pattern of two coherent laser beams. In this case the laser fluence is a periodic function along one of coordinate, so does the temperature distribution as schematically shown in Figure 17.

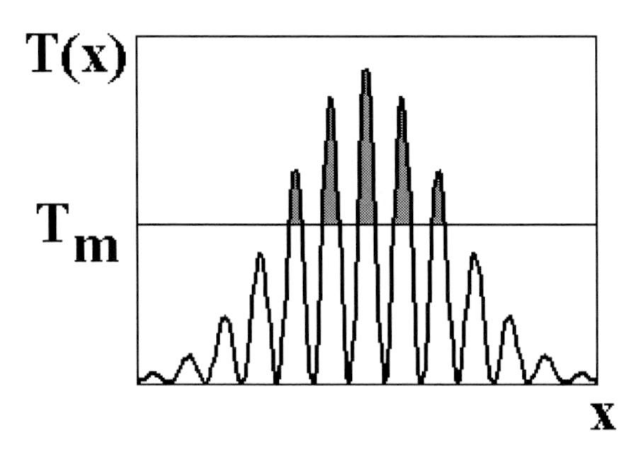

Figure 17. Temperature distribution on a target exposed to two coherent laser beams. Solid line $T(x)=T_m$ indicates melting temperature of the target.

If the fluence in the maxima of interference pattern are sufficient to melt the target, then the melt baths will have elongated shape on the target, which in turn may affect the shape of ejected NP.

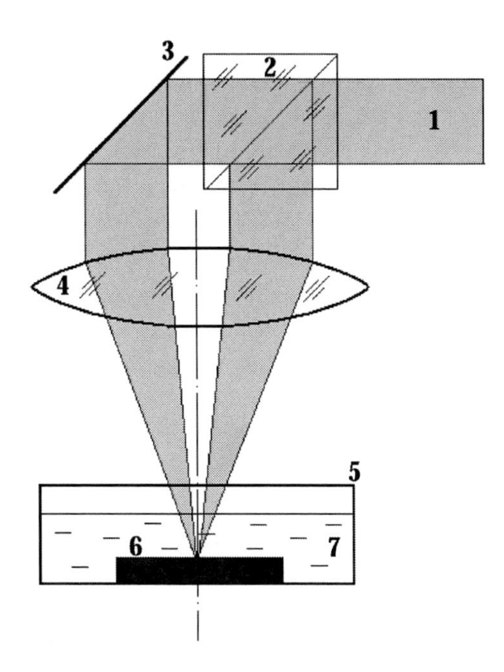

Figure 18. Experimental setup on laser-assisted synthesis of elongated Au NP. 1 – laser beam (Cu vapor laser, 510.6 nm laser output), 2 – splitting cube, 3 – dielectric mirror, 4 – focusing lens, 5 – cell, 6 – Au target, 7 – liquid ethanol.

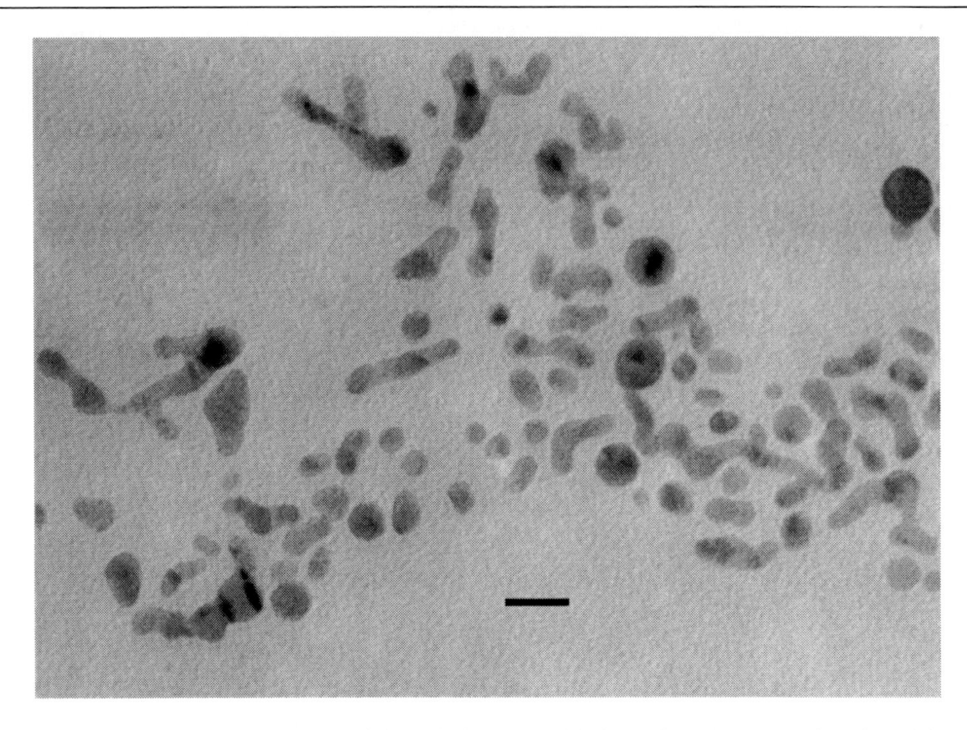

Figure 19. TEM view of Au nanoparticles obtained by ablation of an Au target in ethanol by two interfering laser beams of a Cu vapor laser. Scale bar denotes 50 nm.

The experimental setup is shown in Figure 18. A Cu vapor laser emits at two laser lines, either green (510.6 nm) or yellow (578.2 nm). In this particular experiment only one output was used, namely, the green one. The green laser beam was split into two beams of approximately equal intensities using dividing glass cube. These two beams were superimposed on the surface of a gold target placed into liquid ethanol. The angle between two beams corresponded to the interference pattern on the target surface with period of 4 μm. The cell with gold sample was moved under the laser beams in the direction parallel to the maxima of interference pattern.

Ablation of the Au target in these conditions results in visible coloration of the liquid, and the color of the solution has bluish tint typical of elongated Au NP. TEM examination of generated Au colloid confirms the presence of elongated Au NP, as demonstrated in Figure 19. Some Au NP are indeed elongated, and their aspect ratio (ratio of length to diameter) exceeds 10.

Nanostructuring of Solids under their Laser Ablation in Liquids

The total laser fluence absorbed by a target should be sufficient to melt it in order to produce NP in the liquids. Ideally smooth surface is characterized by definite threshold laser fluence needed for melting. In practice the surface of a target has certain micro-roughness, and this surface may not melt simultaneously. Indeed, any protrusion which is weakly thermally coupled to the bulk of the target is molten at lower fluence than the smooth target. Therefore, at certain fluence a rough surface will not be molten only in some protruding areas. The vapors of surrounding liquid that expands from the target will then carry away the molten fragments thus forming spikes in its surface. The rate of NP generation in the liquid is very low, so at threshold laser fluence the material tends to leave the target in the form of NP but remains on it solidified as nano-spikes or nano-bumps. Period of nano-spikes does not depend on the laser wavelength and is determined mostly by characteristics of the target material.

Formation of NS on the target is accompanied by modification of its absorption spectrum. Indeed, the electrons that oscillate within single NP are also confined, and therefore the spike edges act like the edge of NP as soon as their dimensions are compared with mean free path of electrons at Fermi level (see Figure 20).

This confinement causes the coloration of some metals with formed NS. For instance, Ag plate with NS looks yellow in appearance [66]. A macro view of Ag surface with NS is shown in Figure , a. Yellowish coloration in Ag is typical of silver NP.

Absorption spectrum of both initial Ag target and of Ag with NS is shown in Figure 21.

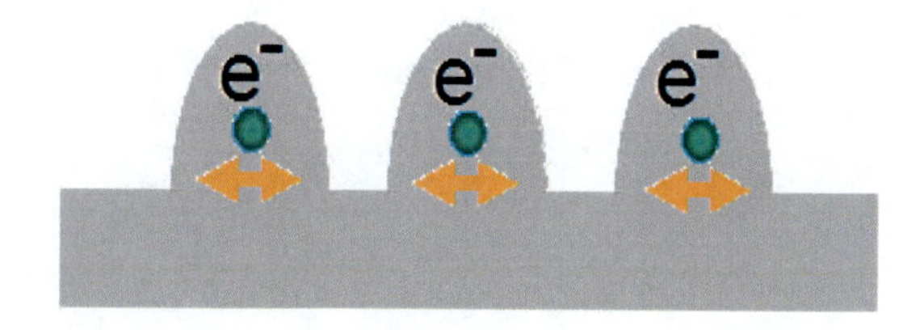

Figure 20. Origin of the surface coloration. Schematic view of oscillations of free electrons in metallic nanostructures. Plasmon resonance of these electrons is close to that of nanoparticles of the same lateral dimensions.

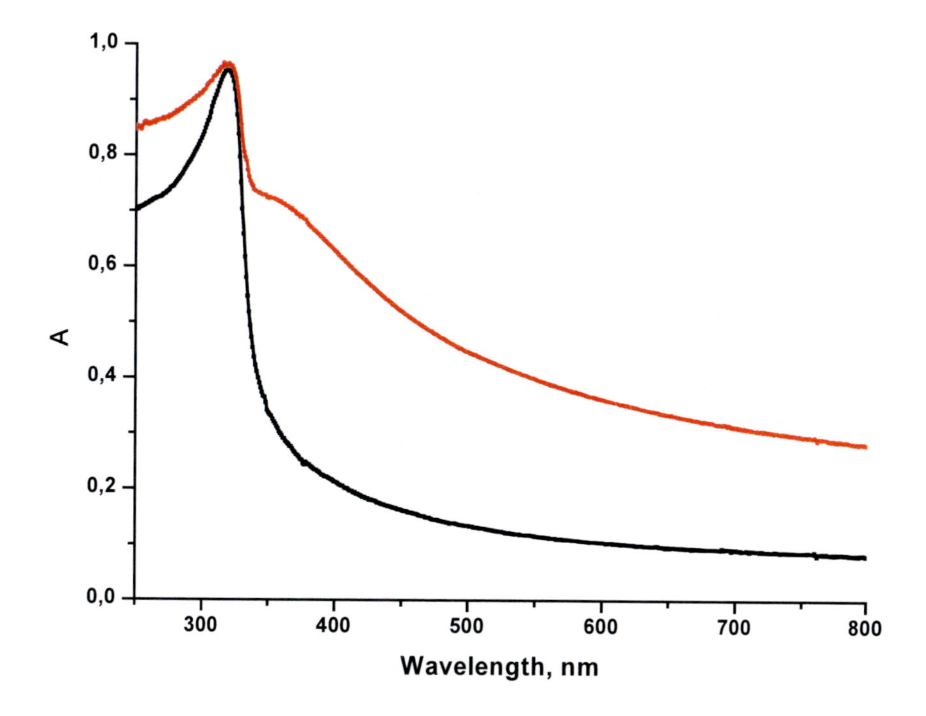

Figure 21. Absorption of an Ag target ablated under water layer with 5 ps laser pulses at wavelength of 248 nm. Black curve corresponds to initial Ag target while the red one to Ag with NS.

Initial spectrum corresponds to plasmon resonance of electrons in bulk Ag. Formation of NS manifests itself in broadening of this resonance. A new maximum appears near 360 nm. This maximum shifts to red with oxidation of Ag upon storage in air during several days. Similar modifications of absorption have also been observed in large variety of metals, e.g., Ti, W, Cu, etc., subjected to laser ablation in liquids at threshold laser fluence.

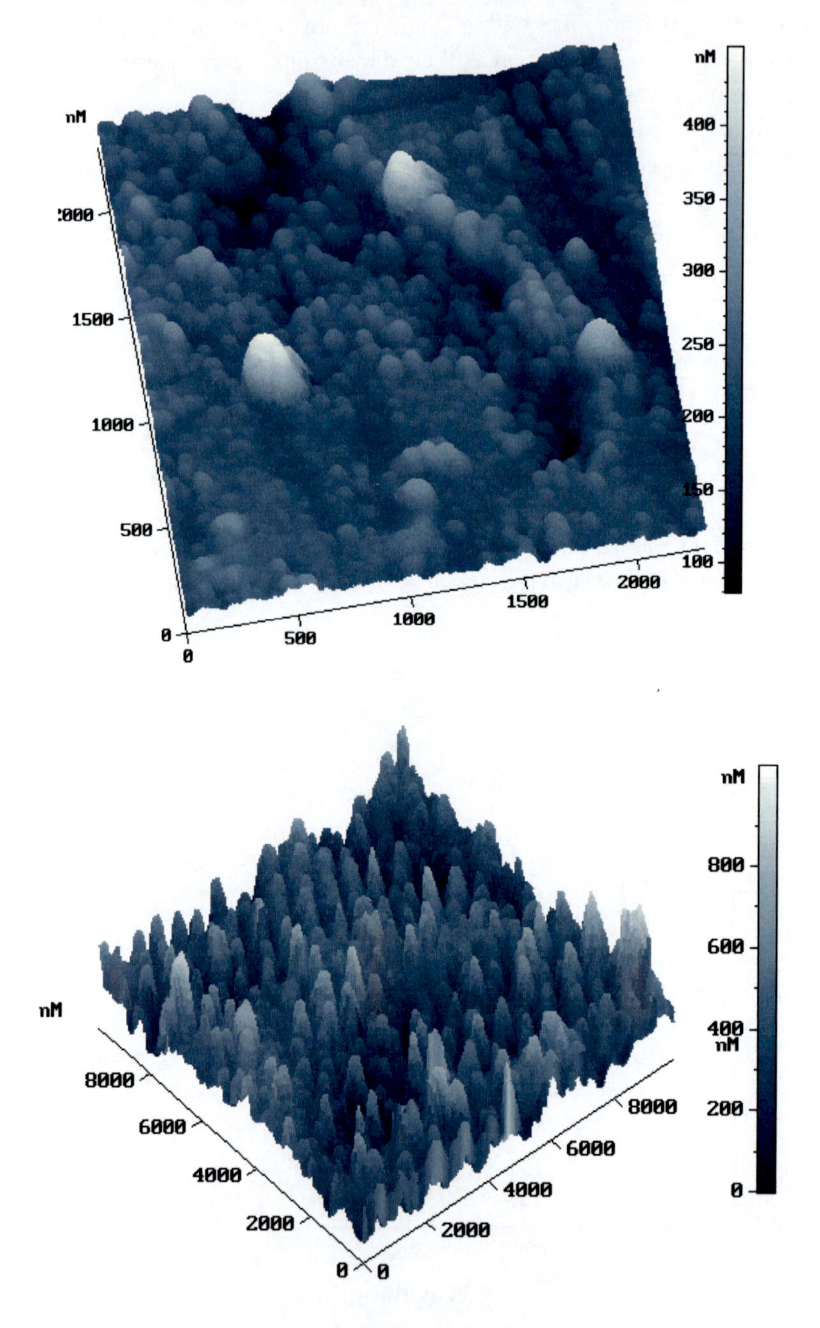

Figure 22. AFM view of NS on Ag (top) and Ta (bottom). NS were generated by ablation of corresponding metal target in water, pulse width of 350 ps at wavelength of 1.06 μm.

Figure 23. Schematic picture of detachment of nanodrops from the melt on the surface of a solid target.

Typical morphology of both Ag and Ta target with NS formed by laser ablation in water is presented in Figure 16. One can see that the period of NS is about 50 nm for Ag and 300 nm for Ta, which is much smaller than laser spot. In this sense NS are self-organized. NS on Ta are characterized by high aspect ratio, while on Ag they are rather nano-bumps.

Formation of NS on metallic targets ablated in liquids requires sufficiently short laser pulses. Indeed, no NS are formed with nanosecond pulse duration. The longest tested pulses that produce NS are 350 ps delivered by a Nd:YAG laser at wavelength of 1.06 μm. Another necessary condition for formation of NS is certain initial roughness of the target surface. The local roughness required for NS formation should exceed 30-50 nm. These two conditions allow suggesting local melting of the target in the micro-protrusions of the target relief, which develops with further increase of the number of laser shots into an array of self-organized nano-spikes.

The process of NS formation under laser ablation of solids immersed into liquids requires further studies. However the experimental data available so far indicate relation of NS to some instability that develops at the interface "melt-vapor of liquid" within the laser spot. In typical conditions the pressure of vapor of the liquid exceed by at least one order of magnitude the capillary pressure stipulated by curvature of the melt surface, so the recoil pressure of vapor that expands from the target plays dominant role.

Nanostructures on Ag generated by its laser ablation in water show the SERS effect of adsorbed organic molecules [67]. The estimated enhancement factor is about 10^5. Unlike Ag colloid in a liquid, Ag plate with nanostructures can be used for SERS measurements many times, since the NS do not degrade upon measurements.

Different stages of NS development are schematically shown in the Figure 23. These structures correspond to morphology that can be found within the laser spot on the target. At sufficiently high laser fluence the nano-drop of the melt is detached from the target surface into surrounding liquid as nanoparticles.

It is also clear that probe microscopes fail to image correctly the mushroom-like structures. This type of nanostructures is imaged by an AFM as a cone.

Small size of NS implies another mechanism of their formation, which should not involve capillary waves. Indeed, capillary waves with period of hundreds of nanometers would be suppressed since capillary pressure at this scale tends to smooth the melt.

Thus, formation of nanostructures under laser ablation of metals in liquid environment is a general conformity. It is characterized by several features. First, nanostructures are observed only in case of sufficiently short laser pulses. Second. Their period foes not depend on the laser wavelength. Nanostructures are merely different from periodic ripples that are generated owing to interference of a Surface Electromagnetic Wave with laser beam.

The reviewed results on laser-assisted generation of nanoparticles via laser ablation in liquids represent only one point of view. Recent review [69] and recent book [70] consider another processes and approaches that have been developed during last decade.

Excitation of High Energy Levels

In typical conditions of experiments on laser ablation of solids in liquid environment the nanoparticles (NP) sputtered into liquid are optically thin at the wavelength of most common lasers. Despite their small size (about 10 nm), the NP can efficiently absorb the laser radiation. The efficiency of this interaction is a function of numerous experimental parameters, such as particle size, detuning of laser frequency from the position of the plasmon resonance of NP, etc. However in most cases, NP are optically thin, that is almost transparent at the laser wavelength. Hence, the time required to reach their temperature under laser exposure is shorter than the duration of laser pulse as soon as the latter exceeds the characteristic time of electron-phonon relaxation. The temperature of NP inside the laser beam is proportional to the peak power of the laser radiation. Estimations show that the temperature of Au nanoparticles inside Cu vapor laser beam with peak intensity of 10^8 W/cm^2 is several kK, and this temperature scales linearly with laser intensity. Therefore, exposure of NP suspension in liquid environment is a novel approach that allows excitation of high energy levels of both the NP material and its close environment. The liquid surrounding the NP in this case turns into a vapor having high temperature and pressure, and eventually into plasma that perfectly surrounds it. The present paper describes several proofs that laser exposure of NP suspension in liquid may lead to excitation of high energy levels including those of nuclei and neutron release. Laser initiation of neutron emission may open new possibilities for transformation of elements, synthesis of desired isotopes, disposal of nuclear waste, etc. The suspension of Hg in D_2O is a model system for laser-assisted transmutation of Hg into Au since the neutron binding energy in D has the lowest value [71,72].

Two kinds of Hg were used for a model system of transmutation of Hg into Au: (i) Hg of analytical purity with natural isotopic composition or (ii) enriched Hg containing 55.6% of ^{196}Hg and 41.4 % of ^{199}Hg, with much lower content of other isotopes, obtained by selective photochemical reaction of Hg with oxygen [73].

Exposure of the colloidal solution of Au NP in water to radiation of a 350 ps Nd:YAG laser is accompanied by visible emission from the solution. The spectrum of this emission contains several spectral features. The first one corresponds to anti-Stokes Raman scattering of H_2O vapors of laser radiation excited at 1060 nm. The second broad peak coincides with the emission of atomic Au(I) at 627.8 nm during transition from the resonant to the meta-stable level. The peak at 530 nm coincides with the position of plasmon resonance of Au NP in H_2O. One can also see that the laser-produced plasma is characterized by intense UV radiation at least up to 200 nm where water itself and Au NP have strong absorption [74].

If the laser spark is placed in the vicinity of a Be window of the cell one can get an image of a metallic grid on Be using an X-ray photo film, though the exposure time has to be at least 2 hours. Therefore, both deep UV and X-ray radiation is produced during exposure of Au NP colloid in H_2O to laser radiation with rather moderate intensity. This result is in agreement with previous observations of X-ray emission under exposure of aqueous solutions to femtosecond laser radiation [75,76].

The energy of X-ray photons is sufficient to cause the modification of the NP environment. This has been confirmed in a model experiment on exposure of Au NP in D_2O to radiation of a 350 ps Nd:YAG laser at peak intensity of 10^{10} W/cm^2. Raman spectrum of

D_2O with Au NP subjected to 2 hours laser exposure indicates the formation of HDO molecules.

Suspension of Hg in D_2O is characterized by an absorption peak in the range 270-290 nm even prior to laser exposure. This peak is assigned to the plasmon resonance of Hg NP, which is close to the theoretical position of 290 nm reported for 10 nm NP of Hg in H_2O [18]. Therefore, in the initial stages, exposure of Hg suspensions to laser radiation leads to the formation of Hg NP (size of the order of 10 nm). Using either a Cu vapor laser (8 hours of exposure) or a 100 femtosecond Ti:sapphire laser (2 hours of exposure), no Au content was detected within the accuracy of measurements.

Table 1. Analysis of Hg samples of natural isotope composition exposed to radiation of a 90 ps Nd:YAG laser, exposure time of 60 min

Sample	Hg in D_2O initial	Hg in D_2O sample 1	Hg in D_2O sample 2
Hg, mg/l	35	23	8
Au, mg/l	0.009	0.1	0.024
Au/Hg ratio	0.00026	0.0043	0.003

Table 2. Exposure of suspension of ^{196}Hg in D_2O to radiation of a 350 ps Nd :YAG laser. The results are averaged over 4 different exposures

Sample	$^{196}Hg/D_2O$ initial	$^{196}Hg/D_2O$ exposed Hg drop	$^{196}Hg/D_2O$ 4 hours of exposure (averaged over 4 samples)	$^{196}Hg/D_2O$ after sedimentation
Au, mg/l	0.0073	0.38	0.23	0.17
Hg, mg/l	20	8.94	2.31	12.6
Au/Hg	0.00036	0.0425	0.10	0.0135

After exposure to a 90 ps Nd:YAG laser radiation, the suspension precipitates very slowly, unlike a freshly prepared suspension in ultrasonic bath. After 1 day of sedimentation the liquid separated into two parts, the first close to the bottom and the other remaining in suspension. Analytic data presented in Table 1 show the formation of Au.

Experiments with a 350 ps Nd:YAG laser were carried out with two types of Hg samples, either of natural isotopic composition or enriched with ^{196}Hg. Results of analysis of Hg suspension of natural isotope composition in D_2O are similar to those obtained with a 90 ps laser. The formation of Au in exposed suspensions of Hg in D_2O is confirmed by analysis, though in general the Au/Hg ratio is lower.

The results of analysis for Hg enriched with ^{196}Hg are presented in Table 2. Note that Au content is found under laser exposure of a Hg drop immersed into D_2O for about 10 min in order to disperse the metal. Four hours of laser exposure increases the Au content to almost 10%.

The mechanism of transmutation is still under discussion. The effect is observed only under sufficiently long laser pulses of picosecond range. One may suggest that the beginning of a laser pulse ionizes the nanoparticles within the laser beam while the remaining laser energy is spent for acceleration of electrons as it is observed in the case of interaction of strong laser beams with plasma [77]. These electrons provide X-ray photons needed for release of a neutron from Deuterium.

Conclusion

The research on the processes of laser ablation in liquids is still in progress. There is a certain hope that the simplicity of the experimental setup on laser-matter interaction will allow developing new cost-effective technologies. Several specific features, such as the absence of a vacuum, may provide wide use of liquid-assisted generation of nanoparticles in medicine, catalysis, etc. The novel process of laser-assisted nanostructuring of solids via their laser ablation in a liquid environment is also of high interest for various applications in medicine, biology, aero- and hydrodynamics.

References

[1] J. Nedersen, G. Chumanov, and T.M. Cotton, *Appl. Spectrosc.*, 1993, 47, 1959.
[2] M. S. Sibbald, G. Chumanov, and T.M. Cotton, *J. Phys. Chem.*, 1996, 100, 4672.
[3] A. Fojtik, A. Henglein, *Ber. Busenges. Phys.Chem.*, 1993, 97, 252.
[4] A. Henglein, *J. Phys.Chem.*, 1993, 97, 5457.
[5] M. Procházka, Štepánek, B. Vickova, I. Srnova, P. Maly, *J. Mol. Struct.* 1997, 410/411, 213.
[6] M. Procházka, P. Mojzeš, J. Štepánek, B. Vlèková, P.-Y. Turpin, *Anal. Chem.*, 1997, 69, 5103.
[7] M. Procházka, J. Štepánek, B. Vlèková, I. Sronová, P. Malŷ, *J.Mol. Struc.*, 1997, 410, 213.
[8] I. Srnová, M. Procházka, B. Vlèková, J. Štepánek, P. Malŷ, *Langmuir,* 1998, 14, 4666.
[9] J.S. Jeon, C.H. Yeh, *J. Chin. Chem. Soc.*, 1998, 45, 721.
[10] M.-S. Yeh, Y.-S. Yang, Y.-P. Lee, H.-F. Lee, Y.-H. Yeh, and C.-S. Yeh, *J. Phys.Chem.*, 1999, B 103, 6851.
[11] A. Takami, H. Kurita, and S. Koda, *J. Phys.Chem.,* 1999, B103, 1226.
[12] S. Link, C. Burda, B. Nikoobakht, and M.A. El-Sayed, *J. Phys.Chem.*, 2000, B104, 6152.
[13] F. Mafuné, J. Kohno, Y. Takeda, T. Kondow, H. Sawabe, *J. Phys. Chem.*, 2000, B 104, 9111.
[14] F. Mafuné, J. Kohno, Y. Takeda, T. Kondow, H. Sawabe, *J. Phys. Chem.*, 2001, B 105, 5114.
[15] A.V. Simakin, V.V. Voronov, G.A. Shafeev, R. Brayner, and F. Bozon-Verduraz, *Chem. Phys. Lett.*, 2001, 348, 182.
[16] S.I. Dolgaev, A.V. Simakin, V.V. Voronov, G.A. Shafeev, F. Bozon-Verduraz, *Appl.Surf.Sci.*, 2002, 186, 546.
[17] P. V. Kamat, M. Flumiani, and G.V. Hartland, *J. Phys.Chem.*, 1998, B 102, 3123.
[18] J.A. Creighton and D.G. Eadon, *J. Chem. Soc. Faraday Trans.*, 1991, 87, 3881.
[19] P. Mulvaney, *Langmuir*, 1996, 12, 788.

[20] R.A. Ganeev, A.I. Ryasnyansky, R.I. Tugushev, and T. Usmanov, *J. Opt. A: Pure Appl. Opt.* 2003, 5, 409.

[21] G.A. Shafeev, E. Freysz, and F. Bozon-Verduraz, *Appl. Phys.,* 2003, A78, 307.

[22] V.A. Karavansky, A.V. Simakin, V.I. Krasovsky, and P.V. Ivanchenko, *Quantum Electronics*, 2004, 34(7), 644.

[23] P.V. Kazakevich, A.V. Simakin, and G.A. Shafeev, *Quantum Electronics*, 2005, 35, 831-834.

[24] S. Link and M. El-Sayed, *J. Phys. Chem.,* 1999, B103, 8410.

[25] A.V. Simakin, V.V. Voronov, N.A. Kirichenko, and G.A. Shafeev, *Appl. Phys.,* 2004, A79, 1127-1132.

[26] F. Mafune, J. Y. Kohno, Y. Takeda and T. Kondow, *J.Phys. Chem.*, 2000, *104*, 8333.

[27] T. Tsuji, T. Kakita, and M. Tsuji, *Appl.Surf.Sci.*, 2003, 206, 314.

[28] B. Nikoobakht and M.A. El-Sayed, *Chem. Mater.*, 2003, 15, 1957.

[29] F. Mafuné, J.-Y. Kohno, Y. Takeda, T. Kondow, and H. Sawabe, *J. Phys.Chem.*, 2000, B104, 9111.

[30] F. Mafuné, J.-Y. Kohno, Y. Takeda, T. Kondow, and H. Sawabe, *J. Phys.Chem.*, 2001, B105, 5144.

[31] Y.-H. Chen and C.-S. Yeh, *Colloids Surf.*, 2002,197, 133.

[32] A.V. Kabashin and M. Meunier, *J. Appl. Phys.*, 2003, 94, 7941.

[33] G. Compagnini, A. A. Scalisi, and O. Puglisi, *J. Appl. Phys.*, 2003, 94(12), 7874.

[34] J.-P. Sylvestre, A. V. Kabashin, E. Sacher, M. Meunier, and J. H. T. Luong, *J. Am. Chem. Soc.*, 2004, 126, 7176.

[35] F. Mafuné, J-Y. Kohno, Yo. Takeda, and T. Kondow, *J. Phys. Chem. B* 2003, 107, 4218-4223.

[36] J.S. Jeon and C.H. Yeh, *J. Chin. Chem. Soc.*, 1998, 45, 721.

[37] T. Tsuji, K. Iryo, N. Watanabe, M. Tsuji, *Appl. Surf. Sci.*, 2002, 202, 80.

[38] X.Y. Zhang, L.D. Zhang, Y. Lei, L.X. Zhao, and Y.Q. Mao, *J. Mat. Chem.*, 2001, *11*, 1732.

[39] B.M.I. van der Zande, M. R. Bohmer, L.G.J. Fokkink, and C. Schonenberger, *Langmuir*, 2000, 16, 451.

[40] S.-S. Chang, C.-W. Shih, C.-D. Chen, W.-C. Lai, and C.R. C. Wang, *Langmuir*, 1999, 15, 701.

[41] Y.-Y. Yu, S.-S. Chang, C.-L. Lee, and C.R. C. Wang, *J. Phys. Chem.,* 1997, B101, 6661.

[42] P.V. Kazakevich, A.V. Simakin, G.A. Shafeev, *Chem. Phys. Lett.,* 2006, 421, 348-350. See also in: P.V. Kazakevich, A.V. Simakin, G.A. Shafeev, G. Viau, Y. Soumare, F. Bozon-Verduraz, *Appl. Surf. Sci.,* 2007, 253 7831–7834.

[43] S. Link and M. El-Sayed, *J. Phys. Chem.* 1999, B103, 8410.

[44] Y.-H. Chen, Y.-H. Tseng, and C. S. Yeh, *J. Mater Chem.*, 2002, 12, 1419.

[45] Y. Kim, R.C. Johnson, J. Li, J.T. Hupp, G.C. Schatz, *J. Phys. Chem.,* 2002, 352, 421.

[46] J.-P. Abid, H.H. Girault, and P.F. Brevet, *Chem.Comm.*, 2001, 829.

[47] A.T. Izgaliev, A.V. Simakin, and G.A. Shafeev, *Quantum Electronics*, 2004, 34(1), 47-52.

[48] A.T. Izgaliev, A.V. Simakin, and G.A. Shafeev, and F. Bozon-Verduraz, *Chem.Phys.Lett.*, 2004, 390, 467.

[49] P.V. Kazakevich, A.V. Simakin, V.V. Voronov, and G.A. Shafeev, *Appl.Surf.Sci.*, 2006, *252*, 4373.
[50] P.V. Kazakevich, V.V. Voronov, A.V. Simakin, and G.A. Shafeev, 2004, *Quantum Electronics*, 34(10) 951.
[51] P.V. Kazakevich, V.V. Voronov, A.V. Simakin, and G.A. Shafeev, *Appl. Surf. Sci.*, 2006, 252 4373.
[52] Y.-H. Yeh, M.-S. Yeh, Y.-P. Lee, C.-S. Yeh, *Chemistry Letters*, 1998, 1183-1184.
[53] M.-S. Yeh, Y.-S. Yang, Y.-P. Lee, H.-F. Lee, Y.-H. Yeh, C.-S. Yeh, *J. Phys. Chem.* 1999, *B* 103 6851-6857.
[54] V. V. Voronov, P. V. Kazakevich, A. V. Simakin, and G. A. Shafeev, 2004, *JETP Letters,* 80(11) 684.
[55] M. Wautelet, J. P. Dauchot, and M. Hecq, *Nanotechnology* 2000, 11, 6–9; M. Wautelet, J.P. Dauchot, M. Hecq, *Materials Science and Engineering* 2003, C 23 187–190; R. Vall´ee, M. Wautelet, J. P. Dauchot, and M. Hecq, *Nanotechnology*, 2001, 12 68–74.
[56] K.Y.Lo, J.T. Lue, *Phys. Rev.*, 1995, *B51(4)* 2467. O. A. Aktsipetrov, P. V. Elyutin, A. A. Fedyanin, A. A. Nikulin, A. N. Rubtsov, *Surface Science*, 1995, *325(3)*, 343.
[58] G. Compagnini, A. A. Scalisi, and O. Puglisi, *Phys. Chem. Chem. Phys*., 2002, 4, 2787–2791.
[59] T. Sasaki, C. Liang, W.T. Nichols, Y. Shimizu, N. Koshizaki, *Appl. Phys*. 2004, A79 1489.
[60] C. Liang, Y. Shimizu, T. Sasaki, and N. Koshizaki, *J. Phys. Chem.,* 2003, B107*,* 9220-9225.
[61] M.S.F. Lima, F.P. Ladàrio, and R. Riva, *Appl. Surf. Sci*., 2006, 252, 4420–4424.
[62] G.X. Chen, M.H. Hong, Q. He, W.Z. Chen, H.I. Elim, W. Ji, T.C. Chong, 2004, *Appl. Phys*. A79 1079.
[63] A. Kruusing, *Optics and Lasers in Engineering*, 2004,41, 307.
[64] A. Iwabuchi, C.-K. Choo, and K. Tanaka, *J. Phys. Chem.,* 2004, B108*,* 10863.
[65] F. Bozon-Verduraz, R. Brayner, V.V. Voronov, N.A. Kirichenko, A.V. Simakin, and G.A. Shafeev, *Quantum Electronics*, 2003, 33(8), 714-720.
[66] E.V. Zavedeev, A.V. Petrovskaya, A.V. Simakin, and G.A. Shafeev, *Quantum Electronics*, 2006, 36(10) 978-980.
[67] S. Lau Truong, G. Levi, F. Bozon-Verduraz, A.V. Petrovskaya, A.V. Simakin, and G.A. Shafeev, *Appl.Phys*. 2007, A89 (2) 373 – 376.
[68] S. Lau Truong, G. Levi, F. Bozon-Verduraz, A.V. Petrovskaya, A.V. Simakin, G.A. Shafeev, *Applied Surface Science*, 2007, 254, 1236–1239.
[69] G.W. Yang, *Progress in Materials Science*, 2007, 52, 648–698.
[70] A. Kruusing, *Handbook of Liquids-assisted laser processing*, Elsevier, 2008, 454 p.
[71] G.A. Shafeev, F. Bozon-Verduraz, and M. Robert, *Preprint of GPI, No45*, Moscow, April 12, 2006.
[72] G.A. Shafeev, F. Bozon-Verduraz, and M. Robert, *Physics of Wave Phenomena*, 2007,15(3), 131–136.
[73] Yu.V. Viazovetsky and A.P. Senchenkov, *Journal of Technical Physics*, 1998, 68(1) 67.
[74] G.A. Shafeev, A.V. Simakin, F. Bozon-Verduraz, M. Robert, *Applied Surface Science* 2007, 254 1022–1026.
[75] K. Hatanaka, T. Miura, and H. Fukumura, *Appl.Phys.Lett*., 2002, 80(21) 3925.

[76] K. Hatanaka and H. Fukumura, X-ray generation from optical transparent materials by focusing ultra-short laser pulses, Chapter 9 in: *Three-Dimensional Laser Microfabrication: Principles and Applications*, H. Misawa and S. Juodkazis, Eds. Wiley-VCH, (2006) (199-238).
[77] A. Pukhov and J. Meyer-ter-Vehn, *Appl. Phys.*, 2002, B74, 355.

In: Nanoparticles: New Research
Editor: Simone Luca Lombardi, pp. 39-107

ISBN: 978-1-60456-704-5

Chapter 2

CARBON NANOPARTICLES AS SUBSTRATES FOR CELL ADHESION AND GROWTH

Lucie Bacakova[1,*], Lubica Grausova[1], Marta Vandrovcova[1], Jiri Vacik[2,†], Aneta Frazcek[3,‡], Stanislaw Blazewicz[3], Alexander Kromka[4,§], Bohuslav Rezek[4], Milan Vanecek[4], Milos Nesladek[5,||], Vaclav Svorcik[6,#], Vladimir Vorlicek[7,&] and Milan Kopecek[1]

[1]Institute of Physiology, Academy of Sciences of the Czech Republic, Videnska 1083, 142 20 Prague 4-Krc, Czech Republic
[2]Nuclear Physics Institute, Academy of Sciences of the Czech Republic, 250 68 Rez near Prague, Czech Republic
[3]AGH University of Science and Technology, Faculty of Materials Science and Ceramics, Department of Biomaterials, Al. Mickiewicza 30, 30-059, Cracow, Poland
[4]Institute of Physics, Academy of Sciences of the Czech Republic, Cukrovarnicka 10, 162 53 Prague 6, Czech Republic
[5]CEA-LIST, Centre d'Etudes Saclay, Bat. 451, p. 84, 91191 Gif Sur Yvette, France;
[6]Department of Solid State Engineering, Institute of Chemical Technology, Technicka 5, 166 28 Prague 6, Czech Republic
[7]Institute of Physics, Academy of Sciences of the Czech Republic, Na Slovance 2, 182 21 Prague 6, Czech Republic

[*] E-mail address: lucy@biomed.cas.cz; grausova@biomed.cas.cz; vandrovcova@biomed.cas.cz; kopecek@biomed.cas.cz
[†] E-mail address: vacik@ujf.cas.cz
[‡] E-mail address: afraczek@op.pl; blazew@agh.edu.pl
[§] E-mail address: kromka@fzu.cz; rezek@fzu.cz; vanecek@fzu.cz
[||] E-mail address: Milos.NESLADEK@cea.fr
[#] E-mail address: vaclav.svorcik@vscht.cz
[&] E-mail: vorlicek@fzu.cz

Abstract

Carbon nanoparticles, such as fullerenes, nanotubes and nanodiamonds, have been considered as promising building blocks in nanotechnology for broadscale nanostuctured devices and materials, e.g., microchips, nanorobots and biosensors, carriers for controlled drug and gene delivery or tracers for novel imaging technologies. Soon after macroscopic production was established, fullerenes, as donors and acceptors of electrons, started to be targeted for use in photodynamic therapy against tumor and microbial cells and also for quenching oxygen radicals. It has been discovered that nanodiamonds can also act as antioxidant and anti-inflammatory agents.

The interaction of carbon nanoparticles with cells and tissues has been investigated mainly using suspensions of these particles in cell culture media or other fluids. Relatively little is known about the influence of layers consisting of carbon nanoparticles on cell-substrate adhesion. Carbon nanoparticles can be advantageously used for surface modification of various artificial materials developed for the construction of tissue replacements and other body implants. In the form of films deposited on the surface of implants, carbon nanoparticles can not only improve the mechanical and other physical properties of a body implant, but can also enhance the attractiveness of the implant for cell colonization. The latter effect is probably mediated by the surface nanostructure of the films, which, at least to a certain degree, mimics the architecture of physiological cell adhesion substrates, such as extracellular matrix molecules. On nanostructured substrates, the cell adhesion-mediating molecules are adsorbed in the appropriate spectrum, amount and spatial conformation that make specific sites on these molecules (e.g. amino acid sequences like RGD) accessible to cell adhesion receptors. In addition, the nanostructured surfaces have been reported to enhance the adsorption of vitronectin, preferred by osteoblasts for their adhesion. The electrical conductivity of carbon nanotube films has also had beneficial effects on the growth and maturation of osteoblasts. Thus, the nanoparticles deposited on the bone-anchoring parts of bone, joint or dental replacements can improve the integration of these devices with the surrounding bone tissue. In addition, carbon nanoparticles admixed into polymers, designed for the fabrication of three-dimensional scaffolds for bone tissue engineering, could decorate the walls of the pores in these materials, and thus promote the ingrowth of bone cells. The interaction of cells with substrates modified with carbon nanoparticles could be further intensified by functionalizing these particles with various chemical functional groups or biomolecules, including KRSR-containing adhesion oligopeptides, recognized preferentially by osteoblasts.

This interdisciplinary review, based mainly on our own results, deals with the interaction of bone-derived cells with glass, polymeric and silicon substrates modified with fullerenes C_{60}, nanotubes, nanocrystalline diamond as well as other carbon allotropes, such as amorphous hydrogenated carbon or pyrolytic graphite. In general, these modifications have resulted in improved adhesion, growth, viability and osteogenic differentiation of these cells. The use of carbon and/or metal-carbon composite nanoparticles for the construction of patterned surfaces for regionally-selective adhesion and directed growth of cells is also discussed in this work.

1. Introduction

Carbon nanoparticles, such as fullerenes, nanotubes and nanodiamonds, are promising building blocks for the construction of novel nanomaterials for emerging industrial technologies, such as molecular electronics, advanced optics, and storage of hydrogen as a potential source of energy. In addition, they are considered as promising materials for a wide range of biomedical applications, including photodynamic therapy against tumors and infectious agents, quenching oxygen radicals, construction of microchips, nanorobots and biosensors, advanced imaging technologies, controlled drug or gene delivery, and also in the simulation of cellular components such as membrane pores or ion channels.

Fullerenes are spheroidal molecules made exclusively of carbon atoms (e.g., C_{60}, C_{70}), present in both terrestrial and extraterrestrial materials [1]. From these materials, fullerenes can be purified by chromatographic methods, selective complexation and fractional crystallization. For large-scale industrial production, various more efficient methods have been developed, such as filtration of a fullerene extract through a thin layer of activated carbon [2]. Fullerenes were discovered, prepared and systemized by Kroto *et al.* [3, 4]. They exhibit a wide range of unique physical and chemical properties, e.g., extreme stability, ability to withstand high temperatures and pressures, a high proclivity to react with other species while maintaining their spherical geometry, the ability to entrap other smaller species such as helium, at the same time not reacting with the fullerene molecule. Potential technical applications for fullerenes include superconductors, lubricants, optical devices, chemical sensors, photovoltaics, polymer additives, polymer electronics such as Organic Field Effect Transistors (OFETS). Due to their high reactivity, fullerenes can serve as catalysts. They also represent an advantageous medium for hydrogen storage, because almost every carbon atom in C_{60} can absorb a hydrogen atom without disrupting the buckyball structure, which could lead to applications in fuel cells. Fullerenes have been also used as precursors to produce diamond films.

In addition, fullerene molecules display a diverse range of biological activity. Their unique hollow cage-like shape and structural analogy with clathrin-coated vesicles in cells support the idea of the potential use of fullerenes as drug or gene delivery agents [5-8]. Fullerenes are able to accept and release electrons. Thus, when irradiated with ultraviolet or visible light, they can convert molecular oxygen into highly reactive singlet (i.e., atomic) oxygen [8, 9]. Thus, they have the potential to inflict photodynamic damage on biological systems, including damage to cellular membranes, inhibition of various enzymes or DNA cleavage [10-13]. This harmful effect can be advantageously exploited for photodynamic therapy against tumors, viruses and bacteria resistant to multiple drugs [14, 15]. On the other hand, C_{60} is considered to be the world's most efficient radical scavenger. This is due to the relatively large number of conjugated double bonds in the fullerene molecule, which can be attacked by radical species. Thus, fullerenes would be suitable for applications in quenching oxygen radicals in medicine, as well as in cosmetics [9, 16-22].

In their pristine unmodified state, fullerenes are highly hydrophobic and water-insoluble. On the other hand, they are relatively highly reactive, which enables them to be structurally modified. Fullerenes can form complexes with other atoms and molecules, e.g. metals, porphyrins, nucleic acids, proteins, as well as other carbon nanoparticles, e.g., nanotubes [6, 7, 23-26]. In addition, fullerenes can be functionalized with various chemical groups, e.g., the hydroxyl, aldehydic, carbonyl, carboxyl, ester or amine groups, as well as amino acids and peptides. This usually renders them soluble in water and intensifies their interaction with biological systems [10, 11, 13, 27, 28]. For example, derivatized fullerenes decrease neuronal degeneration and death, protect the brain against ischemia [19, 20], attenuate ischemia-reperfusion-induced lung injury [22, 29], interact with receptors and inhibit various enzymes, such as HIV protease, HIV-reverse transcriptase, enzymes synthesizing cholesterol, redox enzymes or nitric oxide synthase [8, 17, 30]. Fullerenes C_{60} derivatized with monomalonic acid showed an inhibitory effect on nitric oxide-dependent relaxation of rabbit aortic smooth muscle [16].

Carbon nanotubes were discovered, synthesized and systemized by Iijima and Ichihashi [31-33]. These tubular structures are formed by a single cylindrically-shaped graphene sheet (single-wall carbon nanotubes, SWCNT or SWNT) or several graphene sheets arranged concentrically (multi-wall carbon nanotubes, MWCNT or MWNT). Their diameter is on the nanometer scale (e.g., from 0.4 nm to 2-3 nm in single-walled nanotubes), but their length can reach several micrometers. Related structures, with differences in shape but with a similar architectural principle, known as carbon nanorings, nanohorns, graphene nanoribbons or nanowires, have also been observed [23, 34, 35].

Carbon nanotubes can be prepared by arc-discharge and related methods [36], by laser ablation, laser evaporation or a laser beam pulse [37] or by chemical vapor deposition (CVD) methods [38]. The preparation of nanotubes usually requires the presence of metal catalysts, e.g., Fe, Co, Ni or Y, which can have adverse effects on cells and tissues [39-42]. In addition, in their pristine unmodified state, carbon nanotubes are hydrophobic, and therefore they are prone to form agglomerates in a water environment. To disperse these aggregates, various organic solvents or surfactants, such as Triton X and sodium dodecylbenzene sulfonate have been used [38, 43, 44], but these can alter biomolecules and damage cells. Thus, a more appropriate approach seems to be derivatization of nanotubes, especially with oxygen-containing chemical functional groups. Treatment of carbon nanotubes with acids (usually nitric and sulphuric, also hydrochloric or hydrofluoric acid), which attack the reactive ends of the carbon nanotubes, resulted in the creation of oxygen-containing groups, e.g. carboxylic. At the same time, this treatment purified the nanotubes from metal catalysts [36, 42, 45, 46]. However, Bottini *et al.* [47] and Magrez *et al.* [48] reported increased toxicity of nanotubes after oxidation in acids. Oxidized nanotubes were shorter and straighter than their pristine counterparts, thus they could penetrate more easily inside the cells. In addition, they were better dispersed in an aqueous solution, and therefore reached higher concentrations of free nanotubes at similar weight per volume [47].

Carbon nanotubes have unique mechanical, physical and chemical properties in comparison with conventional carbon fibers [49]. Due to their sp^2 bonds, the tensile strength of SWCNT is about one hundred times higher than that of steel, while their density (i.e., specific weight) is about six times lower [50, 33]. The excellent mechanical properties of carbon nanotubes could be utilized in hard tissue surgery, e.g., to reinforce bone implants, particularly scaffolds for bone tissue engineering made of synthetic polymers or chitosan [51, 52]. In addition, nanotubes can resemble nanofibres of collagen and other extracellular proteins of the bone, and together with other carbon nanoparticles, they can also resemble hydroxyapatite and other inorganic crystals in the bone [53, 54]. In addition, carbon nanotubes can be used effectively as substrates for growth and electrical stimulation of osteoblasts, which has been reported to promote their proliferation, differentiation, production of mineralized bone matrix and thus healing of the damaged bone [55-57]. Carbon nanotubes in the form of layers or as a component of 3D scaffolds can also be used as substrates for attachment, growth and electrical stimulation of neuronal cells [58]. For example, layers of multiwalled carbon nanotubes functionalized with 4-

hydroxynonenal were successfully used as substrates for growing embryonic rat brain neurons *in vitro* [59].

Similarly as fullerenes, carbon nanotubes are relatively highly reactive, able to be functionalized, to form complexes with various atoms and molecules, and to emit photoluminescence. Due to these properties, both fullerenes and carbon nanotubes can be used in a wide range of advanced imaging, diagnostic and therapeutic strategies, such as transmission electron microscopy of DNA [60], labeling and tracer studies [1, 61-63], magnetic resonance [25, 26, 64], radiotherapy of tumors, including advanced methods of boron neutron capture therapy [65, 66], controlled drug and gene delivery [5-8, 67] and biosensor technology [23, 68-70]. The elongated shape of carbon nanotubes, resembling "tiny drinking straws", predisposes them to serve as tips for ultra-high-resolution atomic-force microscopy, and scanning tunneling microscopy [33, 71, 72]. In addition, carbon nanotubes and related materials, such as carbon nanohorns and nanotube-derived carbon foam, are promising substrates for storage of hydrogen as a potential source of energy (34, 73-75], and also for storage of methane, a key technology for producing natural gas vehicles [76].

Carbon nanotubes also represent a novel class of antimicrobial agents [77], ion (e.g., K^+) channel blockers [78], artificial ion and water channels and components of nanoporous membranes [79-81].

On the other hand, the reactivity of nanotubes, their ability to emit electrons from their ends and to generate free radicals, can induce oxidative stress, inflammation and alterations of various biomolecules and cellular structures, such as genes, transcription factors, enzymes, extracellular matrix molecules, mitochondria and cell membranes [40, 82-85], which should be taken into account in the design of nanotube-based materials for biomedical applications.

Nanocrystalline diamond (NCD) has also attracted considerable attention as a promising material for advanced biomedical applications. In the form of a powder, NCD has been prepared by detonation synthesis from a mixture of trinitrotoluene and hexogene [86]. Thin NCD films can also be deposited using microwave plasma-enhanced chemical vapor deposition (MW PECVD) from methane or fullerene C_{60} precursors [86, 87]. Similarly as fullerenes and nanotubes, NCD can also be relatively easily functionalized with various chemical groups and biomolecules (e.g., nucleic acids, enzymes, antibodies and other proteins) with strong covalent bonding [88-90]. At the same time, in comparison with fullerenes and nanotubes, NCDs are practically non-toxic and thus they are the most advantageous among all carbon nanoparticles for biomedical applications [88, 91]. In the form of a powder, NCD could serve as a carrier for controlled drug and gene delivery, for advanced imaging technologies or as an antioxidant and anti-inflammatory agent [91-93]. NCD films exhibit a wide range of unique physicochemical properties, such as mechanical hardness, chemical and thermal resistance, excellent optical transparency and controllable electrical properties (e.g., insulator/semiconductor/metal-like). Thus, biocompatible NCD films could markedly improve the mechanical and other physical properties of medical implants, particularly those designed for hard tissue surgery. Nanocrystalline diamond has already been used for coating the head and cup of artificial joint replacements, e.g., prostheses of temporomandibular joints [94]. In addition, NCD films could be applied in the bone-anchoring stems of articular prostheses or other permanent bone implants in order to improve their integration with the surrounding bone tissue. In studies *in vitro*, NCD provided excellent substrates for adhesion, growth, metabolic activity and maturation of human osteoblast-like MG 63 cells, human SAOS-2 osteoblasts, mouse MC3T3 osteoblasts, as well as other nonosteogenic cell types, such as human cervical carcinoma HeLa cells, rat pheochromocytoma PC12 cells or bovine pulmonary artery endothelial CPAE cells [57, 88, 95-97]. Type IIa diamond, microcontact-printed with laminin, supported adhesion of mouse cortical neurons and neurite outgrowth [98]. In studies *in vivo*, diamond layers, deposited by an MW CVD method on TiAl6V4 probes implanted into a rabbit femur, showed very high bonding strength to the metal base and also to the surrounding bone tissue, without any problems with corrosion [99]. Last but not least, similarly as fullerenes and nanotubes, NCDs are promising building blocks for constructing biomarkers, microchips, nanorobots or biosensors [88-91, 100].

Layers of *other carbon allotropes*, e.g., a-C:H, graphite or hydrocarbon on pristine and modified polymers, are also characterized by nanostructure (i.e., the presence of particles less than 100 nm), in addition to their atomic structure. These layers are represented by carbon structures formed on the surface of synthetic polymers after irradiation, e.g., with plasma, by an ion beam or laser beam, or after modification of the material surface by magnetron sputtering, vacuum evaporation or chemical vapor deposition (CVD). These carbon allotropes could be used for surface modification of various materials designed for the construction of body implants or tissue engineering, e.g., synthetic polymers, metals or carbon-based composites [101-103]. These allotropes can be enriched with metals in order to improve their mechanical properties and to enhance their bioactivity [103-105].

Among all these exciting findings, we have concentrated on the influence of carbon nanoparticles, particularly when arranged into layers and used for biomaterial coating, on cell-substrate adhesion, subsequent growth, differentiation and viability of cells, especially bone-forming cells. A very interesting finding has been reported that nanostructured surfaces can promote preferential adhesion and growth of osteoblasts over other "competitive" cell types, including fibroblasts, and thus they can prevent fibrous encapsulation and loosening of bone implants. It is considered that the underlying mechanism is higher adsorption to the nanostructured surfaces of vitronectin, an extracellular matrix protein preferred by osteoblasts. Therefore, it can be expected that carbon nanoparticles may serve as novel building blocks for creating artificial bioinspired nanostructured surfaces for bone tissue engineering [53, 54, 106-111]. The preferential adhesion of osteoblasts could be further enhanced by the functionalization of carbon nanoparticles by the amino acid sequence Lys-Arg-Ser-Arg (KRSR), a ligand for adhesion receptors on osteoblasts [112-115]. In addition, carbon nanoparticles could be admixed to synthetic polymers, chitosan and other matrices designed for the fabrication of three-dimensional porous scaffolds for tissue engineering. The nanoparticles would form prominences and create a nanopattern on the pore walls, and thus promote the ingrowth of cells inside the scaffolds [109, 116].

Therefore, in our interdisciplinary studies, the following types of materials modified with carbon nanoparticles were investigated:

(1) Continuous layers of fullerenes C_{60} deposited on microscopic glass coverslips or composites with a carbon matrix reinforced with carbon fibers, i.e., a material applicable in hard tissue surgery.
(2) Micropatterned layers created by deposition of fullerenes C_{60} through metallic grids on microscopic glass coverslips.

(3) Hybrid layers created by co-deposition of fullerenes C_{60} and metals on microscopic glass coverslips.
(4) Polymeric materials (i.e., a terpolymer of polytetrafluoroethylene, polyvinyldi fluoride and polypropylene, or polysulphone) mixed with single- or multi-walled carbon nanotubes.
(5) Nanostructured or hierarchically micro- and nanostructured nanocrystalline diamond layers deposited on silicon substrates.
(6) Layers of amorphous hydrogenated carbon (a-C:H), pyrolytic carbon, pyrolytic graphite as well as nanocomposite metal-hydrocarbon plasma polymer films.

The adhesion, growth, viability, metabolic activity and maturation of human osteoblast-like MG 63 cells were then studied in cultures on these materials. The cell behavior correlated with the physical and chemical properties of these materials, and also with results obtained by other authors.

2. Cell Growth on Fullerene C_{60} Layers

2.1. Continuous C_{60} Layers

In our experiments, fullerenes C_{60} (purity 99.5 %, SES Research, U.S.A.) were deposited on to microscopic glass coverslips (Menzel Glaser, Germany; diameter 12 mm) by evaporation of C_{60} in the Univex-300 vacuum system (Leybold, Germany) in the following conditions: room temperature of the substrates, C_{60} deposition rate $\leqslant$ 10 Å/s, temperature of C_{60} evaporation in the Knudsen cells about 450° C, time of deposition up to 50 minutes (1 A). The thickness was measured by atomic force microscopy (AFM). A scratch was made in the layer and its profile was measured in contact mode. The thickness of the layers increased proportionally to the temperature in the Knudsen cell and the time of deposition. We prepared two continuous films of different thicknesses, i.e., a thin layer of 505 ± 43 nm and a thick layer of 1090 ± 8 nm. Optical microscopy revealed that the fullerene layers were brownish in color (Figure 1 B), and the color intensity increased with layer thickness. Raman spectra of the deposited C_{60} films were measured in the back-scattering geometry at room temperature using a Ramascope 1000 Raman microscope (Renishaw, UK) with a 514.5 nm excitation wavelength Ar-ion laser. Immediately after deposition, the Raman spectra showed that the prepared fullerene films were of high quality, confirmed by a high peak $A_g(2)$ at wavenumber 1468 cm^{-1}, low peaks $H_g(7)$ and $H_g(8)$ and the absence of D (disorder, ~ 1350 cm^{-1}) and G (graphitic, ~ 1600 cm^{-1}) bands, which are signs of fragmentation and graphitization of C_{60}, respectively (Figure 2). The surface wettability of the films was estimated from the contact angle measured by a static method in a material-water droplet system using a reflection goniometer. The fullerene C_{60} films were relatively hydrophobic, having contact angles 100.6 ± 6.8° for thin and 96.7 ± 1.7° for thick continuous layers, respectively [117, 118].

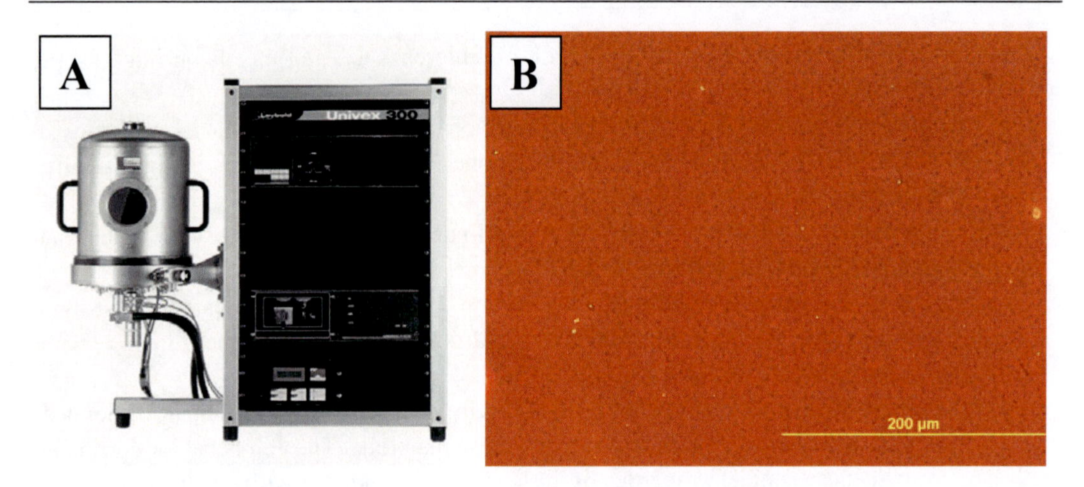

Figure 1. A: Univex-300 vacuum system (Leybold, Germany); B: continuous fullerene C_{60} layer (thickness of 1090 ± 8 nm) deposited on to a microscopic glass coverslip in this system [117, 118].

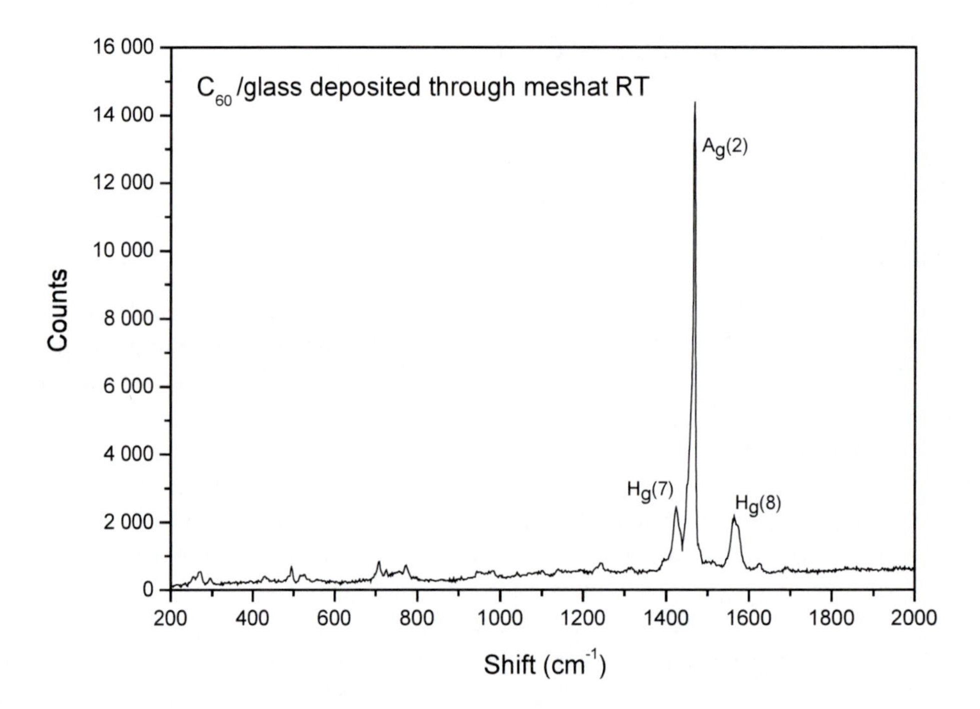

Figure 2. Raman analysis of a fullerene C_{60} layer immediately after deposition on glass coverslips. The quality of the fullerene layer is confirmed by a high peak $A_g(2)$ at wavenumber 1468 cm^{-1}, low peaks $H_g(7)$ and $H_g(8)$ and absence of D (disorder, ~ 1350 cm^{-1}) and G (graphitic, ~ 1600 cm^{-1}) bands, which are signs of fragmentation and graphitization of C_{60}, respectively [117, 118].

To cultivate the cells, the fullerene-coated glass coverslips were sterilized with 70% ethanol for 1 hour, inserted into 24-well polystyrene multidishes (TPP, Switzerland; diameter 15 mm) and seeded with human osteoblast-like MG 63 cells (European Collection of Cell Cultures, Salisbury, UK). Each dish contained 5,000 cells (i.e., about 2,830 cells/cm^2) and 1.5 ml of Dulbecco's modified Eagle's Minimum Essential Medium supplemented with 10% fetal bovine serum and gentamicin (40 µg/ml). The cells were cultured for 1, 3 and 5 or 7 days at 37° C in a humidified air atmosphere containing 5% of CO_2. On day 1 after seeding, the cells on both continuous thin and continuous thick fullerene layers adhered in similar numbers (3,420 ± 420 cells/cm^2 and 2,880 ± 440 cells/cm^2, respectively), which was comparable to the values found on standard cell culture substrates, represented by the tissue culture polystyrene dish (3,080 ± 290 cells/cm^2) and the microscopic glass coverslip (2,560 ± 310 cells/cm^2). In addition, the cells on the fullerene layers and on the control substrates were of similar morphology, i.e., mostly polygonal and spread over a similar area (Figure 3). From days 1 to 5 after seeding, the cells on both continuous C_{60} films, polystyrene and glass also proliferated with similar cell population doubling times ranging from 18.6 ± 0.3 h to 19.6 ± 0.5 hours, and on day 5 they reached similar cell population densities (from 91 600 ± 9500 cells/cm^2 to 112 700 ± 9800 cells/cm^2; Figure 4). In addition, as revealed by staining the adhered cells with calcein and ethidium homodimer using the LIVE/DEAD kit, the percentage of viable cells on days 1 to 7 (after seeding on all fullerene layers) ranged from 80 ± 10 % to 100 ± 15 %, and was similar to the values obtained on the polystyrene culture dishes (99 ± 24 % to 100 ± 15 %), and also on the microscopic glass coverslips (88 ± 21 % to 99 ± 26 %). The trypan-blue exclusion test (performed on the trypsinized cells while they were being counted in the ViCell Analyser) also showed similar proportions of viable cells on the fullerene layers (82 ± 11 % to 89 ± 13 %), as well as on standard cell culture surfaces (80 ± 13 % to 95 ± 13 %). The average viability of the cells tended to increase with time of cultivation.

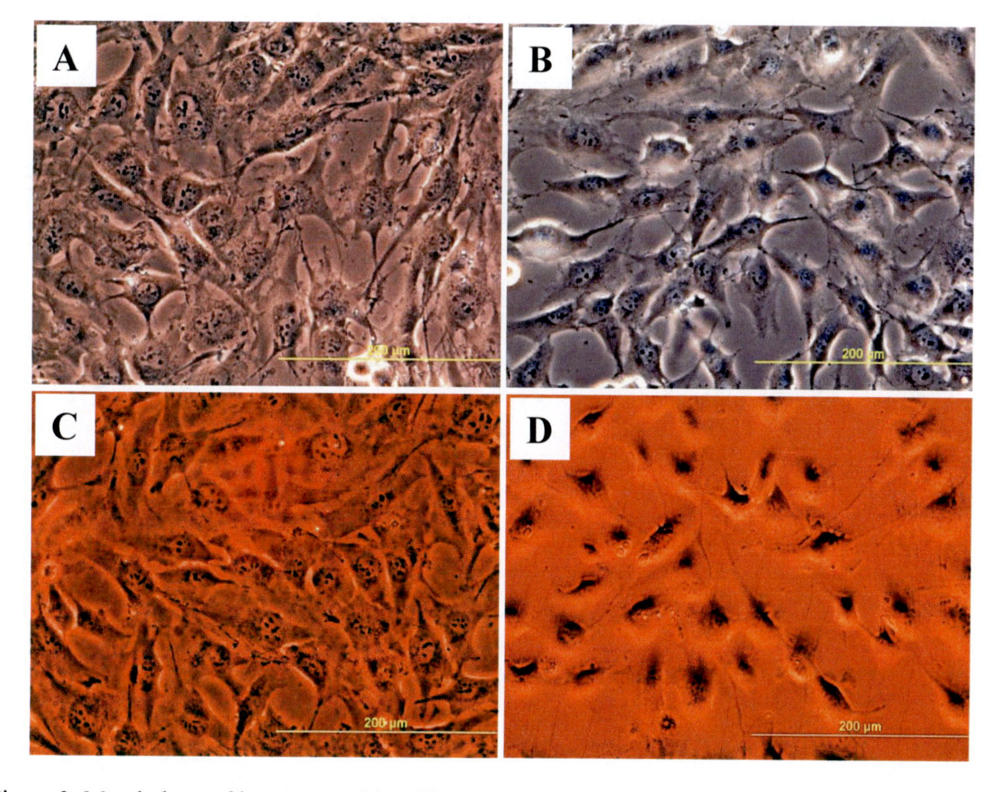

Figure 3. Morphology of human osteoblast-like MG 63 cells in 7-day-old cultures on a polystyrene dish (A) a microscopic glass coverslip (B), thin continuous (C) and thick continuous (D) fullerene C_{60} layers. Native cultures, Olympus IX 50 microscope, DP 70 digital camera, obj. 20x, bar = 200 µm [117].

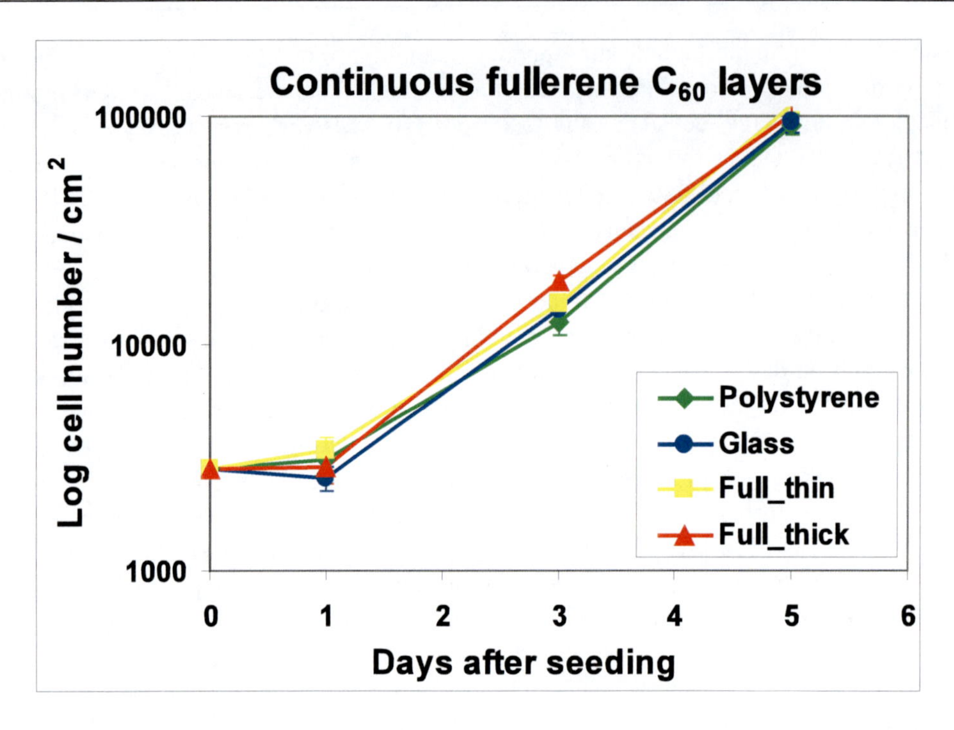

Figure 4. Growth curves of human osteoblast-like MG 63 cells in cultures on polystyrene dishes, microscopic glass coverslips or continuous thin and thick fullerene C_{60} layers. Mean ± S.E.M. (Standard Error of Mean) from 9 to 30 measurements. No statistically significant differences were found by ANOVA, Student-Newman-Keuls Method [117].

The supportive effects of fullerene C_{60} layers on cell colonization could be explained by their surface nanostructure, mimicking the nanoarchitecture of the natural extracellular matrix (ECM), basal lamina, as well as the cell membrane. The nanostructure of the cell adhesion substrate has been reported to enhance the adsorption of cell adhesion-mediating ECM molecules, such as fibronectin, collagen and especially vitronectin, provided by the serum of the culture medium or synthesized by the cells. The spatial conformation of the adsorbed molecules on our films may also be more appropriate for the accessibility of specific sites on these molecules by cell adhesion receptors, e.g., integrins [53, 54, 106-108, 110, 116, 121, 122]. As revealed by immunofluorescence staining of the MG 63 cells on our samples, β_1 integrins (i.e., receptors for collagen, fibronectin and vitronectin), and talin (an integrin-associated protein) were assembled in dot- or streak-like focal adhesion plaques, mainly visible on the cell periphery (Figure 5), and these processes were accompanied by the formation of a fine mesh-like β-actin cytoskeleton and by the presence of a considerable amount of osteopontin, a marker of osteogenic cell differentiation (Figure 6). All these events can be considered signs of active cell-substrate interaction, followed by signal transduction inside the cells. Moreover, β_1 integrins have been reported to enhance osteogenic cell differentiation, which is manifested by higher levels of osteocalcin, higher activity of alkaline phosphatase and a more pronounced response to 1,25-dihydroxyvitamin D_3 [111].

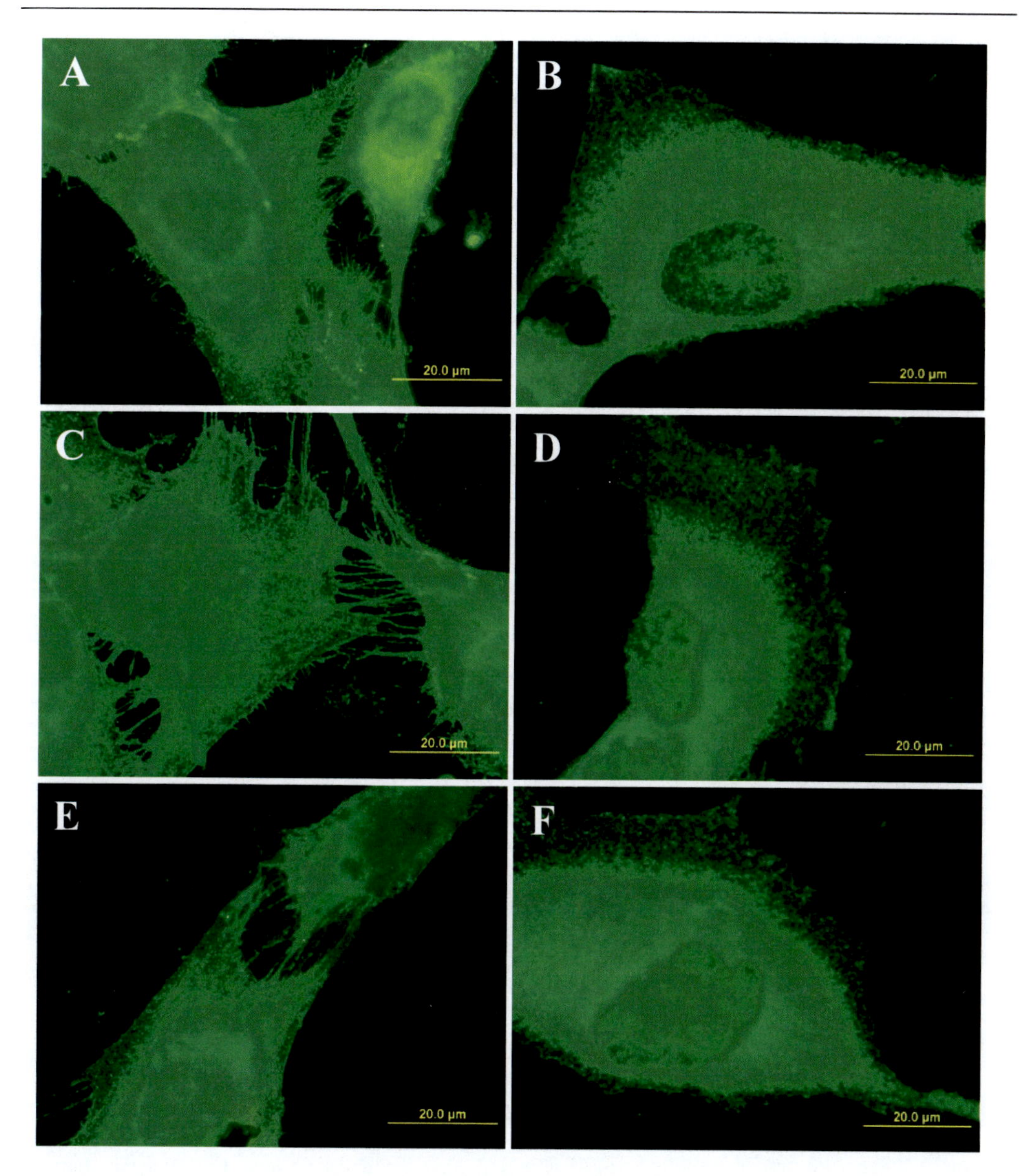

Figure 5. Immunofluorescence staining of β_1 integrins (A, C, E) and talin (B, D, F) in human osteoblast-like MG 63 cells on day 3 after seeding on microscopic glass coverslips (A, B) and thin (505 ± 43 nm; C, D) or thick (1090 ± 8 nm; E, F) continuous fullerene C_{60} layers. Olympus IX 50 epifluorescence microscope, DP 70 digital camera, obj. 100x, bar = 20 μm [117].

The proliferation and differentiation of chondrocytes (another cell type important in bone reconstruction) were also promoted by fullerenes. C_{60} added in cell culture enhanced the production of specific large proteoglycan ECM molecules, typical for cartilage, in rat embryonic limb bud cells in culture [123].

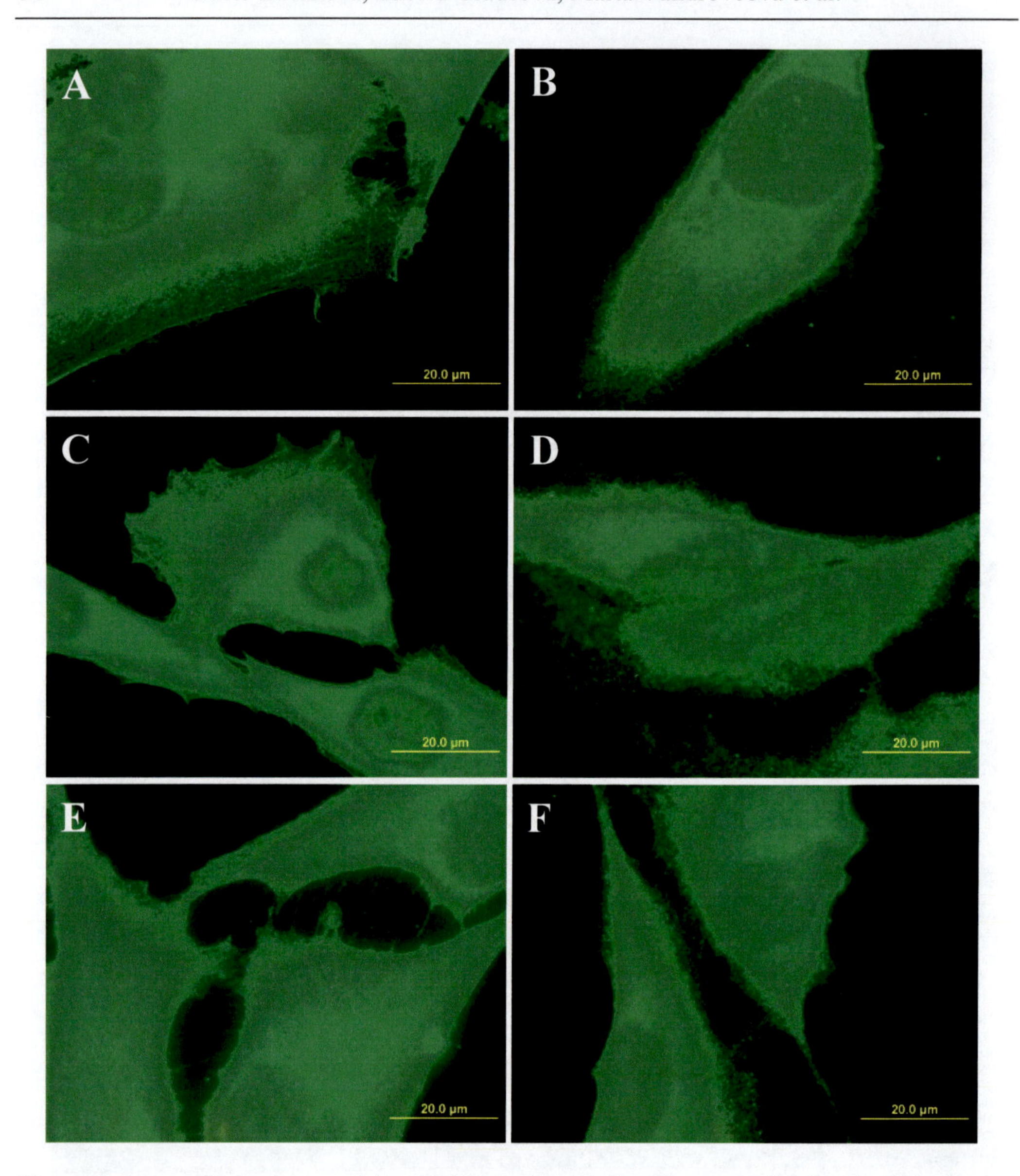

Figure 6. Immunofluorescence staining of β-actin (A, C, E) and osteopontin (B, D, F) in human osteoblast-like MG 63 cells on day 3 after seeding on microscopic glass coverslips (A, B) and thin (505 ± 43 nm; C, D) or thick (1090 ± 8 nm; E, F) continuous fullerene C_{60} layers. Olympus IX 50 epifluorescence microscope, DP 70 digital camera, obj. 100x, bar = 20 µm [117].

Similarly, C_3-fullero-*tris*-methanodicarboxyllic acid (a water-soluble fullerene derivative), supported the formation of focal adhesion plaques, an assembly of an actin cytoskeleton and cell spreading in human epidermoid carcinoma cells exposed to UV light, probably due to its antioxidative action [124]. The free radical scavenger activity of fullerenes and their derivatives (e.g., hydroxylated, carboxylated, butylated or sulphonated C_{60}) also protected keratinocytes from apoptosis induced by ultraviolet light B [27], minimized the oxidative damage of small bowel transplants [18], attenuated ischemia-reperfusion-induced lung injury [22], prevented focal cerebral ischemia and decreased neuronal degeneration and death [19, 20]. However, hydroxylated fullerene derivatives protected cells against oxidative stress only at lower concentrations (10-50 μM in cell culture media), while at higher concentrations (1 mM and 1.5 mM), they caused cell death [22]. Hydroxylated fullerenes usually induced cell death characterized by signs of apoptosis, such as increased caspase activity, DNA fragmentation and annexin V binding to the phosphatidylserine, exposed on the outer side of the cell membrane, without increased membrane permeability [28]. On the other hand, a fullerene derivative $C_{60}(OH)_{24}$ (concentrations from 1 to 100 μg/ml) did not seem to induce apoptosis but caused the accumulation of polyubiquitinated proteins and facilitated autophagic cell death in human umbilical vein endothelial cells [125]. Fullerene colloids in water, also called nano-C_{60}, i.e. fullerene aggregates that readily form when pristine C_{60} is added to water, mainly induced cell necrosis characterized by oxidative damage to the cell membrane (e.g., peroxidation of the membrane lipids) and loss of membrane integrity [10], though the genotoxicity of these agents has also been demonstrated on human lymphocytes [11]. The harmful effects of fullerenes C_{60}, usually potentiated by the light, especially ultraviolet light, could be utilized for antimicrobial therapy or photodynamic therapy against tumor cells [126, 127]. In addition, a nanocrystalline suspension of fullerenes C_{60} and C_{70} prepared by the solvent exchange method using tetrahydrofuran (THF/n$C_{60/70}$) augmented the cytotoxicity of the tumor necrosis factor [128, 129].

Continuous layers of fullerenes C_{60}, experimentally developed on glass, were also tested on composites with the carbon matrix reinforced with carbon fibers (i.e., carbon fiber-reinforced carbon composites, CFRC), which have been considered as promising materials applicable in hard tissue surgery [104, 130-133]. These composites were prepared at the Institute of Rock Structure and Mechanics, Acad. Sci. CR, Prague. Briefly, a commercially available woven fabric (made of Toray T 800 carbon fibers) was arranged in layers, infiltrated with a carbon matrix precursor (phenolic resin UMAFORM LE, Synpo Ltd., Pardubice, CR), pressed, cured, carbonized at 1000°C, and finally graphitized at 2200°C [104, 130-133]. Prior to the deposition of fullerenes, the CFRCs were ground using a metallographic paper of 4000 grade in order to decrease the material surface micro-scale roughness, which has been shown to be less appropriate for spreading and subsequent growth of bone-derived and also vascular smooth muscle cells [104, 130, 131]. This grinding lowered the surface roughness of CFRC about twice. As measured by a profilometer (Rank Taylor Hobson Ltd., England), the departures of the roughness profile from the mean line (i.e., R_a parameter) decreased from 6.5 ± 1.8 μm to 3.5 ± 0.6 μm, and the mean spacing of the adjacent local peaks (parameter S) lengthened from 38 ± 11 μm to 96 ± 49 μm. The fullerene coating did not significantly change this surface microroughness, but created a nanostructured pattern on the pre-existing microarchitecture of the CFRC surfaces. The fullerene C_{60} layers were deposited on the CFRC by evaporation of C_{60} in the Leybold Univex-300 vacuum system (room temperature of the substrates, deposition rate $\leqslant$ 1 Å/s, temperature of C_{60} evaporation in the Knudsen cells 450°C, time of deposition about 15 min, thickness of the fullerene layers < 100 nm). Similarly as in the continuous fullerene layers (deposited on microscopic glass coverslips), the Raman analysis confirmed again that the prepared fullerene films were of high quality and with no fragmentation or graphitization of C_{60} (Figure 7). One third of the CFRC samples (3 x 3 cm) were protected against fullerene coating by a mask in order to achieve regionally-selective growth of the fullerene layer. The other completely uncoated CFRC samples, as well as tissue culture polystyrene (TCPS) dishes were used as control materials [95].

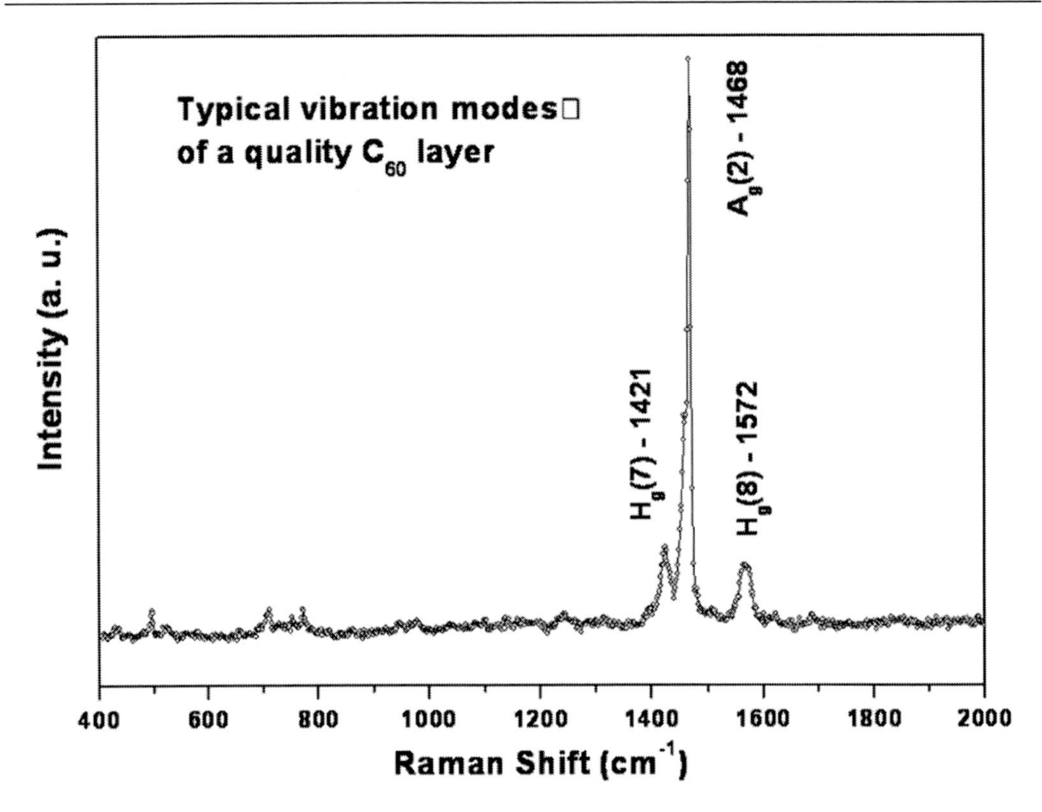

Figure 7. Raman analysis of a thin C_{60} layer deposited on a carbon-fibre reinforced composite (CFRC) substrate. The quality of the fullerene layer is confirmed by a high peak $A_g(2)$ at wavenumber 1468 cm^1, low peaks $H_g(7)$ and $H_g(8)$ and absence of D (disorder, ~ 1350 cm^{-1}) and G (graphitic, ~ 1600 cm^{-1}) bands, which are signs of fragmentation and graphitization of C_{60}, respectively [95].

The material was then seeded with human-osteoblast-like MG 63 cells. Unlike the fullerene layers on glass, the cell population density on day 2 after seeding (Figure 8 A) on C_{60}-coated CFRC was significantly lower (about 19,000 ± 700 cells/cm^2) than that on the control uncoated CFRC and polystyrene dishes (44,000 ± 4,000 cells/cm^2 and 67,000 ± 7,000 cells/cm^2, respectively). This finding could be attributed to a combination of at least two factors less favorable for cell adhesion, such as the remaining surface microroughness and relatively high hydrophobicity of the non-functionalized fullerenes (static water drop contact angle about 100°, see above). It is known that hydrophobic materials promote preferential adsorption of cell non-adhesive proteins from the serum of the culture media, such as albumin. In addition, the cell adhesion-mediating extracellular matrix proteins may be adsorbed in a relatively rigid state, thus their specific amino acid sequences were less accessible for cell adhesion receptors [121, 122]. On the other hand, the water drop contact angle, measured by a static method in a material-water droplet system using a reflection goniometer, was not able to reveal any hydrophobicity of the fullerene layer deposited on CFRCs. The contact angle was non-measurable due to complete absorption of the water drop into the fullerene layer, which suggested a certain non-compactness or porosity of this layer on CFRC. The contact angle of the non-coated CFRC was 99.5 ± 1.0°.

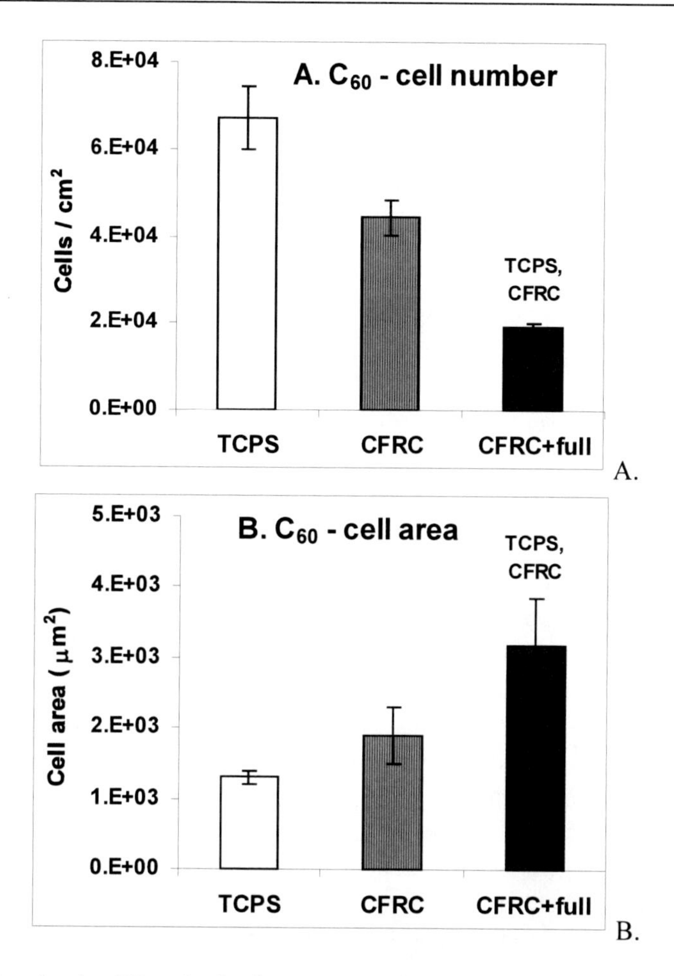

Figure 8. Population density (A) and adhesion area (B) of osteoblast-like MG 63 cells on day 2 after seeding on a tissue culture polystyrene dish (TCPS), carbon fibre-reinforced carbon composites (CFRC) and CFRC coated with a fullerene layer (CFRC+full). Mean ± S.E.M. (Standard Error of Mean) from 4-12 measurements, ANOVA, Student-Newman-Keuls method. Statistical significance: [TCPS, CFRC]: $p \leq 0.05$ compared to the values on tissue culture polystyrene and pure CFRC [95].

The release of fullerenes into the culture media and their cytotoxic action seemed to be less probable in our experiments as an explanation of the lower number of cells adhered to C_{60}-coated CFRC. The cell attachment and spreading on the uncoated regions of the fullerene-modified CFRC, and also on the bottom of the polystyrene dishes containing fullerene-coated samples, were similar as in the control polystyrene dishes without fullerene samples (Figure 9). At the same time, the fullerene layer was resistant to mild wear, represented by swabbing with cotton, rinsing with liquids (water, phosphate-buffered saline, culture media) and exposure to cells and proteolytic enzymes (trypsin) used for cell harvesting. After these procedures and/or one-year storage at room temperature in a dark place, the Raman spectra did not change significantly. The peak $A_g(2)$ at wavenumber 1468 cm^{-1} still remained the highest, though some of the peaks $H_g(7)$ and $H_g(8)$ increased slightly, and additional small D and G bands appeared, which indicated some fragmentation, graphitization, polymerization or oxidation of C_{60} molecules.

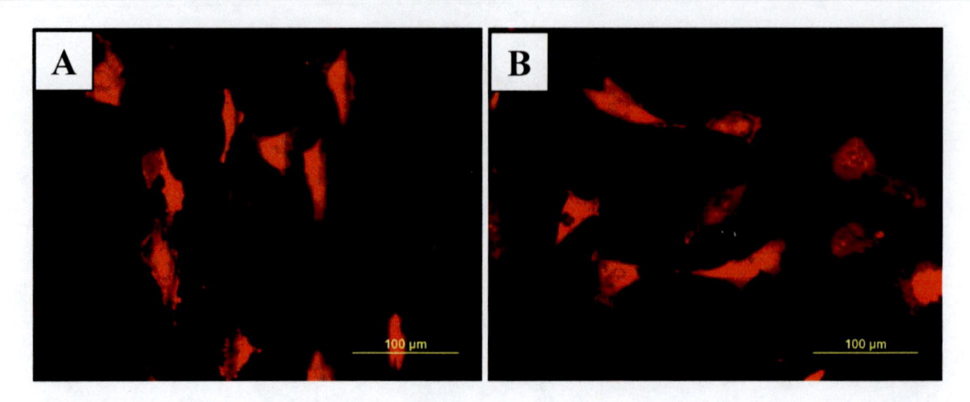

Figure 9. Human osteoblast-like MG 63 cells on day 1 after seeding on the uncoated region of CFRC partially coated by fullerenes (A) and the bottom of a polystyrene dish containing a fullerene-coated CFRC sample (B). Cells fixed with 70% ethanol and stained with propidium iodide. Olympus IX 50 microscope, DP 70 digital camera, obj. 20, bar=100 μm [95].

In addition, the fullerene-coated CFRC surfaces were stronger and less prone to release carbon particles, which is an important limitation to the potential biomedical use of CFRC [104, 130, 131]. Moreover, the spreading area of cells on the fullerene-coated samples amounted to 3,182 ± 670 μm^2, while on both control surfaces it was only 1,888 ± 400 and 1,300 ± 102 μm^2 (Figure 8 B). This could be explained by the low cell population density on the fullerene layer, which provided the cells with more space on which they could spread. On the other hand, the nanostructure of the fullerene layer might enhance the adsorption of vitronectin, i.e., an extracellular matrix protein promoting adhesion of osteoblasts [106-108]. The MG 63 cells on the fullerene-coated CFRC formed dot-like vinculin-containing focal adhesion plaques and a fine network of actin microfilaments (Figure 10), which is a sign of cell vitality and effective binding between cell adhesion receptors and extracellular matrix molecules adsorbed on the material surface [122].

2.2. Micropatterned C_{60} Layers

An interesting idea is that fullerenes might be used for constructing patterned surfaces for regionally-selective adhesion and directed growth of cells. Domains adhesive for cells on the material surface have been created using a wide range of physico-chemical and biochemical approaches. These approaches involve attachment of moderately hydrophilic molecules (e.g., acrylic acid, [134]) or attachment of various chemical functional groups, particularly those containing oxygen. The latter can also be created by irradiation of the material surface, especially the surface of synthetic polymers, by plasma, ions or ultraviolet light [121, 135-140]. In addition, the material surface can be functionalized by various biomolecules, particularly ligands for cell adhesion receptors, such as oligopeptides containing amino acid sequences recognized by all cell types (sequence RGD) or preferentially by a certain cell type (e.g., REDV by endothelial cells, VAPG by smooth muscle cells or KRSR recognized by osteoblasts) [112-114, 122, 141]. Regions repulsive to cells can be prepared using hydrophobic compounds, such as octadiene or polysiloxanes, or extremely hydrophilic molecules, e.g. polyethylene oxide [134, 141, 142].

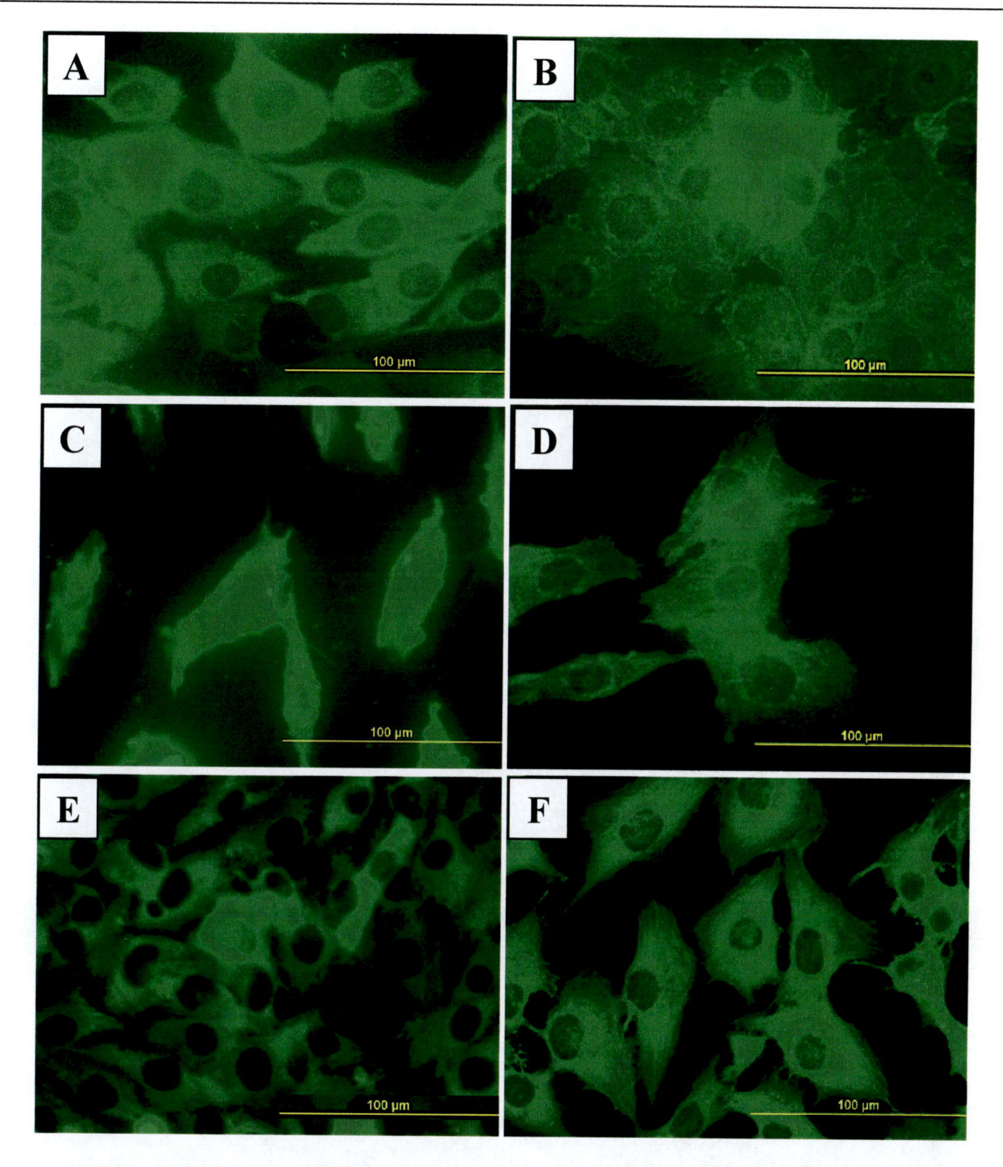

Figure 10. Immunofluorescence staining of vinculin (A, C, E) and β-actin (B, D, F) in osteoblast-like MG 63 cells on day 2 after seeding on CFRC (A, B), CFRC coated with a fullerene layer (C, D) and tissue culture polystyrene dish (E, F). Olympus IX 50 microscope, DP 70 digital camera, obj. 40, bar = 100 µm [95].

Patterned surfaces can also be created by changing the surface roughness and topography, i.e., by creating prominences and hollows of different size, shape and depth [116, 143]. In our experiments, micropatterned layers were prepared by deposition of fullerenes C_{60} on microscopic glass coverslips through contact metallic masks with rectangular holes with an average size of 128 ± 3 µm per 98 ± 8 µm (about 12 430 μm^2) and 50 µm in distance, or 470 x 440 µm (about 207,000 μm^2), distance 200 µm (Figure 11). The fullerenes formed bulge-like prominences below the openings of the metallic grid. Similarly as in the continuous fullerene C_{60} films, the thickness of these prominences increased proportionally to the temperature in the Knudsen cell and the time of deposition. As revealed by AFM, the thickness of the layers on sites underlying the openings of the grid was 128 ± 8 nm, 238 ± 3 nm, 484 ± 5 nm or 1043 ± 57 (data presented as mean ± standard deviation). Fullerene layers were also formed below the metallic part of the grid, where their thickness was either within the size of the standard deviations (the 1st, 2nd and 4th layer) or it amounted to 158 ± 5 nm (the 3rd layer). Therefore, in the 3rd layer, the effective height of the fullerene prominences was 326 ± 5 nm [117, 118].

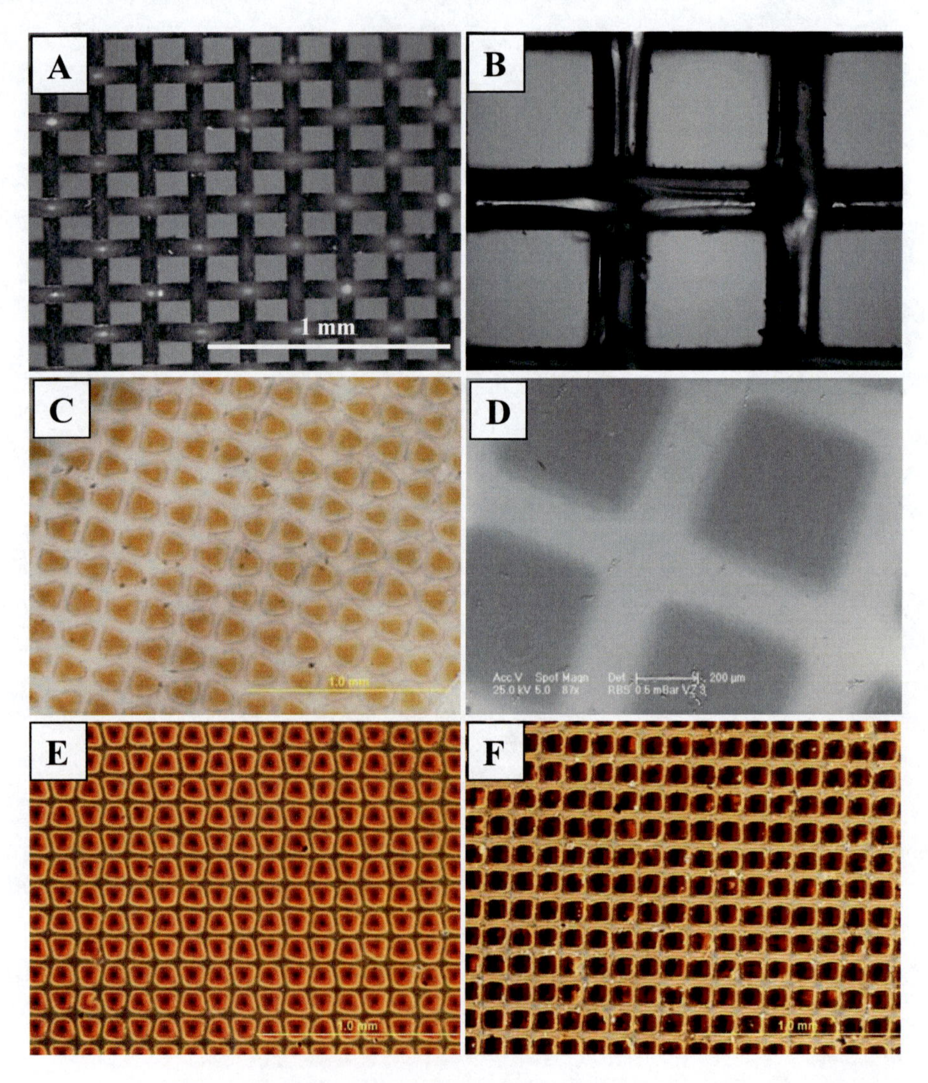

Figure 11. A, B: Metallic grids used for creating micropatterned fullerene C_{60} layers on microscopic glass coverslips. A: openings of 128 ± 3 µm per 98 ± 8 µm (about 12,500 µm^2) and 50 µm in distance, B: openings of 470 x 440 µm (about 207,000 µm^2), distance 200 µm. C-F: micropatterned layers created by the deposition of fullerenes C_{60} through a mesh with smaller (C, E, F) and larger (D) openings. The height of the fullerene protrusions was 128 ± 8 nm (C), 238 ± 3 nm (D), 484 ± 5 nm (E) and 1043 ± 57 nm (F). Olympus BX41 microscope (A, B), Olympus IX 50 microscope (C, E, F) and SEM-EDS scanning electron microscope (JOEL JSM-5600) (D). Objective 10x (A, B) or 4x (C, E, F); bar = 1mm (A, B, C, E, F) or 200 µm (D) [118].

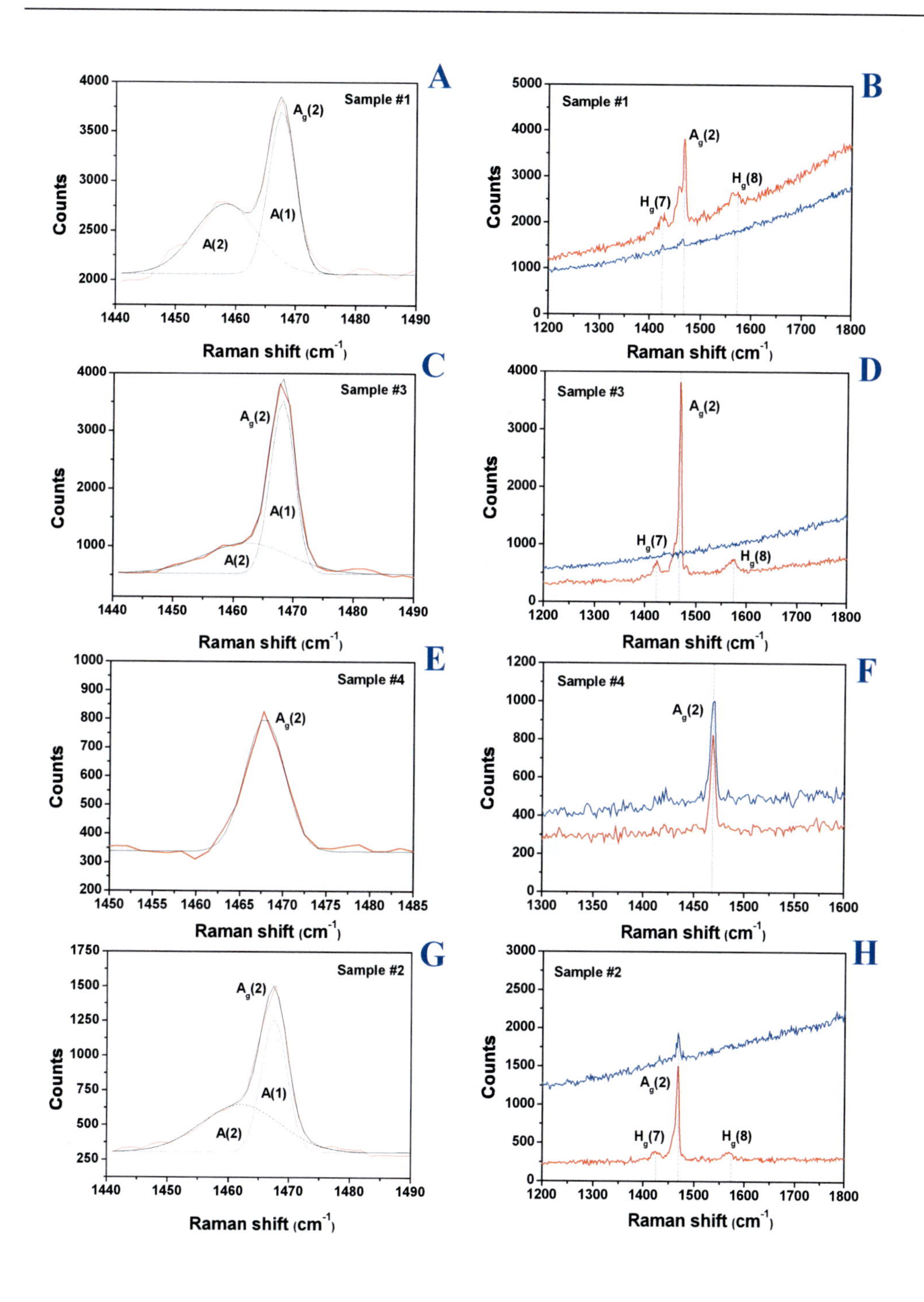

Figure 12. Raman analysis of microstructured C_{60} layers after 1-hour-sterilization with 70% ethanol. A, C, E, G: Gaussian peak fit analysis; B, D, F, H: Raman spectra below the openings (red line) or bars (blue line) of the metallic lattice. As revealed by AFM, the layer thickness below the openings and bars was 128 ± 8 and ≤ 8 nm, respectively, in the sample on B, 238 ± 3 nm and ≤ 3 nm on D, 484 ± 5 nm and 158 ± 5 nm on F, and 1043 ± 57 nm and ≤ 57 nm on H [117, 118].

For cell culture, the fullerene-modified samples were sterilized by 70% ethanol in water (1 hour, at room temperature). Analysis of the vibration mode $A_g(2)$ on the Raman spectra showed that, after sterilization, some of the micropatterned fullerene layers were almost intact (the 3rd layer, Figure 12 E), whereas in the other, some of the C_{60} molecules reacted with oxygen or polymerized, as indicated by the satellite peak A(2). The proportion of C_{60} molecules involved in these chemical changes was determined by the ratio A(2)/A(1), and reached about 30% in the 2nd layer (Figure 12 C) and 50% in the 1st and 4th layer (Figure 12 A, G). Similarly as in AFM, fullerenes were also found below the metallic bars of the grid, though in the thinner layers 1 and 2, the amount was very low and barely detectable (Figure 12 B, D, F, H).

Reflection goniometry showed that similarly to the continuous fullerene C_{60} layers, all of the micropatterned C_{60} layers were relatively highly hydrophobic. All micropatterned layers had similar water drop contact angles, i.e., 93.6 ± 3.6° (layer 1), 94.5 ± 5.5° (layer 2), 95.3 ± 3.1° (layer 3) and 95.6 ± 3.8° (layer 4).

Optical microscopy revealed that below the openings of the grids, the fullerene prominences were brownish in color, and the color intensity increased with layer thickness. Up to a thickness of 484 ± 5 nm, the layers were relatively well transparent in a conventional light microscope (Figure 11). Thus, the cells growing on the layers were well observable in native or hematoxylin- and eosin-stained cultures, and the fullerene bulges and grooves were also well distinguishable. However, the fullerene prominences 1043 ± 57 nm in thickness were less transparent, thus the cells were better observable after fluorescence staining (Figure 13).

The micropatterned layers were seeded with human osteoblast-like MG 63 cells. From day 1 to 7 after seeding, the cells on the layers with fullerene prominences up to 326 ± 5 nm were distributed almost homogeneously over the entire material surface (Figure 13), thus the cell population densities on the prominences and in the grooves were similar (Figure 14). At the same time, the cells on these layers adhered and grew to an extent comparable to that on tissue culture polystyrene dishes and microscopic glass coverslips. Similarly as in continuous fullerene C_{60} layers, these beneficial effects on cell colonization could be explained by their surface nanostructure, mimicking the nanoarchitecture of the natural extracellular matrix [53, 54, 106-110]. Moreover, due to the micropatterning, the fullerene layers in our study were constructed with a hierarchically-organized micro- and nanostructure, i.e., with a similar architectural principle as natural tissues [116]. The presence of oxygen, suggested in some of our fullerene layers by Raman spectroscopy, might also contribute to the supportive effects of these films on cell adhesion and growth. Oxygen-containing chemical functional groups have been repeatedly shown to enhance the colonization of various materials with cells [121, 135-137, 144]. In the case of fullerene-grafted polyurethane, the presence of oxygen-containing and amine functional groups (due to pre-treatment of the polymer with oxygen plasma and the use of aminosilanes as coupling agents) was also attributed to the enhanced adhesion and activation of platelets [120].

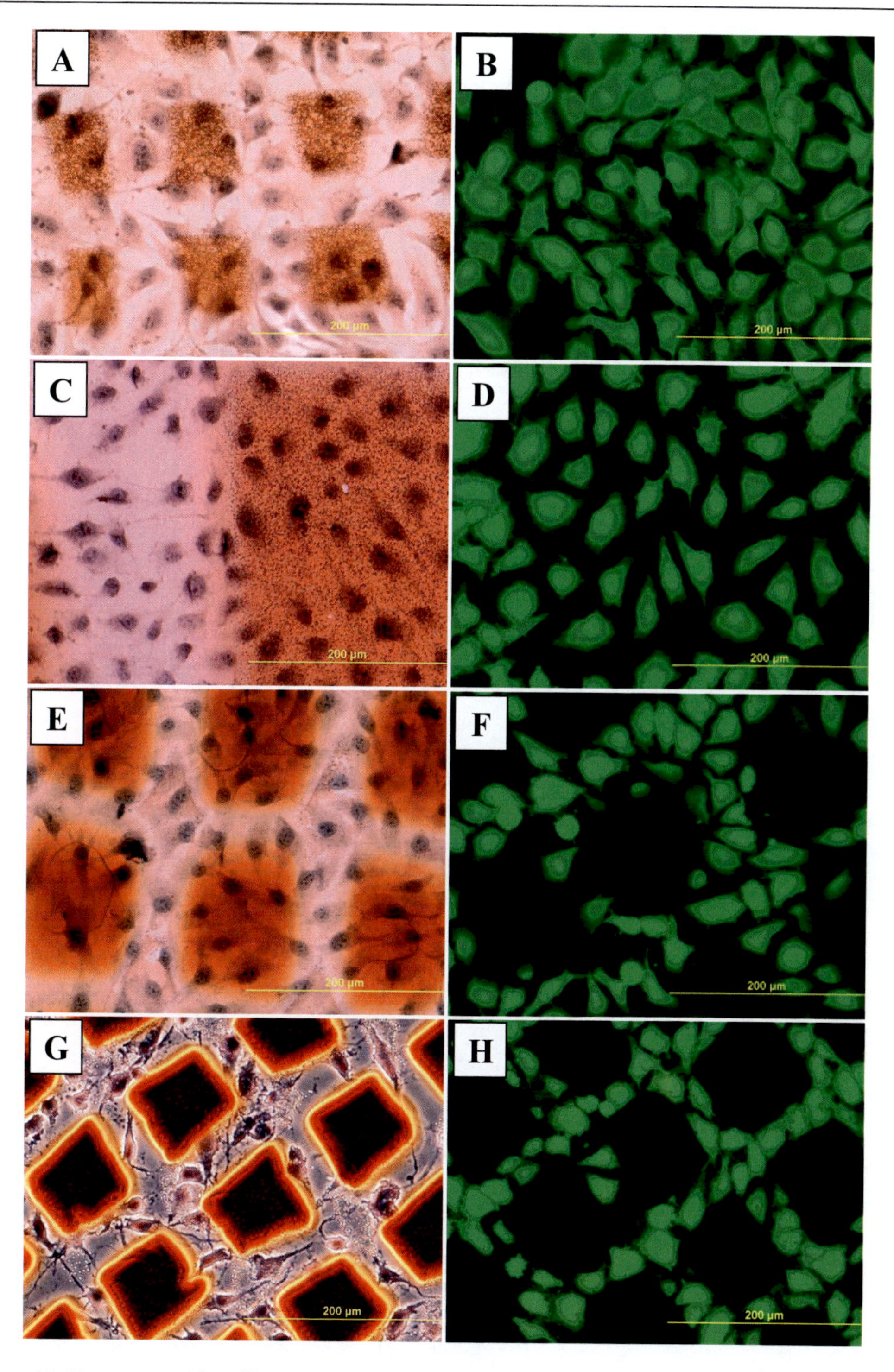

Figure 13. Human osteoblast-like MG 63 cells on day 7 after seeding on fullerene layers micropatterned with prominences 128 ± 8 nm in height (A, B), 238 ± 3 nm in height (C, D), 326 ± 5 nm in height (E, F) or 1043 ± 57 nm in height (G, H). Stained with hematoxylin and eosin (A, C, E, G) or LIVE/DEAD viability/cytotoxicity kit (B, D, F, H). Olympus IX 50 microscope, DP 70 digital camera, obj. 20x, bar = 200 µm [117, 118].

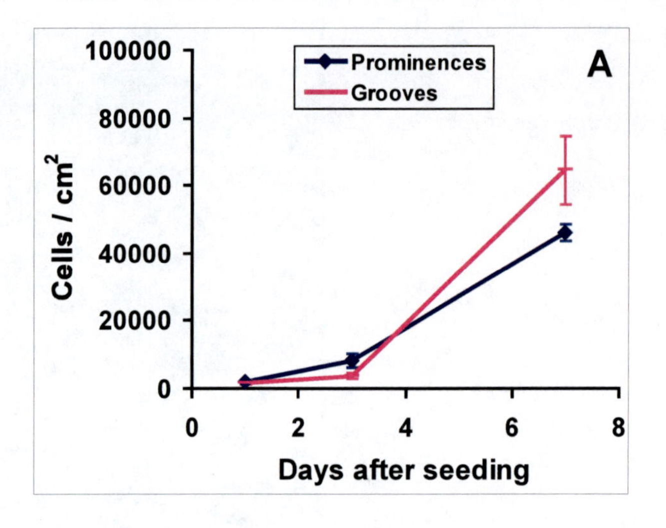

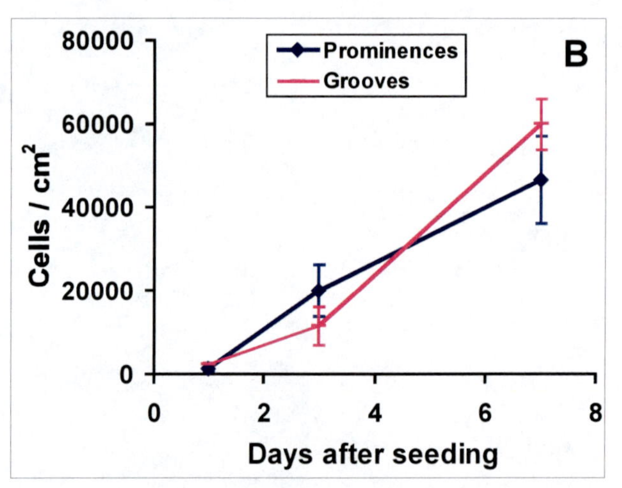

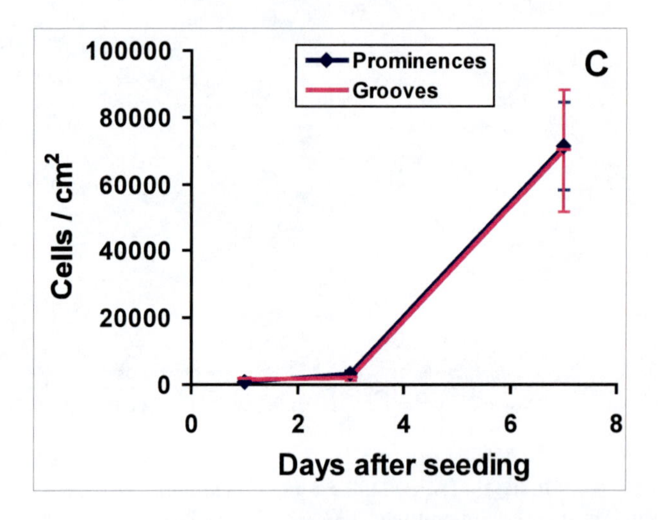

Figure 14. Continued on next page.

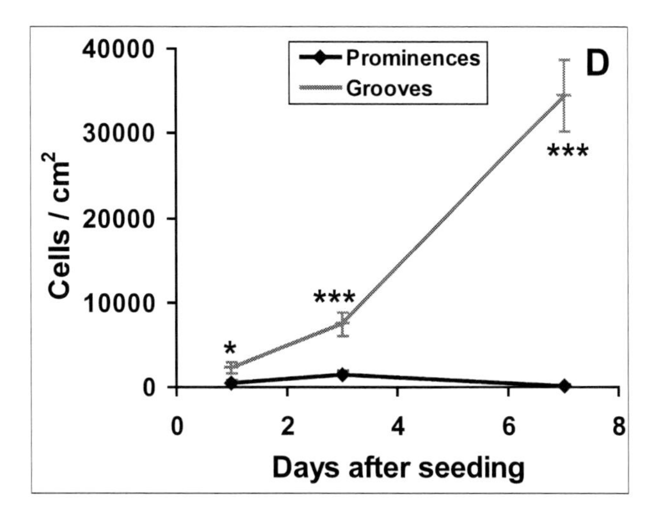

Figure 14. Population densities of MG 63 cells on prominences and grooves on fullerene layers micropatterned with prominences 128 ± 8 nm in height (A), 238 ± 3 nm in height (B), 326 ± 5 nm in height (C) or 1043 ± 57 nm in height (D). Mean ± Standard Error of Mean (S.E.M.) from 9 to 21 microphotographs, Student t-test, statistical significance: *$p \leq 0.05$, ***$p \leq 0.001$ [118].

However, the cells on the layers with prominences of 1043 ± 57 nm were found preferentially in the grooves (Figure 13). These grooves contained from 75.0 ± 13.4 % to 99.6 ± 0.3 % of cells, though they occupied only 44 ± 3 % of the material surface. The cell population density in the grooves was about 5 to 192 times higher than on the bulges, and these differences increased proportionally to the time of cultivation (Figure 14). Preferential growth of bone cells in grooves, pits and other types of hollows has been observed on various artificially microfabricated or primary rough polymeric, metallic or carbon-based materials [104, 116, 130, 143, 145-148]. However, the cells were also able to colonize the prominences on these surfaces, though the prominences were usually much higher (i.e., several micrometers or even tens of μm) than those on the thick micropatterned fullerene layers in our present study (only about 1 μm). Surprisingly, on our fullerene C_{60} layers, the MG 63 cells were not able to "climb up" relatively low prominences only about 1 μm in height, even at a relatively late culture interval of 7 days after seeding. This may be due to a synergetic action of certain physical and chemical properties of the fullerene bulges less appropriate for cell adhesion, such as their hydrophobicity, a steep rise of the prominences, as well as the tendency of spherical ball-like fullerene C_{60} molecules to diffuse out of the prominences [149].

Fluorescence staining of cells using the LIVE/DEAD kit revealed that, on day 1 after seeding, the percentage of viable cells on all micropatterned fullerene layers ranged between 80 ± 10 % and 87 ± 20 %, and was similar to the values obtained on the control tissue culture polystyrene dishes (96 ± 21%) and microscopic glass coverslips (67 ± 16 %). It can be assumed that the presence of a certain number of dead cells was due to the stress to which the cells were submitted during the cell seeding procedure, particularly during trypsinization and keeping the anchorage-dependent cells in suspension, rather than due to a cytotoxic action of the fullerenes. Accordingly, the percentage of viable cells increased with time of cultivation, reaching 97 ± 25 % to 99 ± 15 % on day 7 after seeding on all tested surfaces.

Thus, micropatterned fullerene films could be used as templates for regionally-selective cell adhesion and growth. Moreover, osteogenic cells growing in hollows have been reported to be more active in the phosphorylation of various kinases and transcription factors, signal transduction and differentiation [143, 145-148]. From this point of view, the microstructured fullerene layers (if they adhere strongly to the underlying materials and are resistant to the release of fullerene molecules) could be used as a bioactive coating for bone implants in order to improve their integration with the surrounding bone tissue.

2.3. Fullerene C_{60} Layers Modified by Ion- or Laser-irradiation

Both continuous and micropatterned fullerene C_{60} layers were irradiated with ion beams (Au^+ ions with energy of 1.8 MeV and fluence up to 2 x 10^{15} cm^{-2}) with the aim (i) to convert C_{60} into another biocompatible carbon allotrope, and/or (ii) to induce pattern formation in various scales and dimensions (in order to promote bio-compatibility and bio-attractivity of the micro- and/or nano-patterened carbon substrates with respect to the inspected MG 63 cells).

Figure 15 A shows that using ion irradiation (1.8 MeV Au^+, 2 x 10^{14} cm^{-2}) both goals can be achieved – i.e., conversion of C_{60} into bio-compatible amorphous carbon and selforganization of the as-deposited carbon allotrope layer (i.e., alteration of the original array of rectangular bulges into a system of periodically repeating islands of trapezoidal shape). The form of the trapezoids changes in one direction as a mirror-reflection recurrence; in the perpendicular direction the change is more complex, again combining a mirror reflection with a flip-flop turn. The bulge array form alteration may be due to a coordinated stress relaxation that is accumulated in the system (of a thin C_{60} backing layer supporting the bulge prominences) during ion irradiation. The continuous transformation of C_{60} into amorphous carbon (during progressive ion irradiation) results in the gradual inner stress growth that has (by contrast) a tendency to be relieved. This tendency grows with the accumulation of stress energy, and it results in a sudden stress relaxation. The consequence is a different sample configuration with a distinct topography that has lower area (stress) energy. The observed trapezoidal form and repeating spatial arrangement of the bulge prominences is an interesting self-shaping pattern formation that deserves further attention.

A similar effect can be found when a continuous layer of fullerenes (a thin film fullerite) is irradiated with energetic ions. In Figure 15 B, a dramatic alteration of the originally smooth (as-deposited) C_{60} layer is demonstrated after 1.8 MeV Au^+ ion irradiation to the fluence 2 x 10^{14} cm^{-2}. It can be seen that ion irradiation leads to the formation of ridges and grooves in the surface morphology that could advantageously be controlled by a selected fluence of the ions. As mentioned above, C_{60} is converted during ion irradiation into amorphous carbon, which (due to a different molecular arrangement) requires expansion of the transformed layer into a larger area (than the original C_{60} film holds). This results in an interesting topography of a periodic-like corrugation of the amorphous carbon film. It might be expected that irradiation of the fullerite film through a metallic mask will create a periodic C_{60} - amorphous carbon system with original spatial patterns. Based on the effect of higher bio-compatibility of the micro- and nano-structured surface morphologies, micro-corrugated amorphous carbon films might become attractive for cell adhesion and growth.

The first experiments confirmed that the ion-irradiated fullerene layers enable homogeneous colonization with MG 63 cells to an extent similar to the non-irradiated layers, as well as reference cell culture substrates, represented by microscopic glass coverslips and standard polystyrene dishes. The vast majority of the cells were viable, of normal polygonal or spindle-shaped morphology, well-spread and with a well-assembled fine mesh-like beta-actin cytoskeleton (Figure 16).

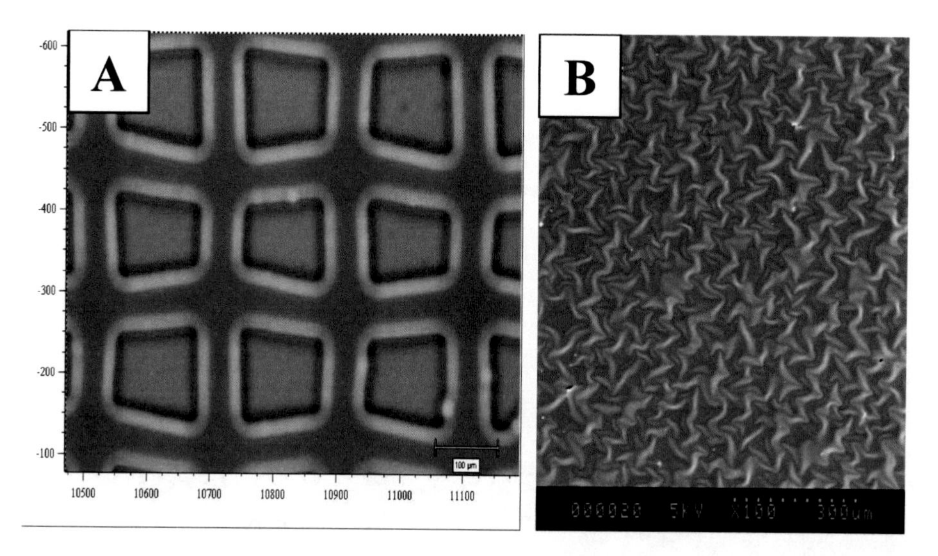

Figure 15. A. Fullerene C_{60} microdomains, created by deposition through a metallic mesh and irradiated with Au^+ ions (energy of 1.8 MeV and fluence of $2x10^{14}$ cm^{-2}). After irradiation, the original system of symmetrical (rectangular) C_{60} bulges was converted into a system of amorphous carbon bulges of trapezoid shapes varying (as a mirror image) periodically. B. A continuous fullerene C_{60} layer irradiated with Au^+ ions (1.8 MeV, $2x10^{15}$ cm^{-2}). The irradiation led to corrugation of the originally smooth fullerene layer.

2.4. Metal-fullerene C_{60} Composites

Another novel and interesting material with potential use in biotechnology and medicine is represented by binary metal/C_{60} composites. In our experiments, several different kinds of these hybrid (metallic-organic) systems, mainly thin films of Ni-C_{60} and Ti-C_{60}, have been inspected. These composites combine important properties of both phases and are assumed also to exhibit satisfying bio-compatible qualities.

In our earlier studies, nanocomposite Ti/hydrocarbon plasma polymer films, created by DC magnetron sputtering and containing amorphous hydrogenated carbon-like material with nanoclusters of Ti, supported the adhesion, growth and maturation of human bone-derived MG 63 cells to an extent comparable to that found on standard cell culture materials, represented by polystyrene dishes, microscopic glass slides or coverslips. In the case of vascular endothelial cells of the line CPAE, derived from the bovine pulmonary artery, the cell adhesion, growth and maturation was even improved in comparison with the standard culture materials and hydrocarbon plasma polymer films without Ti [103]. The beneficial effects of Ti on the colonization of carbon fibre-reinforced carbon composites (with MG 63 cells, as well as primary or low-passaged vascular smooth muscle cells, derived from the rat aorta) were also observed after the CFRC was coated with a carbon-titanium layer [104].

In the case of metal-C_{60} hybrids, both inspected systems (Ni-C_{60} and Ti-C_{60}) were deposited on various supports (e.g., Si and MgO; Figure 17). They exhibited a number of interesting properties, mainly a tendency toward spontaneous self-organization, either already during hybrid synthesis (i.e., simultaneous co-deposition of both metal and organic phases) or during post-deposition processing (e.g., thermal annealing). A typical example of pattern formation of an as-deposited Ni-C_{60} hybrid is shown in Figure 18 A. Under certain deposition kinetics a periodic system of stripe domains was formed. The domains consist of sub-half-micrometer large droplets encompassed with a thin C_{60}-based rind with a polymeric structure. The stripes are embedded in the double-layer platform falling into the epitaxial Ni and amorphous carbon thin films. It has been clarified that during co-deposition the C_{60} molecules partly disintegrate and transform toward amorphous carbon. Due to the accumulation of amorphous carbon (with limited solubility in Ni), thermodynamic instability arises in the system, and after it reaches a certain level it triggers spontaneous partitioning and self-organization of the deposited material. A sequential drift and coordinate release of the thermodynamic instability has been proposed as a principal pattern formation mechanism [150].

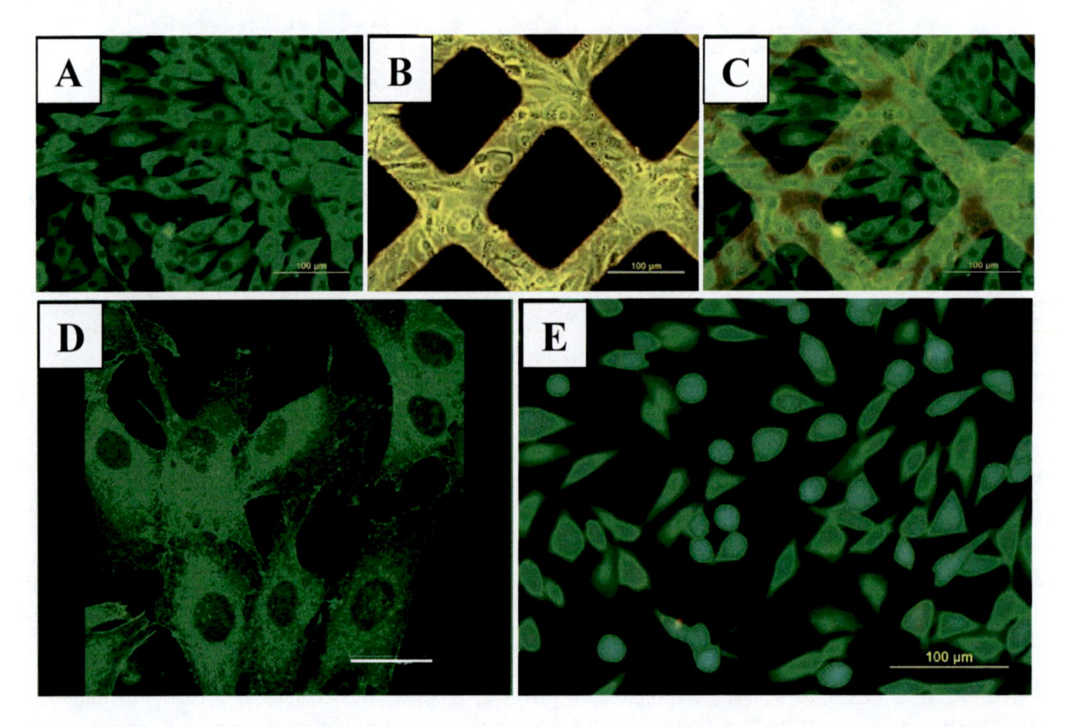

Figure 16. Human osteoblast-like MG 63 cells in cultures on a fullerene C_{60} layer micropatterned by deposition through metallic grid and then irradiated with Au ions of the energy of 1635 keV and fluence of $2x10^{14}$ cm^{-2}. A-D: immunofluorescence staining of beta-actin, day 3 after seeding; E: staining with LIVE/DEAD viability/cytotoxicity kit, day 1 after seeding. A, B, C, E: Olympus IX 50 microscope, DP 70 digital camera. A: fluorescence; B: conventional light; C: combination of both fluorescence and conventional light. Bar = 100 µm. D: Leica DM 2500 confocal microscope, obj. 100, bar = 30 µm.

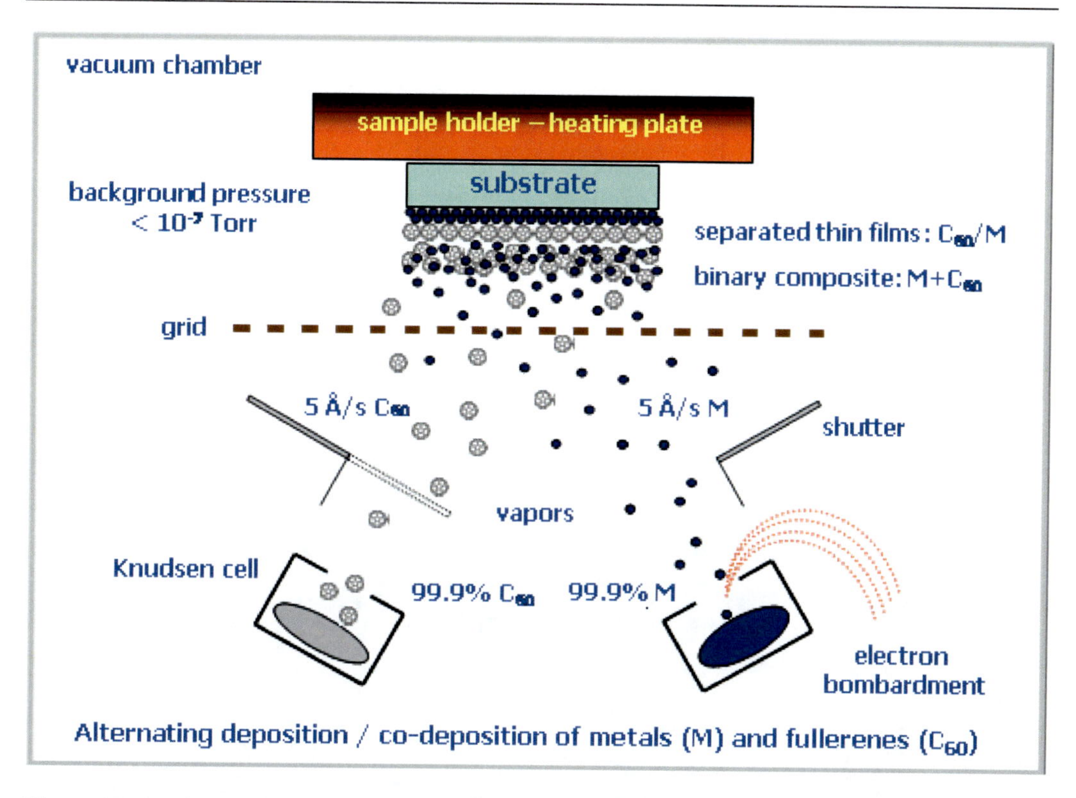

Figure 17. A scheme of hybridization of the metal/C_{60} composites in an Univex-300 vacuum system (Leybold, Germany; see Figure 1). Deposition rates: DR(M) = DR(C_{60}) ~ 5 Å/s; temperatures during deposition: 500°C (M), 120°C (C_{60}); total thickness of the hybrid composites 100 - 1000 nm.

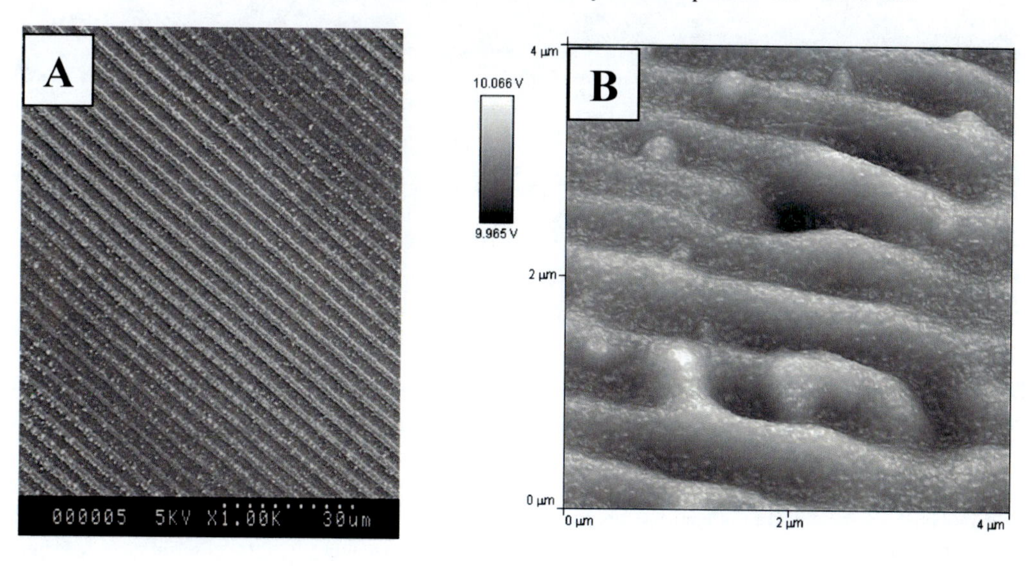

Figure 18. A. SEM micrograph of a stripe system spontaneously formed during co-deposition of Ni and C_{60} on the MgO (001) substrate. SEM-EDS (JOEL JSM-5600), bar = 30 µm. B. MFM (Magnetic Force Microscopy) analysis of the Ni+C_{60} composite synthesized at room temperature on the Si (001) substrate. Veeco CP-II microscope, scan area of 5 x 5 µm.

Another interesting manifestation of the pattern formation proclivity of the hybrid system is shown in Figure 18 B. Here, we observe an interesting property of Ni-C_{60} hybrids deposited on Si at RT - i.e., a buried self-organized morphology, consisting of periodic stripe-like magnetic domains, that has been revealed using the MFM (Magnetic Force Microscopy) method. The 'hidden' system of magnetic domains indicates that even at RT (when only a very small additional thermal energy may enter the system) a certain coordinated rearrangement (relaxation) of the stressed microstructure may appear [151].

An interesting issue is how these hybrid composites and their micro- and nano-pattern forms may affect the bio-activity of the cells. Preliminary results (obtained only recently) have demonstrated that the composites are bio-compatible and support adhesion and growth of the human osteoblast-like MG 63 cells (Figure 19 A-F).

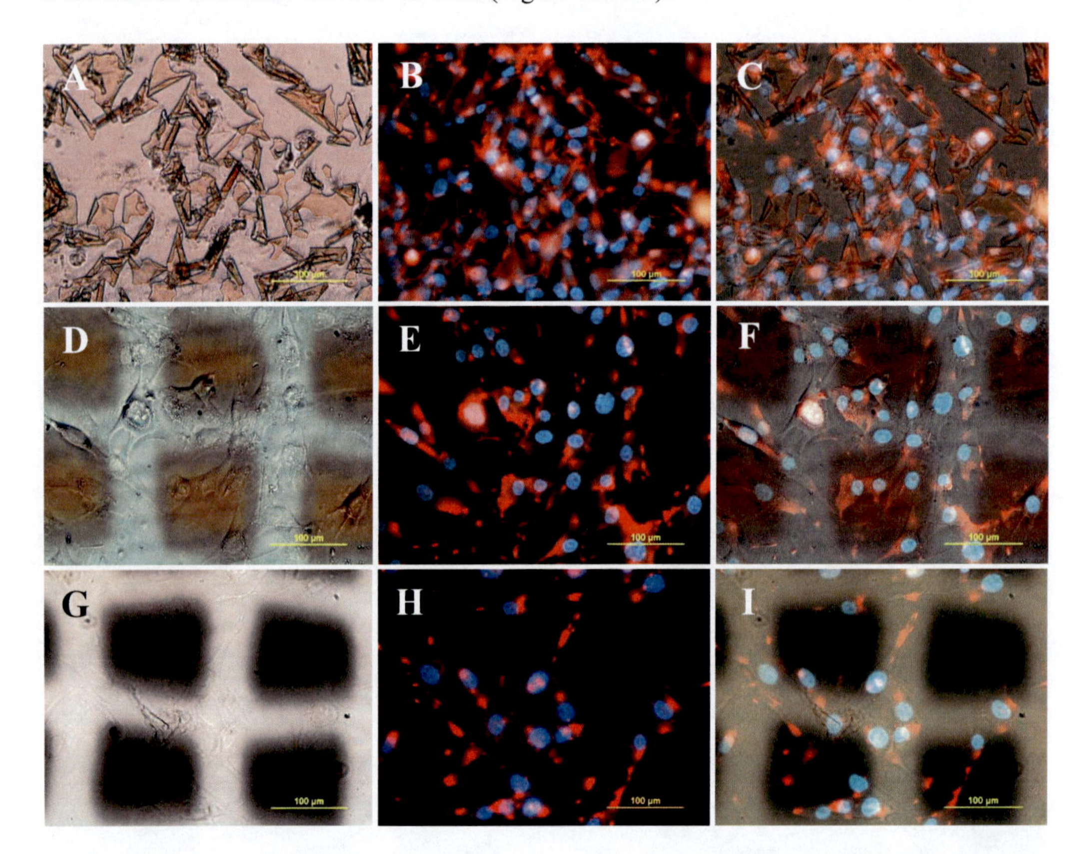

Figure 19. Human osteoblast-like MG 63 cells on day 4 after seeding on hybrid Ti-C_{60} composites deposited on microscopic glass coverslips in the form of continuous films (A-C) or films micropatterned by deposition through a metallic grid (D-F) and irradiated with Au^+ ions of the energy of 1635 keV and fluence of $2x10^{14}$ cm^{-2} (G-I). The cells were fixed with paraformaldehyde and stained with Hoechst 33342 and Texas Red C_2-maleimide fluorescence dyes. Olympus IX 50 microscope, DP 70 digital camera. A,D,G: conventional light; B,E,H: fluorescence; C,F,I: combination of both conventional light and fluorescence. Bar = 100 μm.

A study of ion-irradiated composites has shown that, similarly as in the case of plain fullerite films, also in the case of hybrids the fullerene molecules are converted to amorphous carbon. Because of the metallic phase (which keeps the integrity of the composites firm) no manifestation of the spatial geometry alteration has been observed during the C_{60} →

amorphous carbon transformation (as it is in the case of the plain fullerites). A thorough analysis of the systems showed that the hybrids are (after ion irradiation) more stable than the original structures (which are highly stressed due mainly to low miscibility of the C_{60} and metallic phases). More significant activity of the phase components (e.g., phase separation) can be observed only when the system is exposed to an energy flow, e.g., during thermal annealing or laser illumination. The preliminary results of our study showed that colonization of the human osteoblast-like MG 63 cells may also be possible on ion-irradiated and/or thermally processed hybrids, especially of the Ti-C_{60} type (Figure 19 G-I).

3. Cell Growth on Polymer-Carbon Nanotube Composites

Our studies on cell interaction with nanotube-modified materials were carried out on polymer-nanotube composites, i.e., synthetic polymers mixed with single- or multi-walled carbon nanotubes. As the polymeric matrix, we used a terpolymer of polytetrafluoroethylene, polyvinyldifluoride and polypropylene (PTFE/PVDF/PP) or polysulphone (PSU), i.e., polymers which were considered to be promising for the construction of bone implants. The nanotubes were expected to reinforce the relatively soft and elastic polymeric material and thus to improve its mechanical properties for hard tissue surgery. The other desired function of nanotubes was the formation of prominences in nanoscale on the material surface, which could promote its colonization with bone cells.

3.1. Composites of Carbon Nanotubes and PTFE/PVDF/PP

For the preparation of composites consisting of carbon nanotubes and a PTFE/PVDF/PP terpolymer matrix, a commercially available terpolymer of polytetrafluoroethylene, polyvinyldifluoride and polypropylene (density of 1600 g/dm^3, Aldrich Chemical Co., U.S.A.) was dissolved in acetone at a concentration of 0.1 g/ml. Single-wall carbon nanohorns (SWNH) or high crystalline electric arc multi-wall nanotubes (MWNT-A; both from NanoCraft Inc., Renton, U.S.A.) were mixed with acetone in a sonicator for 5 minutes, and then with the terpolymer solution for 15 minutes to a concentration of 2, 4, 6 or 8 wt. %. The mixtures were then poured on to Petri dishes and left to evaporate the solvent [95].

Despite sonication, both SWNH and MWNT-A formed aggregates in the polymeric matrix, which created microscale irregularities on the material surface (Figure 20). The surface microscale roughness, measured by surface profilometry, increased proportionally to the nanoparticle concentration. In the pure terpolymer, the R_a parameter, i.e., the arithmetic mean of the departures of the roughness profile from the mean line, was only 0.20 ± 0.04 μm, whereas in the SWNH- and MWNT-A-modified material, the R_a parameter ranged from 0.41 ± 0.01 μm to 2.22 ± 0.36 μm in terpolymer samples with 2 to 8 wt. % of nanoparticles (Figure 21 A). The R_a parameter was usually similar in materials containing the same concentrations of SWNH or MWNT-A.

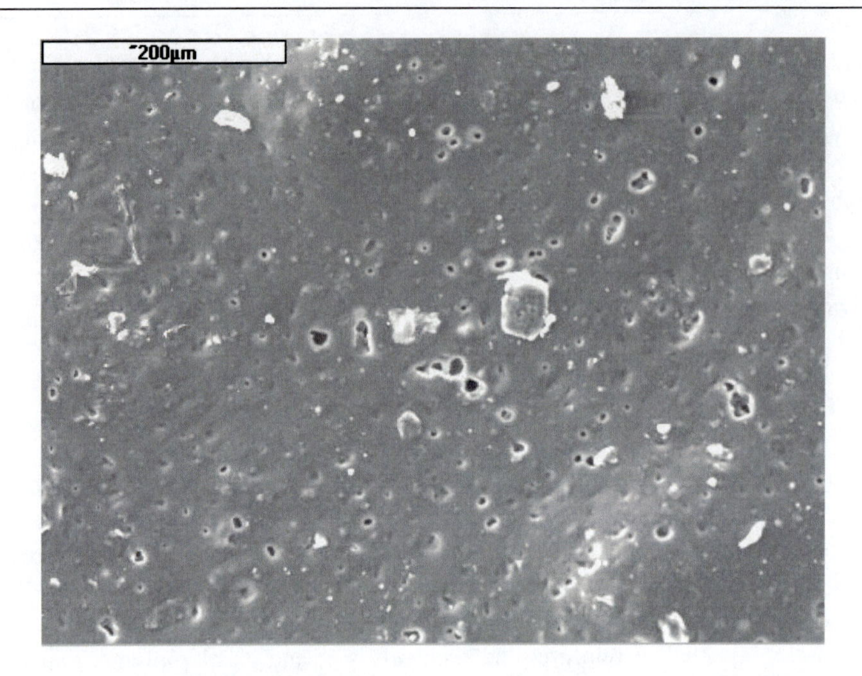

Figure 20. An example of the surface morphology of the terpolymer of polytetrafluoroethylene, polyvinyldifluoride and polypropylene mixed with 4 wt.% of single-wall carbon nanohorns. Jeol JSM 5400 Scanning Electron Microscope, obj. 2000x. Bar = 200µm [95].

At the same time, the surfaces of the terpolymer-nanotube composites contained irregularities in nanoscale, as revealed by AFM (Figure 21 B). Therefore, in fact, similarly to the fullerene-coated CFRC and micropatterned surfaces (created by deposition of C_{60} through metallic grids), the surface of the carbon nanotube-polymer composites was of a hierarchically organized micro- and nanostructure.

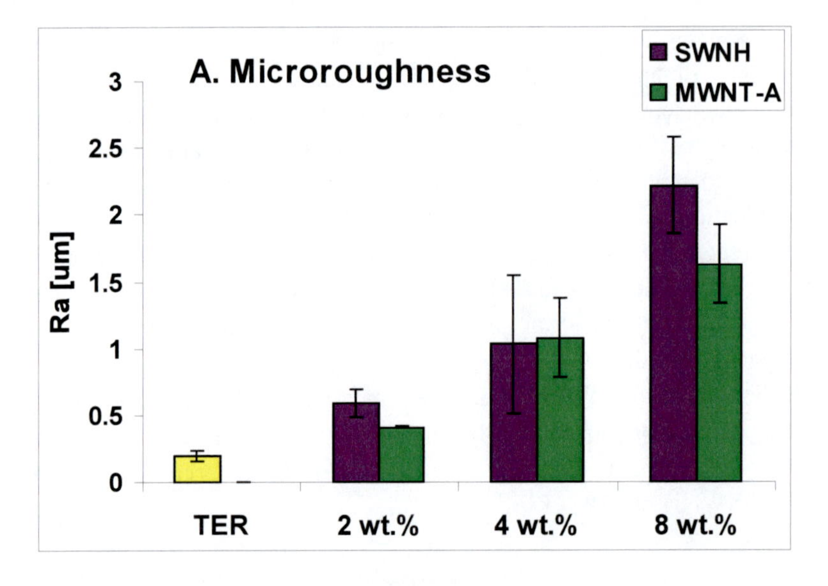

Figure 21. Continued on next page.

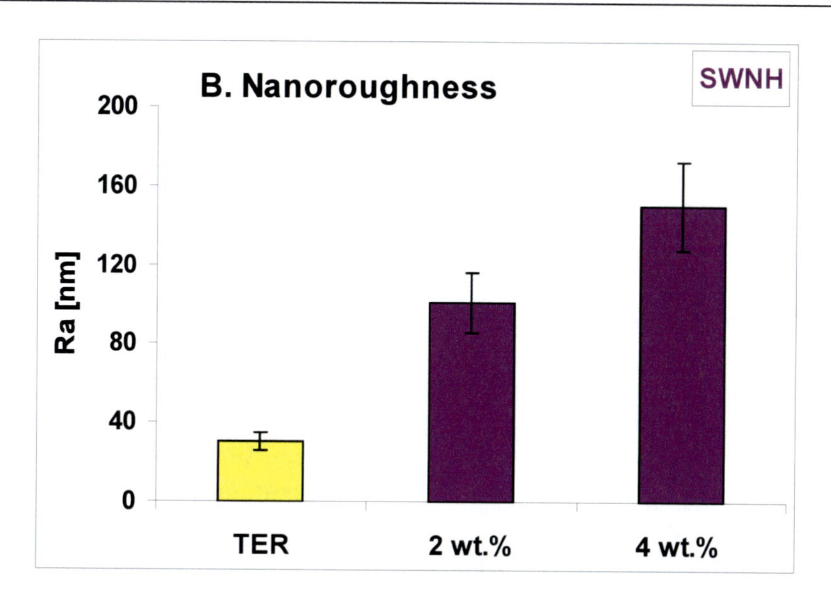

Figure 21. Surface roughness in microscale (A) and nanoscale (B) of unmodified terpolymer of polytetrafluoroethylene, polyvinyldifluoride and polypropylene (TER) and terpolymer samples mixed with 2, 4, 6 and 8 wt.% of single-wall carbon nanohorns (SWNH) or multi-wall carbon nanotubes (MWNT-A). R_a is an average deviation of the roughness profile from the mean line. Measured with a T500 Hommel Tester (Hommelwerke Co., Germany; A) or by atomic force microscopy using SPM LAB 5.01 software (B). Mean ± S.D. (Standard Deviation) from 3 measurements [95].

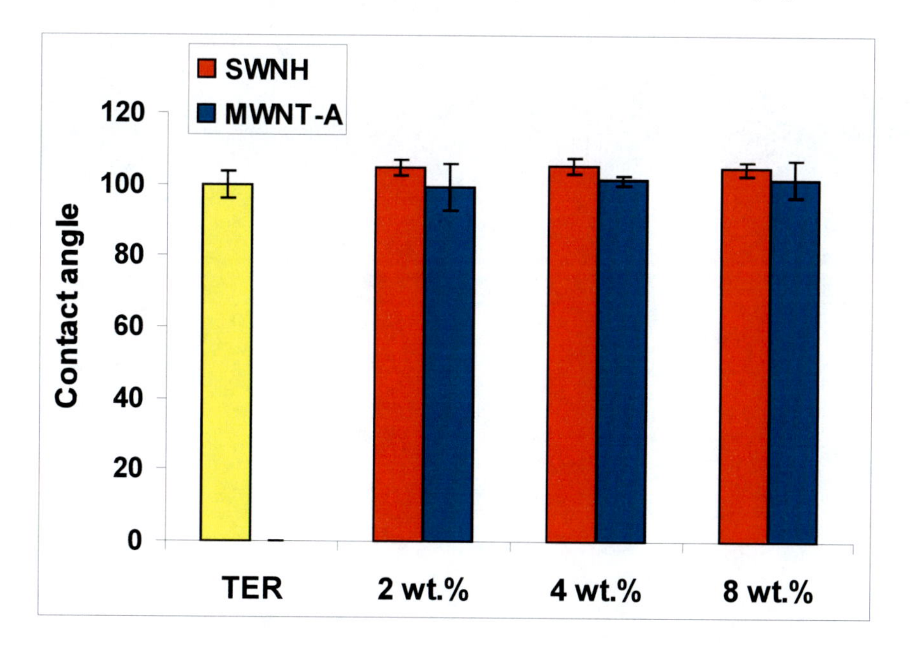

Figure 22. Static water drop contact angle measured on unmodified surface terpolymer of polytetrafluoroethylene, polyvinyldifluoride and polypropylene (TER) and terpolymer samples mixed with 2, 4, 6 and 8 wt.% of single-wall carbon nanohorns (SWNH) or multi-wall carbon nanotubes (MWNT-A). Measurement by the sessile drop method using the DSA 10 Mk2 automatic drop shape analysis system (Kruss, Germany). Mean ± S.D. (Standard Deviation) from 6-10 measurements [95].

The composites of PTFE/PVDF/PP and carbon nanotubes were relatively highly hydrophobic. The water drop contact angle, measured in a static water drop-material contact system, was in the range from 99.3 ± 6.56° to 105.2 ± 2.12°, and was similar in the unmodified and all carbon-nanoparticle-modified terpolymer samples (Figure 22). The surface energy, measured in terpolymers modified with 4 wt. % of SWNH or MWNT-A, was also similar in pure terpolymer and both types of composite samples. Only in the sample with 4 wt. % of SWNH was the polar component of the surface energy significantly lower (Table 1).

Table 1. The surface energy and its dispersion and polar components in pure terpolymer and terpolymer modified with 4 wt. % of SWNH or MWNT-A, measured by contact angle for polar liquid (water) and for non-polar liquid (diodomethane)

Sample name	Surface energy [mN/m]	Dispersion component [mN/m]	Polar component [mN/m]
Terpolymer	22.63 ± 0.61	21.18 ± 0.47	1.45 ± 0.13
4 wt. % SWNH	21.12 ± 0.43	20.47 ± 0.38	0.65 ± 0.04
4 wt. % MWNT-A	22.61 ± 1.08	21.50 ± 0.94	1.11 ± 0.14

Mean ± S.D. (Standard Deviation) from 12 measurements.

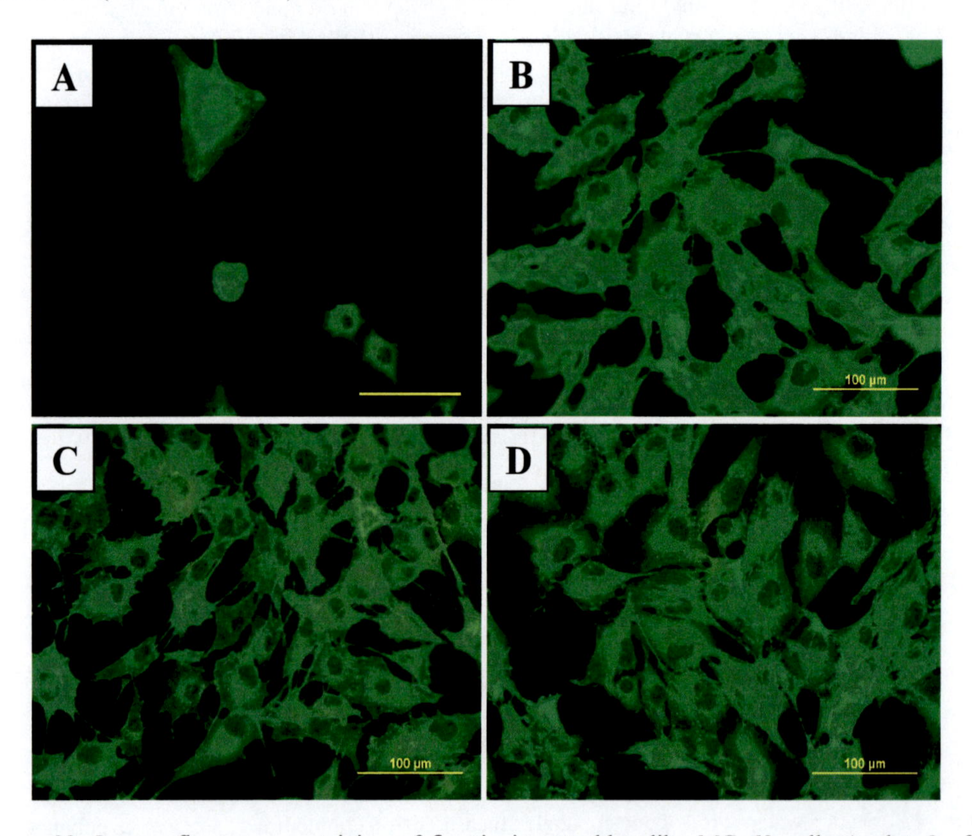

Figure 23. Immunofluorescence staining of β-actin in osteoblast-like MG 63 cells on day 3 after seeding on a terpolymer of polytetrafluoroethylene, polyvinyldifluoride and polypropylene (A), terpolymer mixed with 4 wt. % of single-wall carbon nanohorns (B) or 4 wt. % of multi-wall carbon nanotubes (C) and tissue culture polystyrene dish (D). Olympus IX 50 microscope, DP 70 digital camera, obj. 20, bar = 100 μm [95].

The addition of carbon nanotubes to the PTFE/PVDF/PP terpolymer markedly improved several parameters of cell adhesion, growth and maturation. The human osteoblast-like MG 63 cells in cultures on PTFE/PVDF/PP mixed with SWNH or MWNT-A were well-spread, polygonal, and contained distinct beta-actin filament bundles, whereas most cells on the pure terpolymer were less spread or even round and clustered into aggregates (Figure 23). The enzyme-linked immunosorbent assay (ELISA) revealed that the cells on the material with 4 wt. % SWNH contained a higher concentration of vinculin, a component of focal adhesion plaques, in comparison with the values in cells on the pure terpolymer and also on the tissue culture polystyrene (Figure 24 A). The concentration of talin, another important integrin-associated focal adhesion protein, was also higher in cells grown on terpolymer samples with 4 and 8 wt.% of SWNH (Figure 24 C). However, the presence of MWNT-A did not affect the concentration of vinculin and talin in cells cultured on these samples (Figure 24 B, D). The concentration of osteocalcin, a marker of osteogenic cell differentiation, was similar in cells on all SWNH-modified and pure terpolymer samples, and also TCPS (Figure 24 E). However, in cells on samples with 4 and 6 wt. % of MWNT-A, the concentration of osteocalcin was significantly lower than the concentration in cells on the pure terpolymers and TCPS (Figure 24 F). This could be explained by a higher proliferation activity of these cells (Figure 25), which may delay the process of cell differentiation [122]. Another factor affecting the osteogenic maturation of cells is the softness of the adhesion substrate. On softer substrates, human mesenchymal stem cells differentiated towards muscle or neuronal cells, whereas osteogenic differentiation required matrices of relatively high rigidity [152]. The terpolymer used in our studies, though reinforced with carbon nanoparticles, could still be considered as a relatively soft material in comparison with the metals and ceramics that are also used in bone tissue engineering. Nevertheless, the concentration of intercellular adhesion molecule-1 (ICAM-1), an adhesion molecule of the immunoglobulin superfamily and an important marker of cell immune activation, was not increased in cells on terpolymers modified with SWNH or MWNT-A (Figure 24 G, H), which suggests that the material could be tolerated by the host tissue.

In summary, the improved colonization of the PTFE/PVDF/PP terpolymer with human osteoblast-like MG 63 cells after it has been modified with carbon nanotubes could be attributed to changes in its surface roughness and topography, especially the creation of the surface nanostructure, rather than to its surface wettability, which remained unchanged and relatively low. Only the increased proliferation activity of cells on samples with 4 wt. % of MWNT-A (Figure 25) could be attributed to a favorable combination of the material surface nanostructure and a relatively high polar component of the free surface energy (Table 1) [153-156]. Such a combination was not observed on the other samples, which were either nanostructured (terpolymer with SWNH) or only had a higher polar component of the surface energy (pure terpolymer). On the other hand, the terpolymer modified with 4 wt. % of SWNH seemed to be the most appropriate substrate for cell adhesion, spreading and formation of focal adhesion plaques.

3.2. Composites of Carbon Nanotubes and PSU

The composites of polysulphone (PSU) and carbon nanotubes were prepared by a similar approach as the constructs with a PTFE/PVDF/PP matrix. The PSU (Aldrich Chemical Co.,

U.S.A.) was dissolved in dichloromethane at a concentration of 0.1 g/ml, mixed with 0.5, 1.0 or 2.0 wt.% of SWNH or MWNT-A, sonicated, poured on to Petri dishes and left to evaporate freely. The PSU-carbon nanotube composites also had a combined micro- and nanostructure of the material surface. The microroughness was similar in all nanotube-containing PSU samples, but was higher in nanotube-modified PSU than in unmodified polymer. The surface nanoroughness, measured by AFM, increased proportionally with the concentration of carbon nanotubes in the polymer matrix (Figure 26) [157, 158].

In comparison with the terpolymer PTFE/PVDF/PP, the PSU used in our studies was more wettable, having a sessile water drop contact angle of 84.8 ± 5.4°. Unlike the terpolymer, PSU in its pristine unmodified state gave good support for the adhesion and growth of bone-derived MG 63 cells. PSU proved to be an appropriate growth substrate for human periodontal ligament fibroblasts [159], freshly isolated rat hepatocytes [160], as well as a suitable component of three-dimensional scaffolds for the construction of bioartificial liver assist devices [161]. After the addition of MWNT-A, the surface wettability of the composite remained similar (contact angle in the range from 85.0 ± 4.6° to 90.1 ± 2.7°), while after addition of SWNH, the wettability decreased slightly (contact angle from 93.2 ± 4.3° to 91.2 ± 3.0°).

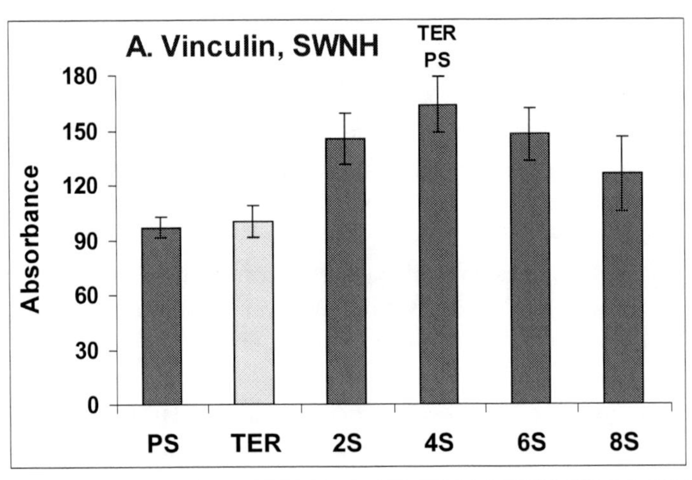

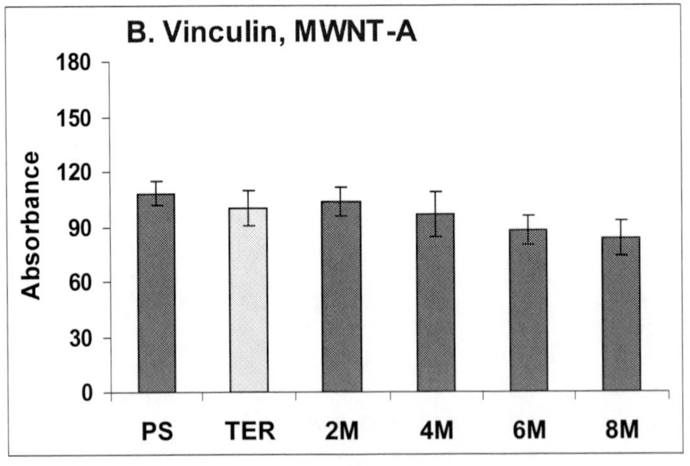

Figure 24. Continued on next page.

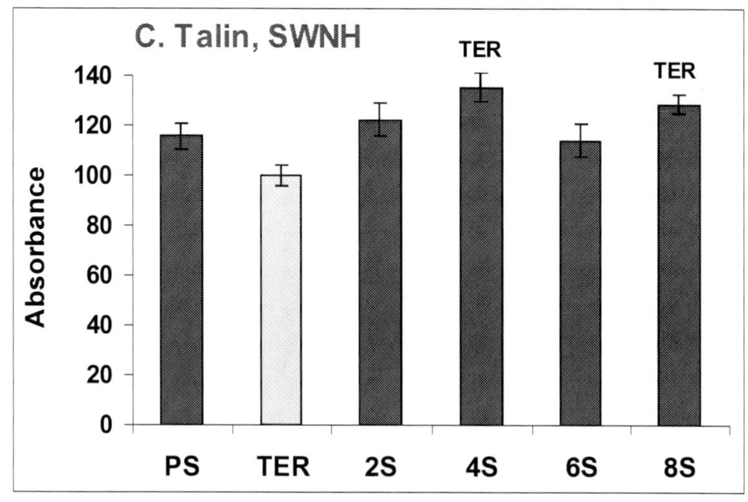

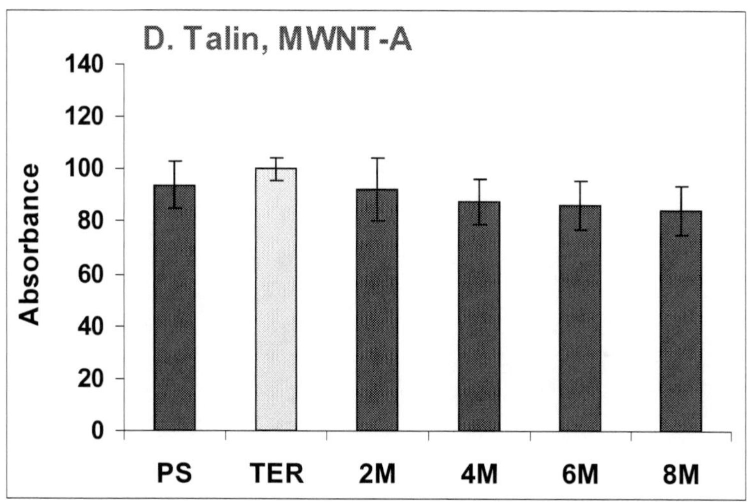

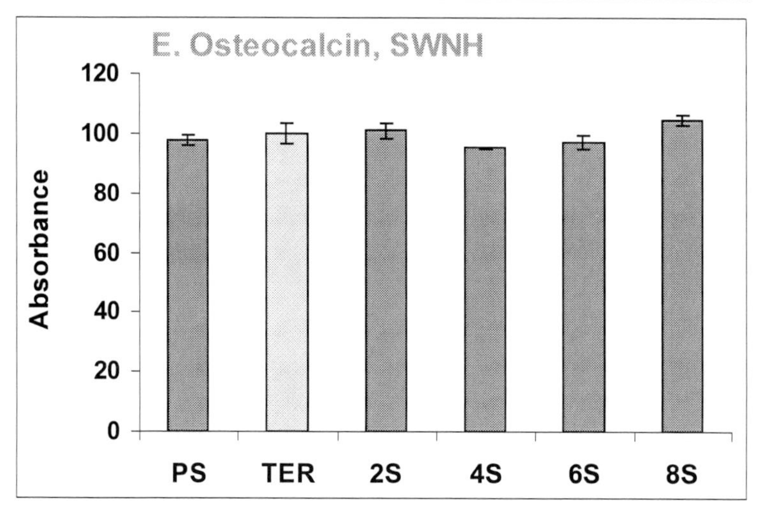

Figure 24. Continued on next page.

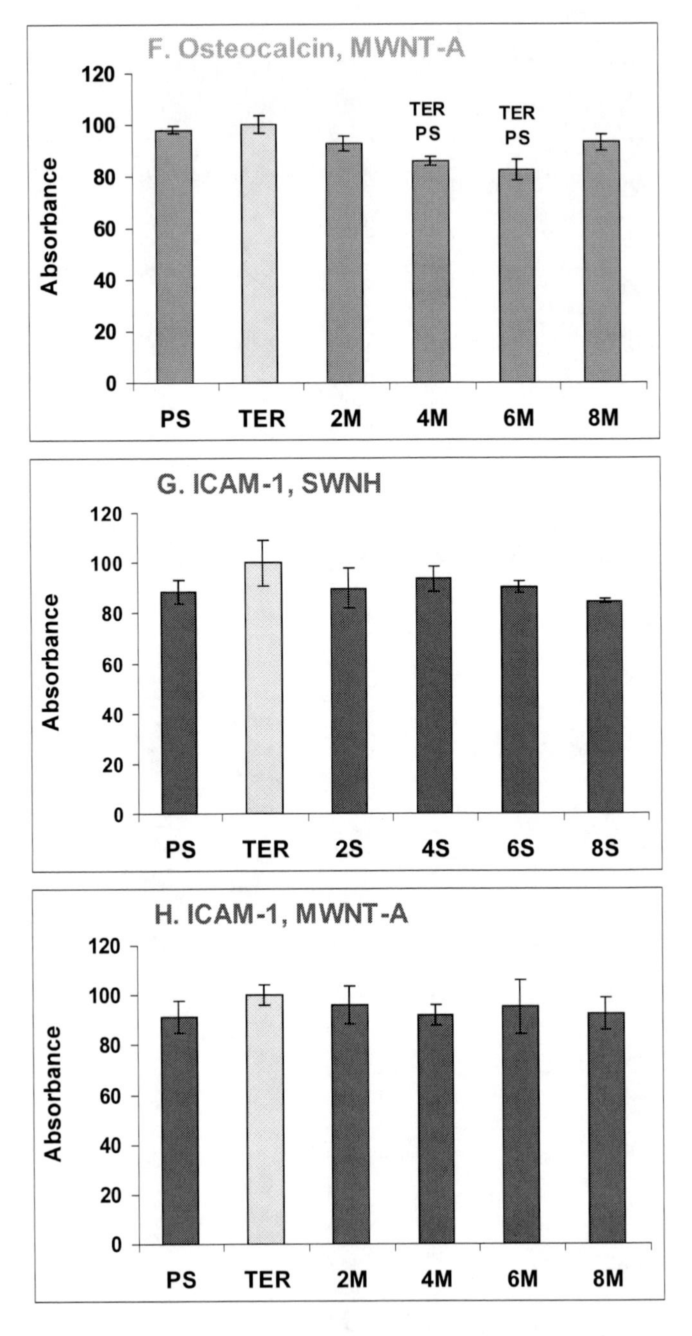

Figure 24. Concentration of vinculin (A, B), talin (C, D), and osteocalcin (E, F) in osteoblast-like MG 63 cells on day 8 after seeding on the pure terpolymer of polytetrafluoroethylene, polyvinyldifluoride and polypropylene (TER), terpolymer mixed with 2, 4, 6 or 8 wt.% (2S, 4S, 6S, 8S) of single-wall carbon nanohorns (SWNH) or 2, 4, 6 or 8 wt.% (2M, 4M, 6M, 8M) of multi-wall carbon nanotubes (MWNT-A) and tissue culture polystyrene (PS). Measured by ELISA per mg of protein; the absorbance values of cells from the modified terpolymers were expressed in % of the values obtained from the control cells grown on the unmodified terpolymer. Mean ± S.E.M. (Standard Error of Mean) from 4 measurements, ANOVA, Student-Newman-Keuls method. Statistical significance: $^{TER,\ PS}$: $p \leq 0.05$ compared to the values on the pure terpolymer and tissue culture polystyrene, respectively [95].

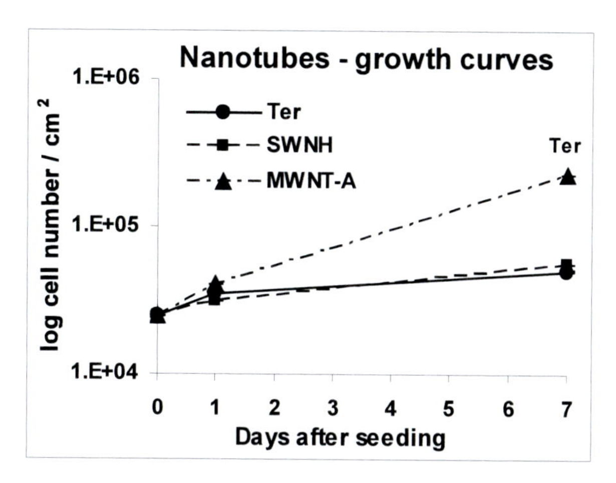

Figure 25. Growth curves of MG 63 cells on a terpolymer of polytetrafluoroethylene, polyvinyldifluoride and polypropylene (Ter), terpolymer mixed with 4 wt.% of single-wall carbon nanohorns (SWNH) or 4 wt.% of multi-wall carbon nanotubes (MWNT-A) [95].

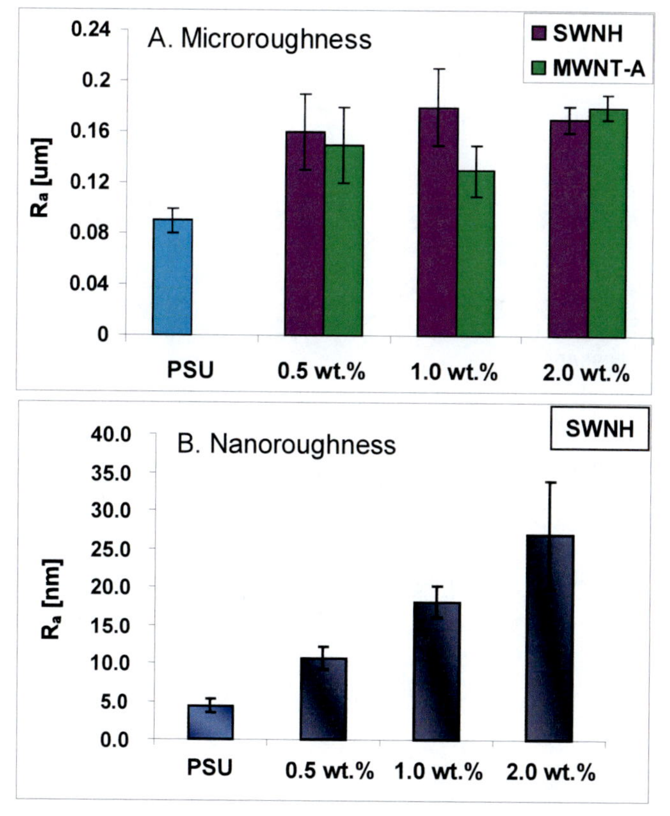

Figure 26. Surface roughness in microscale (A) and nanoscale (B) of unmodified polysulphone (PSU) and PSU mixed with 0.5, 1.0 or 2.0 wt.% of single-wall carbon nanohorns (SWNH) or multi-wall carbon nanotubes (MWNT-A). R_a is an average deviation of the roughness profile from the mean line. Measured using the T500 Hommel Tester (Hommelwerke Co., Germany; A) or by or by atomic force microscopy using SPM LAB 5.01 software (B). (B). Mean ± S.D. (Standard Deviation) from 6 measurements [157, 158].

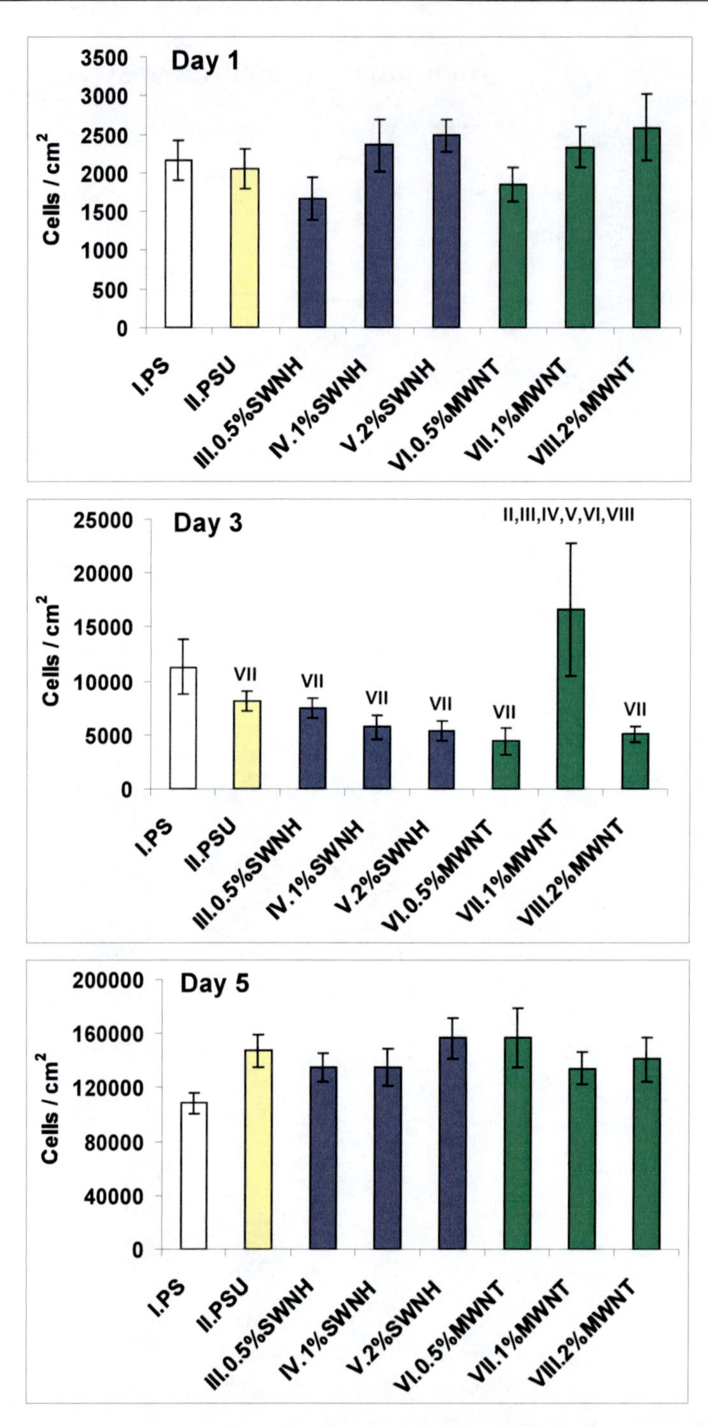

Figure 27. Cell population density (cells/cm^2) of human osteoblast-like MG 63 cells on day 1, 3 and 5 after seeding on cell culture polystyrene dishes (PS), polysulphone (PSU) and polysulphone mixed with 0.5, 1 or 2 wt.% of single-wall carbon nanohorns (SWNH) or multi-wall carbon nanotubes (MWNT). Mean ± S.E.M. (Standard Error of Mean) from 15-50 measurements. ANOVA, Student-Newman-Keuls method. Statistical significance: $^{I, II, III, IV, V, VI, VII, VIII}$: $p \leq 0.05$ compared to the values on the sample labeled with the same number [157, 158].

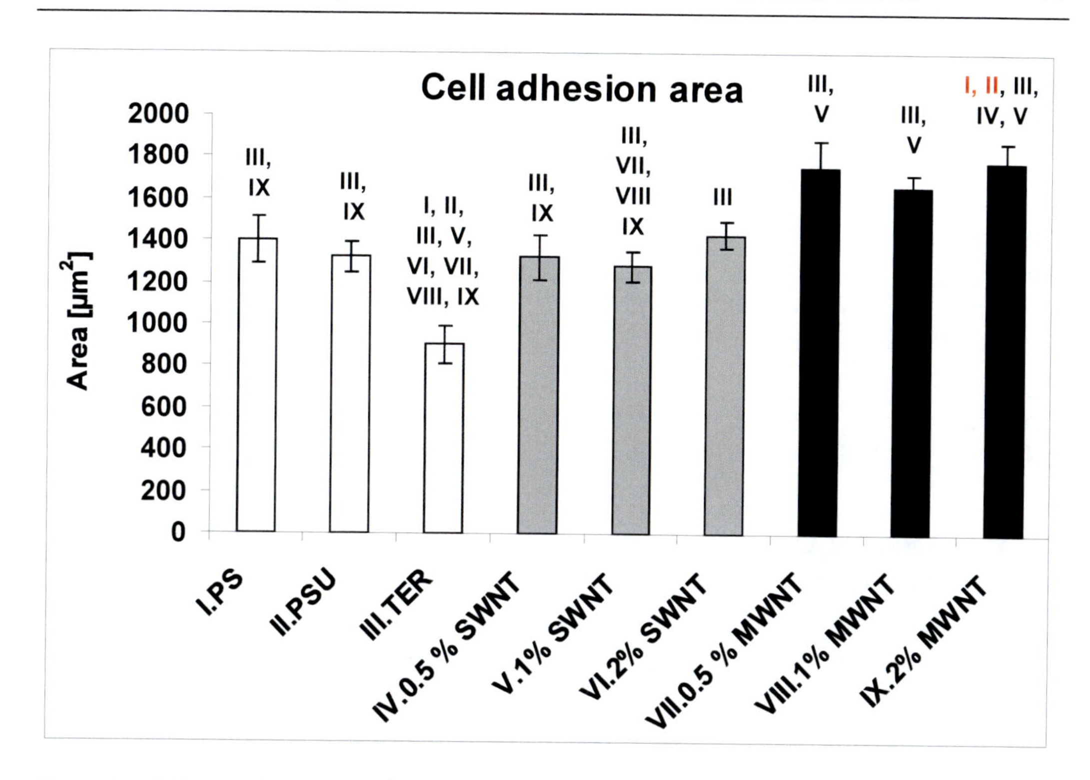

Figure 28. Cell spreading area (μm^2) of human osteoblast-like MG 63 cells on day 1 after seeding on cell culture polystyrene dishes (PS), polysulphone (PSU), terpolymer of polytetrafluoroethylene, polyvinyldifluoride and polypropylene (TER) and polysulphone mixed with 0.5, 1 or 2 wt.% of single-wall carbon nanohorns (SWNH) or multi-wall carbon nanotubes (MWNT). Mean ± S.E.M. (Standard Error of Mean) from 26-124 cells for each experimental group. ANOVA, Student-Newman-Keuls method. Statistical significance: I, II, III, IV, V, VI, VII, IX: $p \leq 0.05$ compared to the values on the sample labeled with the same number [157, 158].

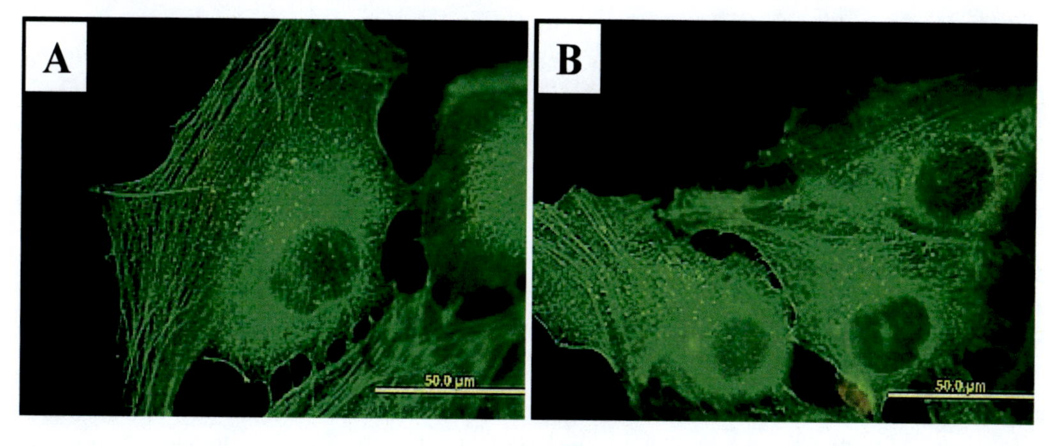

Figure 29. Immunofluorescence of beta-actin in human osteoblast-like MG 63 cells on day 3 after seeding on polysulphone mixed with 2 wt.% of single-wall carbon nanohorns (A) or 2 wt.% of multi-wall carbon nanotubes (B). Olympus IX 50 microscope, DP 70 digital camera, obj. 100, bar = 50 µm [157, 158].

As for the colonization with cells, the numbers of MG 63 cells achieved on PSU-nanotube composites were usually similar to those obtained on unmodified PSU and standard polystyrene cell culture dishes. Only on day 3 after seeding on PSU with 1 wt. % of MWNT-A was the cell population density significantly higher (Figure 27). However, the addition of carbon nanotubes improved the cell spreading. The cell adhesion area, i.e., the cell area projected on the material surface, and measured in cells stained against beta-actin using the immunofluorescence technique, was the largest in cells grown on PSU samples with 0.5, 1 and especially 2 wt. % of MWNT-A (Figure 28). Moreover, the cells on PSU-carbon nanotube composites developed thick beta-actin cables oriented in parallel and resembling those observed in cells of muscle type (Figure 29). However, similarly as in cells on the composites PTFE/PVDF/PP and MWNT-A, this phenotype was often associated with a loss of markers of osteogenic differentiation, particularly osteocalcin [157, 158].

3.3. Interaction of Cells with Carbon Nanotubes Suspended in the Culture Media

Carbon nanotubes can be added to polymers designed for the construction of three-dimensional scaffolds for tissue engineering. These nanoparticles would form nanoscale prominences on the pore walls and thus improve the ingrowth of cells inside the scaffolds. In addition, they would reinforce the scaffolds and improve their mechanical properties in order to be more appropriate for bone tissue engineering, and they would also enable electrical stimulation of cells, which has been reported to promote their growth and maturation, manifested by collagen synthesis and the formation of mineralized bone tissue matrix [55, 56]. An important question is what would happen with the nanotubes if the scaffolds were made of degradable polymers, such as polylactide, polyglycolide, polycaprolactone and their copolymers, which are often used in tissue engineering [162]. While in stable terpolymer PTFE/PVDF/PP, as well as PSU, the nanotubes were permanently bound in the polymeric matrix, they would be released from degradable materials. The nanotubes could then accumulate in adjacent and remote tissues, penetrate inside the cells and have adverse effects on the cells, e.g., a cytotoxic, immunogenic or even genotoxic, mutagenic and carcinogenic action [83-85]. On the other hand, a recent paper showed that there is a real chance of clearing the nanotubes from the organism relatively quickly by glomerular filtration [163].

In order to simulate less or more massive release of carbon nanotubes from the scaffolds and to evaluate their direct effects on cells, we suspended SWNH or MWNT-A in the culture media (DMEM, Sigma, Cat. No. D5648, supplemented with 10% of fetal bovine serum) at concentrations of 0.004, 0.04, 0.4, 4 and 40 mg/ml. This concentration range was chosen because it involves similar concentrations of nanotubes present in our polymer-nanotube composites, as well as concentrations lower or higher by several orders. The nanotube-containing media were then used for seeding human osteoblast-like MG 63 cells into standard polystyrene 24-well multidishes (TPP, Switzerland). Each well contained 30,000 cells (about 17,000 cells/cm^2 and 1.5 ml of the nanotube-containing culture medium). The medium without nanotubes served as a control sample [164]. The cell number and morphology on days 1, 3 and 7 after seeding were then evaluated.

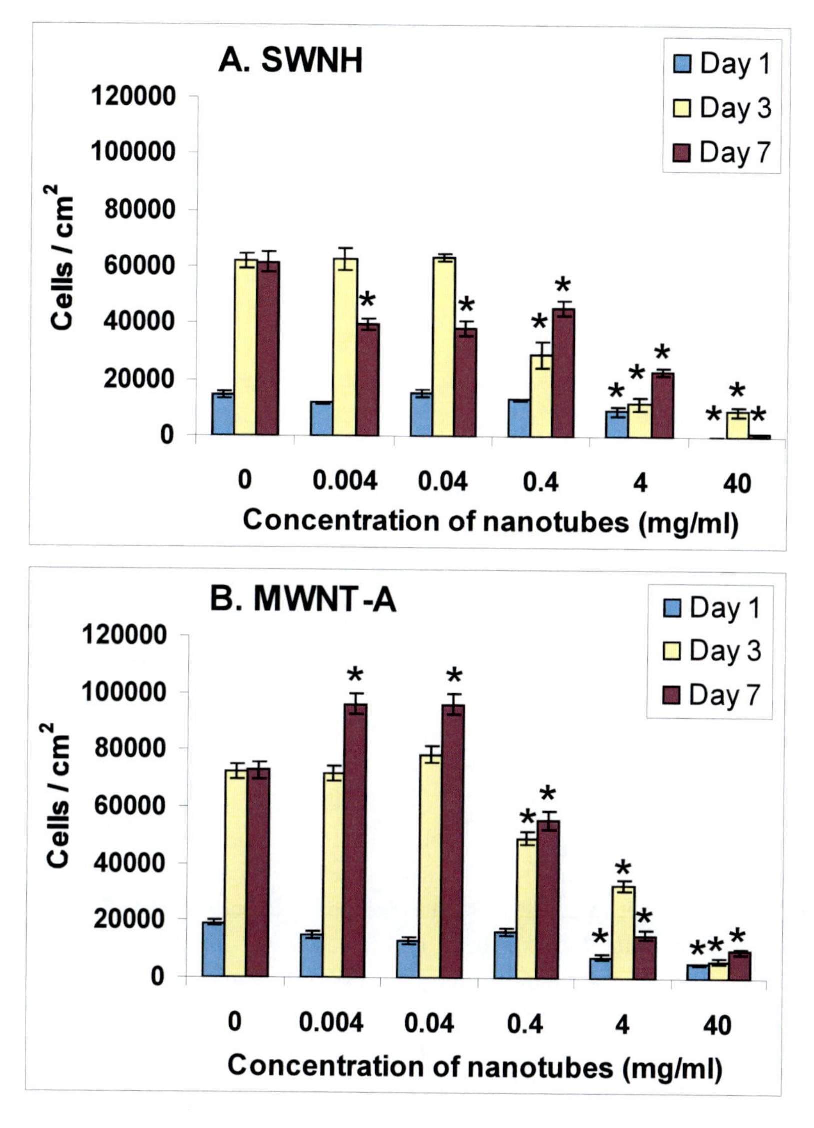

Figure 30. Cell population density (cells/cm^2) of human osteoblast-like MG 63 cells on day 1, 3 and 7 after seeding in the culture medium containing 0-40 mg/ml of single-wall carbon nanohorns (SWNH) or multi-wall carbon nanotubes (MWNT-A). Mean ± S.E.M. (Standard Error of Mean) from 3 samples for each experimental group. ANOVA, Student-Newman-Keuls method. Statistical significance: *: $p \leq 0.05$ compared to the corresponding values obtained in the medium without nanotubes [164].

On day 1 after seeding, the cells in both media adhered in numbers similar to the control value obtained in the medium without nanotubes up to a concentration of 0.4 mg/ml, where the numbers of initially adhering cells were significantly lower (Figure 30). This may be due not only to a cytotoxic action of carbon nanotubes but also to the fact that these nanoparticles, at higher concentrations (4 and 40 mg/ml), covered most of the bottom of the culture well and left only a limited space for cell attachment and spreading (Figures 31 and 32). On day 3 after seeding, the cell number in the media with nanotubes in concentrations of 0.004 and 0.04 mg/ml was similar (SWNH) or even higher (MWNT-A) than in the control nanotube-free medium (Figure 30). A significant decrease in the cell number was apparent from the concentration of 0.4 mg/ml of both types of nanotubes. However, on day 7 after seeding, the cell numbers in the media with SWNH were already significantly decreased from the lowest nanotube concentration of 0.004 mg/ml; whereas in the MWNT-A-containing media, the reduction was only from 0.4 mg/ml, i.e., a concentration 100 times higher (Figure 30). Therefore, it can be concluded from these experiments that the cell growth was inhibited more in the media containing single-walled nanotubes than in media containing multi-walled nanotubes. Similar results were also obtained in cultures of human fibroblasts, where single-wall carbon nanotubes induced stronger toxic effects, manifested by cellular apoptosis or necrosis, than other carbon nanoparticles represented by active carbon, carbon black, multi-wall carbon nanotubes and graphite [165]. In alveolar macrophages, the single-wall carbon nanotubes also impaired phagocytosis and induced cell degeneration and death to an extent more profound than multi-wall carbon nanotubes and fullerenes C_{60} [166]. In addition, single-wall carbon nanotubes stimulated the aggregation of human platelets *in vitro* and vascular thrombosis in rat carotid arteries *in vivo* more than their multi-wall counterparts [167].

The more pronounced adverse effects of single-wall carbon nanotubes on cells could be attributed to their relatively small size, which allows them to penetrate inside the cells more easily than bigger particles. In our experiments, the SWNH were only 2 to 3 nm in diameter and 30 to 50 nm in length, while the diameters of MWNT-A were about 5-20 nm and they were from 300 to 2000 nm in length. In addition, the SWNH had a relatively sharp closed horn-like end (19 degrees), which might further facilitate their penetration inside the cells. Recently, the more pronounced cytotoxicity of single-wall carbon nanotubes has been attributed to their relatively small surface area [165]. The cytotoxicity of both single- and multi-wall carbon nanotubes can be enhanced by the presence of impurities, particularly metal catalysts used during their preparation [39-42], the formation of big, stiff and solid agglomerates resembling asbestos [42, 84, 168] or some types of nanotube derivatization, e.g., oxidation by treatment in acids, as mentioned in the Introduction [47, 48]. On the other hand, other types of nanotube functionalization, such as by 1,3-dipolar cycloaddition reaction, oxidation/amidation treatment or side-wall fuctionalization with phenyl-SO_3H, SO_3Na or phenyl-$(COOH)_2$ has been reported to attenuate the toxicity of carbon nanotubes [44, 169].

In cultures of human embryo kidney HEK 293 cells, single-wall carbon nanotubes in concentrations from 0.001 mg to 0.25 mg/ml, i.e., a concentration range corresponding, at least partly, to that used in our studies, inhibited cell growth by alterations in the expression of cell-cycle associated genes, such as cdk2, cdk4, cdk6 and cyclin D3, leading to an arrest in the G1 phase of the cell cycle, down-regulation of several signal transduction-associated genes (mad2, jak1, ttk, pcdha9 and erk), as well as a lower expression of cell adhesion-associated proteins such as laminin, fibronectin, collagen IV, cadherin and focal adhesion kinase. High nanotube concentration of 0.25 mg/ml induced cell death within 24 hours [82]. In a concentration of 0.1 mg/ml, single-walled carbon nanotubes significantly decreased the number of rat aortic smooth muscle cells relative to the cultures in the control pure medium [168]. In human epidermal keratinocytes, 6-aminohexanoic acid (AHA)-derivatized single-wall carbon nanotubes, applied to the culture media in concentrations from 0.00005 to 0.05 mg/ml, decreased the activity of mitochondrial enzymes, measured by MTT assay. At the same time, the cells treated with 0.05 mg/ml of AHA-SWNTs increased their production of interleukins IL-6 and IL-8 [84]. On the other hand, in human lung A549 cells, the EC_{50} value after 24 hours of exposure to a medium with single-wall carbon nanotubes exceeded the concentration of 8 mg/ml [83]. Relatively safe concentrations of single-walled carbon nanotubes, recommended by Bottini *et al.* [47], did not exceed 40 μg (0.04 mg) per ml of cell culture media. As for the multi-wall carbon nanotubes, when these nanoparticles of high purity were applied to the culture media in concentrations of 16 μg/ml and 20 μg/ml, the growth rate, vitality and shape of Crandell feline kidney fibroblasts remained unchanged, though the nanotube aggregates showed high adhesion to the cell membrane, and were also observed beneath the cell membrane [41].

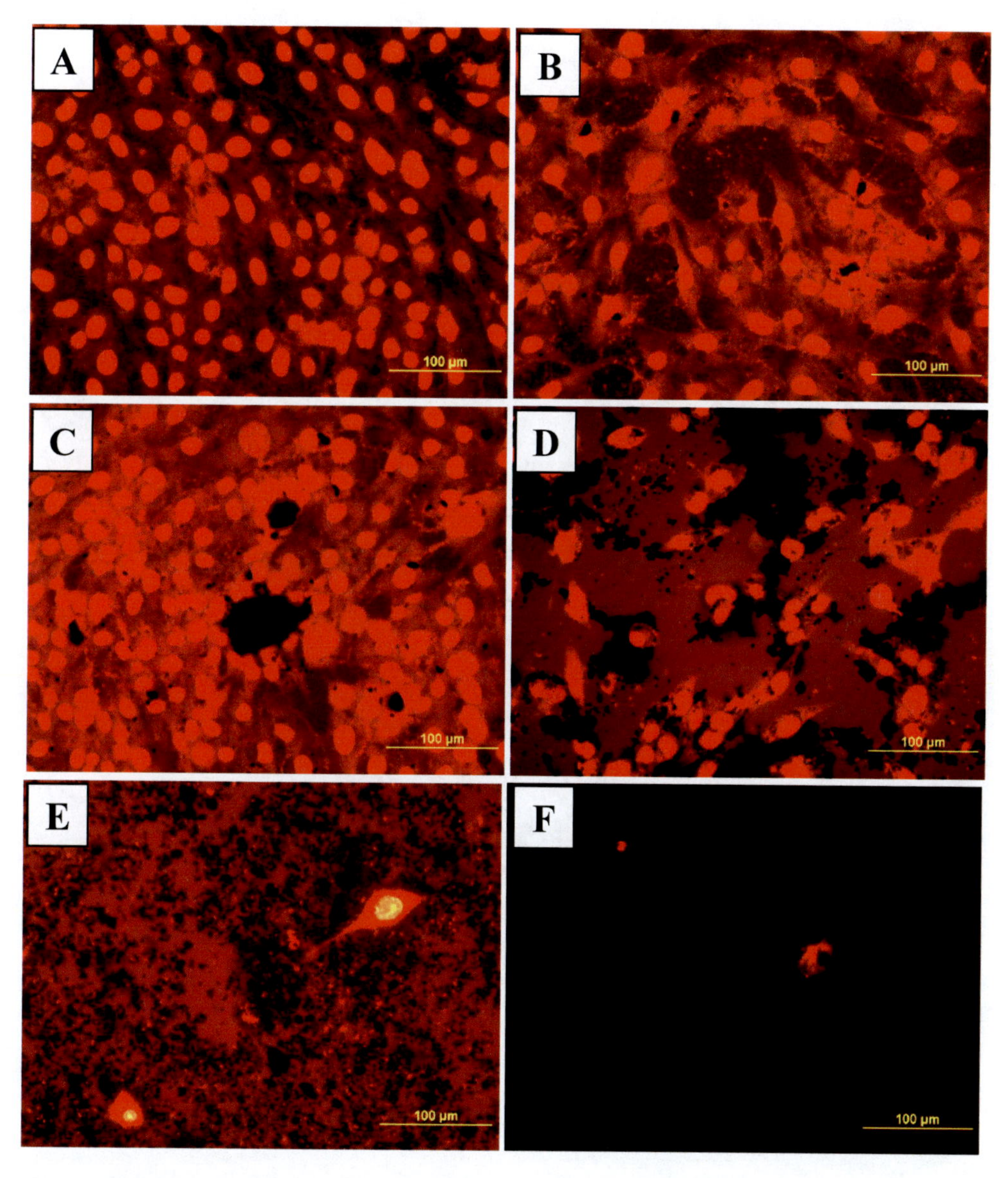

Figure 31. Morphology of human osteoblast-like MG 63 cells on day 3 after seeding in the culture medium containing single-wall carbon nanohorns in concentrations of 0 (A), 0.004 (B), 0.04 (C), 0.4 (D), 4 (E) or 40 (F) mg/ml. Cells fixed with 70% ethanol, stained with propidium iodide. Olympus IX 50 microscope, DP 70 digital camera, obj. 20, bar = 100 μm [164].

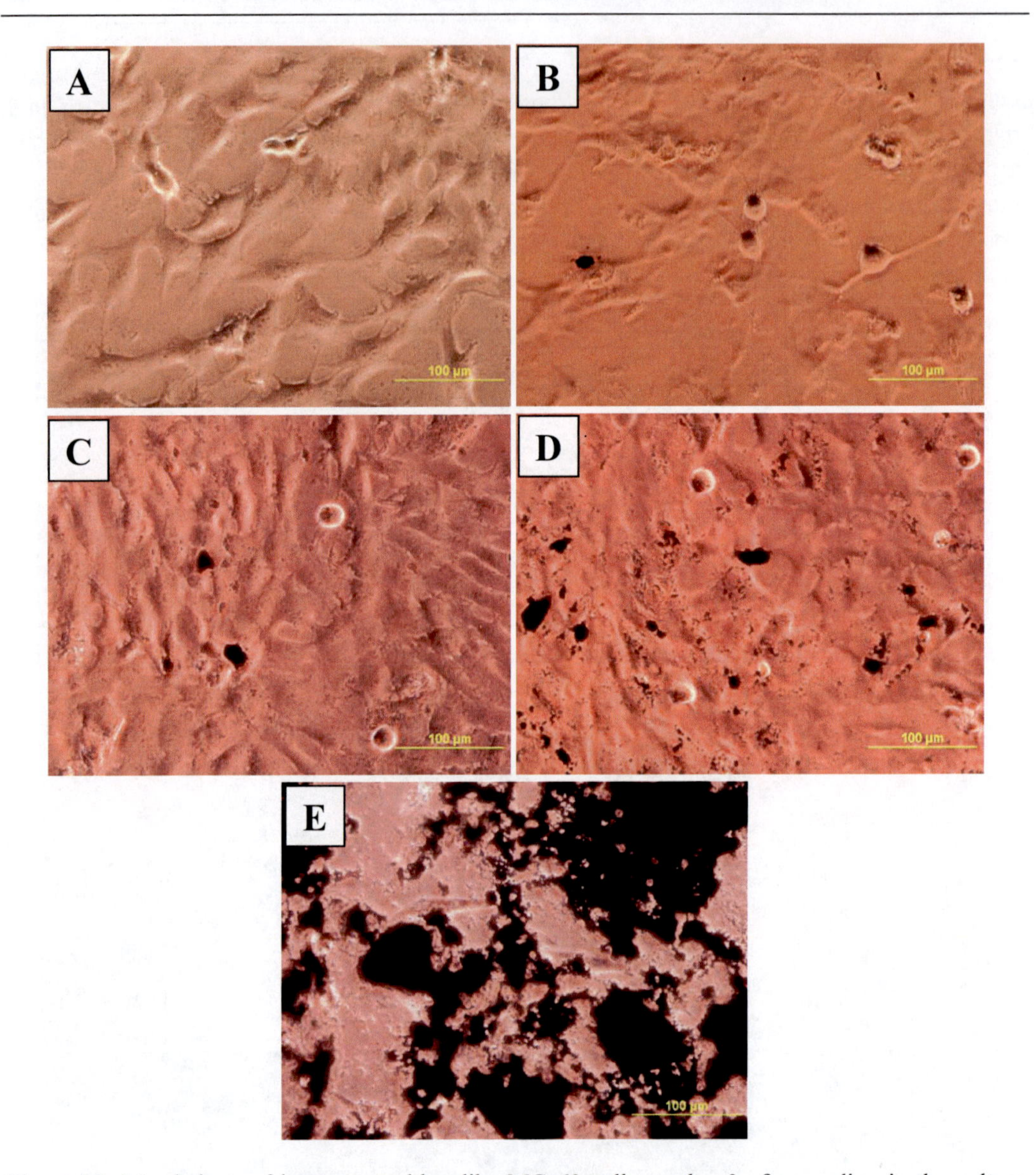

Figure 32. Morphology of human osteoblast-like MG 63 cells on day 3 after seeding in the culture medium containing multi-wall carbon nanotubes in concentrations of 0 (A), 0.004 (B), 0.04 (C), 0.4 (D) or 4 (E) mg/ml. Living cells in native cultures, Olympus IX 50 microscope, DP 70 digital camera, obj. 20, bar = 100 μm [164].

4. Nanocrystalline Diamond Layers

In our studies, nanocrystalline diamond (NCD) films were grown on silicon substrates by a microwave plasma enhanced chemical vapor deposition (MW PECVD) method in an ellipsoidal cavity reactor (Figure 33) [95-97, 170]. The silicon substrates were either mechanically lapped to root mean square (*rms*) roughness up to 300 nm or polished to atomic flatness (*rms* roughness about 1 nm). Prior to the deposition process, the substrates were mechanically seeded in an ultrasonic bath, using a 5 nm diamond powder for 40 minutes. The nucleation procedure was followed by PECVD growth using H_2/CH_4 gas mixtures. A constant methane concentration (1 % CH_4 in H_2), at a total gas pressure of 30 mbar was used. The substrate temperature was 860 °C, as measured by a two-color pyrometer working at wavelengths of 1.35 and 1.55 μm. The silicon substrates were overcoated with an NCD film on both sides, i.e., on the top and bottom side, respectively. Thus, hermetic sealing of the Si substrate minimized any unwanted biochemical reaction, as an opened Si area could result in pollution and disturbances of the subsequent cell experiments. As calculated from optical measurements [171], the film thickness was 330 nm on the atomically polished side.

Prior to the biological test, all NCD films were wet-cleaned in a solution of H_2SO_4 + KNO_3 for 30 min at 200 °C to remove any non-carbon phases from the surface of the deposited NCD films. Finally, the wet-cleaned NCD films were treated in oxygen plasma for 3 minutes at a total power of 300 W and gas pressure 10 mbar to enhance the hydrophilic character of the diamond surface. As mentioned above, moderately wettable surfaces have repeatedly been shown to promote adhesion and growth of cells [57, 121, 134-140]. The wetting angle, measured by a static method in a material-water droplet system using a reflection goniometer, was 35° for both types of NCD surfaces deposited on flat or rough Si substrates. Generally, the NCD surfaces were stable without significant changes of the wetting angle. Virgin silicon substrates treated in oxygen plasma also showed a low contact angle value (~17°), but this value was significantly less stable than for the NCD surface. A Si sample, standing in air for several hours showed local non-stabilities, and the wetting angle varied up to 42° [96, 97].

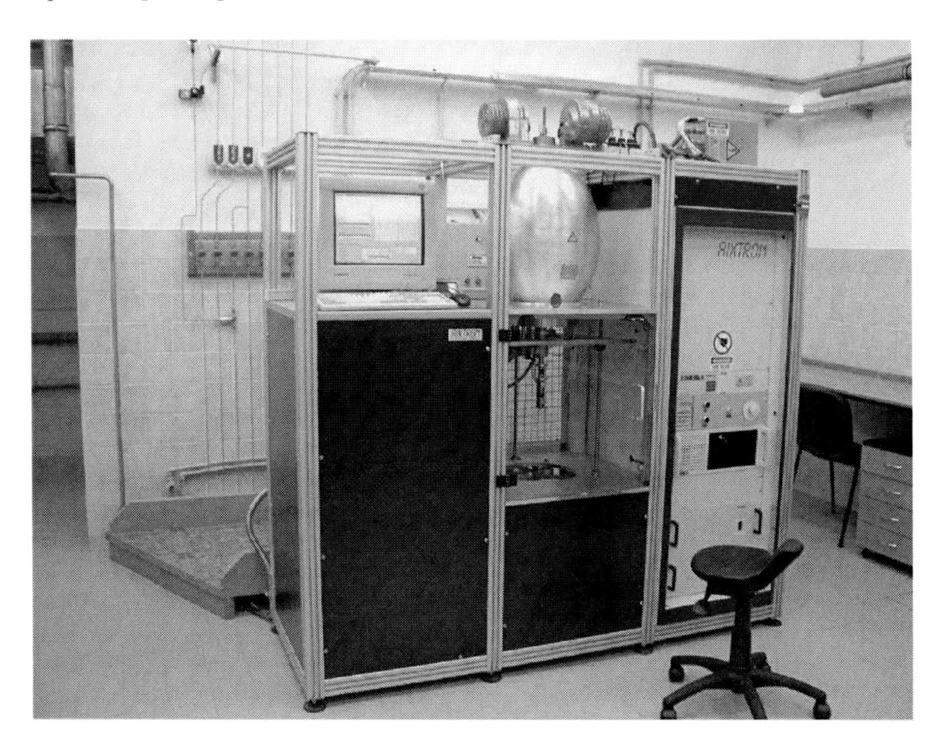

Figure 33. An ellipsoidal cavity reactor (AIXTRON - P6, Germany), characterized by the following parameters: microwave frequency 2.45 GHz, maximum power 6 kW, substrate size 2" or 3", rotating substrate stage, stable plasma at high pressure, fully PC controlled.

The Raman spectra of the deposited NCD films were measured using a 514.5 nm excitation wavelength laser, which enabled us to determine the diamond character of the deposited films (i.e., the sp^3 hybridization). The typical Raman spectrum of an NCD film displayed one dominant peak centered at wavenumber 1333 cm^{-1} (optical phonon in diamond, Figure 34) [172]. In addition, X-ray photoelectron spectroscopy (XPS) determined that the ratio of carbon in sp^3 hybridization, characteristic for diamond, to the sp^2 C was more than 95 % [173].

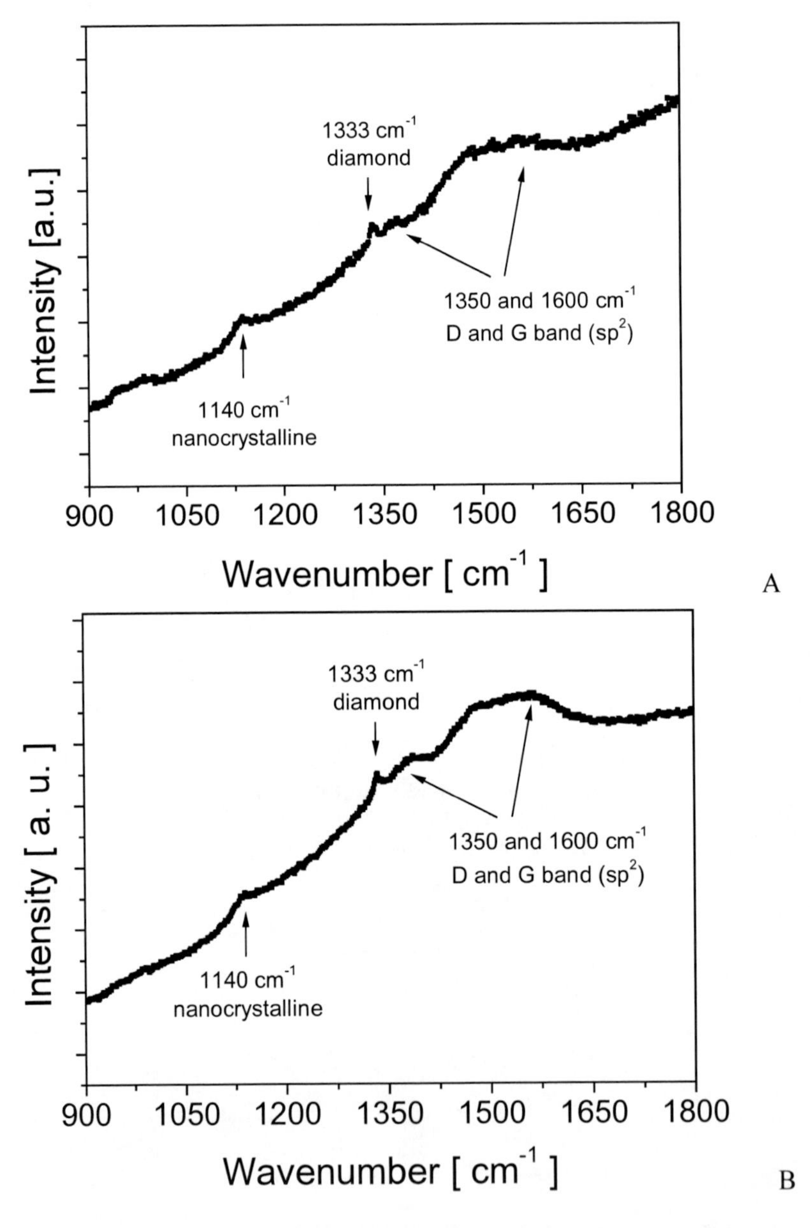

Figure 34. Raman spectroscopy of nanostructured (A) and hierarchically micro- and nanostructured (B) nanocrystalline diamond (NCD) films. The peak 1333 cm^{-1}, characteristic for diamond, and 1140 cm^{-1}, often attributed to NCD, are well visible in both types of films [96].

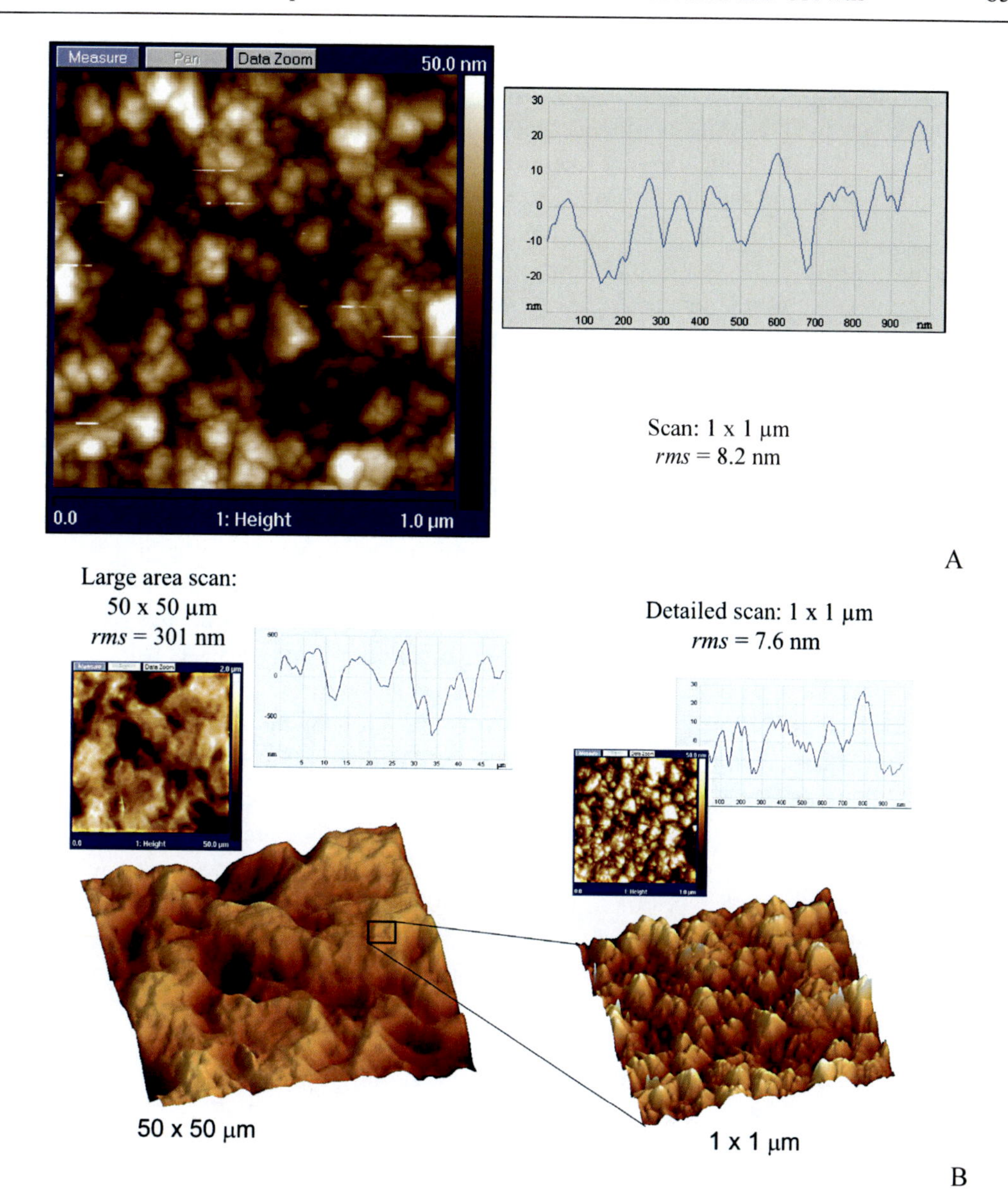

Figure 35. Atomic force microscopy of nanostructured (A) and hierarchically micro-nanostructured (B) NCD films. Root mean square (*rms*) roughness is 8.2 nm (A) or 301 nm and 7.6 nm (B). AFM Microscope Dimension 3100, Veeco [96].

As confirmed by AFM, two types of NCD layers were prepared. The first was nanostructured with a relatively flat surface (Figure 35A). On a detailed AFM scan of 1 x 1 µm, the development of crystal faceting was observable. This NCD film consisted of small crystals up to 30 nm in size. The calculated *rms* roughness was 8.21 nm. The surface features were similar and independent of the scanned area. The second type of layer was hierarchically micro- and nanostructured (Figure 35B). A scan over a large area (50 x 50 µm) exhibited features of relatively high, i.e. submicron, roughness up to ~300 nm. As shown by the detailed scan of 1 x 1 µm, the surface of these irregularities contained nano-scale features with *rms* roughness of 7.6 nm. Both nanostructured and hierarchically micro-nanostructured films thus differ mainly in surface roughness on a large scale [96, 97].

The reason for preparing the hierarchically micro- and nanostructured NCD films was that not only the nanoroughness of the material surface, but also its microstructure, is important for its colonization with osteoblasts and for regeneration of the well functioning bone tissue. Surface roughness on the submicron or micrometer scale (R_a parameter, i.e., departures of the roughness profile from the mean line, in the range approx. from hundreds of nanometers to several micrometers) has repeatedly been shown to enhance the strength of osteoblast adhesion, their spreading and differentiation [111, 154, 174-176], including the deposition of mineralized bone extracellular matrix [177], and the acquisition of osteoblast phenotype in mesenchymal and osteoprogenitor cells [178, 179]. At the same time, increasing surface microroughness reduced the activity of osteoblasts, and thus bone resorption [180]. Microstructured surfaces also modulated the number of attached cells, cell shape, presence and activity of cell adhesion receptors, the assembly of focal adhesion plaques and cytoskeleton, locomotion, proliferation, reactivity to hormones, vitamins and growth factors, as well as the production of various bioactive molecules in cells [111, 154, 174, 175, 177-182]. In addition, as mentioned several times, hierarchically-organized structures of graduated sizes represent the main architectural principle of natural tissues and organs [116].

NCD films were also deposited on microscopic glass slides using the MW PECVD method (26.7 mbar, 710 °C, 0.8-1% CH_4). After the growth, the films were chemically oxidized and subsequently hydrogenated. Some NCD films were also doped with boron (NCD-B; 3 000 ppm B:C; $\rho \sim 1\ 10^{-1}\ \Omega$ cm). The reason was that boron-doped NCD can be used in active sensing applications *in vivo* [175, 176, 183], and the presence of charged states of the ionized B acceptors on the diamond film surface can also influence cell colonization. The wetting angle of both types of films was 85-90°, and the root mean square roughness (*rms*), measured by AFM, was 18 nm.

All samples were sterilized in 70% ethanol, inserted in culture wells in polystyrene multidish plates and seeded with human osteoblast-like MG 63 cells. The number of initially adhered cells, evaluated 24 hours after seeding, was usually similar on all tested NCD layers, microscopic glass slides and cell culture polystyrene dishes. Only on silicon substrates was the number of initially adhering viable cells markedly lower, and this effect was more pronounced in CPAE cells than in MG 63 cells. On the other hand, the percentage of viable CPAE cells on silicon substrates improved with time of cultivation, though their absolute number still remained lower than on NCD-coated surfaces, glass slides and polystyrene wells. In MG 63 cells, the number of viable cells decreased with time of cultivation, and on day 5 after seeding, no cells were detected on Si substrates (Figure 36). It has been reported that the cytotoxicity of silica nanoparticles is dependent on the metabolic and growth activity of a given cell type [184]. For example, fibroblasts with relatively long doubling times were more susceptible to injury induced by silica exposure than tumor cells with short doubling times [184]. Silicon-containing materials can damage cells by the induction of reactive oxygen species, damage to cell membrane and mitochondrial metabolism, as well as inflammatory reactions [184, 185]. Lower adhesion and spreading on silicon substrates was also observed in human cervical carcinoma HeLa cells, rat pheochromocytoma PC12 cells or mouse MC3T3 osteoblasts [88]. Silicon substrates implanted into rabbit eyes behaved in a highly corrosive manner [186].

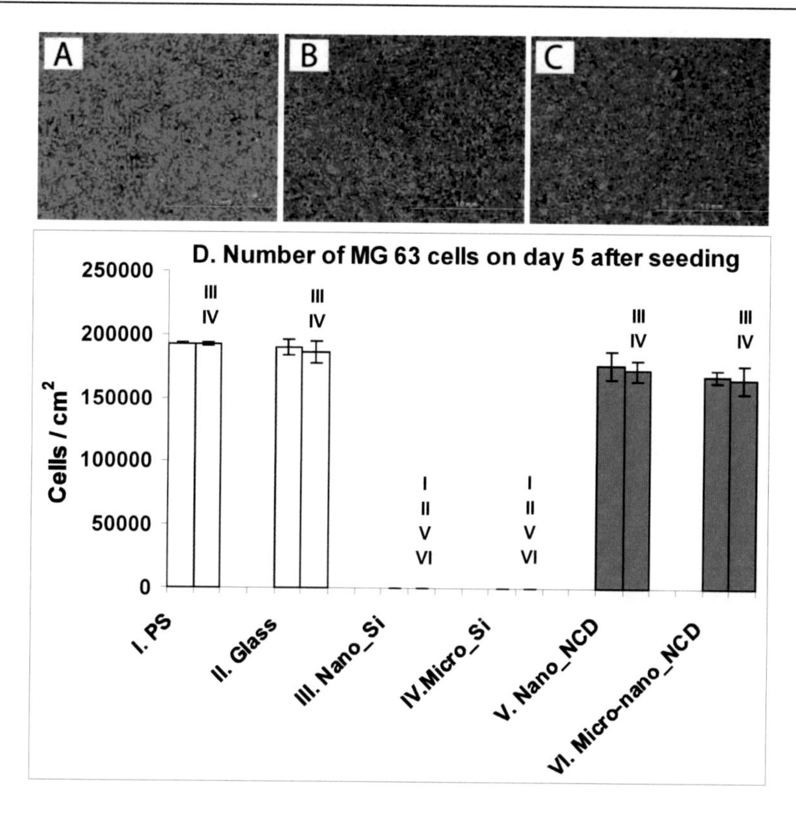

Figure 36. A-C: Fluorescence staining of MG 63 cells with a LIVE/DEAD kit on day 5 after seeding on a control tissue culture polystyrene dish (A), nanostructured NCD films (B) and hierarchically organized micro- and nanostructured NCD (C). D: Number of MG 63 cells on polystyrene culture dishes (PS), microscopic glass coverslips (Glass), nanostructured silicon substrates (Nano_Si), microstructured silicon substrates (Micro_Si), nanostructured diamond films (Nano_NCD) and diamond films with hierarchically organized micro- and nanostructure (Micro-nano_NCD). First column: total cell number, second column: number of viable cells. Mean ± S.E.M. (Standard Error of Mean) from 3 samples. ANOVA, Student-Newman-Keuls Method. Statistical significance: I, II, III, IV, V, VI: $p \leq 0.05$ in comparison with the sample of the same number [97].

The number of MG 63 and CPAE cells on days 3 and 5 after seeding on nanostructured and hierarchically micro-nanostructured O-terminated NCD was usually higher, or at least similar, in comparison with the values obtained on glass and polystyrene dishes (Figures 36, 37). Similarly as in the case of fullerene layers and nanotube-polymer composites, this favorable cell behavior was probably due to the surface nanostructure of both NCD layers. As mentioned above, the material nanostructure has been reported to be highly supportive for colonization with cells, preferentially osteoblasts. Nanostructured surfaces improve the adsorption of cell adhesion-mediating extracellular matrix molecules, e.g., fibronectin, vitronectin and collagen, from the serum of the culture medium or body fluids. These proteins are adsorbed in an appropriate amount and probably also in a favorable spatial conformation, enabling the accessibility of specific amino acid sequences by cell adhesion receptors, e.g., integrins. Preferential adhesion of osteoblasts over other cell types has been explained by preferential adsorption of vitronectin on nanostructured surfaces, due to its relatively small molecule, and its preferential recognition by osteoblasts [53, 54, 106-111, 122].

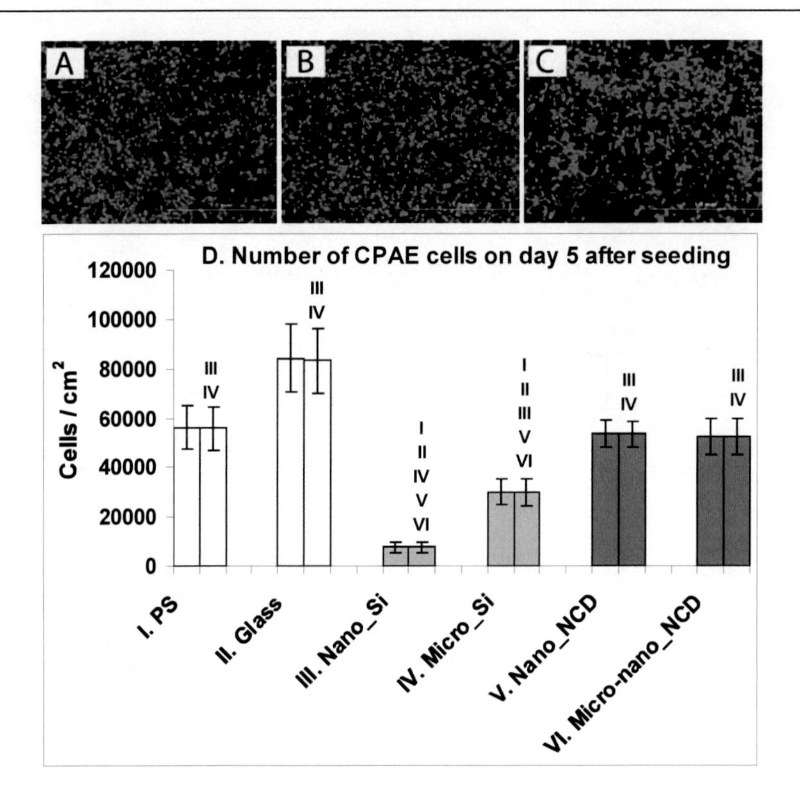

Figure 37. A-C: Fluorescence staining of CPAE cells with a LIVE/DEAD kit on day 5 after seeding on a control tissue culture polystyrene dish (A), nanostructured NCD films (B) and hierarchically organized micro- and nanostructured NCD (C). D: Number of MG 63 cells on substrates described in Figure 36. First column: total cell number, second column: number of viable cells. Mean ± S.E.M. (Standard Error of Mean) from 3 samples. ANOVA, Student-Newman-Keuls Method. Statistical significance: I, II, III, IV, V, VI: $p \leq 0.05$ in comparison with the sample of the same number [97].

In MG 63 cells, the microstructure of the NCD layers also seemed to have a positive effect on their adhesion. On hierarchically micro- and nanostructured surfaces, the MG 63 cells were better spread, i.e., they adhered over a significantly larger area than on the other tested substrates. This result is in good correlation with an earlier study reporting a supportive effect of the surface microroughness of dental titanium implants on the spreading of rat osteoblasts in primary cultures [181], as well as the development of a polygonal shape in osteogenic cells [182]. In addition, as mentioned above, the hierarchically organized micro- and nanostructure of a cell adhesion substrate is considered as a factor mimicking the architecture of natural tissues, and thus beneficial for cell colonization [116].

However, on micro- and nanostructured surfaces, the cell number tended to be lower than in purely nanostructured NCD, especially in endothelial CPAE cells. For example, on day 3 after seeding, the cell population density obtained on nanostructured NCD amounted to 44,000 ± 6,900 cells/cm^2, while on micro- and nanostructured NCD it was only 17,400 ± 2,400 cells/cm^2. A possible explanation is that on the latter surfaces, the spreading of CPAE cells, which are relatively large and have a flat polygonal morphology, and their subsequent growth, might be hampered by micro-sized surface irregularities [104, 130-133]. It can be hypothesized that the idea of a hierarchically-organized micro- and nanoarchitecture of biomaterials, resembling the arrangement of natural tissues [116], may be useful mainly for constructing three-dimensional porous bioartificial implants rather than for planar two-dimensional surfaces. For example, nanodiamonds could be admixed into a polymeric matrix designed for the production of scaffolds for bone tissue engineering. The nanodiamonds would then decorate the walls of the pores, usually hundreds of micrometers in diameter, and the resulting nanostructure of the pore walls would facilitate the ingrowth of osteoblasts and bone tissue regeneration.

The cells on NCD films, especially those that are purely nanostructured, also displayed higher metabolic activity. As revealed by the XTT test, the absorbance of the formazan dye produced by MG 63 cells on nanostructured and hierarchically micro-nanostructured diamond films was significantly higher (by 260 ± 4 % and 102 ± 2 %, respectively) than on the control polystyrene dishes (absorbance normalized to 100 ± 8 %). The absorbance was not detectable on the silicon substrates, probably due to the very low number of viable cells on these materials. In addition, the color of the XTT reagent solution was altered, which may have been due to some non-specific reaction with the silicon substrates. Higher absorbance of formazane from cells cultivated on the two diamond layers indicates higher activity of mitochondrial enzymes in these cells than in those on the control cell culture surfaces (i.e., glass and polystyrene), and is in good correlation with the cell numbers found on both types of NCD films [96].

On boron-doped NCD films, the numbers of MG 63 cells were often lower than on undoped films modified only by oxidation and hydrogenation, though these cells were well spread and of normal morphology. Similarly as on the other NCD films, the cells on boron-doped NCD assembled dot- or streak-like focal adhesion plaques containing integrins $alpha_v$ and $beta_1$, i.e. receptors for vitronectin, fibronectin and collagen, as well as integrin-associated proteins talin and vinculin. These cells also developed a mesh-like beta actin cytoskeleton (Figure 38). However, the presence of boron might affect their proliferation ability, e.g., through an alteration to the calcium metabolism of these cells. Incubation of MG 63 cells with 2-aminoethoxydiphenylborate (2-APB) resulted in a blockage of store-operated Ca^{2+} channels, followed by lengthening of the S and G2/M phases of the cell cycle [187]. On the other hand, supplementation of rats with boron in drinking water improved the mechanical properties of the bone tissue [188]. In accordance with this finding, the MG 63 cells grown on boron-doped NCD surfaces were more brightly stained for osteocalcin, an important extracellular matrix glycoprotein involved in bone tissue mineralization (Figure 38).

In summary, nanocrystalline diamond films give good support for the adhesion and growth of osteogenic and endothelial cells and can be used in tissue engineering applications and *in vivo* experiments. This is a great advantage over other carbon nanoparticles, such as fullerenes and nanotubes, which, as mentioned above, can have a cytotoxic behavior, especially in a free form dispersed in cell culture media [10, 11, 82, 47]. Virgin silicon substrates (typically used in the semiconductor industry) have been found to be cytotoxic and not suitable for biomedical applications, e.g., for biomaterial coatings or for *in situ* bio-sensor elements. Endothelial cells were usually more sensitive to surface roughness and their adhesion and growth was better on purely nanostructured than on hierarchically organized micro- and nanostructured NCD films. At the same time, MG 63 cells were more responsive to the chemical composition of the adhesion substrate, i.e., Si cytotoxicity [97].

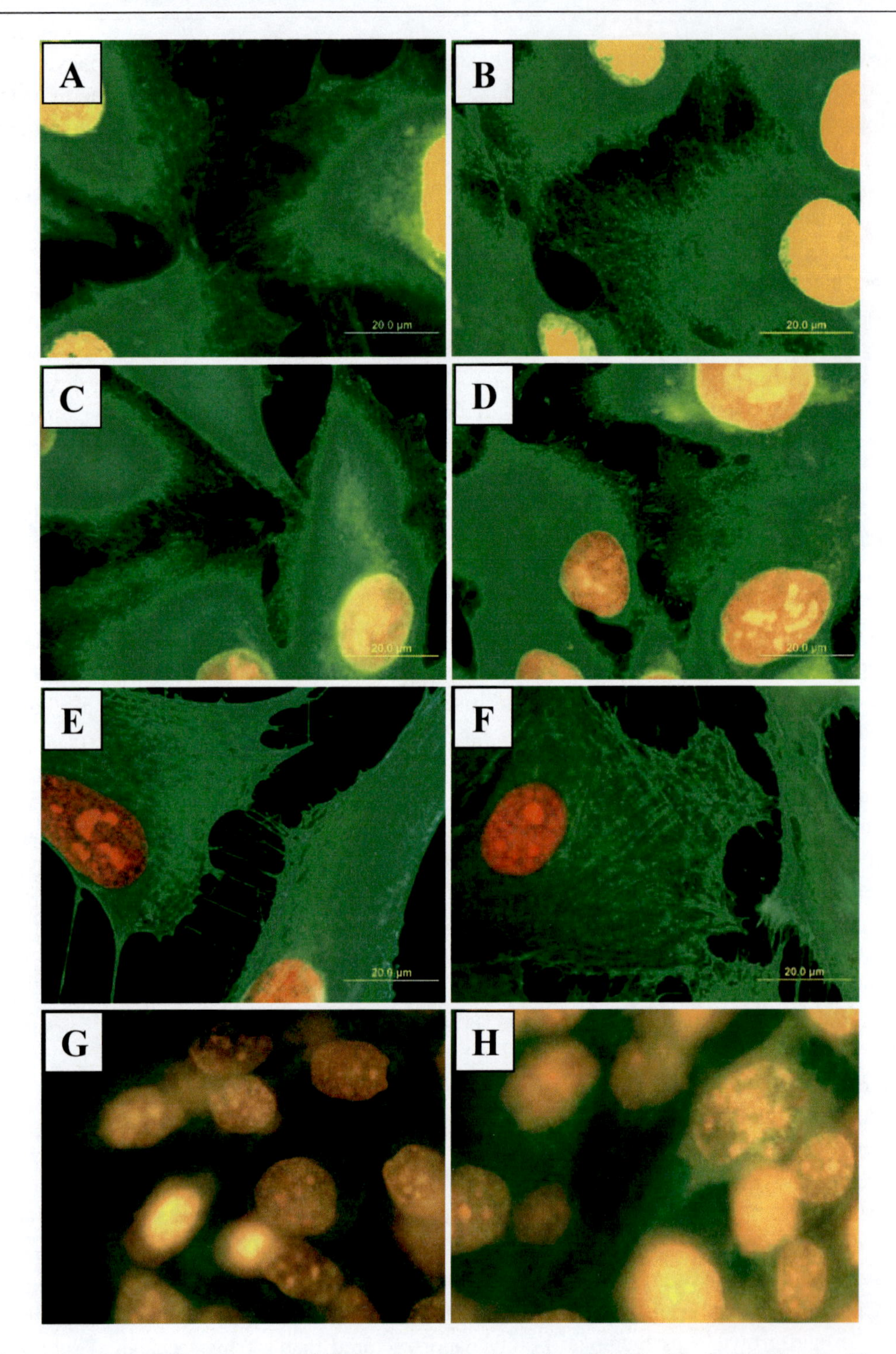

Figure 38. Immunofluorescence of alpha$_v$ integrins (A, B), talin (C, D), beta-actin (E, F) and osteocalcin (G, H) in human osteoblast-like MG 63 cells on day 1 (E, F), 3 (A-D) or 7 (G, H) after seeding on non-doped (A, C, E, G) and boron-doped (B, D, F, H) nanocrystalline diamond films deposited on microscopic glass slides. Olympus IX 50 microscope, DP 70 digital camera, obj. 100, bar = 20 µm.

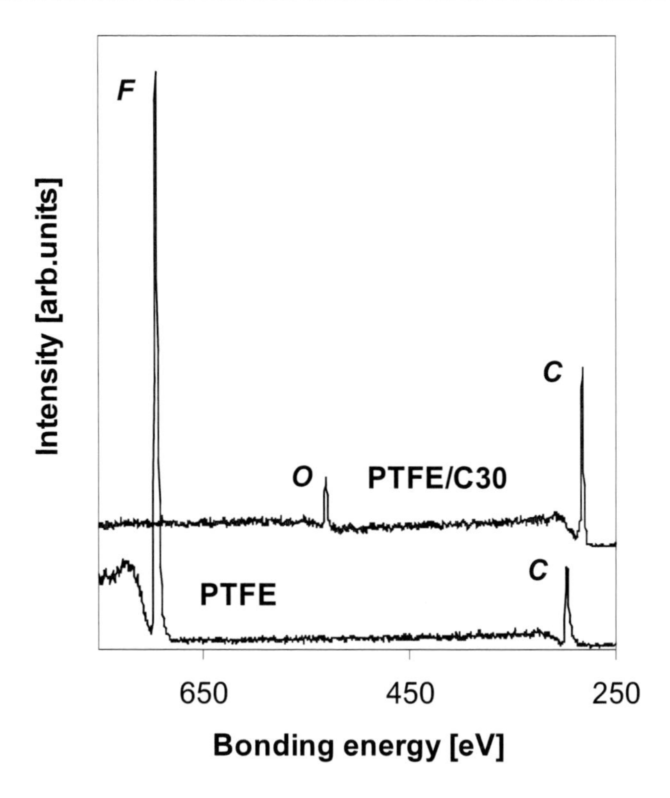

Figure 39. XPS spectra of the pristine polytetrafluoroethylene (PTFE) and PTFE with a carbon layer deposited for 30 min (PTFE/C30). A constant off-set is added to the latter spectrum [102].

5. Other Carbon-Based Layers for Potential Biomaterial Coating

These layers comprised films of amorphous hydrogenated carbon (a-C:H), deposited by by magnetron sputtering on carbon fiber-reinforced carbon composites (CFRC) to strengthen and smooth their surface in order to prevent a potential release of carbon particles from the material, and to enhance its colonization with cells. However, the adhesion and growth of MG 63 cells and primary vascular smooth muscle cells was not significantly increased in comparison with uncoated CFRC and reference cell culture polystyrene. This was most probably due to the relatively high hydrophobicity of these a-C:H films [189]. Similarly, a-C:H films have been used as bioinert coatings for blood-contacting implants and devices, as well as articular surfaces of joint replacements [103, 105, 122, 190, 191]. However, other a-C:H in our experiments, deposited on polyethyleneterephtalate (PET) by RF magnetron sputtering, enhanced adhesion and growth of mouse 3T3 fibroblasts [101]. Similarly, the a-C:H formed on polytetrafluoroethylene (PTFE) by chemical vapor deposition from acetylene induced by a UV-excimer lamp increased the colonization of the polymer with human umbilical vein endothelial cells (HUVEC, line EA.hy926) [102]. The explanation is that these a-C:H films contained additional oxygenic structures (Figure 39) and were moderately wettable (the static water drop contact angle decreased from 90° to approx. 65° with the deposition time, see Figure 40), which rendered them more attractive for cells. In addition, the material surface morphology was changed during the photo-deposition of carbon. The dependence of the surface morphology on the deposition time is shown in Figure 41 in AFM scans of 1×1 μm^2. In the first phase of deposition, the surface is covered by carbon and then the holes present in the material are filled. However, as is apparent from Figure 41, the surface nanoroughness of PTFE then increased up to 20 minutes of photo-deposition of carbon, after which it decreased. The increasing nanoscale surface roughness of the adhesion substrate has been shown to have a positive influence on cell functioning. Human osteoblast-like cells of the line SAOS-2 in cultures on nanocrystalline diamond layers showed increasing activity of cellular dehydrogenases with increasing surface nanoroughness (the root mean square roughness, rms, was in the range from 11 nm to 39 nm) [57]. Similarly, in our studies, the a-C:H layers photo-deposited on PTFE through a metallic mask for 20 minutes, were optimal for micropatterning the material surface with domains adhesive for cells, and thus for induction of regionally-selective cell adhesion and directed cell growth (Figure 42) [102].

The heterogeneous results obtained on a-C:H films indicate that the physical and chemical properties of these films, and their subsequent behavior in a biological environment can vary among different laboratories depending on the conditions of preparation, including the heterogeneity of the pristine substrates used for deposition. For example, due to the production process, the surfaces of pristine PTFE foils are typically not very homogeneous regarding surface roughnesses.

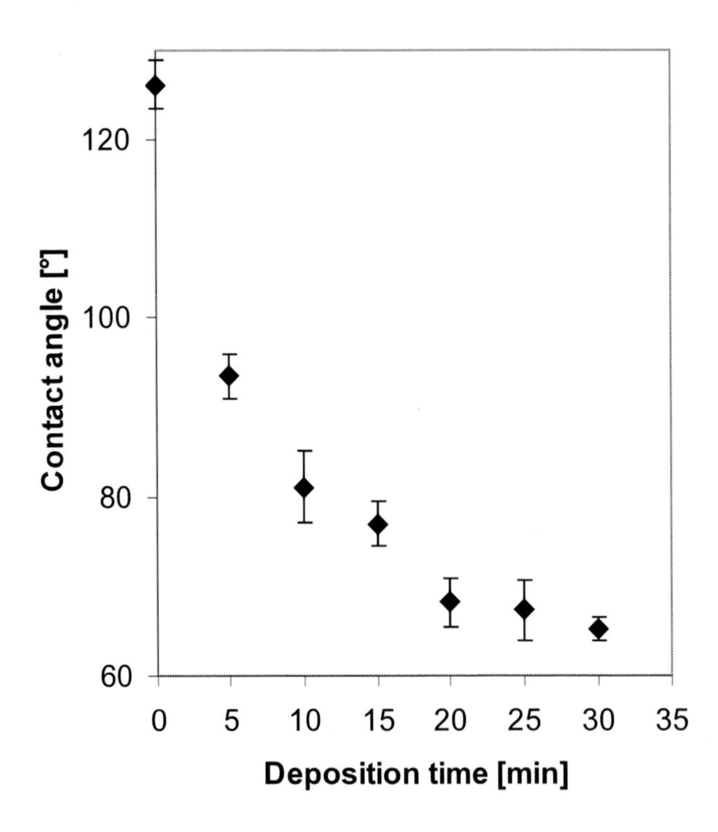

Figure 40. Dependence of water contact angle on CVD carbon deposition time on PTFE [102].

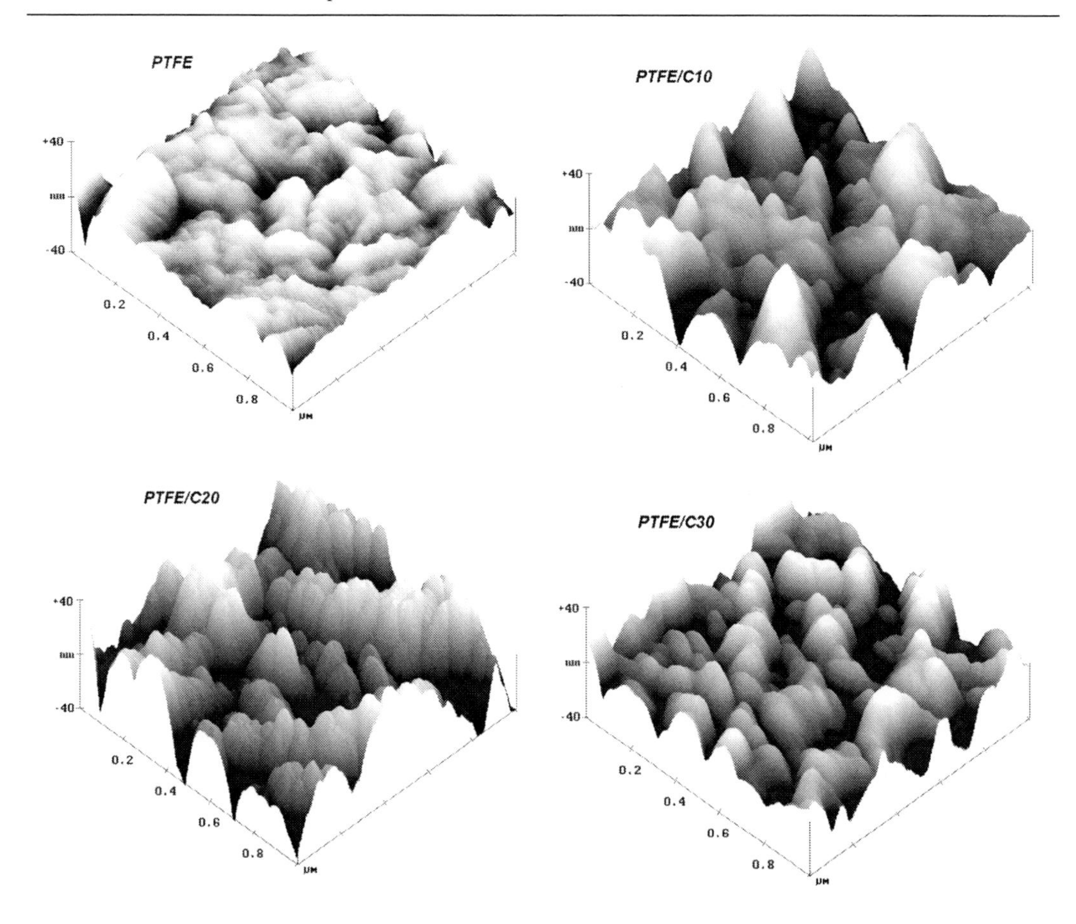

Figure 41. AFM images (scans of 1×1 μm^2) of the surface morphology of pristine polytetrafluoroethylene (PTFE) and PTFE with carbon layers deposited from 10 to 30 min (PTFE/C10 to PTFE/C30) [102].

Enrichment of amorphous carbon films with Ti also stimulated cell adhesion and growth. Coating CFRC with a carbon-titanium layer deposited by the plasma-enhanced physical vapor deposition method (PECVD), using unbalanced d.c. planar magnetron sputtering, stimulated attachment, spreading, proliferation, protein synthesis and an increase in cell volume in MG 63 and vascular smooth muscle cells [104]. Similarly, nanocomposite hydrocarbon plasma polymer films with 20 at. % of Ti, deposited on microscopic glass slides and polystyrene dishes, also had a supportive effect on adhesion, growth and maturation of MG 63 cells and especially bovine vascular endothelial cells of the line CPAE [103]. On nanocomposite hydrocarbon plasma polymer films with 3 at. % of silver, focal adhesion plaques in CPAE cells were more brightly immunostained against vinculin than those in the cells on control Ag-free hydrocarbon films, glass or polystyrene dishes. The endothelial CPAE cells on films adhered, grew, formed a filamentous beta-actin cytoskeleton, developed Weibel-Palade bodies containing the von Willebrand factor, a marker of endothelial cell maturation, and created a confluent layer in an extent similar to that observed in cells growing on control surfaces. At the same time, films with 3at. % of Ag displayed bacteriostatic activity against *Escherichia coli* [105].

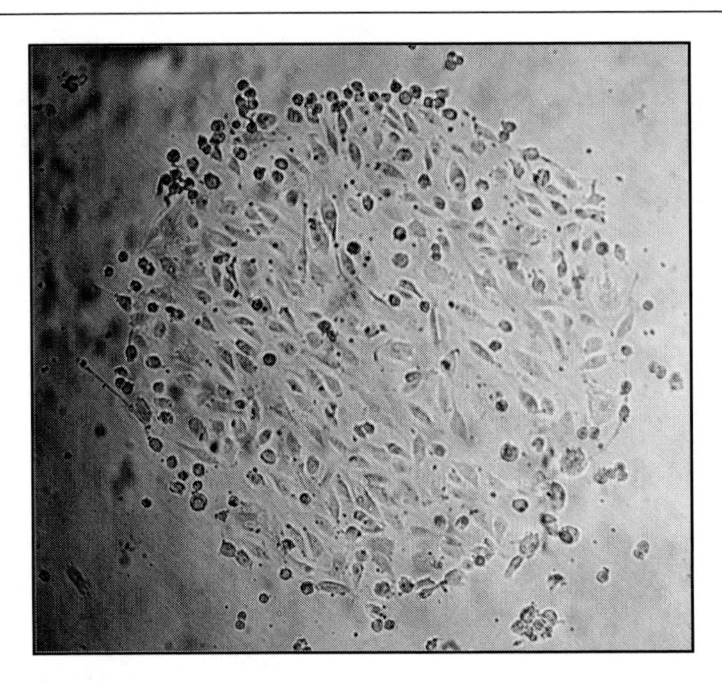

Figure 42. Phase-contrast micrographs of HUVEC (7 days after cell seeding) on a PTFE sample coated selectively with carbon for 20 min by photo-induced CVD through a contact mask (spot diameter 1.5 mm) [102].

Pyrolytic carbon deposited on CFRC in the form of isotropic layers created in a tumbling bed reactor also supported adhesion and growth of rat vascular smooth muscle cells, human osteoblast-like MG 63 cells and human embryonal lung LEP fibroblasts [155, 192, 193]. However, these effects were more pronounced after coating CFRC with pyrolytic graphite prepared by decomposition of butane in a mixture with N_2 (pressure 4 Pa, temperature 1900 ^{o}C, exposure 300–400 min), especially if this treatment was combined with polishing CFRC with metallographic paper and diamond paste [132, 133]. The MG 63 on these surfaces were better spread, grew more quickly, attained higher maximum population densities and contained higher concentrations of osteocalcin and osteopontin, i.e. indicators of osteogenic maturation. At the same time, the concentration of immunoglobulin molecules ICAM-1, markers of cell immune activation, in MG 63 cells remained similar as on reference culture substrates, represented by polystyrene dishes [131].

6. Conclusion and Further Perspectives

Fullerenes C_{60} deposited as continuous films or layers micropatterned with grooves and bulges clearly supported the adhesion, growth, viability and maturation of human osteoblast-like MG 63 cells to a similar extent as standard tissue culture polystyrene dishes and microscopic glass coverslips. In addition, layers with fullerene prominences more than 1000 nm in height promoted regionally-selective adhesion and growth of MG 63 cells in the grooves located among the prominences, and thus they could be used as templates for directed cell adhesion and growth. Composites of C_{60} and metals, particularly Ti, are also promising as growth substrates for bone-derived cells.

The terpolymer of polytetrafluoroethylene, polyvinyldifluoride and polypropylene (PTFE/PVDF/PP), or polysulphone mixed with single- or multi-wall carbon nanotubes also provided good support for the adhesion and growth of MG 63 cells. In the case of PTFE/PVDF/PP, these parameters improved markedly in comparison with cells on the pure terpolymer. Carbon nanotubes could be added into polymeric matrices designed for the construction of porous 3D scaffolds for bone tissue engineering in order to stimulate cell ingrowth inside the material. However, the concentration of osteocalcin and osteopontin, indicators of osteogenic maturation, in cells on polymer-nanotube composites was often lower than on pure terpolymer or polystyrene dishes.

All nanocrystalline diamond films used in this study, showing different levels of surface organization, i.e., nanostructured (*rms* 8.2 nm or 18 nm) and hierarchically micro-nanostructured (*rms* 301 and 7.6 nm, respectively), as well as different chemical composition (i.e. hydrogenated, oxygen-terminated and boron-doped) provided good support for the adhesion, spreading, viability, growth and maturation of human osteoblast-like MG 63 cell in cultures on these materials. Thus, these NCD films could be used for the surface modification of bone implants (e.g., bone-anchoring parts of joint prostheses or bone replacements) in order to improve their integration with the surrounding bone tissue.

Other carbon allotropes and carbon-based materials, such as amorphous hydrogenated carbon, pyrolytic graphite, hydrocarbon plasma polymers, as well as their composites with metals can also serve for coating various materials in order to enhance their bioactivity.

The interactions of carbon nanoparticles with biological systems can be further intensified by the functionalization of these molecules with various atoms, chemical groups or molecules including ligands for cell adhesion receptors, such as RGD-containing oligopeptides or KRSR preferred by bone cells. However, at the same time, thorough studies on the potential cytotoxicity, immunogenicity, mutagenicity and carcinogenicity of carbon nanoparticles are needed for their potentional biomedical applications.

Acknowledgements

This study was supported by the Grant Agency of the Czech Republic (Grants No. 204/06/0225 and 101/06/0226), by the Ministry of Education of the Czech Republic (Research Program No. LC 06041), and the Acad. Sci. CR (Grants No. KAN400480701, KAN200100801, KAN 101120701 and KAN 400100701). Mr. Robin Healey (Czech Technical University, Prague) and Mrs. Sherryl Ann Vacik, Pangrac & Associates, Aransas, Texas, U.S.A. are gratefully acknowledged for their language revision of the manuscript. We also thank Dr. Vera Lisa and Mrs. Ivana Zajanova (Inst. Physiol., Acad. Sci. CR) for their excellent assistance in cell culturing and immunocytochemical methods.

References

[1] Poreda, RJ; Becker L. Fullerenes and interplanetary dust at the Permian-Triassic boundary. *Astrobiology*, 2003, 3, 75-90.

[2] Komatsu, N; Ohe, T; Matsushige, K. A highly improved method for purification of fullerenes applicable to large-scale production. *Carbon*, 2004, 42, 163–167.

[3] Kroto, HW; Heath, JR; O'Brien, SC; Curl, RF; Smalley, RE. C_{60}: Buckminsterfullerene. *Nature*, 1985, 318, 162-163.

[4] Kroto, HW. Smaller carbon species in the laboratory and space. *Int. J. Mass Spectrometry and Ion Processes*, 1994, 138, 1-15.

[5] Foley, S; Crowley, C; Smaihi, M; Bonfils, C; Erlanger, BF; Seta, P; Larroque, C. Cellular localisation of a water-soluble fullerene derivative. *Biochem. Biophys. Res. Commun.*, 2002, 294, 116-119.

[6] Isobe, H; Nakanishi, W; Tomita, N; Jinno, S; Okayama, H; Nakamura, E. Gene delivery by aminofullerenes: structural requirements for efficient transfection. *Chem. Asian. J.*, 2006, 1, 167-175.

[7] Isobe, H; Nakanishi, W; Tomita, N; Jinno, S; Okayama, H; Nakamura, E. Nonviral gene delivery by tetraamino fullerene. *Mol. Pharm.,* 2006, 3, 124-134.

[8] Bakry, R; Vallant, RM; Najam-ul-Haq, M; Rainer, M; Szabo, Z; Huck, CW; Bonn, GK. Medicinal applications of fullerenes. *Int. J. Nanomedicine.*, 2007, 2, 639-649.

[9] Guldi, DM; Prato, M. Excited-state properties of C60 fullerene derivatives. *Acc. Chem. Res.,* 2000, 33, 695-703.

[10] Sayes, CM; Gobin, AM; Ausman, KD; Mendez, J; West JL; Colvin,VL. Nano-C60 cytotoxicity is due to lipid peroxidation. *Biomaterials,* 2005, 26, 7587-7595.

[11] Dhawan, A; Taurozzi, JS; Pandey, AK; Shan, W; Miller, SM; Hashsham, SA; Tarabara, VV. Stable colloidal dispersions of C60 fullerenes in water: evidence for genotoxicity. *Environ. Sci. Technol.,* 2006, 40, 7394-7401.

[12] Isakovic, A; Markovic, Z; Todorovic-Markovic, B; Nikolic, N; Vranjes-Djuric, S; Mirkovic, M; Dramicanin, M; Harhaji, L; Raicevic, N; Nikolic, Z; Trajkovic, V. Distinct cytotoxic mechanisms of pristine versus hydroxylated fullerene. *Toxicol. Sci.*, 2006, 91, 173-183.

[13] Yamawaki, H; Iwai, N. Cytotoxicity of water-soluble fullerene in vascular endothelial cells. *Am. J. Physiol. Cell. Physiol.,* 2006, 290, C1495-C1502.

[14] Yamakoshi, Y; Umezawa, N; Ryu, A; Arakane, K; Miyata, N; Goda, Y; Masumizu, T; Nagano, T. Active oxygen species generated from photoexcited fullerene (C_{60}) as potential medicines: O_2^- versus $^1O_2$. *J. Am. Chem. Soc.*, 2003, 125, 12803-12809.

[15] Tang, YJ; Ashcroft, JM; Chen, D; Min, G; Kim, CH; Murkhejee, B; Larabell, C; Keasling, JD; Chen, FF. Charge-associated effects of fullerene derivatives on microbial structural integrity and central metabolism. *Nano Lett.* 2007, 7, 754-760.

[16] Satoh, M; Matsuo, K; Kiriya, H; Mashino, T; Hirobe, M; Takayanagi, I. Inhibitory effect of a fullerene derivative, monomalonic acid C60, on nitric oxide-dependent relaxation of aortic smooth muscle. *Gen. Pharmacol.*, 1997, 29, 345-351.

[17] Wilson, SR. *Fullerenes: Chemistry, Physics, and Technology*. Kadish KM and Ruoff RS (Eds.), John Wiley and Sons, 2000, 437-465.

[18] Lai, H-S; Chen, Y; Chen, W-J; Chang, K-J; Chiang, L-Y. Free radical scavenging activity of fullerenol on grafts after small bowel transplantation in dogs. *Transplant. Proc.*, 2000, 32, 1272–1274.

[19] Huang, SS; Tsai, SK; Chih, CL; Chiang, LY; Hsieh, HM; Teng, CM; Tsai, MC. Neuroprotective effect of hexasulfobutylated C_{60} on rats subjected to focal cerebral ischemia. *Free Radic. Biol. Med.*, 2001, 30, 643-649.

[20] Yang, DY; Wang, MF; Chen, IL; Chan, YC; Lee, MS; Cheng, FC. Systemic administration of a water-soluble hexasulfonated C(60) (FC(4)S) reduces cerebral ischemia-induced infarct volume in gerbils. *Neurosci. Lett.*, 2001, 311, 121-124.

[21] Nakamura, E; Isobe, H. Functionalized fullerenes in water. The first 10 years of their chemistry, biology, and nanoscience. *Acc. Chem. Res.* 2003, 36, 807-815.

[22] Chen, Y-W; Hwang, KC; Yen, C-C; Lai, YL. Fullerene derivatives protect against oxidative stress in RAW 264.7 cells and ischemia-reperfused lungs. *Am. J. Physiol. Regul. Integr. Comp. Physiol.*, 2004, 287, R21-R26.

[23] Kawase, T; Tanaka, K; Seirai, Y; Shiono N; Oda, M. Complexation of carbon nanorings with fullerenes: Supramolecular dynamics and structural tuning for a fullerene sensor. *Angew. Chem. Int. Ed. Engl.*, 2003, 42, 5597-5600.

[24] Bhattacharya, S; Tominaga, K; Kimura, T; Uno, H; Komatsu, N. A new metalloporphyrin dimer: Effective and selective molecular tweezers for fullerenes. *Chem. Phys. Lett.*, 2007, 433, 395-402.

[25] Simon, F. Studying single-wall carbon nanotubes through encapsulation: from optical methods till magnetic resonance. J. *Nanosci. Nanotechnol.*, 2007, 7, 1197-1220.

[26] Sitharaman, B; Tran, LA; Pham, QP; Bolskar, RD; Muthupillai, R; Flamm, SD; Mikos, AG; Wilson, LJ. Gadofullerenes as nanoscale magnetic labels for cellular MRI. *Contrast Media Mol. Imaging.*, 2007, 2, 139-146.

[27] Fumelli, C; Marconi, A; Salvioli, S; Straface, E; Malorni, W; Offidani, AM; Pellicciari, R; Schettini, G; Giannetti, A; Monti, D; Franceschi, C; Pincelli, C. Carboxyfullerenes protect keratinocytes from ultraviolet-B-induced apoptosis. *J. Invest. Dermatol.*, 2000, 115, 835-841.

[28] Isakovic, A; Markovic, Z; Todorovic-Markovic, B; Nikolic, N; Vranjes-Djuric, S; Mirkovic, M; Dramicanin, M; Harhaji, L; Raicevic, N; Nikolic, Z; Trajkovic, V. Distinct cytotoxic mechanisms of pristine versus hydroxylated fullerene. *Toxicol. Sci.*, 2006, 91, 173-183.

[29] Lai, YL; Murugan, P; Hwang, KC. Fullerene derivative attenuates ischemia-reperfusion-induced lung injury. *Life Sci.*, 2003, 72, 1271-1278.

[30] Wolff, DJ, Papoiu, AD; Mialkowski, K; Richardson, CF; Schuster, DI; Wilson, SR. Inhibition of nitric oxide synthase isoforms by tris-malonyl-C(60)-fullerene adducts. *Arch. Biochem. Biophys.*, 2000, 378, 216-223.

[31] Iijima, S. Helical microtubules of graphitic carbon. *Nature*, 1991, 354, 56-58.

[32] Iijima, S; Ichihashi, T. Cobalt-catalysed growth of carbon nanotubes with single-atomic-layer walls. *Nature*, 1993, 363, 603-605.

[33] Iijima, S. Carbon nanotubes: past, present, and future. *Physica B*, 2002, 323, 1–5.

[34] Fernandez-Alonso, F; Bermejo, FJ; Cabrillo, C; Loutfy, RO; Leon, V; Saboungi, ML. Nature of the bound states of molecular hydrogen in carbon nanohorns. *Phys. Rev. Lett.*, 2007, 98, 215503.

[35] Lee, H; Cho, J. Sn(78)Ge(22)@carbon core-shell nanowires as fast and high-capacity lithium storage media. *Nano Lett.*, 2007, 7, 2638-2641.

[36] Ando, Y; Zhao, X; Inoue, S; Iijima, S. Mass production of multiwalled carbon nanotubes by hydrogen arc discharge. *J. Crystal Growth*, 2002, 237–239, 1926–1930.

[37] Yudasaka, M; Ichihashi, T; Komatsu, T; Iijima, S. Single-wall carbon nanotubes formed by a single laser-beam pulse. *Chemical Physics Letters*, 1999, 299, 91–96.

[38] Okazaki, T; Saito, T; Matsuura, K; Ohshima, S; Yumura, M; Oyama, Y; Saito, R; Iijima, S. Photoluminescence and population analysis of single-walled carbon nanotubes produced by CVD and pulsed-laser vaporization methods. *Chem. Phys. Lett.*, 2006, 420, 286–290.

[39] Kagan, VE; Tyurina, YY; Tyurin, VA; Konduru, NV; Potapovich, AI; Osipov, AN; Kisin, ER; Schwegler-Berry, D; Mercer, R; Castranova, V; Shvedova, AA. Direct and indirect effects of single walled carbon nanotubes on RAW 264.7 macrophages: role of iron. *Toxicol. Lett.*, 2006, 165, 88-100.

[40] Pulskamp, K; Diabaté, S; Krug, HF. Carbon nanotubes show no sign of acute toxicity but induce intracellular reactive oxygen species in dependence on contaminants. *Toxicol. Lett.*, 2007, 168, 58-74.

[41] Pensabene, V; Vittorio, O; Raffa, V; Menciassi, A; Dario, P. Investigation of CNTs interaction with fibroblast cells. *Conf. Proc. IEEE Eng. Med. Biol. Soc.*, 2007, 1, 6620-6623.

[42] Wick, P; Manser, P; Limbach, LK; Dettlaff-Weglikowska, U; Krumeich, F; Roth, S; Stark, WJ; Bruinink, A. The degree and kind of agglomeration affect carbon nanotube cytotoxicity. *Toxicol. Lett.*, 2007, 168, 121–131.

[43] Islam, MF; Rojas, E; Bergey, DM; Johnson, AT; Yodh, AG. High weight fraction surfactant solubilization of single-wall carbon nanotubes in water. *Nano Letters*, 2003, 3, 269-273.

[44] Sayes, CM; Liang, F; Hudson, JL; Mendez, J; Guo, W; Beach, JM; Moore, VC; Doyle, CD; West, JL; Billups, WE; Ausman, KD; Colvin, VL. Functionalization density dependence of single-walled carbon nanotubes cytotoxicity in vitro. *Toxicol. Lett.*, 2006, 161, 135-142.

[45] Dillon, C; Gennett, T; Jones, KM; Alleman, JL; Parilla, PA; Heben, MJ. A simple and complete purification of single-walled carbon nanotube materials. *Adv. Mater.*, 1999, 11, 1354-1358.

[46] Chattopadhyay, D; Galeska, I; Papadimitrakopoulos, F. Complete elimination of metal catalysts from single wall carbon nanotubes. *Carbon*, 2002, 40, 985–988.

[47] Bottini, M; Bruckner, S; Nika, K; Bottini, N; Bellucci, S; Magrini, A; Bergamaschi, A; Mustelin, T. Multi-walled carbon nanotubes induce T lymphocyte apoptosis. *Toxicol. Lett.*, 2006, 160, 121-126.

[48] Magrez, A; Kasas, S; Salicio, V; Pasquier, N; Seo, JW; Celio, M; Catsicas, S; Schwaller, B; Forro, L. Cellular toxicity of carbon-based nanomaterials. *Nano Lett.*, 2006, 6, 1121-1125.

[49] Ebbesen, TW. Carbon nanotubes. *Phys. Today,* 1996, 49, 26-32.

[50] Yakobson, B; Smalley, RE. Fullerene Nanotubes: C[sub1,000,000) and Beyond. *Am. Sci.*, 1997, 85, 324-333.

[51] Shi, X; Sitharaman, B; Pham, QP; Liang, F; Wu, K; Edward Billups, W; Wilson, LJ; Mikos, AG. Fabrication of porous ultra-short single-walled carbon nanotube nanocomposite scaffolds for bone tissue engineering. *Biomaterials*, 2007, 28, 4078-4090.

[52] Abarrategi, A; Gutiérrez, MC; Moreno-Vicente, C; Hortigüela, MJ; Ramos, V; López-Lacomba, JL; Ferrer, ML; del Monte, F. Multiwall carbon nanotube scaffolds for tissue engineering purposes. *Biomaterials*, 2008, 29, 94-102.

[53] Price, RL; Waid, MC; Haberstroh, KM; Webster, TJ. Selective bone cell adhesion on formulations containing carbon nanofibers. *Biomaterials*, 2003, 24, 1877-1887.
[54] Price, RL; Ellison, K; Haberstroh, KM; Webster, TJ: Nanometer surface roughness increases select osteoblast adhesion on carbon nanofiber compacts.*J Biomed. Mater. Res. A*, 2004, 70, 129–138.
[55] Supronowicz, PR; Ajayan, PM; Ullmann, KR; Arulanandam, BP; Metzger, DW; Bizios, R. Novel current-conducting composite substrates for exposing osteoblasts to alternating current stimulation. *J. Biomed. Mater. Res.*, 2002, 59, 499-506.
[56] Zanello, LP; Zhao, B; Hu, H; Haddon, RC. Bone cell proliferation on carbon nanotubes. *Nano Lett.*, 2006, 6, 562-567.
[57] Kalbacova, M; Kalbac, M; Dunsch, L; Kromka, A; Vanecek, M; Rezek, B; Hempel, U; Kmoch, S: The effect of SWCNT and nano-diamond films on human osteoblast cells. *Phys. Stat. Sol.*, 2007, *(b)*, 1–4.
[58] Malarkey, EB; Parpura, V. Applications of carbon nanotubes in neurobiology. *Neurodegener. Dis.*, 2007, 4, 292-299, 2007.
[59] Mattson, MP; Haddon, RC; Rao, AM. Molecular functionalization of carbon nanotubes and use as substrates for neuronal growth. *J. Mol. Neurosci.*, 2000, 14, 175-182.
[60] http://www.cellnucleus.com/TEM.htm.
[61] Ohtsuki, T; Masumoto, K; Shikano, K; Sueki, K; Tanaka, T; Komatsu, K. Direct preparation of radioactive fullerenes as a tracer for applications. *Biol. Trace. Elem. Res.*, 1999, 71-72, 489-498.
[62] Bottini, M; Cerignoli, F; Dawson, MI; Magrini, A; Rosato, N; Mustelin, T. Full-length single-walled carbon nanotubes decorated with streptavidin-conjugated quantum dots as multivalent intracellular fluorescent nanoprobes. *Biomacromolecules*, 2006, 7, 2259-2263.
[63] Hoshino, A; Manabe, N; Fujioka, K; Suzuki, K; Yasuhara, M; Yamamoto, K. Use of fluorescent quantum dot bioconjugates for cellular imaging of immune cells, cell organelle labeling, and nanomedicine: surface modification regulates biological function, including cytotoxicity. *J. Artif. Organs.*, 2007, 10, 149-157.
[64] Hartman, KB; Wilson, LJ. Carbon nanostructures as a new high-performance platform for MR molecular imaging. *Adv. Exp. Med. Biol.*, 2007, 620, 74-84.
[65] Yinghuai, Z; Peng, AT; Carpenter, K; Maguire, JA; Hosmane, NS; Takagaki, M. Substituted carborane-appended water-soluble single-wall carbon nanotubes: new approach to boron neutron capture therapy drug delivery. *J. Am. Chem. Soc.*, 2005, 127, 9875-9880.
[66] Escorcia, FE; McDevitt, MR; Villa, CH; Scheinberg, DA. Targeted nanomaterials for radiotherapy. *Nanomed.*, 2007, 2, 805-815.
[67] Singh, R; Pantarotto, D; McCarthy, D; Chaloin, O; Hoebeke, J; Partidos, CD; Briand, JP; Prato, M; Bianco, A; Kostarelos, K. Binding and condensation of plasmid DNA onto functionalized carbon nanotubes: toward the construction of nanotube-based gene delivery vectors. *J. Am. Chem. Soc.*, 2005, 127, 4388-4396.
[68] Yao, DS; Cao, H; Wen, S; Liu, DL; Bai, Y; Zheng, WJ. A novel biosensor for sterigmatocystin constructed by multi-walled carbon nanotubes (MWNT) modified with aflatoxin-detoxifizyme (ADTZ). *Bioelectrochemistry*, 2006, 68, 126-133.
[69] Du, D; Huang, X; Cai, J; Zhang, A; Ding, J; Chen, S. An amperometric acetylthiocholine sensor based on immobilization of acetylcholinesterase on a

multiwall carbon nanotube-cross-linked chitosan composite. *Anal. Bioanal. Chem.*, 2007, 387, 1059-1065.

[70] Goyal, RN; Gupta, VK; Bachheti, N. Fullerene-C60-modified electrode as a sensitive voltammetric sensor for detection of nandrolone--an anabolic steroid used in doping. *Anal Chim Acta.*, 2007, 597, 82-89.

[71] Woolley, AT. Biofunctionalization of carbon nanotubes for atomic force microscopy imaging. *Methods Mol. Biol.*, 2004, 283, 305-319.

[72] Gibson, CT; Carnally, S; Roberts, CJ; Attachment of carbon nanotubes to atomic force microscope probes. *Ultramicroscopy*, 2007, 107, 1118-1122.

[73] Ding, F; Lin, Y; Krasnov, PO; Yakobson, BI. Nanotube-derived carbon foam for hydrogen sorption. *J. Chem. Phys.*, 2007, 127, 164703.

[74] Kowalczyk, P; Hołyst, R; Terrones, M; Terrones, H. Hydrogen storage in nanoporous carbon materials: myth and facts. *Phys. Chem. Chem. Phys.*, 2007, 9, 1786-1792.

[75] Nikitin, A; Li, X; Zhang, Z; Ogasawara, H; Dai, H; Nilsson, A. Hydrogen storage in carbon nanotubes through the formation of stable C-H bonds. *Nano Lett.*, 2008, 8, 162-167.

[76] Murata, K; Miyawaki, J; Yudasaka, M; Iijima, S; Kaneko, K. High-density of methane confined in internal nanospace of single-wall carbon nanohorns. *Carbon,* 2005, 43, 2817–2833.

[77] Kim, JW; Shashkov, EV; Galanzha, EI; Kotagiri, N; Zharov, VP. Photothermal antimicrobial nanotherapy and nanodiagnostics with self-assembling carbon nanotube clusters. *Lasers Surg. Med.*, 2007, 39, 622-34.

[78] Park, KH; Chhowalla, M; Iqbal, Z; Sesti, F. Single-walled carbon nanotubes are a new class of ion channel blockers. *J. Biol. Chem.*, 2003, 278, 50212-50216.

[79] Hinds, BJ; Chopra, N; Rantell, T; Andrews, R; Gavalas, V; Bachas, LG. Aligned multiwalled carbon nanotube membranes. *Science*, 2004, 303, 62-65.

[80] Majumder, M; Zhan, X; Andrews, R; Hinds, BJ. Voltage gated carbon nanotube membranes. *Langmuir*, 2007, 23, 8624-8631.

[81] van Hijkoop, VJ; Dammers, AJ; Malek, K; Coppens, MO. Water diffusion through a membrane protein channel: a first passage time approach. *J. Chem. Phys.*, 2007, 127, 085101.

[82] Cui, D; Tian, F; Ozkan, CS; Wang, M; Gao, H. Effect of single wall carbon nanotubes on human HEK293 cells. *Toxicol. Lett.*, 2005, 155, 73-85.

[83] Davoren, M; Herzog, E; Casey, A; Cottineau, B; Chambers, G; Byrne, HJ; Lyng, FM. In vitro toxicity evaluation of single walled carbon nanotubes on human A549 lung cells. *Toxicol. In Vitro*, 2007, 21, 438-448.

[84] Zhang, LW; Zeng, L; Barron, AR; Monteiro-Riviere, NA. Biological interactions of functionalized single-wall carbon nanotubes in human epidermal keratinocytes. *Int. J. Toxicol.*, 2007, 26, 103-113.

[85] Kisin, ER; Murray, AR; Keane, MJ; Shi, XC; Schwegler-Berry, D; Gorelik, O; Arepalli, S; Castranova, V; Wallace, WE; Kagan, VE; Shvedova, AA. Single-walled carbon nanotubes: geno- and cytotoxic effects in lung fibroblast V79 cells. *J. Toxicko. Environ. Health.*, 2007, *A* 70: 2071-2079.

[86] Iakoubovskii, K; Adriaenssens, GJ; Meykens, K; Nesladek, M; Vul, AY; Osipov, VY. Study of defects in CVD and ultradisperse diamond. *Diamond Relat. Mater.*, 1999, 8, 1476–1479.

[87] Qin, LC; Zhou, D; Krauss, AR; Gruen, DM. TEM Characterization of nanodiamond thin films. *NanoStructured Materials*, 1998, 10, 649-660.

[88] Bajaj, P; Akin, D; Gupta, A; Sherman, D; Shi, B; Auciello, O; Bashir, R. Ultrananocrystalline diamond film as an optimal cell interface for biomedical applications. *Biomed. Microdevices*, 2007, 9, 787–794.

[89] Hartl, A; Schmich, E; Garrido, JA; Hernando, J; Catharino, SC; Walter, S; Feulner, P; Kromka, A; Steinmüller, D; Stutzmann, M. Protein-modified nanocrystalline diamond thin films for biosensor applications. *Nat. Mater.*, 2004, 3, 736-742.

[90] Christiaens, P; Vermeeren, V; Wenmackers, S; Daenen, M; Haenen, K; Nesládek, M; vandeVen, M; Ameloot, M; Michiels, L; Wagner, P. EDC-mediated DNA attachment to nanocrystalline CVD diamond films. *Biosens. Bioelectron.*, 2006, 22, 170-177.

[91] Schrand, AM; Huang, H; Carlson, C; Schlager, JJ; Osawa, E; Hussain, SM; Dai, L. Are diamond nanoparticles cytotoxic? *J. Phys. Chem.*, 2007, 111, 2-7.

[92] Bakowicz, K; Mitura, S. Biocompatibility of NCD. *J. Wide Bandgap Mater.*, 2002, 9, 261.

[93] Fu, CC; Lee, HY; Chen, K; Lim, TS; Wu, HY; Lin, PK; Wei, PK; Tsao, PH; Chang, HC; Fann, W. Characterization and application of single fluorescent nanodiamonds as cellular biomarkers. *Proc. Natl. Acad. Sci. U. S. A.*, 2007, 104, 727-732.

[94] Papo, MJ; Catledge, SA; Vohra, YK; Machado, C. Mechanical wear behavior of nanocrystalline and multilayer diamond coatings on temporomandibular joint implants. *J. Mater. Sci. Mater. Med.*, 2004, 15, 773-777.

[95] Bacakova, L; Grausova, L; Vacik, J; Fraczek, A; Blazewicz, S; Kromka, A; Vanecek, M; Svorcik, V. Improved adhesion and growth of human osteoblast-like MG 63 cells on biomaterials modified with carbon nanoparticles. *Diamond Relat. Mater.*, 2007, 16, 2133-2140.

[96] Grausova, L; Bacakova, L; Kromka, A; Potocky, S; Vanecek, M; Nesladek, M; Lisa, V. Nanodiamond as a promising material for bone tissue engineering. *J. Nanosci. Nanotechnol.,* 2008, in press.

[97] Grausova, L; Kromka, A; Bacakova, L; Potocky, S; Vanecek, M; Lisa, V. Bone and vascular endothelial cells in cultures on nanocrystalline diamond films. *Diamond Relat. Mater.*, 2008, in press.

[98] Specht, CG; Williams, OA; Jackman, RB; Schoepfer, R. Ordered growth of neurons on diamond. *Biomaterials*, 2004, 25, 4073-4078.

[99] Rupprecht, S; Bloch, A; Rosiwal, S; Neukam, FW, Wiltfang, J. Examination of the bone-metal interface of titanium implants coated by the microwave plasma chemical vapor deposition method. *Int. J. Oral Maxillofac. Implants*, 2002, 17, 778-785.

[100] Xiao, X; Wang, J; Liu,C; Carlisle, JA; Mech, B; Greenberg, R; Guven, D; Freda, R; Humayun, MS; Weiland, J; Auciello, O. In vitro and in vivo evaluation of ultrananocrystalline diamond for coating of implantable retinal microchips. *J. Biomed. Mater. Res. B Appl. Biomater.*, 2006, 77, 273-281.

[101] Kubova, O; Bacakova, L; V. Svorcik, V. Biocompatibility of carbon layer on polymer. *Mater. Sci. Forum*, 2005, 482, 247-250.

[102] Kubova, O; Svorcik, V; Heitz, J; Moritz, S; Romanin, C; Matejka, P; Mackova, A. Characterization and cytocompatibility of carbon layers prepared by photo-induced chemical vapor deposition. *Thin Solid Films*, 2007, 515, 6765-6772.

[103] Grinevich, A; Bacakova, L; Choukourov, A; Boldyryeva, H; Pihosh, Y; Slavinska, D; Noskova, L; Skuciova, M; Lisa, V; Biederman, H. Nanocomposite Ti/hydrocarbon plasma polymer films from reactive magnetron sputtering as growth supports for osteoblast-like and endothelial cells. *J. Biomed. Mater. Res. Part A*, 2008, in press.

[104] Bacakova, L; Stary, V; Kofronova, O; Lisa, V. Polishing and coating carbon fiber-reinforced carbon composites with a carbon-titanium layer enhances adhesion and growth of osteoblast-like MG63 cells and vascular smooth muscle cells *in vitro*. *J. Biomed. Mater. Res.*, 2001, 54, 567-578.

[105] Bacakova, L; Koshelyev, H; Noskova, L; Choukourov, A; Benada, O; Mackova, A; Lisa, V; Biederman, H. Vascular endothelial cells in cultures on nanocomposite silver/hydrocarbon plasma polymer films with antimicrobial activity. *J. Optoelectronics and Advanced Materials*, 2008, in press.

[106] Webster, TJ; Ergun, C; Doremus, RH; Siegel, RW; Bizios, R. Enhanced functions of osteoblasts on nanophase ceramics. *Biomaterials*, 2000 , 21, 1803-1810.

[107] Webster, TJ; Ergun, C; Doremus, RH; Siegel, RW; Bizios, R. Specific proteins mediate enhanced osteoblast adhesion on nanophase ceramics. *J. Biomed. Mater. Res.*, 2000, 51, 475-483.

[108] Webster, TJ; Smith, TA. Increased osteoblast function on PLGA composites containing nanophase titania. *J. Biomed. Mater. Res. A*, 2005, 74, 677-686.

[109] Woo, KM; Chen,VJ; Ma, PX. Nano-fibrous scaffolding architecture selectively enhances protein adsorption contributing to cell attachment. *J. Biomed. Mater. Res. A*, 2003, 67, 531-537.

[110] Wei, G; Ma, PX. Structure and properties of nano-hydroxyapatite/polymer composite scaffolds for bone tissue engineering. *Biomaterials*, 2004, 25, 4749-4757.

[111] Wang, L; Zhao, G; Olivares-Navarrete, R; Bell, BF; Wieland, M; Cochran, DL; Schwartz, Z; Boyan, BD. Integrin $beta_1$ silencing in osteoblasts alters substrate-dependent responses to 1,25-dihydroxy vitamin D_3. *Biomaterials*, 2006, 27, 3716-3725.

[112] Dee, KC; Andersen, TT; Bizios, R. Design and function of novel osteoblast-adhesive peptides for chemical modification of biomaterials. *J. Biomed. Mater. Res.*, 1998, 40, 371-377.

[113] Gobin, S; West, JL. Val-ala-pro-gly, an elastin-derived non-integrin ligand: Smooth muscle cell adhesion and specificity. *J. Biomed. Mater. Res. A*, 2003, 67, 255–259.

[114] Shin, H; Jo, S; Mikos, AG. Biomimetic materials for tissue engineering. *Biomaterials*, 2003, 24, 4353–4364.

[115] Bacakova, L; Noskova, L; Koshelyev, H; Biederman, H; Vascular endothelial cells in cultures on metal/C:H composite films. *Engineering of Biomaterials*, 2004, 7 [37], 18-20.

[116] Tan, J; Saltzman, WM. Biomaterials with hierarchically defined micro- and nanoscale structure. *Biomaterials*, 2004, 25, 3593-3601.

[117] Bacakova, L; Grausova, L; Vacik, J; Svorcik, V. Human bone-derived cells in cultures on fullerene C_{60} layers of continuous and patterned morphology. *Conference NanoSMat* 2007, Algarve, Portugal, Abstract No. BBE20, p. 66.

[118] Grausova, L; Bilkova, P; Bacakova, L; Vacik, J; Svorcik, V.: Regionally-selective adhesion and growth of human osteoblast-like MG 63 cells on micropatterned fullerene C_{60} layers. 14th *International Conference on Plasma Physics and Applications (*14th *CPPA* 2007*)*, Brasov, Romania, p. 127.

[119] Levi, N; Hantgan, RR; Lively, MO; Carroll, DL; Prasad, GL. C_{60}-fullerenes: detection of intracellular photoluminescence and lack of cytotoxic effects. *J. Nanobiotechnol.*, 2006, 4, 14.

[120] Lin, JC; Wu, CH: Surface characterization and platelet adhesion studies on polyurethane surface immobilized with C_{60}. *Biomaterials*, 1999, 20, 1613-1620.

[121] Bacakova, L; Walachova, K; Svorcik, V; Hnatowitz, V. Adhesion and proliferation of rat vascular smooth muscle cells (VSMC) on polyethylene implanted with O(+) and C(+) ions. *J. Biomater. Sci. Polym. Ed.*, 2001, 12, 817-834.

[122] Bacakova, L; Filova, E; Rypacek, F; Svorcik, V; Stary, V. Cell adhesion on artificial materials for tissue engineering. *Physiol Res.*, 2004, 53 *Supp* , S35-45.

[123] Tsuchiya, T; Yamakoshi, YN; Miyata, N. A novel promoting action of fullerene C_{60} on the chondrogenesis in rat embryonic limb bud cell culture system. *Biochem. Biophys. Res. Commun.*, 1995, 206, 885-894.

[124] Straface, E; Natalini, B; Monti, D; Franceschi, C; Schettini, G; Bisaglia, M; Fumelli, C; Pincelli, C; Pellicciari, R; Malorni, W. C3-fullero-tris-methanodicarboxylic acid protects epithelial cells from radiation-induced anoikia by influencing cell adhesion ability. *FEBS Lett.*, 1999, 454, 335-340.

[125] Yamawaki, H; Iwai, N. Cytotoxicity of water-soluble fullerene in vascular endothelial cells. *Am. J. Physiol. Cell. Physiol.*, 2006, 290, C1495-C1502.

[126] Yamakoshi, Y; Umezawa, N; Ryu, A; Arakane, K; Miyata, N; Goda, Y; Masumizu, T; Nagano, T. Active oxygen species generated from photoexcited fullerene (C_{60}) as potential medicines: O_2^- versus $^1O_2$. *J. Am. Chem. Soc.*, 2003, 125, 12803-12809.

[127] Tang, YJ; Ashcroft, JM; Chen, D; Min, G; Kim, CH; Murkhejee, B; Larabell, C; Keasling, JD; Chen, FF. Charge-associated effects of fullerene derivatives on microbial structural integrity and central metabolism. *Nano Lett.*, 2007, 7, 754-760.

[128] Harhaji, L; Isakovic, A; Raicevic, N; Markovic, Z; Todorovic-Markovic, B; Nikolic, N; Vranjes-Djuric, S; Markovic, I; Trajkovic, V. Multiple mechanisms underlying the anticancer action of nanocrystalline fullerene. *Eur J Pharmacol.*, 2007, 568, 89-98.

[129] Harhaji, L; Isakovic, A; Vucicevic, L; Janjetovic, K; Misirkic, M; Markovic, Z; Todorovic-Markovic, B; Nikolic, N; Vranjes-Djuric, S; Nikolic, Z; Trajkovic, V. Modulation of tumor necrosis factor-mediated cell death by fullerenes. *Pharm. Res.*, 2007, in press.

[130] Bacakova, L; Stary, V; Hornik, J; Glogar, P; Lisa, V; Kofronova O. Osteoblast-like MG63 cells in cultures on carbon fibre-reinforced carbon composites. *Inzynieria Biomaterialów-Engineering of Biomaterials*, 2001, 4 [17-19], 11-12.

[131] Bacakova, L; Stary, V; Glogar, P; Lisa, V. Adhesion, differentiation and immune activation of human osteogenic cells in cultures on carbon-fibre reinforced carbon composites. *Inzynieria Biomaterialów-Engineering of Biomaterials*, 2003, 6, 8-9.

[132] Stary, V; Bacakova, L; Hornik, J; Chmelik, V. Bio-compatibility of the surface layer of pyrolytic graphite. *Thin Solid Films*, 2003, 433, 191-198.

[133] Stary, V; Glogar, P; Bacakova, L; Hnilica, F; Chmelik, V; Korinek, Z; Gregor, J; Mares, V; Lisa, V. A study of surface properties of composite materials and their influence on biocompatibility. *Acta Montana AB*, 2003, 11, 19-36.

[134] Haddow, DB; France, RM; Short, RD; MacNeil, S; Dawson, RA; Leggett, GJ; Cooper, E. Comparison of proliferation and growth of human keratinocytes on plasma

copolymers of acrylic acid/1,7-octadiene and self-assembled monolayers. *J. Biomed. Mater. Res.*, 1999, 47, 379-387.

[135] Bacakova, L; Mares, V; Bottone, MG; Pellicciari, C; Lisa, V; Svorcik, V. Fluorine-ion-implanted polystyrene improves growth and viability of vascular smooth muscle cells in culture. *J. Biomed. Mater. Res.*, 2000, 49, 369-379.

[136] Bacakova, L; Mares, V; Lisa, V; Svorcik, V. Molecular mechanisms of improved adhesion and growth of an endothelial cell line cultured on polystyrene implanted with fluorine ions. *Biomaterials*, 2000, 21, 1173-1179.

[137] Mikulikova, R; Moritz, S; Gumpenberger, T; Olbrich, M; Romanin, Ch; Bacakova, L; Svorcik, V; Heitz, J. Cell microarrays on photochemically modified PTFE, *Biomaterials*, 2005, 26, 5572.

[138] Parizek, M; Bacakova, L; Lisa, V; Kubova, O; Svorcik, V; Heitz J. Vascular smooth muscle cells in cultures on synthetic polymers with adhesive microdomains. *Inzynieria Biomaterialow (Engineering of Biomaterials)*, 2006, *IX (*58-60*)*, 7-10.

[139] Parizek, M; Bacakova, L; Kasalkova, N; Svorcik, V; Kolarova, K. Improved adhesion and growth of vascular smooth muscle cells in cultures on Polyethylene modified by plasma discharge. *Inzynieria Biomaterialow (Engineering of Biomaterials)*, 2007, *X (*67-68*)*, 1-4.

[140] Kasalkova, N; Kolarova, K; Bacakova, L; Parizek, M; Svorcik, V. Cell adhesion and proliferation on modified polyethylene, *Mater. Sci. Forum*, 2007, 567-568, 269-272.

[141] Bacakova, L; Filova, E; Kubies, D; Machova, L; Proks, V; Malinova, V; Lisa, V; Rypacek, F. Adhesion and growth of vascular smooth muscle cells in cultures on bioactive RGD peptide-carrying polylactides. *J. Mater. Sci. Mater. Med.*, 2007, 18,1317-1323.

[142] Patrito, N; McCague, C; Norton, PR; Petersen, NO. Spatially controlled cell adhesion via micropatterned surface modification of poly(dimethylsiloxane). *Langmuir*, 2007, 23, 715-719.

[143] Hamilton, DW; Chehroudi, B; Brunette, DM. Comparative response of epithelial cells and osteoblasts to microfabricated tapered pit topographies in vitro and in vivo. *Biomaterials*, 2007, 28, 2281-2293.

[144] Heitz, J; Svorcik, V; Bacakova, L; Rockova, K; Ratajova, E; Gumpenberger, T; Bauerle, D; Dvorankova, B; Kahr, H; Graz, I; Romanin, C. Cell adhesion on polytetrafluoroethylene modified by UV-irradiation in an ammonia atmosphere. *J. Biomed. Mater. Res.*, 2003, 67*A*, 130-137.

[145] Charest, JL; Bryant, LE; Garcia, AJ; King, WP. Hot embossing for micropatterned cell substrates. *Biomaterials*, 2004, 25, 4767-4775.

[146] Kenar, H; Kose, GT; Hasirci, V. Tissue engineering of bone on micropatterned biodegradable polyester films. *Biomaterials*, 2006, 27, 885-895.

[147] Hamilton, DW; Brunette, DM. The effect of substratum topography on osteoblast adhesion mediated signal transduction and phosphorylation. *Biomaterials*, 2007, 28, 1806-1819.

[148] Papenburg, BJ; Vogelaar, L; Bolhuis-Versteeg, LA, Lammertink, RG; Stamatialis, D; Wessling, M. One-step fabrication of porous micropatterned scaffolds to control cell behavior. *Biomaterials*, 2007, 28,1998-2009.

[149] Guo, S; Fogarty, DP; Nagel, PM; Kandel, SA: Scanning tunneling microscopy of surface-adsorbed fullerenes: C_{60}, C_{70}, and C_{84}. *Surf. Sci.*, 2007, 601, 994-1000.

[150] Vacik, J; Naramoto, H; Narumi, K; Yamamoto, Y; Miyashita, K. Pattern formation induced by co-deposition of Ni and C60 on MgO.100. *Journal of Chemical Physics*, 2001, 114 [20], 9115 -9119.

[151] Vacik, J; Lavrentiev, V; Hnatowicz, V; Yamamoto, S; Vorlicek, V; Stadler, H. Spontaneous Partitioning of the Ni+C_{60} Thin Film Grown at RT. *Journal of Alloys and Compounds* 2008, in press.

[152] Engler, AJ; Sen, S; Sweeney, HL; Discher, DE. Matrix elasticity directs stem cell lineage specification. *Cell*, 2006, 126, 677-689.

[153] Hao, L; Lawrence, J; Chian, KS. Osteoblast cell adhesion on a laser modified zirconia based bioceramic. *J. Mater. Sci. Mater. Med.*, 2005, 16, 719-726.

[154] Zhao, G; Schwartz, Z; Wieland, M; Rupp, F; Geis-Gerstorfer, J; Cochran, DL; Boyan, BD. High surface energy enhances cell response to titanium substrate microstructure. *J. Biomed. Mater. Res. A*, 2005, 74, 49-58.

[155] Pesakova, V; Kubies, D; Hulejova, H; Himmlova, L. The influence of implant surface properties on cell adhesion and proliferation. *J. Mater. Sci. Mater. Med.*, 2007, 18, 465-473.

[156] Dos Santos, EA; Farina, M; Soares, GA; Anselme, K. Surface energy of hydroxyapatite and beta-tricalcium phosphate ceramics driving serum protein adsorption and osteoblast adhesion. *J. Mater. Sci. Mater. Med.*, 2008, in press.

[157] Grausova, L; Bacakova, L; Fraczek, A; Blazewicz, S; Blazewicz, M; Pepka, M. Biological effects of polymers modified with carbon nanotubes on human osteoblast-like MG 63 cells. *Engineering of Biomaterials*, 2006, 9[58-60], 14-16.

[158] Grausova, L; Bacakova, L; Fraczek, A; Filova, E; Blazewicz M. Adhesion, proliferation and viability of human osteoblast-like MG 63 cells on carbon nanotube-polysulfone composites. *Eight International Conference on the Science ad Application of Nanotubes, „Nanotubes 2007"*, Ouro Preto, Minas Gerais, Brasil, p. 181.

[159] Mailhot, JM; Sharawy, MM; Galal, M; Oldham, AM; Russell, CM. Porous polysulfone coated with platelet-derived growth factor-BB stimulates proliferation of human periodontal ligament fibroblasts. *J. Periodontol.*, 1996, 67, 981-985.

[160] Jozwiak, A; Karlik, W; Wiechetek, M; Werynski, A. Attachment and metabolic activity of hepatocytes cultivated on selected polymeric membranes. *Int. J. Artif. Organs*, 1998, 21, 460-466.

[161] Hoque, ME; Mao, HQ; Ramakrishna, S. Hybrid braided 3-D scaffold for bioartificial liver assist devices. *J. Biomater. Sci. Polym. Ed.*, 2007, 18, 45-58.

[162] Pamula, E; Bacakova, L; Filova, E; Buczynska, J; Dobrzynski, P; Noskova, L; Grausova, L: The influence of pore size on colonization of poly(L-lactide-glycolide) scaffolds with human osteoblast-like MG 63 cells in vitro. *J. Mater. Sci. Mater. Med.*, 2008, 19, 425-435.

[163] McDevitt, MR; Chattopadhyay, D; Jaggi, JS; Finn, RD; Zanzonico, PB; Villa, C; Rey, D; Mendenhall, J; Batt, CA; Njardarson, JT; Scheinberg, DA. PET imaging of soluble yttrium-86-labeled carbon nanotubes in mice. *PLoS ONE* 2, 2007, e907.

[164] Fraczek, A; Menaszek, E; Czajkowska, B; Bacakova, L; Blazewicz, M. In vitro and in vivo biocompatibility of single wall carbon nanotubes. *Engineering of Biomaterials*, 2007, 10(69-72), 8-11.

[165] Tian, F; Cui, D; Schwarz, H; Estrada, GG; Kobayashi, H. Cytotoxicity of single-wall carbon nanotubes on human fibroblasts. *Toxicol. In Vitro*, 2006, 20, 1202-1212.

[166] Jia, G; Wang, H; Yan, L; Wang, X; Pei, R; Yan, T; Zhao, Y; Guo, X. Cytotoxicity of carbon nanomaterials: single-wall nanotube, multi-wall nanotube, and fullerene. *Environ. Sci. Technol.*, 2005, 39, 1378-1383.

[167] Radomski, A; Jurasz, P; Alonso-Escolano, D; Drews, M; Morandi, M; Malinski, T; Radomski, MW. Nanoparticle-induced platelet aggregation and vascular thrombosis. *Br. J. Pharmacol.*, 2005, 146, 882-893.

[168] Raja, PM; Connolley, J; Ganesan, GP; Ci, L; Ajayan, PM; Nalamasu, O; Thompson, DM. Impact of carbon nanotube exposure, dosage and aggregation on smooth muscle cells. *Toxicol. Lett.*, 2007, 169, 51-63.

[169] Dumortier, H; Lacotte, S; Pastorin, G; Marega, R; Wu, W; Bonifazi, D; Briand, JP; Prato, M; Muller, S; Bianco, A. Functionalized carbon nanotubes are non-cytotoxic and preserve the functionality of primary immune cells. *Nano Lett.*, 2006, 6, 1522-1528.

[170] Potocky, S; Kromka, A; Potmesil, J; Remes, Z; Polackova, Z; Vanecek, M. Growth of nanocrystalline diamond films deposited by microwave plasma CVD system at low substrate temperatures. *Phys. Stat. Sol. (a)*, 2006, 203, 3011-3015.

[171] Poruba, A; Fejfar, A; Remes, Z; Springer, J; Vanecek, M; Kocka, J; Meier, J; Torres, P; Shah, A. Optical absorption and light scattering in microcrystalline silicon thin films and solar cells. *J. Appl. Phys.*, 2000, 88, 148-160.

[172] Buhlman, S; Blank, E; Haubner, R; Lux, B: Characterization of ballas diamond depositions, *Diamond Rel. Mater.*, 1999, 8, 194-201.

[173] Zemek, J; Houdkova, J; Lesiak, B; Jablonski, A; Potmesil, J; Vanecek, M. Electron spectroscopy of nanocrystalline diamond surfaces. *J. Optoelectr. & Adv. Materials*, 2006, 8, 2133.

[174] Kim, HJ; Kim, SH; Kim, MS; Lee, EJ; Oh, HG; Oh, WM; Park, SW; Kim, WJ; Lee, GJ; Choi, NG; Koh, JT; Dinh, DB; Hardin, RR; Johnson, K; Sylvia, VL; Schmitz, JP; Dean, DD.Varying Ti-6Al-4V surface roughness induces different early morphologic and molecular responses in MG63 osteoblast-like cells. *J. Biomed. Mater. Res.* 2005, 74, 366-373.

[175] Zhao, G; Zinger, O; Schwartz, Z; Wieland, M; Landolt, D; Boyan, BD. Osteoblast-like cells are sensitive to submicron-scale surface structure. *Clin. Oral. Implants Res.*, 2006, 17, 258-264.

[176] Zhao, W; Xu, JJ; Qiu, QQ; Chen, HY. Nanocrystalline diamond modified gold electrode for glucose biosensing. *Biosens Bioelectron*, 2006, 22, 649-655.

[177] Boyan, BD; Bonewald, LF; Paschalis, EP; Lohmann, CH; Rosser, J; Cochran, DL; Dean, DD; Schwartz, Z; Boskey, AL. Osteoblast-mediated mineral deposition in culture is dependent on surface microtopography. *Calcif. Tissue Int.*, 2002, 71, 519-529.

[178] Lohmann, CH; Bonewald, LF; Sisk, MA; Sylvia, VL; Cochran, DL; Dean, DD; Boyan, BD; Schwartz, Z. Maturation state determines the response of osteogenic cells to surface roughness and 1,25-dihydroxyvitamin D3. *J. Bone Miner. Res.*, 2000, 15, 1169-1180.

[179] Schneider, GB; Zaharias, R; Seabold, D; Keller, J; Stanford, C. Differentiation of preosteoblasts is affected by implant surface microtopographies. *J. Biomed. Mater. Res. A*, 2004, 69, 462-468.

[180] Lossdorfer, S; Schwartz, Z; Wang, L; Lohmann, CH; Turner, JD; Wieland, M; Cochran, DL; Boyan, BD. Microrough implant surface topographies increase

osteogenesis by reducing osteoclast formation and activity. *J. Biomed. Mater. Res. A,* 2004, 70, 361-369.

[181] Sammons, RL; Lumbikanonda, N; Gross, M; Cantzler, P. Comparison of osteoblast spreading on microstructured dental implant surfaces and cell behaviour in an explant model of osseointegration. A scanning electron microscopic study. *Clin. Oral. Implants. Res.*, 2005, 16, 657-666.

[182] Sader, MS; Balduino, A; Soares, G de A; Borojevic, R. Effect of three distinct treatments of titanium surface on osteoblast attachment, proliferation, and differentiation. *Clin. Oral. Implants Res.*, 2005, 16, 667-675.

[183] Mares, JJ; Hubik, P; Nesladek, M; Kristofik. Boron-doped diamond — Grained Mott's metal revealing superconductivity. *Diamond Relat. Mater.*, 2007, 16, 921-925.

[184] Chang, JS; Chang, KL; Hwang, DF; Kong, ZL. *In vitro* cytotoxicitiy of silica nanoparticles at high concentrations strongly depends on the metabolic activity type of the cell line. *Environ. Sci. Technol.*, 2007, 41, 2064-2068.

[185] Oner, G; Cirrik, S; Bakan, O. Effects of silica on mitochondrial functions of the proximal tubule cells in rats. *Kidney Blood Press Res.*, 2005, 28, 203-210.

[186] Xiao, X; Wang, J; Liu, C; Carlisle, JA; Mech, B; Greenberg, R; Guven, D; Freda, R; Humayun, MS; Weiland, J; Auciello, O. In vitro and in vivo evaluation of ultrananocrystalline diamond for coating of implantable retinal microchips. *J. Biomed. Mater. Res. B Appl. Biomater.*, 2006, 77, 273-281.

[187] Labelle, D; Jumarie, C; Moreau, R. Capacitative calcium entry and proliferation of human osteoblast-like MG-63 cells. *Cell Prolif.*, 2007, 40, 866-884.

[188] Naghii, MR; Torkaman, G; Mofid, M. Effects of boron and calcium supplementation on mechanical properties of bone in rats. *Biofactors*, 2006, 28, 195-201.

[189] Bacakova, L; Stary, V., Glogar P.: Adhesion and growth of cells in culture on carbon-carbon composites with different surface properties. *Engineering of Biomaterials*, 1998, 1 [2], 3-5.

[190] Lappalainen, R; Selenius, M; Anttila, A; Konttinen, YT; Santavirta, SS. Reduction of wear in total hip replacement prostheses by amorphous diamond coatings. *J. Biomed. Mater. Res. B Appl. Biomater.*, 2003, 66, 410-413.

[191] Ma, WJ; Ruys, AJ; Mason, RS; Martin, PJ; Bendavid, A; Liu, Z; Ionescu, M; Zreiqat, H. DLC coatings: Effects of physical and chemical properties on biological response. *Biomaterials*, 2007, 28, 1620–1628.

[192] Bacakova, L; Balik, K; Zizka, S; Adhesion and growth of vascular smooth muscle cells in cultures on carbon fibre-reinforced carbon composites covered with pyrolytic carbon. *Engineering of Biomaterials*, 1998, 1, [-], 19-22.

[193] Pesakova, V; Klezl, Z; Balik, K; Adam, A. Biomechanical and biological properties of the implant material carbon-carbon composite covered with pyrolytic carbon. *J. Mater. Sci.: Mater. Med.*, 2000, 11, 793-798.

In: Nanoparticles: New Research
Editor: Simone Luca Lombardi, pp. 109-142

ISBN: 978-1-60456-704-5

Chapter 3

ORGANIC-SHELL INORGANIC-CORE HYBRID NANOPARTICLES WITH ADVANCED FUNCTIONS DESIGNED BY WET PROCESS

Mami Yamada*

Department of Applied Chemistry, Tokyo University of Agriculture and Technology, 2-24-16 Nakacho, Koganei, Tokyo 184-8588: PRESTO/Japan Science and Technology Agency, 4-1-8, Honcho, Kawaguchi-shi, Saitama 332-0012, Japan

Abstract

This chapter reviews the novel synthesis of inorganic nanoparticles stabilized by an organic shell layer with investigation into their specific characteristics. How to design organic-inorganic hybrid nanoparticles in chemical solution process effectively is indispensable to uncover their unprecedented natures due to the fusion of merits of both organic and inorganic substances. Expressly, the two synthetic parameters are a focus of this chapter: the physical construction of the inorganic core and the chemical constituents introduced. Concretely, the catalytic activities of nanoparticles are controllable by modifying the crystal faces and the surface area of their inorganic cores in connection with their shape and size. In another case, when functional units, as an organic shell, are attached to the surface of nanoparticles, external stimuli such as electric field or light are utilized to transform the characteristics of nanoparticles themselves. It is also interesting to bring metal coordination polymers into an inorganic core, leading to the first isolation of alkyl chain stabilized-metal coordination nano-polymers with ferromagnetism. The nanomaterials presented here are listed as Pt nano-cube, size-selected Au nanoparticles, metal hexacyanoferrate coordination nano-polymers, and metal nanoparticles functionalized by bifferocene, anthraquinone, porphyrin and triphenylene derivatives. The synthetic procedure and their remarkable characteristics with some application examples are demonstrated.

* E-mail address: m-yamada@cc.tuat.ac.jp. Tel./Fax: +81-1-42-388-7379.

Introduction

Nanometer-sized metal particles are generally of mesoscopic areas consisting of dozens to thousands atoms [1]. Since gold colloid sol was synthesized for the first time in the 17th century, the particular natures of metal nanoparticles, which are different from those of bulk metals and molecules, have attracted much attention from researchers. These unique properties of nanoparticles are derived from two characteristic effects—the quantum size effect and the surface effect, leading to the expectation of diverse applications [2] *e.g.* optoelectronic devices, molecular catalysts and chemical sensors. The preparation methods of metal nanoparticles are mainly divided into two categories: physical dry method and chemical wet process. Basically, the former means that bulk metals are downsized mechanically such as with the breaking grinder method, laser ablation, and plasma irradiation, while the latter comprises metal ions dissolved in solution and reduced to neutral atoms, gathering on a nanometer scale. Recently, the improvement of the chemical process is remarkable after Brust *et al.* reported in 1994 the large-scale preparation of mono-dispersed gold nanoparticles which are significantly air-stable in solvent-free form [3]. The nanoparticles obtained by the chemical method commonly possess "the metallic nanocore-organic shell hybrid structure" [4-6]. The organic shell supplies the high stability of nanoparticles, avoiding thermal aggregation among inorganic cores. The chemical process has some benefits over the physical method: the equipment is very convenient and easy (so to speak "synthesis in one vessel"), and the innumerable kinds of nanoparticles could be constructed just by adjusting appropriate synthetic parameters (starting materials, unit ratio, temperature, solvents, etc.) in thereaction mixture. These synthetic variations enable the unique design of nanoparticle materials with unprecedented physical and chemical structures. The other astonishing merit of nanoparticles synthesized by wet process is the re-dispersibility into solvent after purification, which gives the facile fabrication of nanoparticle films onto substrates for application usage. From these points, stable nanoparticle dispersions have been developed for industrial usage by many academic institutes and manufacturing companies.

Against the backdrop noted above, this chapter describes the latest nanoparticle research under the primary theme of 'creation and functionality of organic-shell inorganic-core hybrid nanoparticles by wet process'. There exist numerous patterns in selecting components for organic and inorganic composites. Generally, an organic material excels in the versatility of its physical structure, including flexibility, while an inorganic substance can accumulate multiple electrons in forming various electronic states. First, the creation of organic/inorganic hybrid nanoparticles should be pursued through effective selection of components. It should be also considered that functions of nanoparticles are affected substantially by their physical structure (shape and size) and composition ratio among the units selected. In other words, in order to build up hybrid nanoparticles with novel functions, the preparation method requires the precise control of at least three synthetic parameters of "unit selection", "physical structure" and "component ratio". Simultaneously, the promising development of the wet process would be promoted by exploring the effective synthetic procedure which accompanies the assembling process of the nanoparticles on a substrate for the purpose of further application. Through feedback on the experimental information gathered on the basis of these research concepts, the research presented in this chapter aim to show the preparation of novel organic/inorganic nanoparticles with functions including electrochemical, optical, catalytic, and magnetic properties. The synthesized nanoparticles have been classified into three groups: (1) functionalized organic molecules-attached metal nanoparticles, (2) shape- and size-controlled metal nanoparticles, and (3) nanometer-sized metal coordination polymers. These topics are demonstrated mostly in connection with my related publications.

1. Functionalization of Metal Nanoparticles by Organic Molecules

The first section is presented to prepare and investigate the functional-molecule modified noble metal nanoparticles which possess the prominent properties derived from the combination of the inorganic metal core and organic shell. Functionalization based on a metal core will open an unexpected interesting way to study intra- and intermolecular nanoparticle chemistry and to design new photoelectrochemical systems or multi-step, and electron donor/acceptor reactions using nanoparticles. The fundamental idea of functionalization of metal nanoparticles by thiol exchange reaction was advocated by Murray *et al,.* who proposed the preparation of functionalized gold nanoparticles stabilized by alkylthiolate by exchanging some thiolates on the particle surface with ferrocene thiol derivatives in toluene solution [7]. The method is facile and can be utilized for various nanoparticle functionalizations. In addition, it could create poly-functionalized metal nanoparticles by means of particles modified with mixtures of different thiols [8]. The reactivity of functionalized alkyl thiolate-stabilized metal nanoparticles should be investigated [9-12] for the further synthetic steps and the development of novel functions of nanoparticles. In this section, as functional molecules, redox-active units (biferrocene and anthraquinone), a chlomophore (phorphyrine) and a liquid crystal molecule (triphenylene) are selected. The characters of the prepared organic-inorganic nanoparticles could be controlled by physical and chemical states of the surrounding functional species with outside stimulation, *i.e.,* applied electronic field, light and dielectric strength around the particles.

1.1. Electrodeposition of Metal Nanoparticles Surrounded by Multiple Redox Units

The biferrocene (BFc) unit is one of multiple-redox species, which undergoes a two-step one-electron oxidative reaction of $BFc/BFc^{+}/BFc^{2+}$ in electrolyte solution [13]. The high charge accumulation on a particle/solution interface is expected by the applied electronic field when the BFc units are attached onto a particle surface. The BFc-modifled gold nanoparticles (Au_n-BFc) were synthesized by a thiolate exchange between an alkyl thiolate-stabilized metal nanoparticles (Au_n-AT) and a BFc thiol derivative, (1-(9-thiononyl-1-one)-1', 1''-biferrocene (BFcS) (Scheme 1) [13-15]. Au_n-ATs were obtained by the reduction of Au (III) ions ($HAuCl_4$, $AuCl_3$ *etc.*) in solution mixture with a surface active agent (TOAB, DDAB *etc.*), a reducing agent ($NaBH_4$, Super Hydride *etc.*) and an alkyl thiolate with appropriate chain length [3, 16, 17]. The average diameter of gold core, d_{av}, was changed with the range of 1.5~8.0 nm by the synthetic parameters of Au_n-AT, mainly the molar ratio of Au ions to an alkyl thiolate. The number of the exchanged BFcS on the Au_n-AT surface, θ_{BFc}, was calculated based on the ratio of the integrals of the ^{1}H-NMR signals between BFc (2.9-4.7 ppm) and methyl protons (0.8-0.9 ppm), respectively. Figure 1 shows the transmission electron microscopy (TEM) image of Au_n-BFcs with d_{av} = 2.3 nm and θ_{BFc} = 7.5) [14]. It is seen that the nanoparticles are individually isolated by stable organic shells. Au_n-BFcs could be obtained as a black powdery form after the alcohol purification process, however, they were easily redispersed into several organic solvents, *e.g.*, CH_2Cl_2, chloroform, THF, hexane due to the hydrophobic nature of organic shell.

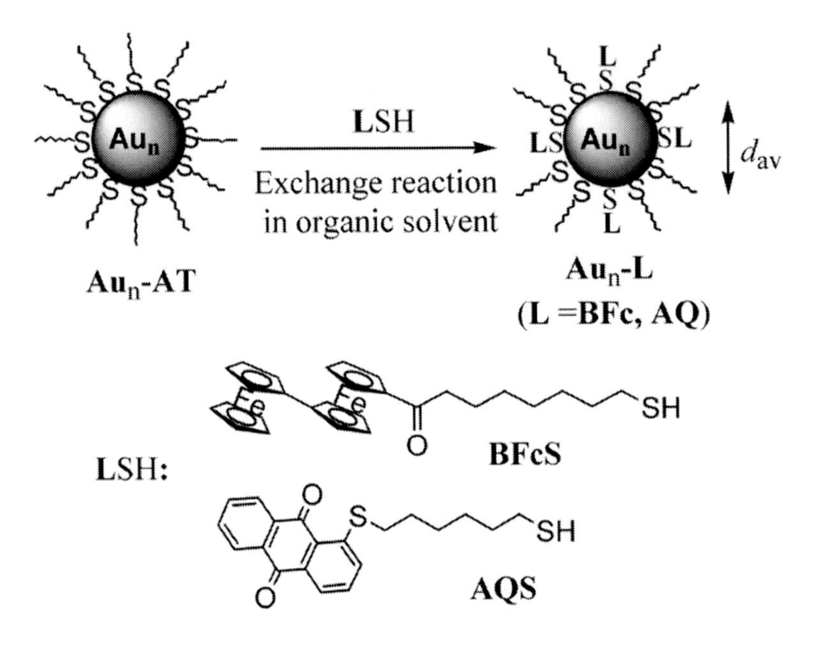

Scheme 1. Illustration of Au_n-BFc and Au_n-AQ.

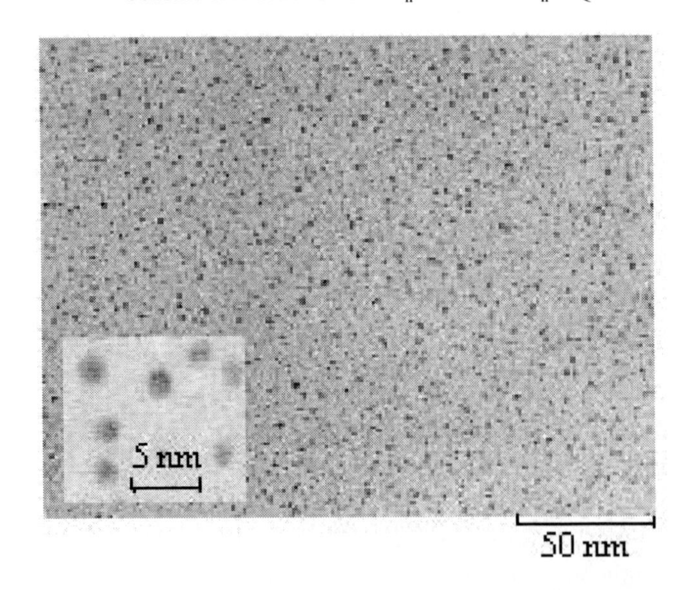

Figure 1. TEM image of Au_n-BFc (d_{av} = 2.3 nm, θ_{BFc} = 7.5).

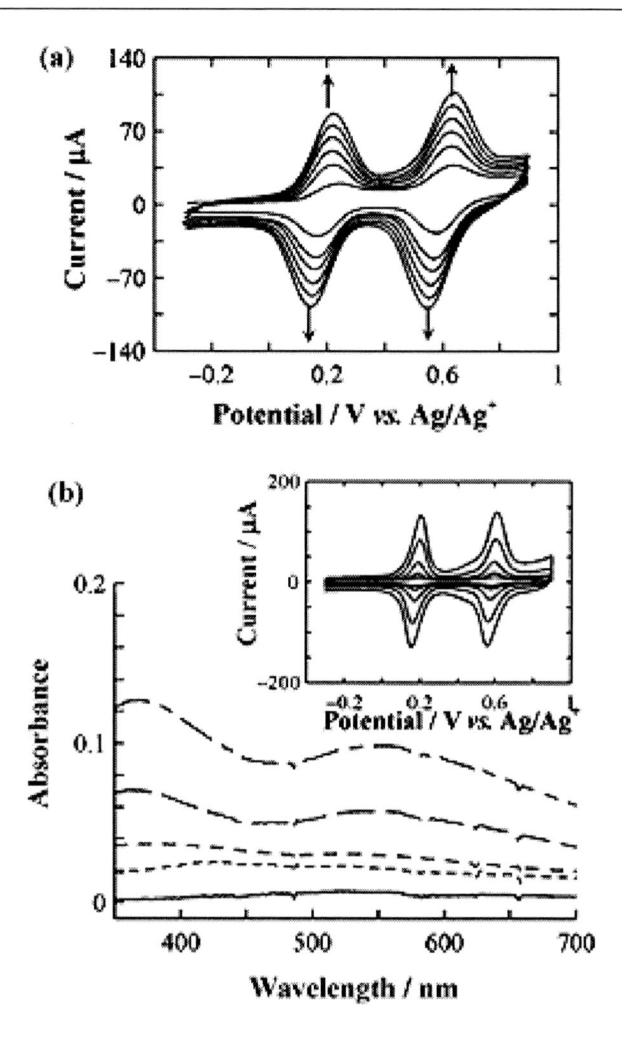

Figure 2. (a) Cyclic voltammograms of 1.7 μM of Au_n-BFc (d_{av} = 2.9 nm, θ_{BFc} = 20.8) at ITO in 0.1 M Bu_4NClO_4–CH_2Cl_2 at 100 mV/s between –0.3 and 0.9 V *vs.* Ag/Ag^+ in the positive direction at the 1st, 10th, 20th, 30th, 40th, and 50th cyclic scan, as shown from the bottom of the figures to the top. (b) UV-Vis spectra and (inset) cyclic voltammograms in 0.1 M Bu_4NClO_4–CH_2Cl_2 of the electrodeposited Au_n-BFc films prepared in the same condition of Figure 2a with the 3, 10, 25, 50, and 75 cyclic scans, as shown from the bottom of the figure to the top [15].

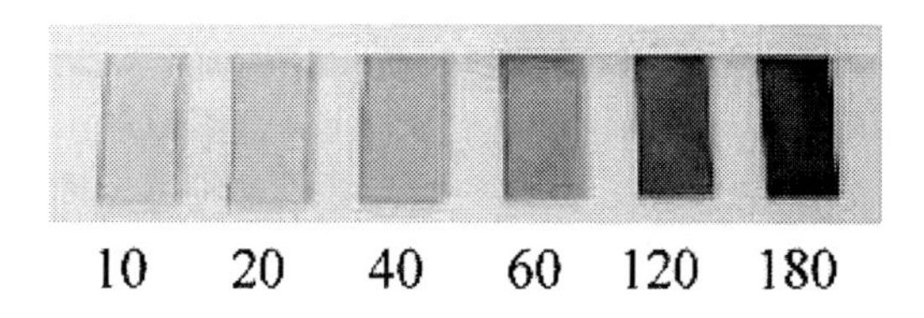

Figure 3. Photographs of electrodeposited Au_n-BFc films prepared in a solution of 5.0 μM Au_n-BFc (d_{av} = 2.3 nm, θ_{BFc} = 15) at ITO 0.1 M Bu_4NClO_4–CH_2Cl_2 at 100 mV/s between –0.3 and 0.9 V *vs.* Ag/Ag^+. The numbers in the figure are those of the cyclic scans [14].

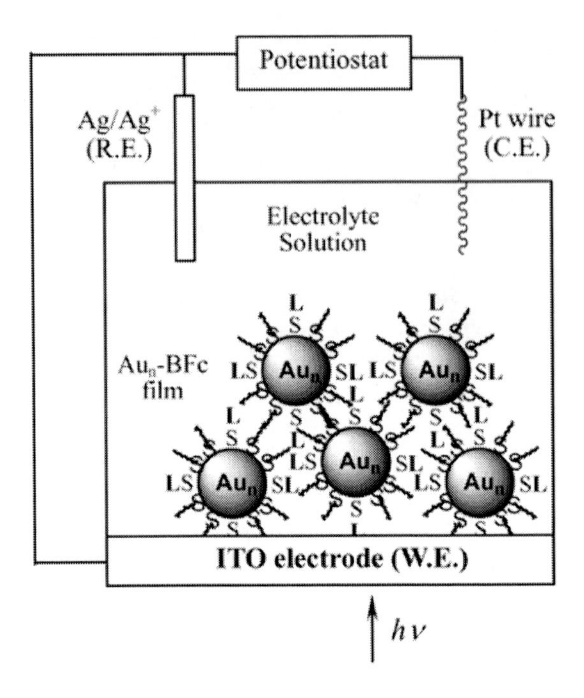

Scheme 2. Electro-optical measurement spectroelectrochemical cell of the Au_n-BFc film [21].

Figure 2a shows the typical cyclic voltammograms (CVs) of Au_n-BFcs, which is measured at an indium tin oxide (ITO) electrode in electrolyte solution of Bu_4NClO_4–CH_2Cl_2 [15]. The peak current increases gradually by consecutive potential scans between –0.3 and 0.9 V *vs.* Ag/Ag^+ where two-step $1e^-$ oxidation due to BFc units on a particle surface occurs at 0.20 and 0.61 V *vs.* Ag/Ag^+, suggesting that Au_n-BFc particles flock on an electrode/solution interface by 2 electron oxidation of BFc sites. UV-Vis spectra of Au_n-BFc films thus prepared exhibit broad absorption bands that grow in intensity with increasing in the number of potential scans (Figure 2b), meaning that the thickness of the film is controlled by changing the number of potential scans. Figure 3 displays the actual photographs of the Au_n-BFc films on ITO, which confirms that the black deposition due to Au_n-BFc is tightly attached onto a transparent ITO electrode [18]. CVs of the electrode films in pure electrolyte solution show two pairs of cathodic and anodic waves (Figure 2b, inset), the peak currents of which are proportional to the potential scan rate, indicating behavior of surface immobilized BFc species in the film. The construction of the Au_n-BFc films with a different core diameter was applicable by using this electrodeposition process, independent of the core size [15]. In turn, this oxidative deposition also occurred by introducing Pd in place of Au, BFc-modified palladium nanoparticles (Pd_n-BFc) (see below) [19].

As seen in Figure 2b, the Au_n-BFc film with relatively large diameter (2.3 nm~) exhibits the broad absorption band at *ca.*550 nm due to the collective surface plasmon (SP) band influenced by the dipole-dipole coupling among the adjacent particles [20]. This collective SP band in the formed film offered special electro-optical features different from a single isolated nanoparticle (Scheme 2) [21]. Figure 4 illustrates that the UV-Vis spectra of the Au_n-BFc film change with the λ_{max} of the SP band at 584 nm by applied potential. The λ_{max} value is extremely shifted to a shorter wavelength by 74 nm with increasing absorbance by potential in the negative direction to –1.6 V. On the contrary, the positive potential shift to 1.6 V gives the longer λ_{max} by 8 nm with decreasing absorbance. These spectral changes means that the direct charging of electrons in the gold core of Au_n-BFc is predominant comparing to the charge accumulation at the immobilized BFc units, because the λ_{max} shift in the negative potential is significantly larger than that in the positive direction where redox reaction of BFc moities occur.

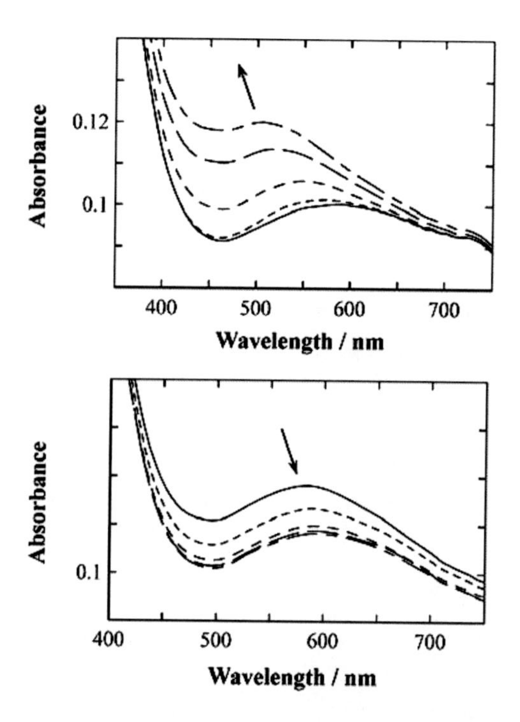

Figure 4. (Top) UV-Vis spectra of the Au_n-BFc film (d_{av} = 6.4 nm, θ_{BFc} = 4.6) on ITO at given potentials of 0, –0.4, –0.8, –1.2, and –1.6 V *vs.* Ag/Ag^+ in 0.1 M Bu_4NClO_4–CH_2Cl_2 in the negative direction, and (bottom) at given potentials of 0, 0.4, 0.8, 1.2, and 1.6 V *vs.* Ag/Ag^+ in 0.1 M Bu_4NClO_4–CH_2Cl_2 in the positive direction [21].

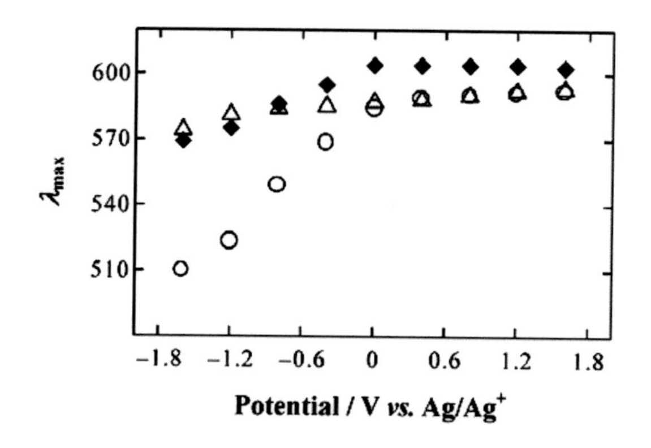

Figure 5. The plots of the absorption maximum (λ_{max}) of the Au_n-BFc film (d_{av} = 2.3 nm, θ_{BFc} = 15) (triangles), the Au_n-BFc film (d_{av} = 4.3 nm, θ_{BFc} = 11.3) (squares), and Au_n-BFc film (d_{av} = 6.4 nm, θ_{BFc} = 4.6) (circles) film *vs.* the applied potential [21].

Table 1. Spectroelectrochemical data of the Au_n-BFc films [21]

Particle Diameter / nm	C_{core} [a] / aF	ΔV [b] / V	Number of Injected Electrons to the Core [c]	λ_{init} [d] / nm	λ_{final} [d] / nm	$\lambda_{final.ideal}$ [e] / nm	$\Delta\lambda_{ex}$ [f] / nm	$\Delta\lambda_{ideal}$ [g] / nm
2.3	1.1	0.15	$11e^-$	587	574	577	13	10
4.3	2.8	0.057	$28e^-$	604	569	601	35	3
6.4	4.2	0.038	$42e^-$	584	510	582	74	2

a) The core capasitance: $C_{core} = 4\pi\varepsilon\varepsilon_0 r\,(r + d)/d$ where ε is the monolayer dielectric constant, r is the core radius, and d is the monolayer thickness. b) The potential spacing of consecutive single-electron transfer processes: $\Delta V = e/C_{core}$. c) Electron provision from 0 V to –1.6 V calculated by the value of ΔV. d) See Figure 20 at 0 V (λ_{init}) and –1.6 V (λ_{final}). e) The value calculated by $\lambda_{final.ideal} / \lambda_{init} = (N_{init} / N_{final})^{1/2}$ where N_{init} and N_{final} are the numbers of free electrons per metal core after and before the charging, respectively., referred to Table 1. f) $\Delta\lambda_{ex} = \lambda_{init} - \lambda_{final}$. g) $\Delta\lambda_{deal} = \lambda_{init} - \lambda_{final.ideal}$

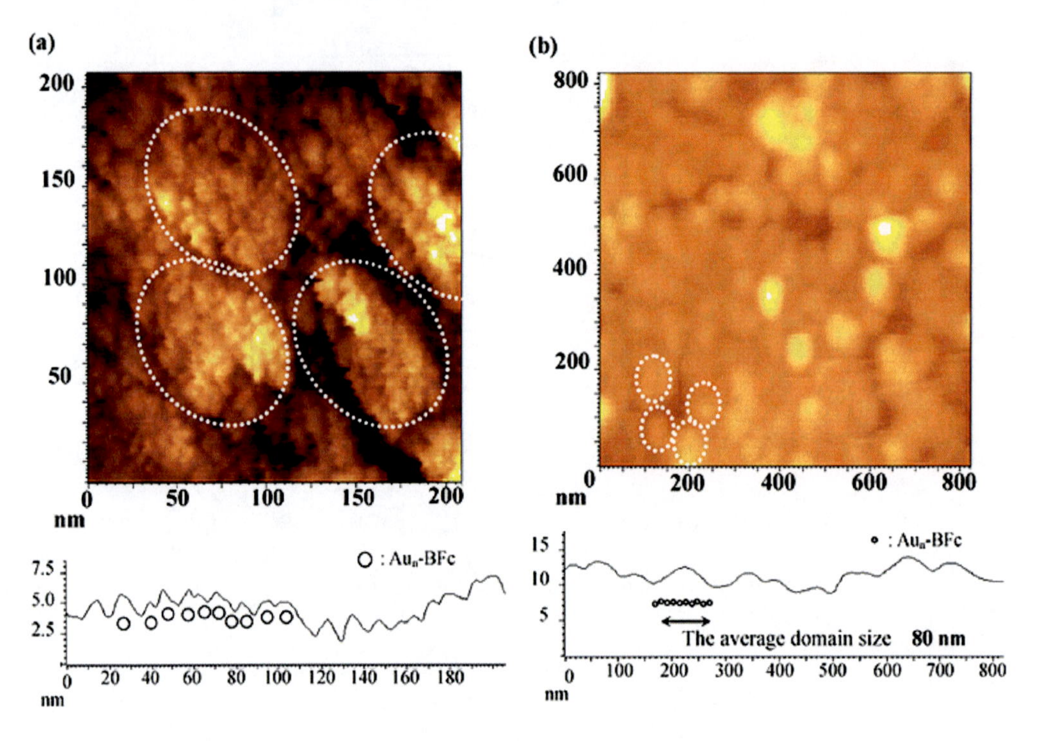

Figure 6. (a) STM and (b) AFM images of the Au_n-BFc film prepared in a solution of 5.2 μM Au_n-BFc (d_{av} = 2.3 nm, θ_{BFc} = 7.5) at HOPG with 5 cyclic scans in 0.1 M Bu_4NClO_4–CH_2Cl_2 at 100 mV/s between –0.3 and 0.9 V *vs.* Ag/Ag^+ in the positive direction, and the typical cross-sectional profile along the cross axis [14].

Au_n-BFc films with a different core diameter could be prepared with the same electrochemical method. Notably, the shift of λ_{max} is more conspicuous for the Au_n-BFc film with larger core size in the negative potential region (Figure 5). The electroscopic data are summarized in Table 1, suggesting that the particle-particle interaction works more effectively to the Au_n-BFc film with larger particle size, considering that the shifts grow with increasing the core size of particles. It is interesting that the order of ideal amount of the λ_{max} shift is opposite to the experimental values; that is, the theoretical calculation [20] predicts that the SP band position of the isolated particles are stimulated more easily with smaller core size by applied potential. These results point out that the collective SP band of the Au_n-BFc films might be shifted particularly with interparticle interaction different to a single particle and/or some other parameters, which affects the SP resonance, *e.g,*. electronic state of the surface ligands [22], chemical adsorbates on the surface, and the number of layers, should be considered for understanding these electro-optical shifts. The properties of nanoaprticles provoked by their interparticle interactions are also filled with many potentialities for nanotechnological science.

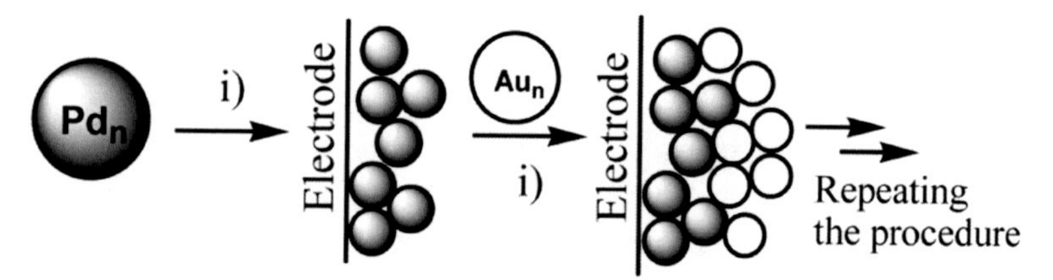

i) Electro-oxidation of biferrocene units on the particle surface in 0.1 M Bu_4NClO_4-CH_2Cl_2

Pdn, Aun : **M_n-BFc :**
M = Pd, Au

Scheme 3. Preparation of the Au_n-BFc and Pd_n-BFc composite film [27].

STM and AFM images of the Au_n-BFc film (d_{av} = 2.3 nm, θ_{BFc} = 7.5) are shown in Figure 6 [14]. The surface maintains a monolayer-level flatness seen in its cross sectional view, whereas it is apparent that domains of particles *ca.* 70–80 nm in diameter (encircled by a dotted line) are constructed. The AFM image of the same sample reveals this peculiar nanostructure of the Au_n-BFc film; round-shaped domains spreading on the whole surface shown in Figure 6b. About the Pd_n-BFc deposition, in Figure 7, tetragonal-like aggregation is observed similar to Au_n-AQ (see below). These specific morphological features testify that these characteristic assembling structures proceed in this deposition system.

The other multiple redox unit, an anthraquinone (AQ) thiol derivatice 1-(1,8-dithiaoctyl)anthracene-9,10-dione (AQS) [23] is next introduced onto a gold nanoparticle to produce Au_n-AQ (Scheme 1) [24, 25]. BFc and AQ are both multiple redox species, however, they are contrastive. Namely, the former undergoes two-step one electron 'oxidative' reaction of $BFc/BFc^{+}/BFc^{2+}$, while the latter receives two-step one electron 'reductive' reaction of $AQ/AQ^{-}/AQ^{2-}$. From the results noted above, the 2 electron oxidation process of BFc units of Au_n-BFc led to the formation of a uniform BFc-active gold nanoparticle film on an electrode. On the contrary, Au_n-AQ is found to be aggregated by 2 electron "reduction" of AQ sites (Figure 8), clarifying that metal nanoparticles functionalized with multiple-redox molecules could assemble by charge accumulation of redox species on a particle/solution interface. Note that the multiple-electron system is indispensable for this deposition phenomenon since it is not observed for particles with a single redox species such as ferrocene [8]. The detailed mechanism of Au_n-BFc is revealed by electrochemical quartz crystal microbalance (EQCM) [26] and summarized in Figure 9 that (1) Au_n-BFcs are apt to aggregate at the electrode/electrolyte interface and adsorb to the electrode by the formation of the BFc^{2+} state on the particle surface, (2) significant desorption of softly assembled Au_n-BFcs from the electrode occurs after BFc sites are returned to the neutral state by reversible reduction, and (3) a small portion of the strongly adsorbed flocks of Au_n-BFcs remains on the electrode [14]. By repeating the potential scans, the Au_n-BFc film is gradually fabricated by the remaining Au_n-BFc domains observed in the STM and AFM images.

About application experiments using this system, a combination of electro-oxidative deposition of Au_n-BFc and Pd_n-BFc forms a thin BFc-active composite film with a layered hybrid structure [27]. First, the electrodeposition of Pd_n-BFc (d_{av} = 3.8 nm) is performed by potential cyclic scans, which is followed by the electrodeposition of Au_n-BFc (d_{av} = 2.9 nm), thus forming the Pd_n-BFc/Au_n-BFc composite film (Scheme 3). The two-step hetero-electrodeposition procedure is repeated in order to increase the number of composite layers (Figure 10a). Figure 10b shows the XPS spectra of the composite film, displaying that the Au $4f_{5/2}$ (87.5 eV) and $4f_{7/2}$ (83.9 eV) peaks are not detected in the first layer of Pd_n-BFc, although they appear after deposition of the second layer of Au_n-BFc. In addition, the formation of the third layer of Pd_n-BFc shelters most of these Au peaks, suggesting that the prepared films build up an alternately layered structure of Pd_n-BFc and Au_n-BFc. It proves that enough electron transfer exists among heterogeneous interfaces of Pd_n-BFc and Au_n-BFc for developing the composite film in gaining the average 5 M_n-BFc (M = Pd, Au) layers [28] above the second layer.

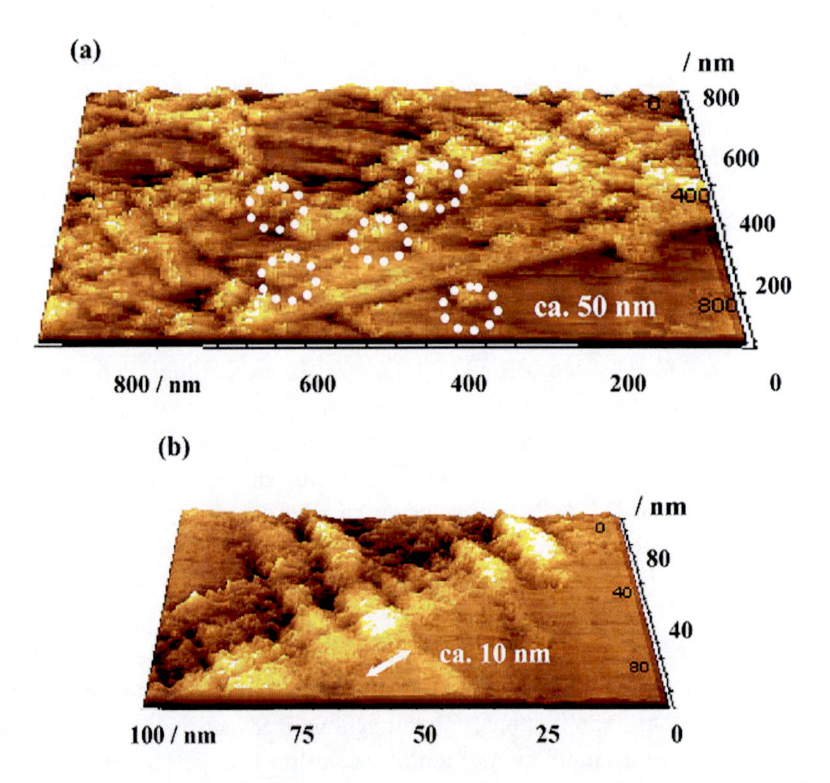

Figure 7. STM image of the Pd_n-BFc film (d_{av} = 3.8 nm, θ_{BFc} = 26.3)prepared in a solution of 1.9 μM Pd_n-BFc at HOPG with 60 cyclic scans in 0.1 M Bu_4NClO_4–CH_2Cl_2 at 100 mV/s between –0.3 and 0.9 V *vs.* Ag/Ag^+ in the area of (a) 800 × 800 nm^2 and (b) 100 × 100 nm^2 [19].

By utilizing scanning electrochemical microscopy (SECM) [29], dot-and line-shape lithographic assemblies of Au_n-BFc particles at a resolution of 50-100 nm were constructed [13]. When the substrate potential is kept at 1.1 V in Au_n-BFc solution where the electro-oxidative deposition of Au_n-BFc occurs, the electronic density due to the faradaic current tends to gather at the area of the substrate where the distance to the counter electrode is the closest. Figure 11 displays the CCD images of the lithographic deposition of Au_n-BFc by using this method under several conditions. We can clearly recognize the black precipitates deposited in a specific area. The simplest diagram is a "dot" described in Figure 11C, which is prepared by maintaining the deposition time for several seconds without moving the Pt counter electrode. The line drawing of the Au_n-BFc deposition is possible with *ca.* 100 μm width deposited by moving the counter electrode, displayed in Figure 11A, B, D, and E. Furthermore, it is expected that the variable parameters, such as the distance between a counter electrode and a substrate electrode, the scan speed, and the diameter of a counter electrode, could be applied to change the line width and the deposition thickness to produce more precisely controlled shapes. These assemblies systems would be also applicable by attempting to use STM and AFM instruments in place of SECM in order to achieve higher lithographic resolution.

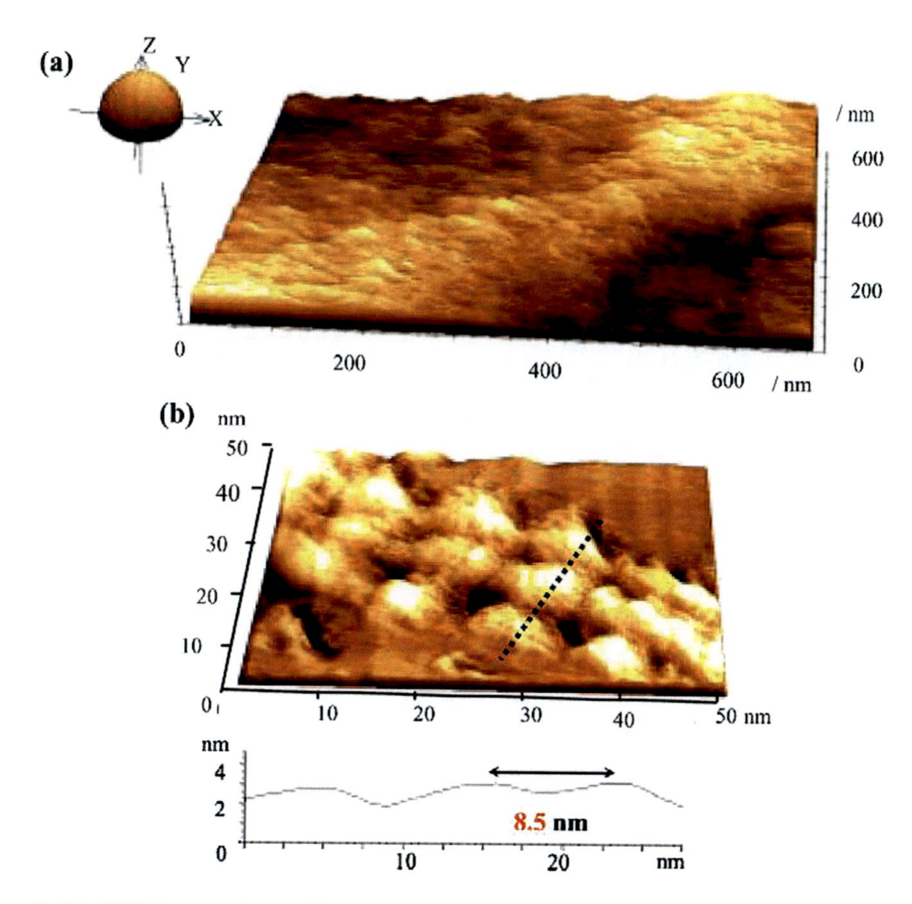

Figure 8. (a) STM image of a multilayer Au_n-AQ (d_{av} = 2.2 nm, θ_{AQ} = 26) film electrodeposited on HOPG in the area of 700 × 700 nm^2. (b) STM image of a submonolayer Au_n-AQ film on HOPG in the area of 50 × 50 nm^2 with the cross-sectional profile along the solid line [25].

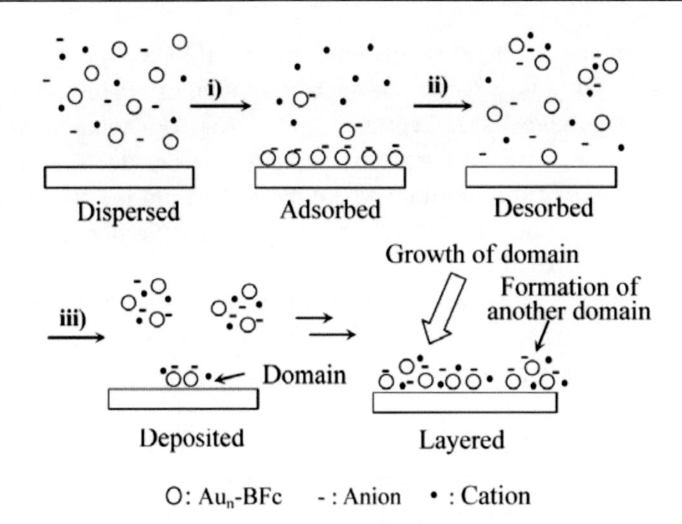

Figure 9. The illustration of the electrodeposition process in Au$_n$-BFc solution at the electrode interface. i) Two-electron oxidation of the BFc sites, ii) set back to the neutral state by two-electron reduction, and iii) repeating the potential sweep [14].

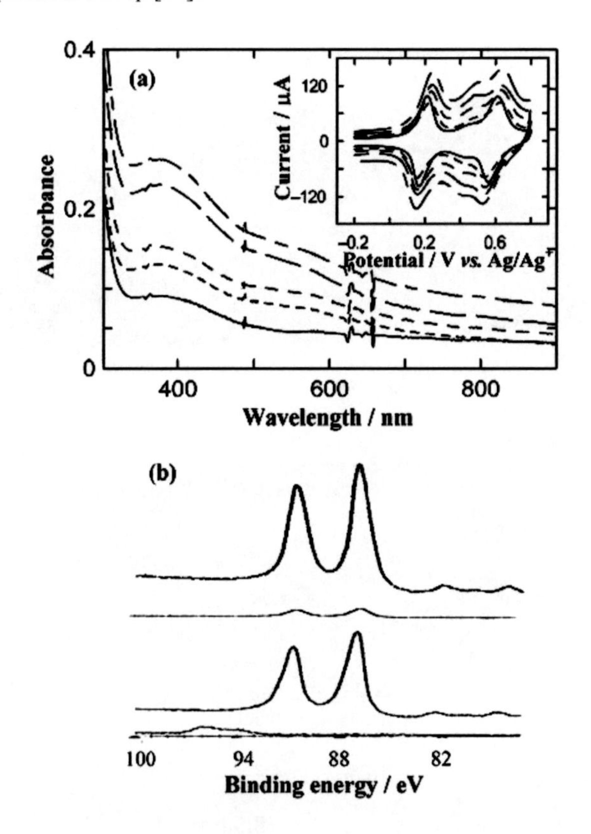

Figure 10. (a) The UV-Vis spectrum, (inset) the cyclic voltammogram, and (b) the XPS spectrum of Pd$_n$-BFc, Pd$_n$-BFc/Au$_n$-BFc, Pd$_n$-BFc/Au$_n$-BFc/Pd$_n$-BFc, [Pd$_n$-BFc/Au$_n$-BFc]$_2$, [Pd$_n$-BFc/Au$_n$-BFc]$_2$Pd$_n$-BFc films, as shown from the bottom of the figure to the top. The films were prepared with 25 cyclic potential scans between –0.3 V and 0.9 V *vs.* Ag/Ag$^+$ for each metal nanoparticle film in a solution of 3.2 μM Pd$_n$-BFc or Au$_n$-BFc, at ITO in 0.1 M Bu_4NClO_4–CH_2Cl_2 at 100 mV/s [27].

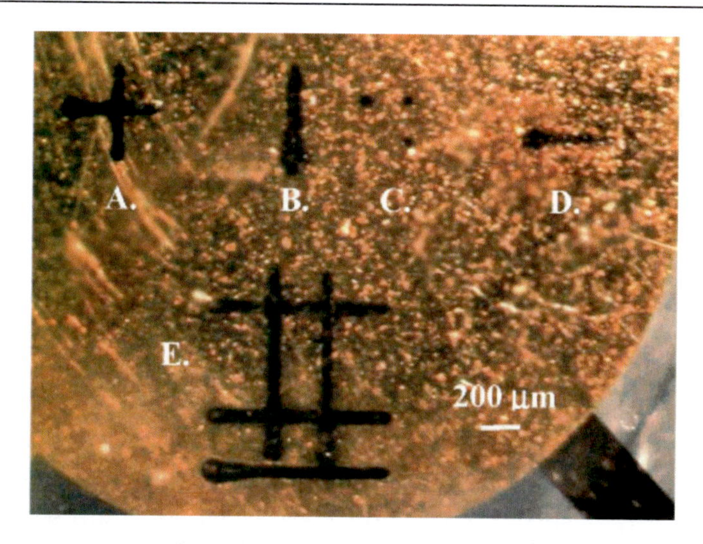

Figure 11. CCD image of the lithographic Au_n-BFc deposition on a gold electrode substrate prepared in 10 μM Au_n-BFc in 0.1 M Bu_4NClO_4-CH_2Cl_2 at the substrate potential 1.1 V by the SECM apparatus. Each line was deposited by moving the Pt counter electrode in a 7 μm diameter at 122.7 μm/s for 500 μm (A, B, and D) or 900 μm (E) with 10 scans. As for part D, the deposition time was maintained for 5 s without moving the counter electrode [18].

1.2. Discotic Liquid Crystalline Molecule Modified Gold Nanoparticles with Controlling their Assemble Structure

Many studies related to the construction of highly-ordered metal nanoparticle assemblies have been reported [2]. Among them, from the viewpoint of electronic energy transport, researchers should especially focus on the one-dimensional (1D) arrangement of nanoparticles. The decrease in interparticle spacing can elicit a dramatic increase in particle-particle electronic interaction, leading to efficient 1D tunneling current as a role for nanowires [30]. Existing methods for fabricating the 1D arrangement of particles include the use of electron beam lithography, templates and self-assembly of nanoparticles. The self-assembly method is the most facile way, by which an appropriate volume of nanoparticle solution is spread over a substrate, spontaneously constructing the thermodynamically stable arrangement of nanoparticles accompanied by solvent vaporization. The development of the 1D assembly is still hard to achieve, unfavourable to that of the hexagonal closed packed (hcp) structure because of the low-symmetric structural isotropy. In order to accomplish the 1D assembly of nanoparticles, their structural design should be refined at best. As a secret, the discotic liquid crystal molecule, hexaalkoxy-substituted triphenylene (TP) [31], is introduced into an organic ligand surrounding a gold nanoparticle surface (Au-TP) (Figure 12) [32, 33]. TP molecules are known to self-assemble into 1D columnar mesophase via π-π interaction [34]. Consequently, it is expected that TP molecules would be a strong trigger for the 1D assembly of metal cores by introducing TP units onto the metal nanoparticle surface.

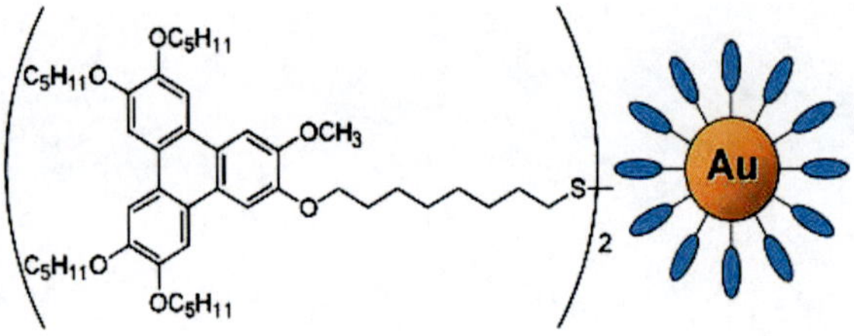

Figure 12. (Left) Structure of TPD. (Right) Schematic representation of Au-TP [32].

Au-TP with 2.4 nm diameter was synthesized in a homogeneous reduction of $HAuCl_4$ in DMF/water with TP thiol derivatice, [8-(3-methoxy-6, 7, 10, 11-tetrakis-pentyloxy-triphenylen-2-yloxy)-octanyl disulfide (TPD, Figure 12). Figure 13 shows the TEM images of the Au-TP assemblies prepared in a mixture of a polar (methanol) and a less polar (toluene) solvent. The solvent polarity is controlled by altering the mixed ratio of methanol to toluene (v/v), R_{MT}, from 0/1 to 3/1. The R_{MT} value is regarded as a kind of polarity index. When the solvent polarity is relatively low (R_{MT} = 0/1, 1/1), a clear hcp structure is formed with a particle spacing, d_{int}, of 4.2 and 3.2 nm, respectively (Figure 13a and 13b). The full molecular length of the TP ligand is *ca.* 2.4 nm calculated by MOPAC described in Figure 14d, where the C8 alkyl chain is 1.1 nm, the aromatic frame of the TP moiety is 0.6 nm, and the surrounding pentyloxy group of the TP moiety is 0.7 nm. These values indicate that the Au-TP in Figure 13a self-assemble by intercalating only the adjacent pentyloxy moiety around the TP ligands (Figure 14a). On the other hand, those in Figure 13b should accompany the *partial* π-π interaction of TP units (Figure 14b).

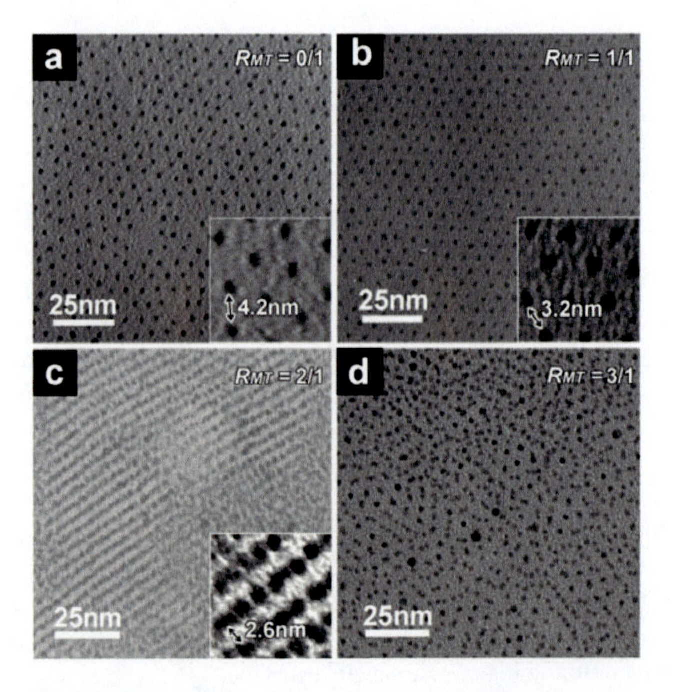

Figure 13. TEM images of Au-TP prepared in (a) R_{MT} =0/1, (b) 1/1, (c) 2/1, and (d) 3/1. The inset shows the enlarged image [32].

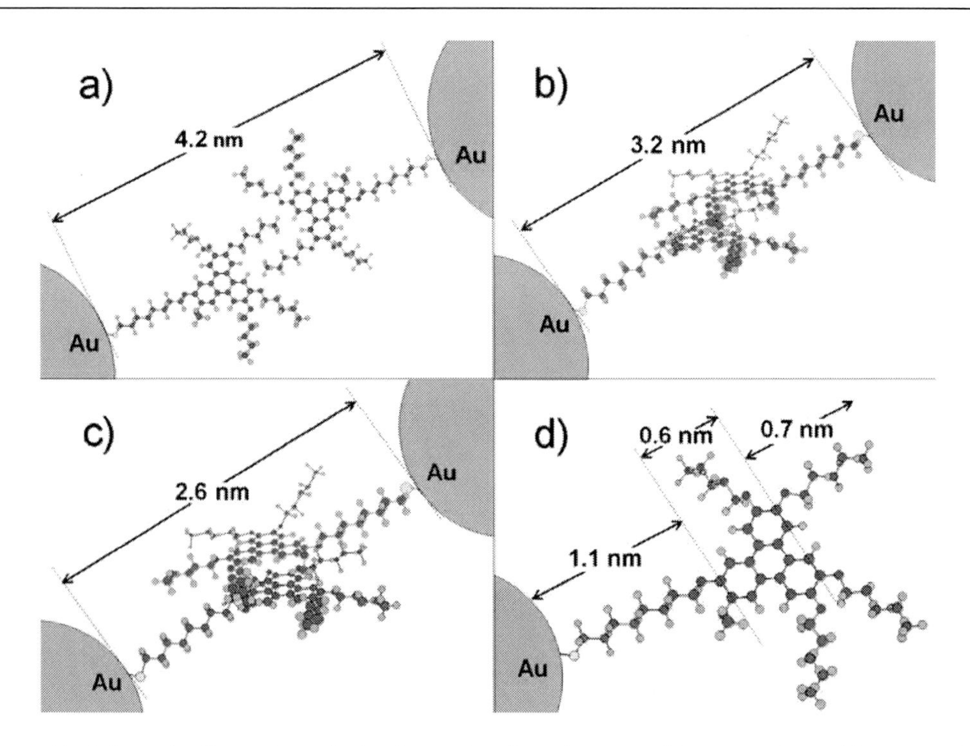

Figure 14. Schematic illustration of (a) adjacent TP ligands on a gold core surface with non-stacking (d_{int} = 4.2 nm), (b) with partial stacking (d_{int} = 3.2 nm), (c) with full stacking (d_{row} = 2.6 nm) of TP units, and (d) a single TP ligand on a gold core surface [32].

Interestingly, when the polarity of the solvent is further increased to R_{MT} = 2/1, the assembly structure is entirely converted from hcp to a 1D arrangement (Figure 13c). The interparticle spacing in a chain is *ca.* 0.7 nm. The row spacing, d_{row}, is 2.6 nm, a value which is consistent with the *full* stacking of the TP units interdigitated among the adjacent Au-TP particles (Figure 14c). The well-defined superstructure of Au-TP disappears with an increase of R_{MT} above 3/1 (Figure 13d). The whole results deduce that the strong interparticle π-π interaction of TP units in solution is inevitable for the 1D arrangement of Au-TPs, while that interaction is supportive, but not indispensable for the hcp ordering of Au-TPs.

The plausible mechanism is attributed to the specific nature of TP ligands that discotic liquid crystalline molecules tend to gather into a 1D column as mentioned at the beginning. When the *inter*particle π-π stacking is promoted in an appropriate mixed solvent, the small columnar phase of TP units will be constructed by the intercalation of TP ligands among Au-TPs. During the ordering process of the collected TP columnar phase, the partial area on a metal core surface, where the protection by TP ligands becomes weak due to the alkyl chain bending of TP ligands turning to the TP column direction, is inevitably produced. Accordingly, the van der Waals attraction among the metal cores is accelerated, causing the gradual aggregation of Au-TPs. At that time, the specific 1D nucleus of Au-TPs are compulsorily constructed on account of the surrounding TP column, resulting in the secondary growth of the 1D Au-TP arrays through directional motion of nanoparticles toward the ordered arrays, accompanied by the solvent evaporation process. The further investigation of the other parameters, *e.g.,* solvent molecular structure, boiling point, Au-TPs on their physical properties (core size, alkyl chain length of the TP ligand) produced more variable assemble structure of Au-TPs [33]. The general relations between the design of organic/inorganic metal nanoparticles and their self-assemble nature should been clarified in detail. From this point, this study is important which demonstrates definitely that both the appropriate physical architecture of nanoparticles and the parametric balance in self-assembly process are essential for the precise control of nanoparticle arrangement.

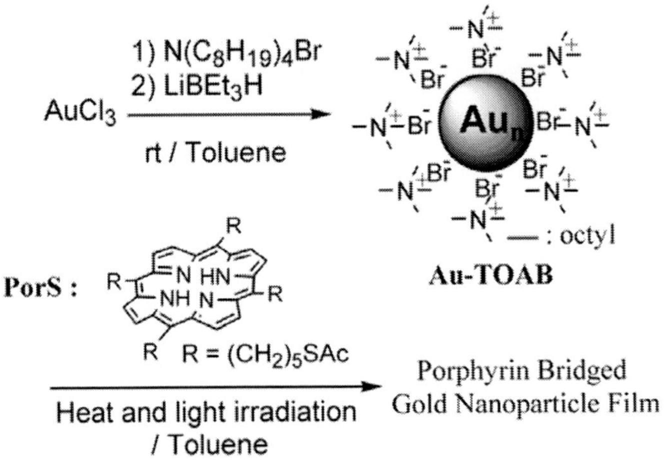

$AuCl_3$ → 1) $N(C_8H_{19})_4Br$ 2) $LiBEt_3H$, rt / Toluene → Au-TOAB (— : octyl)

PorS : R = $(CH_2)_5SAc$

→ Heat and light irradiation / Toluene → Porphyrin Bridged Gold Nanoparticle Film

Scheme 4. Preparation of the PorS-bridged Au nanoparticle film [35].

1.3. Formation of a Porphyrin-Gold Nanoparticle Network

Next, the combination between metal nanoparticles and photo-functional molecules is presented. The system is made by the connection between octylammonium bromide-covered gold nanoparticle (Au-TOAB) and a porphyrin derivative, meso-tetra (5-thioacetylpentyl)porphyrin with four alkyl chains terminated with a thioacetate group (PorS) (Scheme 4) [35]. It is imaginable that the heat activation can produce aggregation of Au-TOAB since the adsorption force of the ammonium salt to gold nanoparticle surface is weak. Actually, heating a toluene solution of Au-TOAB caused enormously grown coagulation, resulting in precipitation at the bottom of the vessel. In this study, an extra perturbation of 'light irradiation' is introduced in addition with heat and PorS in order to control the cohesion process of Au-TOAB to form a rigid network film on a specific part of the substrate successfully.

A toluene solution of Au-TOAB was prepared by reducing $AuCl_3$ by lithium triethylborohydride dissolved in a micelle solution of tetraoctylammonium bromide (TOAB) in degassed toluene under N_2 inert atmosphere based on the literature [4]. The core diameter of the prepared gold particles was 3.8 ± 0.8 nm, and the nanoparticle solution was stable for a few weeks in a condition. The porphyrin derivative, PorS, equipped with four adsorptive thiolate hands around rigid planar porphyrin moiety to generate chemically tight combination with Au-TOAB, since the adsorption strength of thiolate is larger than that of TOAB. Figure 15a displays the time-resolved UV-Vis spectra of the mixed solution of Au-TOAB and PorS in toluene with an incident light of "D_2 + W" lamps simultaneous heating at 50 °C. A formation of a purple thin film on the inner wall of the used UV quartz cell is observed, which corresponds to a gradual increase of the absorption in the range of 500 nm to 900 nm (Figure 15a).The nanoparticle film thus prepared exhibits a red-shifted surface plasmon band of assembled gold nanoparticles at 576 nm, and a sharp Soret band due to the porphyrin moiety at 420 nm, confirming that the film is composed of PorS-gold nanoparticles (Figure 15a inset).

Figure 15b shows the UV-Vis spectra of the films stuck on the inside of the cell prepared using a "Xe" lamp with different irradiation time. The spectra present two bands derived from Au-TOAB and PorS similarly to Figure 15a, and the intensity of the absorption bands grows with increases in irradiation time almost linearly (Figure 15b inset). The molar ratio of Au-TOPB to PorS is 1: 7.5, indicating that the film contains less porphyrin moieties than that prepared using the D_2 + W lamps, in addition the film formation process depends on the light source. Additional experiments were carried out for the detailed elucidation of this networking mechanism, indicating that the main factor for this system is 'the thermal energy of molecular vibration' in substrate (here quartz of the UV cell) by infrared light which generates a specifically heated area of the inner wall. As a result, the isolated and/or partially gathered nanoparticles are stuck on the irradiated spot connected by a thioacetate group of PorS, and the nanoparticle networks grow gradually in three-dimensional direction.

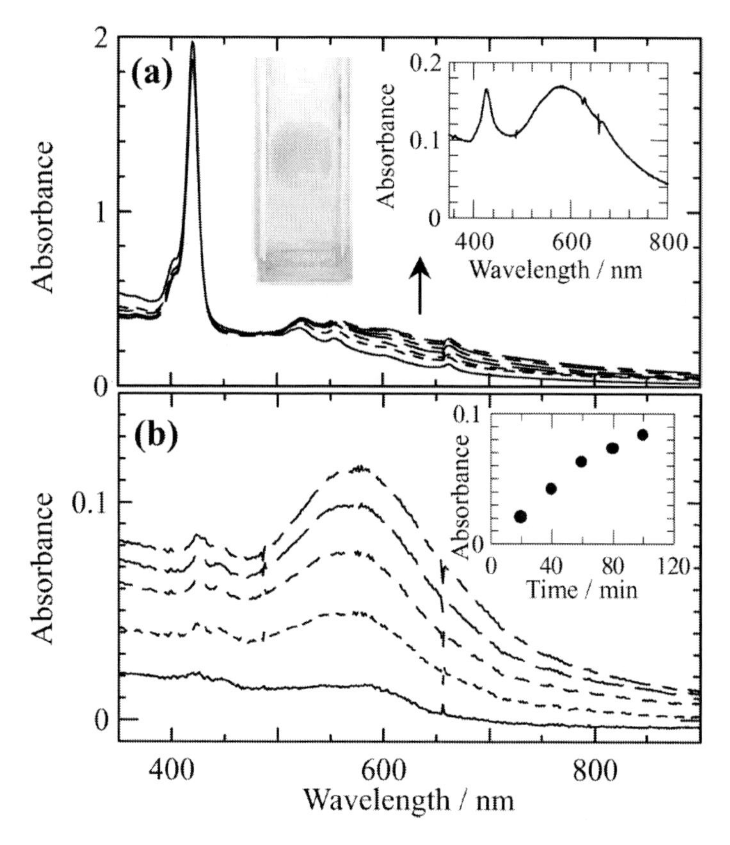

Figure 15. (a) UV-Vis spectral changes of the mixture of Au-TOAB and PorS with an incident light in every 20 min for 2 h with simultaneous heating at 50 °C from the bottom to the top. (Inset) The picture of the prepared film on a cell wall after the measurement and its UV-Vis spectrum. (b) The UV-Vis spectrum of the prepared PorS-Au nanoparticle film on the wall of the UV cell prepared with a Xe lamp from the bottom to the top for 20, 40, 60, 80, and 100 min with simultaneous heating at 50 °C in the mixture of Au-TOAB and PorS. (Inset) The correlation between the absorption intensity at 420 nm of the film and the irradiation time [35].

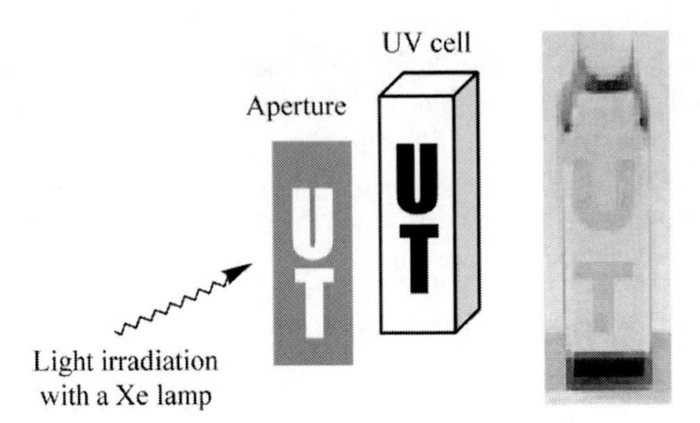

Figure 16. A photograph of porphyrin bridged gold nanoparticles on the inner wall of the UV cell and the illustration of the irradiation process.

This film formation is applicable for the larger gold nanoparticles in 5.6 nm core diameter as shown in Figure 16. It demonstrates that this technique would be possibly utilized for the film formation with appropriate metal nanoparticels in designed area. The photochemical properties of the film was also examined using the interdigitated array (IDA) electrode, which showed that photo-current response was switchable by light irradiation, involving the reliable increase of conductivity in 30 %. The excited electrons of the porphyrin units could induce the faster electron transfer through the film.

2. Physical Structure Control of Metal Nanoparticles as Advanced Catalysts

A large number of researchers up to now have pointed out that the properties of a metal nanoparticle are greatly dependent on its physical structure. For instance, the absorption originated to the surface plasmon oscillation of gold nanoparticles is dramatically red-shifted when the nanoparticle shape changes from spherical structure to a rod like one [37]. In common, the catalytic activity of metal nanoparticles is accelerated with decreasing their diameter, since the ratio of surface atoms increases exceedingly [38]. The focus of this section is to develop advanced catalysts by controlling the shape and size of metal nanoparticles. It is not so simple to arrange their physical structure precisely in solution, since among various synthetic parameters, at least solution temperature, time, an organic ligand, and a free energy change (ΔG) should be precisely controlled in order to obtain the specific shape of nanoparticles [39, 40]. The intelligent synthetic system which controls the accurate crystal growth of metal nanoparticles both kinetically and thermodynamically should be invented by selecting appropriate synthetic parameters. Under consideration of these factors, highly selective synthesis of cubic Pt nanoparticles is accomplished by means of an ionic additive in mixture, and the preparation of gold nanoparticle catalyst with high monodispersity in size is practicable for carbon nanotube synthesis.

2.1. Synthesis and Size Control of Platinum Nanocubes with High Selectivity of Shape

The catalytic properties of Pt nanoparticles are greatly expected for the organic synthesis, the decomposition of exhaust gas, and the polymer electrolyte fuel cell (PEFC). Fuel cells possess high energy conversion efficiency originated from the chemical energy of the combustion reaction between hydrogen and oxygen, without emitting environmental pollutants including CO_2. Among several kinds of fuel cells, PEFC would become the most common source in the general community because of its low operating temperature (below 100 °C). However, to enable its practical usage, the cost of the Pt nanoparticles used as an electrode catalyst in PEFCs must be reduced, which can be carried out by developing Pt nanoparticles with more efficient catalytic activity. Their catalytic activity depends on their shape and size, and cubic Pt nanoarticles are anticipated to exhibit higher catalytic activity compared to spherical particles, since cubic particles are composed of only (100) facets, with more defective and active sites accompanied by dissolution and surface reconstruction than (111) facets [41]. As for particle size, around 5 nm is reported to produce the most efficient catalytic performance [42].

Here, described is the novel preparation and size control of Pt nanocubes stabilized by polyacrylic acid sodium salt (PAA-Pt) with selectivity of max. 84% by the additive effect of NaI [43]. NaI was selected as an additive to control the particle shape because I^- anions are known to adsorb onto a Pt crystal surface most strongly among halogen anions, and was thought to act as a good controller for the growth rate between {100} and {111} faces of PAA-Pt. PAA-Pt could be prepared by the reduction of K_2PtCl_4 with H_2 bubbling in the presence of PAA and NaI in aqueous solution.

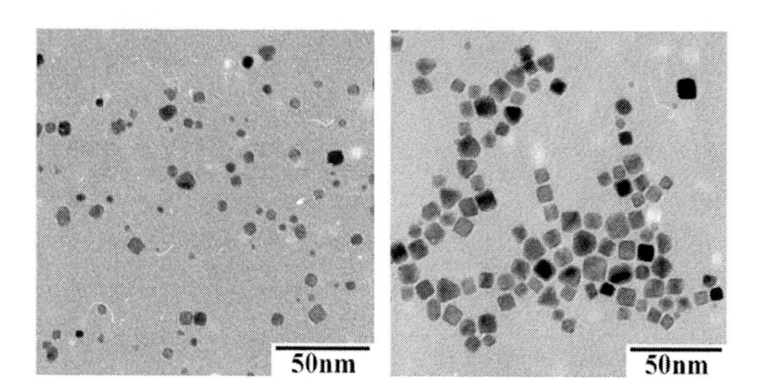

Figure 17. TEM image of PAA-Pt prepared under the reaction condition of NaI/K_2PtCl_4 = (right) 1.0 mM/1.0 mM and (left) 0 mM/1.0 mM.

Figure 17 shows the TEM image of the prepared PAA-Pt with and without NaI. From these images, it is obvious that the shape of PAA-Pt tends to become cubic when NaI coexists in the reaction mixture. The further investigation of the NaI effect on the physical structure (shape and size) of PAA-Pt was carried out by changing the molar ratio of NaI to K_2PtCl_4, as summarized in Figure 18. The ratio of the cubic nanoparticles to all of the produced particles (= cubic ratio) and the average particle size tend to increase with increasing NaI ratio and Pt concentration. This result indicates the formation mechanism of the cubic PAA-Pt: after a particle nucleus is generated at the first stage of the reaction, the particle crystal grows gradually, accompanied by the different growth ratios between the {111} and {100} faces under the additive effect of I^-.

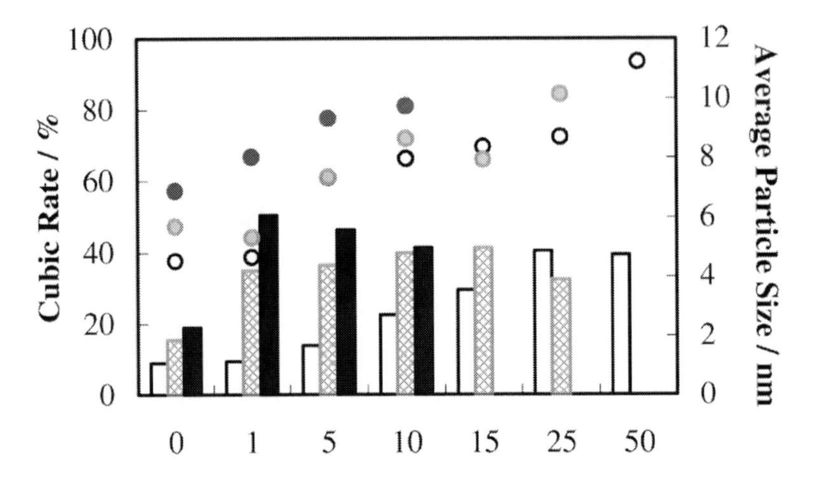

Figure 18. Effect of the NaI molar ratio to K_2PtCl_4 on the average particle size (circles) and the cubic selectivity (bars) of the prepared PAA-Pt. The concentration of K_2PtCl_4 was 0.1 (white), 0.5 (grey) and 1.0 mM (black), respectively [43].

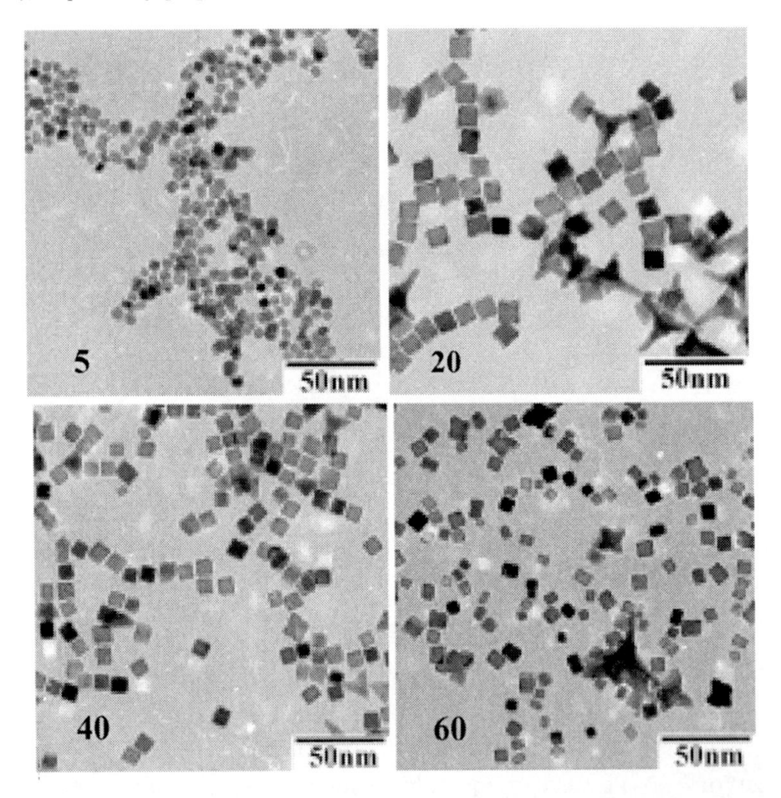

Figure 19. TEM image of PAA-Pt prepared with a different reaction temperature of which the value is noted in the images [43].

Figure 19 shows the TEM images of PAA-Pt synthesized at a different reaction temperature, *T*, from 5 to 60 ° C. They demonstrate that the cubic PAA-Pt can be preferentially obtained between 20 and 60 ° C with high selectivity *ca.* 70-80 %, in addition, the cubic size is controllable between 7.5−10.5 nm by changing *T*. The particle size first increased with rise of *T* until 20 ° C, and then gradually decreased. In this system, the temperature was the most effective parameter to control the cubic size of PAA-Pt, keeping the high shape selectivity. It was preliminarily confirmed that the catalytic performance of the prepared PAA-Pt for the O_2 reduction was higher than the ordinary spherical Pt nanoparticles. Our group would report their detailed catalytic capacity in near future.

2.2. Synthesis and Diameter Control of Multi-walled Carbon Nanotubes over Gold Nanoparticle Catalysts

The characteristics of carbon nanotubes (CNTs) [38, 44] have been energetically studied for decades including electrical behavior and mechanical strength. In general, nanoparticles of 3d transition metals (Fe, Co, Ni *etc.*) are used as catalysts for the synthesis of CNTs in chemical vapor deposition (CVD) [38]. Accordingly, the studies of "noble" metal nanoparticles for a CNT catalyst attract great interest. Bulk gold has long been regarded as inert and less active as a catalyst, a condition attributed to its completely filled 5d shell and the relatively high value of its first ionization; however, gold begins to show interesting catalytic behavior, *e.g,*. CO oxidation, hydrocarbon oxidation, and NO reduction when its size is reduced into a nanoscale regime [45, 46]. In this section, the preparation of multi-walled carbon nanotubes (MWNTs) over a gold nanoparticle catalyst is discussed [47, 48]. The adopted gold nanoparticles have been stabilized by dodecanethiol (DT), used as a catalyst after supported on SiO_2-Al_2O_3.

DT-stabilized gold nanoparticle with a core diameter of 2.1 ± 0.5 nm (Au-DT1) was synthesized on the basis of reducing Au (III) ions in toluene/water solution with DT and tetraoctyl ammonium bromide (TOAB) [49]. The diameter control of Au-DT1 was carried out by the calcinations process of the crude solid of Au-DT1 containing both TOAB and DT, obtained before the purification process of methanol. The size selection of gold nanoparticles could be accomplished by the solid heating method, where the crude Au-DT1 was heated in an electric furnace at fixed temperatures of 150, 180, and 200 °C to produce Au-DT2 (d_{av} = 3.7 nm) Au-DT3 (d_{av} = 5.3 nm) and Au-DT4 (d_{av} = 7.0 nm), respectively [49, 50]. Seen in the TEM images of the prepared samples (Figure 20), the particles are well-assembled due to their very low size distribution. By this step, the particle size could be conveniently adjusted almost linearly to the applied heating temperature.

Each size-controlled Au-DT was supported to produce the Au-DT/SiO_2-Al_2O_3 catalyst (0.1 wt. % gold), obtained after the calcination process for 1 h min at 200 °C. The particle core diameter in Au-DT(1-4)/SiO_2-Al_2O_3 was measured at 3.0 ± 0.7 nm, 5.0 ± 0.7 nm, 8.3 ± 1.0 nm, and 10.0 ± 1.1 nm, respectively. Comparing the values of the particle size before and after being supported, an aggregation among adjacent particles in the SiO_2-Al_2O_3 matrix occurred, leading to the *ca.* 1.5 times crystal growth in size during the calcination. However, the particles remained separately dispersed in SiO_2-Al_2O_3, and reflected the size order of the used Au-DT as a starting particle. This confirms that this catalyst preparation using a size-selective metal nanoparticle prepared in solution is effective for managing the precise particle size in catalysts. Apparently, the particle size distribution is narrower than that prepared by general preparation of nanoparticle catalysts, such as an impregnation method [51] in which metal ions dispersed into a support are reduced in a solid state.

Table 2. Dependence of a used catalyst on the outer diameter of the MWNTs formed by using C_2H_2 as a carbon source [47]

Used Nanoparticle for a Catalyst	Outer Diameter of MWNTs / nm				
	Reaction Temperature / °C				
	450	500	550	600	700
Au-DT1	-[a]	9.1	9.5	10.3	14.0
Au-DT2	-	-	13.0	14.3	16.5
Au-DT3	-	-	23.0	17.0	20.2
Au-DT4	-	-	-	19.3	23.3

a) Dash indicates that no MWNTs were identified.

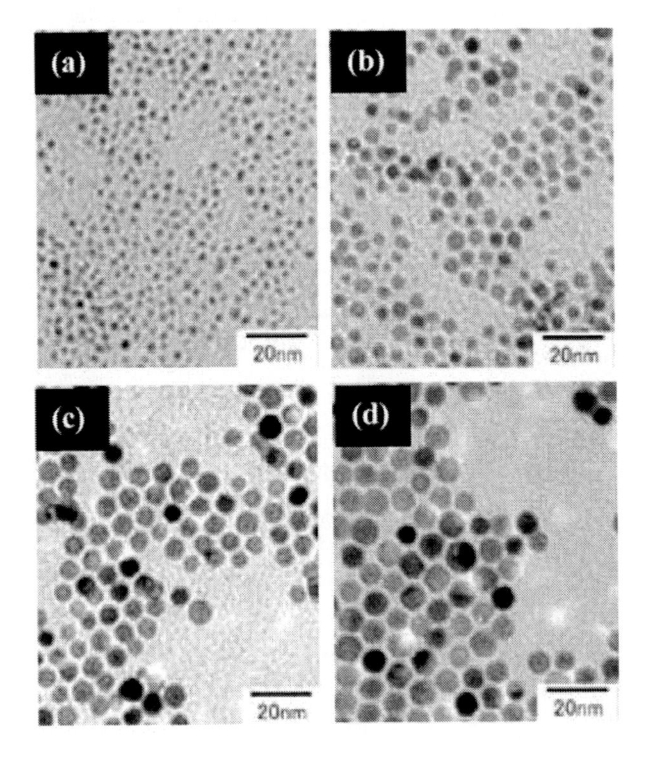

Figure 20. TEM image of (a) Au-DT1, (b) Au-DT2, (c) Au-DT3, and (d) Au-DT4 [47].

Figure 21 shows the relation between the size of Au-DT used as a catalyst and the outer diameter of MWNTs formed at 600 and 700 °C, which displays the discrepancy between C_2H_2 and C_2H_4 used as a carbon source. The diameter of MWNTs produced by C_2H_2 (MWNT-C_2H_2: Figure 22) increases almost proportionally with increasing the size of Au-DT. On the other hand, the reasonable dependence of the MWNT diameter synthesized by C_2H_4 (MWNT-C_2H_4) on Au-DT size cannot be recognized.

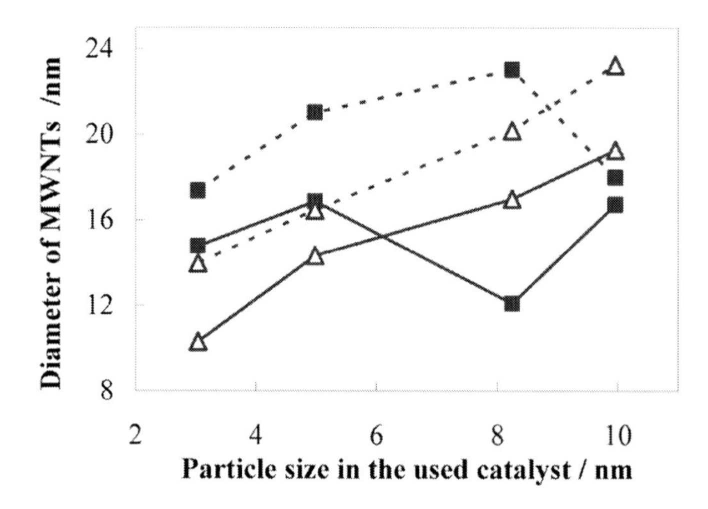

Figure 21. Relation between the particle size in the used catalyst and the outer diameter of MWNTs formed at 600 (solid line) and 700 (dotted line) °C by using C_2H_2 (triangles) and C_2H_4 (squares) as a carbon source [47].

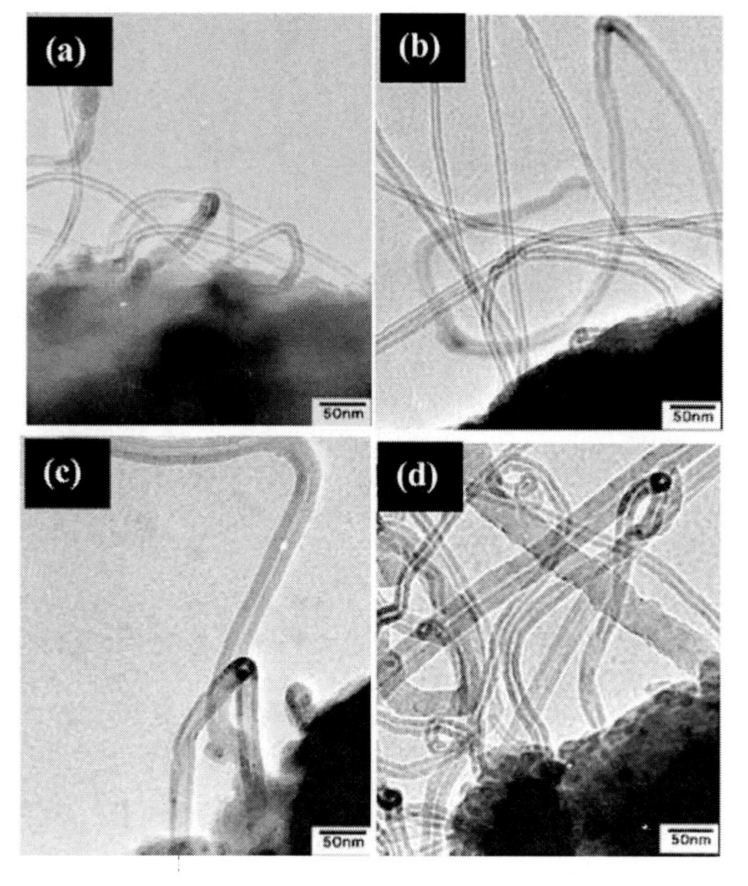

Figure 22. TEM image of MWNTs formed at 600 °C by using C_2H_2 as a carbon source over (a) Au-DT1, (b) Au-DT2, (c) Au-DT3, and (d) Au-DT4/SiO_2-Al_2O_3 [47].

The influence of the reaction temperature and the size of Au-DT on the formation of MWNTs was investigated by using C_2H_2 as a carbon source over Au-DT(1-4)/SiO_2-Al_2O_3, summarized in Table 2. It is obvious that the outer diameter of the formed MWNTs becomes larger with an increase of the size of Au-DT introduced as a catalyst. The diameter of MWNTs also tends to increase at higher reaction temperatures, which is explained by the considerations that (1) the size of the gold nanoparticle catalyst grows gradually by heat-induced aggregation during the reaction among adjacent particles, and (2) the catalytic reaction is accelerated at higher temperatures. Additionally, it should be mentioned that MWNTs can be generated at lower reaction temperatures with decreasing size of the Au-DTs, attributed to higher catalytic ability for smaller nanoparticles.

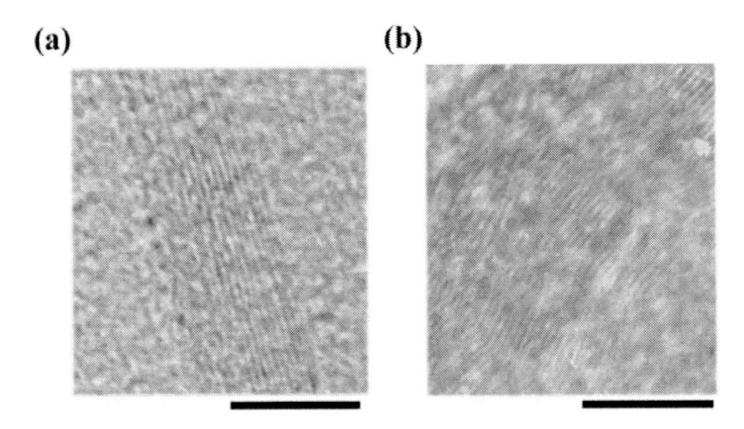

Figure 23. HRTEM images of MWNTS formed at formed at 600 °C by using (a) C_2H_2 and (b) C_2H_4 as a carbon source over Au-DT2/SiO_2-Al_2O_3 [47]. Scale bar : 5 nm.

To elucidate this phenomenon, the HRTEM images of MWNT-C_2H_2 and –C_2H_4 were taken, especially in the part of the tube inner wall (Figure 23); these demonstrate that consecutive carbon layers with interlayer ideal spacing of 0.38 nm [52] are observed in the image of MWNT-C_2H_2, while undulated layers covered by amorphous carbon are identified in that of MWNT-C_2H_4. Many unrevealed points remain concerning the mechanism of CNT formation, even over a general 3d metal catalyst; however, it is proposed that the tip-and base-growth models [53] can be also applied for a gold nanoparticle catalyst. After the nucleation of CNTs at the gold nanoparticle surface, the tube growth process begins. During the growth, the wall layers are created in a uniform axial direction by the introduction of C_2H_2 molecules, possessing a regular growth of the parameter values relating to the CNT diameter. On the other hand, the carbon wall grows more randomly into several axes with C_2H_4 molecules, leading to less dependence of the CNT configurations on the size of Au-DT.

3. Introduction of Metal Coordination Polymers into a Nanoparticle Core

The last section focused on metal coordination compounds as an inorganic core in place of metal. Metal coordination polymers, where some transition metal ions are connected in three-dimensional (3-D) nets by organic ligands, are fascinating organic-inorganic hybrid materials [54]. Variable geometries and connections have been created by selecting proper bridging ligands and metal ions in order to produce novel functional materials with useful wide-ranging properties [55, 56], *e.g.* electronic, magnetic, electrochromic, optical, catalytic, and adsorptive characteristics. The discussion about their specific natures has centered on bulk crystals (μm ~) for a long time. It has not been well clarified yet whether downsizing effects like metallic nanoparticles are observed for the properties of nanometer-sized metal coordination polymers, which we have named as metal coordination "nano"-polymers (MCNPs) [57-59]. For the first step, it is important to popularize the synthetic procedure of MCNPs in order to reveal the relations between the physical structure of MCNPs and their characters experimentally. As a metal coordination polymer, metal cyanometalates, M_1-CN-M_2 (M_1 = Fe, Ru, Co, Cr, Os, *etc.*, M_2 = Fe, Co, Ni, Mn, Cr, Eu, La, Sm, *etc.*), called Prussian blue (M_1 = M_2= Fe) [61] and its analogues, are selected [56]. Because of its convenient synthetic procedure compared to other metal coordination polymers, M_1-CN-M_2 is one of the most widely-explored metal coordination polymers with various outstanding features as a bulk crystal including, *e.g,*. magnetism (photo-induced magnetism), electrochromism, and conductivity, etc. M_1-CN-M_2 MCNPs would be a standard model case to demonstrate the possibility of MCNPs as novel nanomaterials.

3.1. Synthesis and Downsizing Effect of Metal Coordination Nano-polymers

The attempts to synthesize MCNPs have appeared by degrees in the past 5 years using, *e.g.*, the reverse micelles technique [62], the Langmuir-Bloddgett method [63], and template synthesis [64]. Among them, our group succeeded to isolate the Fe/Cr-CN-Co MCNPs (d_{av} = 5~7 nm) stabilized by organic shell of stearylamine (SA), utilizing reverse micelle solution of H_2O/cyclohexane/a nonionic surfactant, polyethylene glycol mono 4-nonylphenyl ether (NP-5: $HO(CH_2CH_2O)_nC_6H_4C_9H_{19}$) [57]. The complexation between $[Fe/Cr(CN)_6]^{3-}$ anions and Co^{2+} occurs in a water nano-droplet to generate nanometer-sized crystals of Fe/Cr-CN-Co MCNPs with narrow size distribution (Scheme 5). The merits of this synthesis deserving special mention is that the prepared Fe/Cr-CN-Co MCNPs can be obtained as an air-stable powdery form, and the excessive organic stabilizers of SA are sufficiently removed during the purification process. The key to this isolation is the stable and strongly coordinated protection by an amine group of SA onto the Co sites around the Fe/Cr-CN-Co inorganic core. The powder X-ray diffraction (XRD) patterns of the samples showed a typical face centered cubic (fcc) structure as the same as their bulk crystal, though the peaks were widely broadened due to the small crystal size [65]. In this synthesis, the metal elemental control among Fe, Cr and Co is also demonstrated by adjusting the ratio among the starting complexes in reaction mixture (Table 3). In particular, compound **1,** comprised of only Co and Fe, exhibits the typical cubic shape stacked together by self-assembly (Figure 24). The lattice constant, *a*, is in good accordance with that of a bulk form (*e.g.*, *a* = 10.06 and 10.54 Å for **1** and **5**, respectively). The value of *a* shifted almost linearly from **1** to **5** with increase of the Cr component, suggesting that the metal units were uniformly dispersed in the nano-polymers.

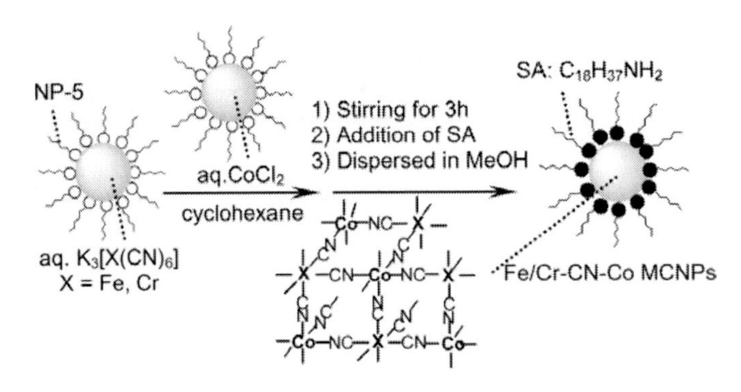

Scheme 5. Preparation of Fe/Cr-CN-Co MCNPs [57].

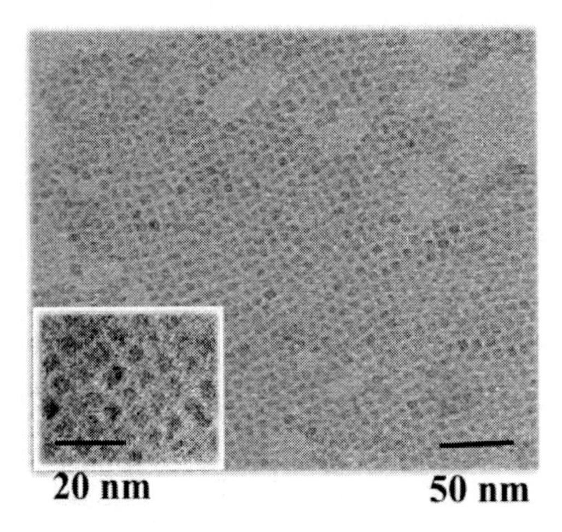

Figure 24. TEM image of the prepared Fe/Cr-CN-Co MCNPs (**1**).

Table 3. Formula of the prepared Fe/Cr-CN-Co MCNP [57]

Compound	Formula[a]
1	$K_{1.04}Co_{1.10}[Fe(CN)_6](SA)_{0.41}{\cdot}2.1H_2O$
2	$K_{0.81}Co_{1.10}[Fe(CN)_6]_{0.72}[Cr(CN)_6]_{0.28}(SA)_{0.53}{\cdot}3.0H_2O$
3	$K_{0.80}Co_{1.01}[Fe(CN)_6]_{0.62}[Cr(CN)_6]_{0.38}(SA)_{0.60}{\cdot}2.4H_2O$
4	$K_{0.82}Co_{1.30}[Fe(CN)_6]_{0.32}[Cr(CN)_6]_{0.68}(SA)_{0.46}{\cdot}3.1H_2O$
5	$K_{0.30}Co_{1.35}[Cr(CN)_6](SA)_{0.69}{\cdot}4.3H_2O$

a) Determined by the thermogravinetric analysis (TGA), the induced couple plasma (ICP) analysis, and the common CHN analysis method. The ratio of the hexacyano complexes is normalized as 1.

All of the samples were redispersed in generally less polar solvents, *e.g.* CH_2Cl_2, $CHCl_3$, THF, pyridine, while their bulk form [66, 67] was insoluble in any solvent). Figure 25 displays the THF solution of **1-5,** which indicates that the solution color is dependent on the elemental ratio in Fe/Cr-CN-Co MCNPs. This color variation reflects the electronic state of metal components in MCNPs, affecting their properties. For example, about their magnetism (Figure 26), field-cooled magnetization versus temperature curves of **1-5** show that compounds **4** and **5**, containing a high ratio of Cr units, exhibit ferromagnetism with spontaneous magnetization at the Curie temperature (*T*c). These findings indicate that the isolated Fe/Cr-CN-Co MCNPs would become a potential substance for the field of coordination chemistry in a nano-scale range.

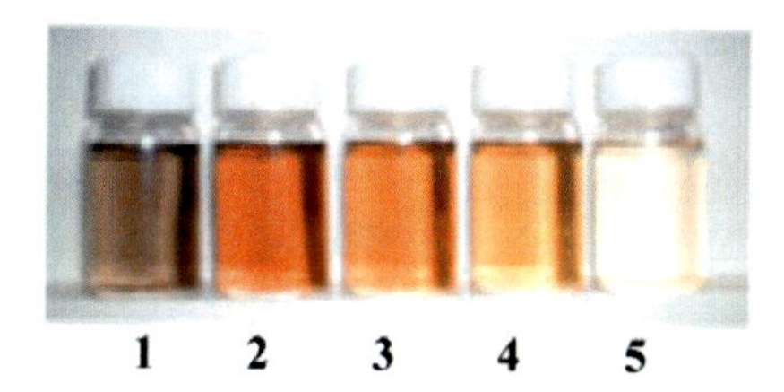

Figure 25. THF solution of the prepared Fe/Cr-CN-Co MCNPs (**1-5**). The number under the figure refers to that of the compound [57].

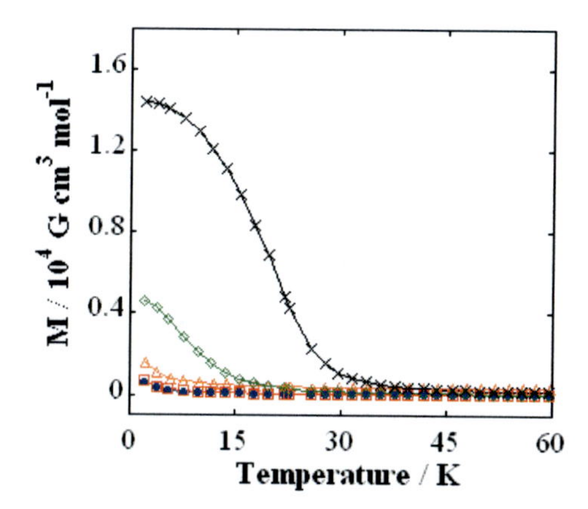

Figure 26. Field-cooled magnetization curves in an external magnetic field of 0.1 T of Fe/Cr-CN-Co MCNPs: blue (**1**), red (**2**), orange (**3**), green (**4**), and black (**5**) [57].

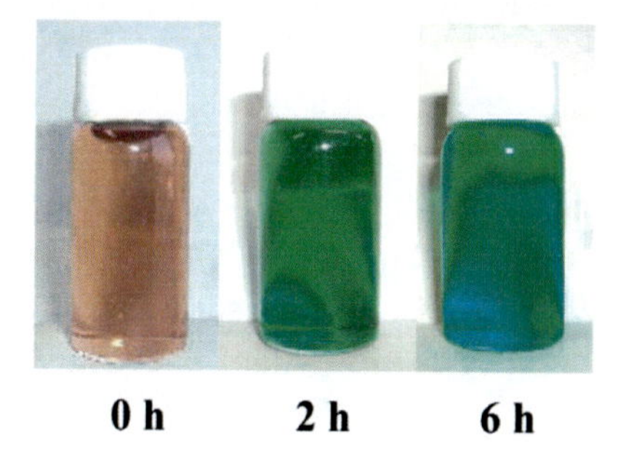

Figure 27. Photographs of Fe-CN-Co MCNPs prepared in the reverse micelle solution of CTAB during the reaction [59]. The reaction times are noted in the figures.

The other surfactants except for nonionic NP-5, of which the hydrophilic group is cationic or anionic, can be applied for the synthesis of MCNPs. Fe-CN-M_2 MCNPs (M_2 = Eu, La) were prepared in an anionic reverse micelle of sodium bis(2-ethylhexyl)sulfosuccinate (AOT) [68, 69]. The crystal size is *ca.* 50 ~ 800 nm, indicating that the size tends to be larger when rare earth metal ions are introduced into MCNPs. In another case, Fe-CN-Co MCNPs with the average diameter of *ca.* 3 nm in cationic reverse micelle solution of cetyltrimethyl ammonium halides [CTAX, X = B(bromide), C(chloride)] [59]. It is interesting to note that the material color dramatically changes from *red* to *green* with increasing reaction time when Fe-CN-Co MCNPs are synthesized in CTAB reverse micelles (Figure 27). This is attributed to the temporal evolution of elemental composition and crystal structure of Fe-CN-Co nanoparticle core, of which the phenomenon is not observed for its bulk crystal. Various measurement revealed that the driving force was the interaction between the Fe-CN-Co MCNP surface and cetyltrimethyl ammonium species in solution, since the ratio of surface area in MCNPs was exceedingly increased compared to bulk material. This is one piece of evidence that metal coordination polymers possess a characteristic size effect on their properties.

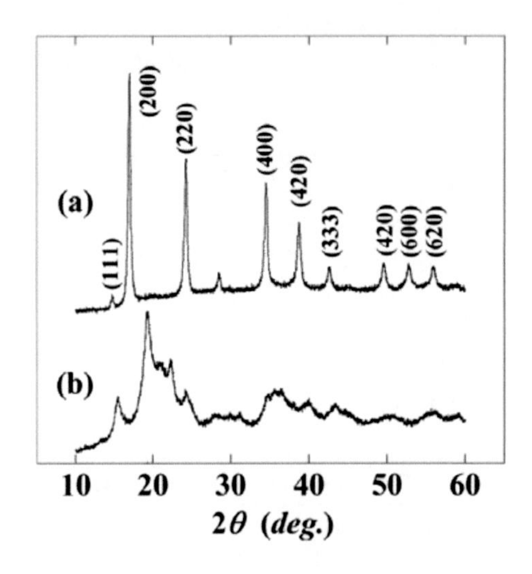

Figure 28. XRD patterns for (a) Cr-CN-Cu bulk crystals and (b) Cr-CN-Cu MCNPs [67].

In another experiment, these downsizing effects were also demonstrated by investigating the crystal structure and magnetic properties of Cr-CN-M_2 (M_2 = Co, Fe, Cu) MCNPs stabilized by pyridine derivatives [67]. For example, the XRD pattern of Cr-CN-Cu MCNPs with 15 nm diameter showed unexpected diffraction peaks, entirely different from the typical fcc structure of its bulk crystal originated from the complexation between a $[Cr(CN)_6]^{3-}$ anion and an octahedral Cu^{2+} cation (Figure 28). The crystal disorder of Cr-CN-Cu MCNPs is possibly attributed to some tetrahedral Cu sites which are provoked by the release of water molecules from Cr-CN-Cu crystals. This is also the surface effect of MCNPs derived from the interaction between "the organic species in solution" and "Cu sites in MCNPs". Numbers of unrevealed questions have been left about MCNPs. Further investigation is very significant not only for the coordination chemistry but also for the nanotechnology science.

3.2. Applicable Usage of Metal Coordination Nano-polymers

M_1-CN-M_2 MCNPs thus prepared become novel materials based on the original properties of their bulk crystal, *e.g.* nano-magnets, sensors (H_2O_2, glucose), electrochromism *etc.* For example, Fe-CN-Cr MCNPs can act as a nano-magnet material as noted above. In another case, the toluene dispersion of Fe-CN-Fe MCNPs [70] was spread over an ITO electrode to form a thin film of Fe-CN-Fe MCNPs. The film displays the reversible electrochromism between blue (+0.6 V *vs.* SCE) and coloress (–0.4 V *vs.* SCE) in 0.1 M potassium hydrogen phthalate [71].

Generally, the electrochromic film of Fe-CN-M_2 (M_2 = Fe, Co, Ni) are made by electrodeposition process of starting metal ions, while versatile electrochromic patterns can be fabricated more conventionally by utilizing the combination between Fe-CN-M_2 MCNP dispersion and appropriate film formation methods (*e.g.* ink jet printing, spincoating, LB, *etc.*)

From the other point, it is also perceptive to use MCNPs as a novel precursor of metal alloy or ceramic nano-materials [58, 60]. It is proposed that a novel synthetic approach to create metal alloy or ceramic nanoparticles by transforming MCNPs through the decomposition of the bridging organic ligand of MCNPs. Metal coordination polymers have a great advantage where the various metal elements are connected by an organic ligand with uniform composition [54]. In M_1-CN-M_2 MCNPs, M_1 and M_2 ions are bridged in advance by CN ligands with a regular elemental composition; therefore, the M_1M_2 alloy or M_1M_2 oxide nanoparticles obtained after the transformation reaction hold the uniform dispersion of the metal constituents with almost the same metal elemental ratio in the M_1-CN-M_2 MCNPs precursor. This method has the advantages, first, that the kinds of the metal elements can be accurately introduced before hand, and second, the electronic state of the nano-materials finally obtained is controllable just by altering the decomposition condition between reductive (H_2) and oxidative (O_2) atmosphere.

The transforming process of cobalt tetracyanoplatinate (Pt-CN-Co) MCNPs through a gas phase reduction in the H_2 atmosphere is demonstrated as an example. Pt-CN-Co MCNPs (Pt/Co = 0.9) with *ca.* 15 nm diameter were prepared by the complexation between $[Pt^{II}(CN)_4]^{2-}$ and Co^{2+} as the same as Fe/Cr-CN-Co MCNPs (Figure 29a). From thermal gravimetry analysis (TGA) of Pt-CN-Co MCNPs under H_2 atmosphere, the removal of the CN ligands occurred from 350 to 450 ° C with a weight loss of 24 %. The X-ray photoelectron spectroscopy (XPS) spectra of Pt-CN-Co MCNPs before and after the gas-phase reduction at 350 ° C show that when the reaction time is increased, the Pt4f signals at 73.1 eV and 76.4 eV, due to Pt^{II} sites in Pt-CN-Co MCNPs, decrease in intensity, while new signals appear at 71.2 eV and 74.5 eV assigned to the metal Pt^0 (Figure 30, left). The intensity of the CN stretching signal of Pt-CN-Co MCNPs in an IR spectrum was reduced with increasing reaction time. It is understood that the transformation reaction of Pt-CN-Co MCNPs gradually occurs by the reduction of Pt and Co ions accompanied by the removal of the bridging CN ligands. When reaction time is fixed as 3 h, the higher reaction temperature in the range of 350 to 450 ° C promotes the metallization reaction of Pt^{II} (Figure 30, right).

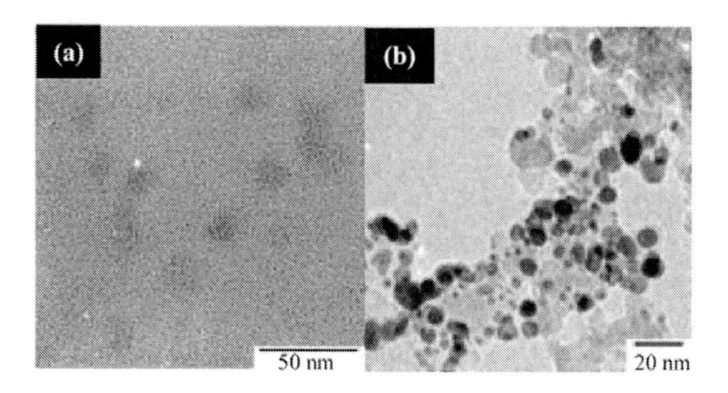

Figure 29. TEM image of Pt-CN-Co MCNPs (Pt/Co = 0.9) (a) before and (b) after the transforming reaction in H_2 atmosphere at 400 °C for 3h [58].

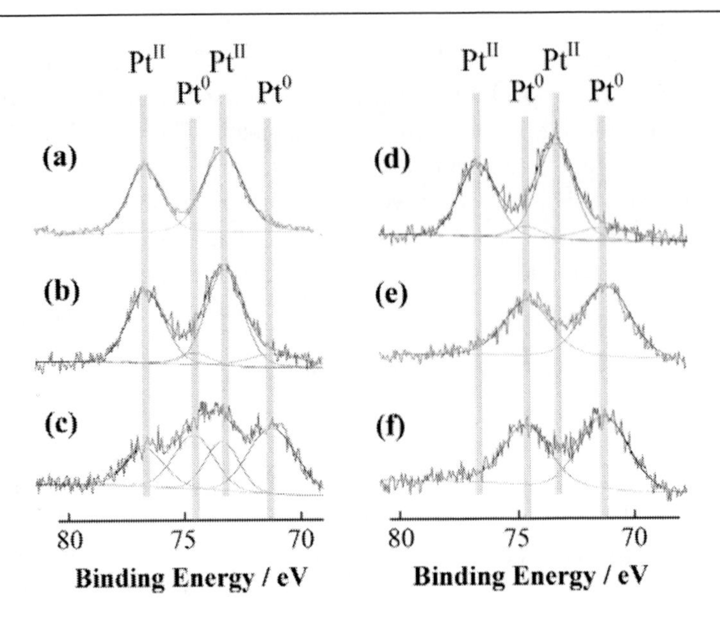

Figure 30. XPS curve in the Pt4f energy range of Pt-CN-Co MCNPs (Pt/Co = 0.9) before (a) and after the transformation reaction in H_2 atmosphere at 350 °C for 3h (b)(d), at 350 °C for 5h (c), at 400 °C for 3h (e), and at 450 °C for 3h (f) [58].

Table 4. Pt/Co ratio and the lattice constant of Pt-CN-Co MCNPs [58]

Pt/Co (cal.)[a]	Pt/Co (anal.)[b]	Lattice Constant, *a* /Å[c]
1	0.9	3.76
2	1.6	3.83
3	2.9	3.86
5	3.6	3.88

a) The mixed ratio of the starting complexes. b) Determined by EDX. c) Calculated from the XRD results.

The XRD pattern of Pt-CN-Co MCNPs after the transformation reaction at 400°C with 3h exhibits a fcc structure consistent with metallic PtCo. In the TEM image of Pt-CN-Co MCNPs after the reaction, most particles have a below 10 nm diameter (Figure 29b). The particle size is reduced compared to that of Pt-CN-Co MCNPs (Figure 29a). The size shrinkage happened by the removal of the bridging CN ligand between Pt and Co ions. The value of Pt/Co in Pt-CN-Co MCNPs is changeable by inserting $Pt^{IV}Cl_6^{2-}$ ions in place of Co^{2+} in reaction solution (Pt/Co = 0.9, 1.6 , 2.9, and 3.6). In the XRD patterns, the diffraction peaks are shifted to lower diffraction angles almost linearly, meaning that the lattice constant, *a*, calculated by the XRD patterns increases almost linearly with an increase of the Pt component in the compound, suggesting that the metal components are well dispersed in an PtCo alloy nanoparticle (Table 4). Using the same concept, GeFe nanoparticles with a $B8_2$ type structure (η phase) could be composed from iron tetraoxalategermanium MCNPs and exhibited the unique ferromagnetism [60]. The uniformity of metal units in MCNPs as a precursor plays an important role for the fabrication of nanomaterials.

Conclusion

The novel organic-shell inorganic-core hybrid nanoparticles were synthesized by introducing metal or metal coordination compounds as an inorganic core. In favor of withdrawing the advantage of the combination between organic substances and inorganic ones, the suitable design of their construction is inevitable by tuning the principal parameters of the unit selection, the physical structure and the constituent ratio between inorganic and organic units. In the first section, metal nanoparticles with added functionalities have been created by attaching molecules with functions including a multiple redox moiety (biferrocene and anthraquinone), a photo-active unit (porphyrin) and a liquid crystalline molecule (triphenylene) onto the surface core through coordination bonds by thiolate. External stimuli such as light and electric field were utilized to transform the properties of functionalized molecules. Second, the catalytic activities of noble metal nanoparticles were investigated from the points of controlling these activities by their physical structure of shape and size. PAA-protected Pt nanoparticles (PAA-Pt) with cubic shape were selectively prepared by addition of I^- anions to give a Pt nanocube surrounded by (100) facets with the high catalytic activity for O_2 reduction. The other experiment clarified the correlation between the diameter of an Au particle and that of the synthesized carbon nanotube (CNT) by changing the size of Au nanoparticle catalysts. The last section was allocated to the achievements with novel nanomaterials using metal coordination polymers as an inorganic core. The metal coordination nano-polymers (MCNPs) were stable in air, obtained as a solid powdery form by organic stabilizers, and they exhibited characteristic properties different from their bulk crystals derived from a size effect. Additionally, they could work as a novel intelligent precursor of an alloy or ceramic nanomaterial after the decomposition of organic ligands under an appropriate gas atmosphere. Nanoparticle science in wet process could progress further from the point of the development of the synthetic technique, the investigation of specialized natures, and the construction of their dimensional network, which could lead to practical widely-used nanoparticle materials in the future.

Acknowledgements

The described studies were partially supported by Grants-in-Aid for Precursory Research for Embryonic Science and Technology (PRESTO) from Japan Science and Technology Agency (JST) and for Scientific Research (No. 20710069) from the Ministry of Culture, Education, Sports, Science and Technology, Japan. I thank Prof. H. Nishihara in the University of Tokyo for the research contents.

References

[1] Schimid, G. *Chem. Rev.* 1992, 92, 1709-1727.
[2] Daniel, M.-C.; Astruc, D. *Chem. Rev.* 2004, 104, 293-346.
[3] Brust, M.; Walker, M.; Bethell, D.; Schiffrin, D. J.; Whyman, R. *J Chem Soc, Chem. Commun.* 1994, 801-802.

[4] Quiros, I.; Yamada, M.; Kubo, K.; Mizutani, J.; Kurihara, M.; Nishihara, H. *Langmuir,* 2002, 18, 1413-1418.

[5] Kiely, C. J.; Fink, J.; Brust, M.; Bethell, D.; Schiffrin, D. J. Nature 1998, 396, 444-446.

[6] Chen, S.; Ingram, R. S.; Hostetler, M. J.; Pietron, J. J.; Murray, R. W.; Schaaff, T. G.; Khoury, J. T.; Alvarez, M. M.; Whetten, R. L. *Science* 1998, 280, 2098-2101.

[7] Hostetler, M. J.; Green, S. J.; Stokes, J. J.; Murray, R. W. J. Am. Chem. Soc. 1996, 118, 4212-4213.

[8] Ingram, R. S.; Hostetler, M. J.; Murray, R. W. *J. Am. Chem. Soc.* 1997, 119, 9175-9178.

[9] Templeton, A. C.; Hostetler, M. J.; Warmoth, E. K.; Chen, S.; Hartshorn, C. M.; Krishnamurthy, V. M.; Forbes, M. D. E.; Murray, R. W. *J. Am. Chem. Soc.* 1998, 120, 4845-4849.

[10] Gittins, D.; Bethell, D.; Schiffrin, D. J.; Nishols, R. J. Nature 2000, 408, 67-69.

[11] Templeton, A. C.; Cliffel, D. E.; Murray, R. W. *J. Am. Chem. Soc.* 1999, 121, 7081-7089.

[12] Kamat, P. V.; Barazzouk, S.; Hotchndani, S. Angew. *Chem. Int. Ed.* 2002, 41, 2764-2767.

[13] Horikoshi, T.; Itoh, M.; Kurihara, M.; Kubo, K.; Nishiahra, H. *J. Electroanal. Chem.* 1999, 473, 113-116.

[14] Yamada, M.; Tadera, T.; Kubo, K.; Nishihara, H. *J. Phys. Chem. B* 2003, 107, 3703-3711.

[15] Yamada, M.; Nishihara, H. *Eur. Phys. J. D* 2003, 24, 257-260.

[16] Hostetler, M. J.; Wingate, J. E.; Zhong, C.-J.; Harris, J. E.; Vachet, R. W.; Clark, M. R.; Londono, J. D.; Green, S. J.; Stokes, J. J.; Wignall, G. D.; Glish, G. L.; Porter, M. D.; Evans, N. D.; Murray, R. W. *Langmuir* 1998,14, 17-30.

[17] Lin, X. M.; Sorensen, C. M.; Klabunde, K. *J. Chem. Mater.* 1999, 11, 198-202.

[18] Yamada, M.; Nishihara, H. *Langmuir* 2003, 19, 8050-8056.

[19] Yamada, M.; Quiros, I.; Mizutani, J.; Kubo, K.; Nishihara, H. *Phys.Chem. Chem. Phys.* 2001, 3, 3377-3381.

[20] Templeton, A. C.; Pietron, J. J.; Murray, R. W.; Mulvaney, P. *J. Phys. Chem. B.* 2000, 104, 564-570.

[21] Yamada, M.: Nishihara, H. *ChemPhysChem* 2004, 5, 555-559.

[22] Alvarez, M. M.; Khoury, J. T.; Schaaff, T. G.; Shafigullin, M. N.; Vezmar, I.; Whetten, R. L. *J. Phys. Chem.* B 1997, 101, 3706-3712.

[23] Nishiyama, K.; Kubo, A.; Taniguchi, I., Yamada, M.; Nishihara, H. *Electrochemistry* 2001, 69, 980-983.

[24] Yamada, M.; Kubo, K.; Nishihara, H. *Chem. Lett.* 1999, 1335-1336.

[25] Yamada, M.; Tadera, T.; Kubo, K.; Nishihara, H. *Langmuir* 2001, 17, 2363-2370.

[26] Sauerbrey, G. Z. *Phys.* 1959, 155, 206-222.

[27] Yamada, M.; Nishihara, H. *Chem. Commun.* 2002, 2578-2579.

[28] The coverage of Au_n-BFc on an electrode, Γ_{BFc}, can be estimated from the optical measurements by equation: $\Gamma_{BFc} = A/1000\varepsilon$ [mol cm^{-2}] where A is the absorbance at the SP band and ε is the molar extinction coefficient at the SP band for each particle (see ref. 22). The number of layers is estimated by assuming that the electrodeposited nanoparticles are hexagonaly packed in the film.

[29] Bard, A. J.; Denuault, G.; Lee, H.; Mandeler, D.; Wipf, D. O. *Acc. Chem. Res.* 1990, 23, 357-363.
[30] Berry, V.; Saraf, R. F. Angew. *Chem., Int. Ed.* 2005, 44, 6668-6611.
[31] D. Adam, D.; Schuhmacher, P.; Simmerer, J.; Haussling, L.; Siemensmeyer, K.; Etzbach, K. H.; Ringsdorf, H.; Haarer, D. *Nature* 1994, 371, 141-143.
[32] Yamada, M..; M.; Shen, Z.; Miyake, M. *Chem. Commun.* 2006, 2569-2571.
[33] Shen, Z.; Yamada, M.; Miyake, M. *J. Am. Chem. Soc.* 2007,129, 14271-14280.
[34] Markovitsi, D.; Germain, A.; Millie, P.; Lecuyer, P.; Gallos, L.; Argyrakis, P.; Bengs, H.; Ringsdorf, H. *J. Phys. Chem.* 1995, 99, 1005-1017.
[35] Yamada, M.; Kuzume, A.; Kurihara, M.; Kubo, K.; Nishihara, H. *Chem. Commun.* 2001, 2476-2477.
[36] Fink, J.; Kiely, C. J.; Bethell, D.; Schiffrin, D. *J. Chem. Mater.* 1998, 10, 922-926.
[37] Link, S.; El-Sayed, M. A. *J. Phys. Chem. B.* 1999, 103, 8410-8426.
[38] Saito, T.; Ohshima, S.; Xu, W.-C.; Ago, H.; Yumura, M.; Iijima, S. *J. Phys. Chem. B.* 2005, 109, 10647-10652.
[39] Burda, C.; Chen, X.; Narayanan, R.; El-Sayed, M. A. *Chem. Rev.* 2005, 105, 1025-1102.
[40] Shen, Z.; Yamada, M.; Miyake, M. *Chem. Commun.* 2007, 245-247.
[41] Narayanan, R.; El-Sayed, M. A. *J. Am. Chem. Soc.* 2004, 126, 7194-7195.
[42] Stonehart, P. *J. Appl. Electrochem.* 1992, 22, 995-1001.
[43] Yamada, M.; Kon, S.; Miyake, M. *Chem. Lett.* 2005, 34, 1050-1051.
[44] Tans, S. J.; Verschueren, A. R. M.; Dekker, C. Nature 1998, 393, 49–52.
[45] Okumura, M.; Akita, T.; Haruta, M. *Catal. Today* 2001, 74, 265–269.
[46] Jie, J.; Haraki, K.; Kondo, J. N.; Domen, K.; Tamaru, K. *J. Phys. Chem.* B 2000, 104, 11153-11156.
[47] Yamada, M.; Kawana, M.; Miyake, M. *Appl. Catal. A-Gen.* 2006, 302, 201-207.
[48] Lee, S. Y.; Yamada, M.; Miyake, M. *Carbon* 2005, 43, 2654-2663.
[49] T. Teranishi, S. Hasegawa, T. Shimizu, M. Miyake, *Adv. Mater.* 2001, 13, 1699-1701.
[50] Kyung-Hoon Kim, K. –H.; Yamada, M.; Park, D. –W.; Miyake, M. *Chem. Lett.* 2004, 33, 344-345.
[51] Merenyi, G.; Lind, J.; Shen, X.; Eriksen, T. E. *J. Phys. Chem.* 1990, 94, 748-752.
[52] Lyu, S. C.; Lee, T. J.; C.W. Yang, C. W.; Lee, C. *J. Chem. Commun.* 2003, 1404-1406.
[53] Amelinckx, S.; Zhang, X. B.; Bernaerts, D.; Zhang, X. F.; Ivanov, V.; Nagy, J. B. *Science* 1994, 265, 635-639.
[54] Veciana, J.; Rovira, C.; Amabilino, D. B. *Supramolecular Engineering of Synthetic Metallic Materials*; Kluwer Academic Publishers: Dordrecht,The Netherlands, 1998.
[55] Ferlay, S.; Mallah, T.; Ouahe´s, R.; Vellet, P.; Verdaguer, M. *Nature 1*995, 378, 701-703.
[56] Sato, O.; Hayami, S.; Einaga, Y.; Gu, Z.-Z. *Bull. Chem. Soc. Jpn.* 2003, 76, 443-470.
[57] Yamada, M.; Arai, M.; Kurihara, M.; Sakamoto, M.; Miyake, M. *J. Am. Chem. Soc.* 2004, 126, 9482-9483.
[58] Yamada, M.; Maesaka, M.; Kurihara, M.; Sakamoto, M.; Miyake, M. *Chem. Commun.* 2005, 4851-4853.
[59] Yamada, M.; Sato, T.; Miyake, M.; Kobayashi, Y. *J. Colloid. Intef. Sci.* 2007, 315, 369-375.
[60] Yamada, M.; Ohkawa, R.; Miyake, M. *IEEJ Trans.* 2007, 127, 1342-1343.

[61] Itaya, K.; Ataka, T.; Toshima, S. J. *Am. Chem. Soc.* 1982, 104, 4767-4772.
[62] Vaucher, S.; Fielden, J.; Li, M.; Dujardin, E.; Mann, S. *Nano Lett.* 2002, 2 225-229.
[63] Culp, J. T.; Park, J. H.; Benitez, I. O.; Huh, Y. D.; Meisel, Y. D.; Talham, D. R. *Chem. Mater.* 2003, 15, 3431-3436.
[64] D-Vera, J. M.; Colacio, E. *Inorg. Chem.* 2003, 42, 6983-6985.
[65] Cullity, B. D. *Elements of X-Ray Diffraction,* 2nd ed.; Addison-Wesley Publishers: Reading, MA, 1978.
[66] The bulk crystal of M_1-CN-M_2 is generally prepared by mixing diluted solution of M_1 cyano complexes and M_2 cations in water. See ref. 56.
[67] Arai, M.; Miyake, M.; Yamada, M. *J. Phys. Chem.* C. 2008, 112, 1953-1962.
[68] Kondo, N.; Nakajima, A.; Sasaki, Y.; Kurihara, M.; Yamada, M.; Miyake, M.; Mizukami M.; Sakamoto, M. *Chem. Lett.* 2006, 35, 1302-1303.
[69] Kondo, N.; Yokoyama, A.; Kurihara, M.; Sakamoto, M.; Yamada, M.; Miyake, M.; Ohsuna, T.; Aono, H.; Sadaoka, Y. *Chem. Lett.* 2004, 33, 1182-1183.
[70] Gotoh, A.; Uchida, H.; Ishizaki, M.; Satoh, T.; Kaga, S.; Okamoto, S.; Ohta, M.; Sakamoto, M.; Kawamoto, T.; Tanaka, H.; Tokumoto, M.; Hara, S.; Shiozaki, H.; Yamada, M.; Miyake, M.; Kurihara, M. *Nanotechnology* 2007, 18, 345609.
[71] Hara, S.; Tanaka, H.; Kawamoto, T.; Tokumoto, M.; Yamada, M.; Gotoh, A.; Uchida, H.; Kurihara, M.; Sakamoto, M. *Jpn. J. Appl. Phys.* 2007, 46, L945-L947.

In: Nanoparticles: New Research
Editor: Simone Luca Lombardi, pp. 143-165
ISBN: 978-1-60456-704-5

Chapter 4

HIGHLY STABILIZED GOLD NANOPARTICLES SYNTHESIZED AND MODIFIED BY PEG-*B*-POLYAMINE

Daisuke Miyamoto[1,2,3] and Yukio Nagasaki[1,2,3,4,5,6,*]
[1]Graduate School of Pure and Applied Sciences; Japan
[2]Tsukuba Research Center for Interdisciplinary Materials Science; Japan
[3]Center for The Tsukuba Advanced Research Alliance; Japan
[4]College of Engineering Sciences; Japan
[5]Master's School of Medical Sciences, Graduate School of Comprehensive Human Sciences, University of Tsukuba; Japan
[6]Satellite Laboratory, International Center for Materials Nanoarchitectonics, Japan National Institute of Materials Science, Japan

Introduction

Colloidal gold nanoparticles (GNPs) with a size range of several to hundreds of nanometers show a bright-pinkish color due to plasmon resonance and have been widely utilized for preparation of stained glasses for several hundred years. Recently, much attention has been given to gold nanoparticles (GNPs) because of their unique characteristics, and GNPs have been applied in many areas, such as biomedical materials science, optics and electronics [1].

In the 20th century, various methods for the preparation of gold colloids were reported and reviewed [2-8]. Most of these methods were based on the reduction of tetrachloroauric acid ($HAuCl_4$). The most popular one for a long time has been that using citrate reduction of $HAuCl_4$ in water [3]. It leads to GNPs of ca. 20 nm. To obtain GNPs of prechosen size between 16 and 147 nm via their controlled formation, a method was proposed where the ratio between the reducing and stabilizing agents was varied [4]. GNPs thus prepared are dispersed in the solution by ionic repulsion of adsorbed ions such as citric acid on their surface [9-12]. Under physiological condition, however, the ionically stabilized GNPs tend to

* E-mail address: nagasaki@nagalabo.jp. 1-1-1 Tennoudai, Tsukuba, Ibaraki 305-8573, Japan

aggregate because of the charge shielding. To improve the dispersion stability of ionically stabilized GNPs, much kind of polymer has been used as GNPs-stabilizer [11].

Although there are a variety of ways to achieve nanoparticle-polymer composites[13,14], most of them divided two different approaches. The first one consist of the *in situ* synthesis of the nanoparticles in the polymer matrix either by reduction of the metal salts dissolved in that matrix [15], or by evaporation of the metals on the heated polymer surface [16]. The second one involves polymerization of the matrix around the nanoparticles [17-19]. Blending of *pre*made GNPs into a *pre*synthesized polystyrene (PSt) bound to a thiol group was also reported [20]. The physical process involving mechanical crushing or pulverization of bulk metals and arc discharge yielded large nanoparticles with a wide size distribution, while nanoparticles prepared by reduction of metal salts tends to give small and a narrow size distribution. For this reduction process, several kinds of reducing agents such as $NaBH_4$ [21] and alchol [22] are utilized. Reduction of metal ions in the presence of the polymer is the most often chosen because the complexation of the metal cations by the ligand atoms of the polymer is crucial before reduction. In particular, it dramatically limits the particles size [23]. The most important role of the stabilizing polymer is to protect the nanoparticles from coagulation.

To apply GNPs for biomedical materials science, it is important to modify the surface of GNPs in order to improve their biocompatibility and colloidal stability under physiological conditions. Thiol chemistry has been widely used for the modification on gold surfaces with chemical species ranging from small molecules [24-26] to biomacromolecules [27-30]; synthetic polymers [31,32], especially, thiol-ended PEG (PEG-SH), have been widely used [33-38], and they have shown the inhibition of the adsorption of nonspecific molecules. The S-Au linkage is strong enough (47 kcal/mol) [39] for the construction of a dense PEG brush on the colloid surface, however, the oxidative stability of thiolate species was limited [40-42], as well as exchanged with thiolated compounds inside the body. Thiolate-modified surfaces are also damaged by exposure to light, high temperature and oxygen. In this way, the stability of PEGylated GNPs via a thiol-gold bond is not always sufficient to maintain its functionality for long-term, especially *in vivo*. For versatile applications, alternative methodology must be developed in terms of surface modification chemistry.

One of the best ways to solve these problems is usage of a water soluble polymer possessing multiple-coordination ability with metals, such as poly(2-vinylpyridine) (PVP) and poly(ethleneimine) (PEI), as a stabilizer for metal colloids [43,44]. Although the N-Au linkage is not as strong (6 kcal/mol) [66] as the linkage, these polymers were strongly interacted with gold surface by the multiple-coordination. For example, Antonietti and co-workers reported the preparation of PEG-stabilized gold colloids through the mixing of $AuCl_3$ with PEI-PEG graft copolymers [44]. The amino groups in the PEI segment of the PEI-PEG graft copolymer was considered to have the ability to reduce auric cations to form a gold colloid in the $AuCl_3^-$ incorporated PEI/PEG micelles, however, the surface modified with PEG/polyamine graft copolymers was not stable [45,46].

In this study, we used PEG/polyamine block copolymers and highly stabilized PEGylated GNPs (PEG-GNPs) was facilely synthesized by autoreduction of $HAuCl_4$ using poly(ethylene glycol)-*b*-poly(2-(*N,N*-dimethylamino)ethyl methacrylate) (PEG-*b*-PAMA) without addition of the further reduction reagents. The condition of GNP synthesis, such as pH and the block copolymer concentration, was investigated. Furthermore, a function of target molecule recognition was added to the PEG-GNPs and the ability of the target molecule recognition was investigated. In order to investigate the interaction between the GNPs surface and PEG-*b*-PAMA, the technique of the GNPs stabilization was applied for the surface modification of GNPs using PEG-*b*-PAMA and commercial GNPs. The GNP modification conditions, such as pH, block copolymer concentration and PAMA chain length, were investigated, and the dispersion stability of the PEG-GNPs under various conditions was evaluated in detail.

2.1. Synthesis of α-acetal-poly(ethylene glycol)-*block*-poly(2-(*N,N*-dimethylamino)ethyl methacrylate) [47]

3,3-Diethoxy-1-propanol (1.00 mmol) and potassium naphthalene (1.00 mmol) were added to 45 mL of tetrahydrofuran (THF) in a 100 mL flask with a three-way stopcock under a nitrogen atmosphere [48], then, ethylene oxide (EO) was added to the solution. After the mixture was stirred for 2 days at room temperature, the appropriate amount of AMA was added in the solution and stirred for 20 min at 0 ℃. The reaction was stopped by adding 5 mL of methanol. The solution was poured to excess cold 2-propanol to precipitate the polymer. The precipitate was dissolved in methanol and precipitated again in cold 2-propanol. This procedure was repeated again. The precipitate were dried *in vacuo* and freeze-dried with benzene. The recovered polymers were protonated by hydrochloric acid and purified by soxhlet extraction for 2 days using THF to remove unreacted acetal-PEG. The white precipitate was dissolved water and the polymer was obtained by freeze-dry. The chemical structure of PEG-*b*-PAMA [49] was confirmed by ^{1}H-NMR (EX-400, JEOL, Japan), (400 MHz, $CDCl_3$): δ (ppm) 0.85-1.23 (d, 3H, α-CH_3), 1.81-1.87 (d, 2H, -C(COO-)(CH_3)-CH_2-), 2.52-2.57 (d, 6H, -N(CH_3)$_2$), 3.00-3.06 (d, 2H, -CH_2-CH_2-N-) 3.27 (d, 3H, -O-CH_3), 3.45-3.74 (d, 4H, -CH_2-CH_2-O-), and 4.15-4.21 (d, 2H, -CH_2-CH_2-N-). The polymerization results are summarized in Table 1.

Table 1. Results of acetal-poly(ethylene glycol)-*block*-poly(2-(*N,N*-dimethylamino)ethyl methacrylate)

Run	Code	Mn ($\times 10^{-3}$)$^{a)}$	
		PEG	PAMA
1	6k/16k	5.7	15.7
2	5k/4k	4.9	3.8

a) Determined by SEC data. Column: TSK-Gel SUPER HZ3000, HZ2500; carrier: THF with 0.5 wt% triethylamine; detecter: RID, flow rate: 0.35 mL/min; Temperature: 40 ℃.

2.2. Facile Synthesis of PEGylated Gold Nanoparticles [47]

Five mL of acetal-PEG-*b*-PAMA (Run 1, Table 1) aqueous solution (6.8 mg/mL) was added in the 1 mL of $HAuCl_4 \bullet 4H_2O$ aqueous solution (2.5 mg/mL) and stirred at room temperature. This reaction was monitored by the measurement of UV spectrum. A schematic representation of PEG-*b*-PAMA synthesis and preparation of PEG-GNPs is shown in Scheme 1.

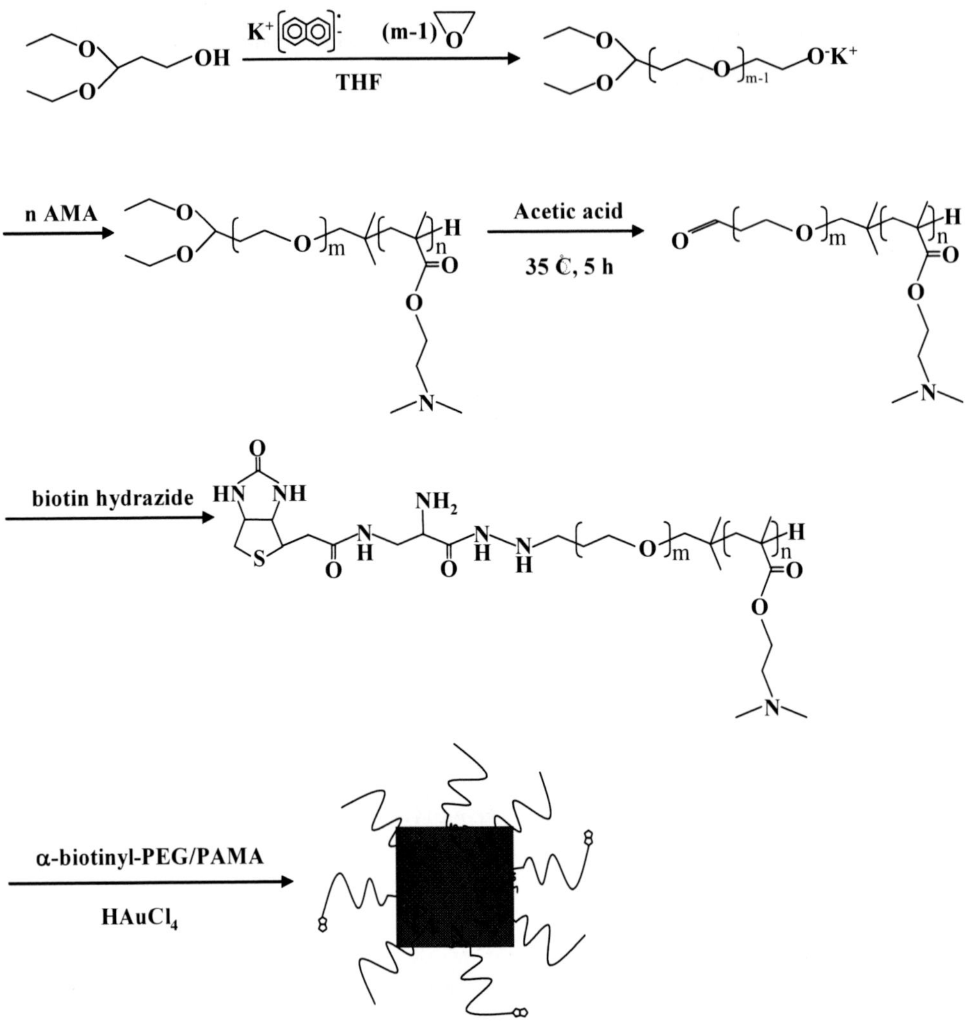

Scheme 1. Synthesis of PEG-*b*-PAMA and facile synthesis of PEGylated nanoparticles.

Since amino groups are known to show coordination ability on a metal surface, acetal-PEG-*b*-PAMA was expected to function as a stabilizer of GNPs. An unprecedented finding is that the block copolymer even facilitated autoreduction of the auric cations to obtain GNP without any additional reducing agent. The tetrachloroauric acid solution was changed from colorless to bright red due to the formation of GNPs when the block copolymer was added to the solution. The tertiary amino groups in the PAMA sgment play a crucial role in the reduction of auric cations as well as the anchoring of PEG on the surface of the GNPs. Figure 1 shows the change in the plasmon absorbance at 520 nm of tetrachloroaurate aqueous solution with different polymer addition. In the presence of acetal-PEG-OH at room temperature, no color change was observed. On the contrary, both PAMA homopolymer and acetal-PEG-*b*-PAMA induced a change in the color of the solution due to a gradual increase in the absorbance at 520 nm over several hours, indicating the reduction of auric cations to form GNPs. Nevertheless, the GNPs obtained by the PAMA homopolymer were rather unstable under a high salt concentration, and the characteristic absorption of GNPs was gradually reduced with time (data not shown). Worth nothing is that the reduction process finished within ca. 3 h in the case of acetal-PEG-*b*-PAMA at room temperature to form stable nanoparticles having a narrow size distribution. Figure 2 shows transmission electron microscope (TEM) image of (a) GNPs synthesized by acetal-PEG-*b*-PAMA, (b) GNP synthesized by PAMA homopolymer. The GNPs synthesized by acetal-PEG-*b*-PAMA showed homogeneous particles size. On the contrary, the GNPs synthesized by PAMA homopolymer showed large particle size distribution. The acetal-PEG-*b*-PAMA-stabilized GNPs have a substantially narrower distribution than those prepared by the PEI-PEG graft copolymers [44], PAMA homopolymer, and comparable to conventional method. This is obviously an advantage of using block copolymers as stabilizers because they can form a multimolecular micelle structure with a definite association number and well-defined core-shell architecture: an aurate-complex core segregated and surrounded by hydrophilic shell layer.

2.3. Characterization of Facilely Synthesized PEGylated Gold Nanoparticles [47]

GNPs were synthesized using the acetal-PEG-*b*-PAMA (Table 1, Run 2). The obtained GNPs were purified by centrifuge at 45,000 g, for 30 min at 20 ℃. The precipitated particles were resuspended in water or various buffer solutions with a wide range of pH. The centrifugations were repeated several times to remove free block copolymers. The average size and the distribution index (D_w/D_n) of the obtained GNPs were 11.8 nm and 1.01, respectively, which was determined by dynamic light-scattering (DLS) measurement.

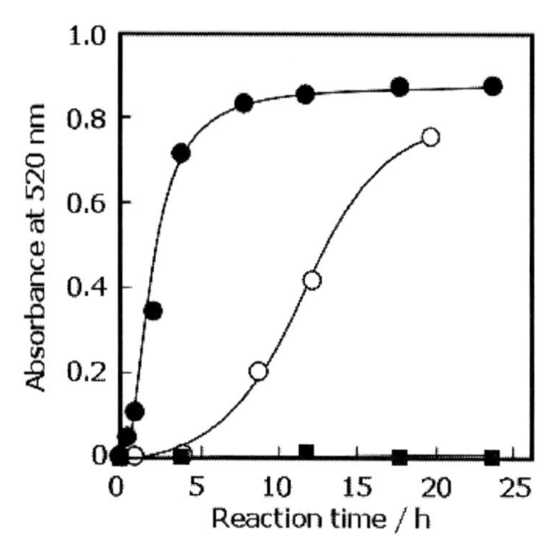

Figure 1. Change in the plasmon absorbance at 520nm of the tetrachloroaurate aqueous solution in the presence of R-acetal-PEG-OH: open circle, PAMA: closed triangle, and R-acetal-PEG-*b*-PAMA: closed square. Reproduced from ref. 61 by courtesy of publishers, American Chemical Society, U.S.A.

Commercially available GNPs, prepared by citrate, were dispersed in an aqueous solution through the ionic repulsion of the surface-adsorbed ions. Thus, the citrate-reduced gold nanoparticles have an appreciably negative ζ-potential (<-20 mV) at neutral pH, yet they undergo aggregation in the acidic region (pH <5) due to the neutralization of the surface negative charge. On the other hand, as can be seen in Figure 3, the ζ-potential of the GNPs prepared by the acetal-PEG-*b*-PAMA was close to zero regardless of the environmental pH. This result indicates that the PAMA segment in the block copolymer coordinates with the gold surface, allowing the PEG segment to be tethered from the surface into the aqueous exterior to shield the surface charge. The clear phase separation of the PEGlayer and the anchored PAMA segment should be an important factor in shielding the cationic charge of the PAMA segment.

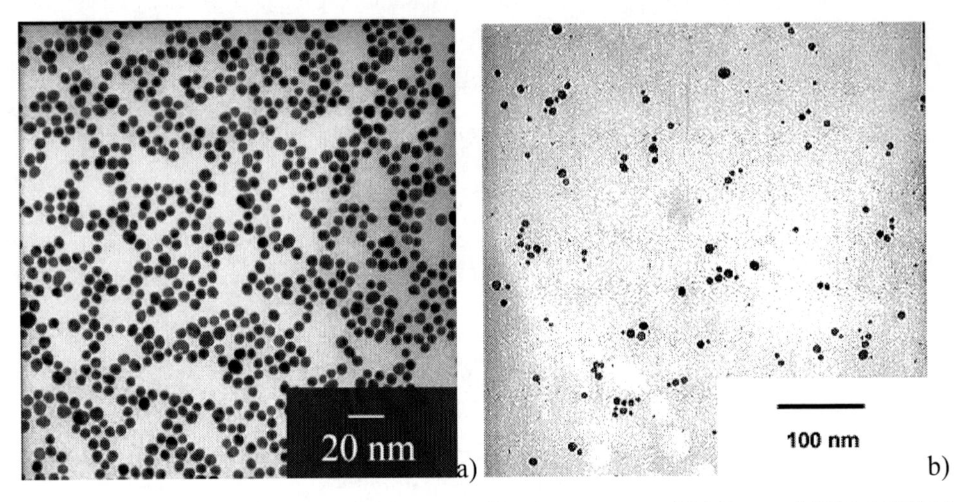

Figure 2. TEM image of gold nanoparticles prepared in the presence of (a) R-acetal-PEG-*b*-PAMA and (b) PAMA. Reproduced from ref. 61 by courtesy of publishers, American Chemical Society, U.S.A.

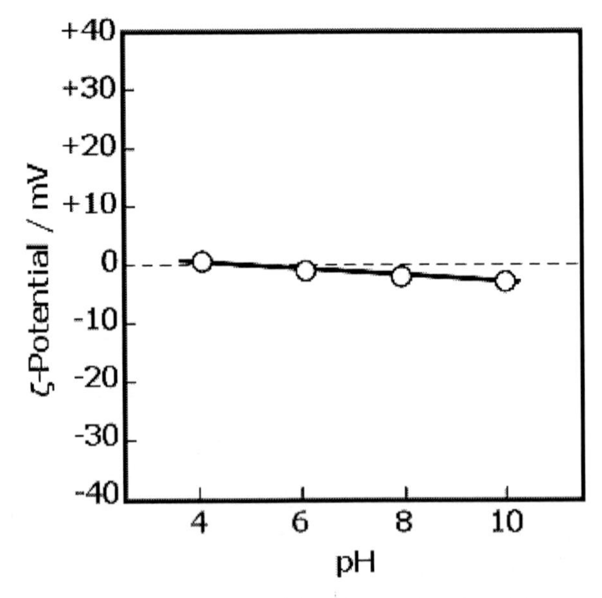

Figure 3. Change in ζ-potential of R-acetal-PEG-*b*-PAMA-anchored gold nanoparticles with pH ranging from 4 to 10 at 25 °C in 10mM NaCl solution. Reproduced from ref. 61 by courtesy of publishers, American Chemical Society, U.S.A.

Although the gold nanoparticles prepared by the acetal-PEG-*b*-PAMA showed an almost neutral surface charge, they maintain a fairly high dispersion character even after repeated centrifugation. To evaluate the stability of the block-copolymer-anchored nanoparticles, the time course of the change in the plasmon absorption was monitored under very high salt concentration (1.0 M NaCl). Note that the characteristic plasmon band at 520 nm is lowered, and an absorption band in a longer wavelength region appears when the GNPs are close to each other due to the electric dipole-dipole interaction between plasmons of neighboring particles[50-52]. Figure 4 shows the time course of plasmon absorbance at 520 nm as a function of time in 1.0 M NaCl. It should be noted that citrate stabilized GNPs immediately aggregate even under physiological salt concentration (0.15 M NaCl), while gold nanoparticles anchored with acetal-PEG-*b*-PAMA showed almost no decrease in the plasmon absorbance at 520 nm even under high ionic strength (1.0 M NaCl). The appreciable stability of the PEG-GNPs should be due to the steric stabilization effect of the hydrophilic PEG palisade tethered from the surface of the nanoparticles. The nanoparticles were also stable at a low pH region for several days even though the amino groups in acetal-PEG-*b*-PAMA should protonate completely under this condition.

The PEG density on the GNPs surface was also evaluated by thermal gravimetric analysis (TGA). GNPs was synthesized using the acetal-PEG-*b*-PAMA (Table 1, Run 2). After the solution of the PEG-GNPs was substituted with water through centrifugation and the PEG-GNPs was completely dried *in vacuo*, TGA was carried out. The average size was 8.16 nm, which was determined by TEM measurement.

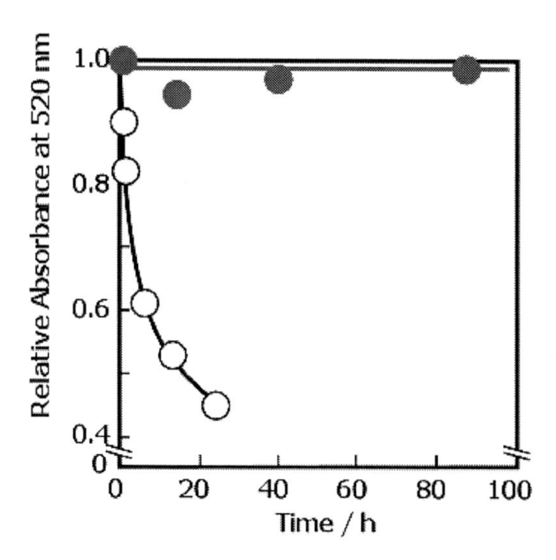

Figure 4. Change in the relative absorbance at 520nm of gold nanopaticles R-acetal-PEG-*b*-PAMA-anchored gold nanoparticles in 1.0 M NaCl solution: closed circle, citrate-stabilized gold nanoparticles in 0.1 M NaCl solution: closed square. Reproduced from ref. 61 by courtesy of publishers, American Chemical Society, U.S.A.

The amounts of PEG-*b*-PAMA on the GNPs are summarized in Table 2. PEG chain density on the GNP surface was 0.49 chains/nm^2. This PEG chain density was slightly low, comparing to the references [33, 53]. The reason was not clear. This result may be caused by the different chain length of terminal segment of PEG or the difference of the curvature of

GNPs. However, the PEG chain density was enough to disperse the PEG-GNPs under very high salt concentration (1.0 M NaCl).

2.4. Facile Synthesis of Biotinyl-PEGylated Gold Nanoparticles Using α-biotinyl-poly(ethylene glycol)-*block*-poly(2-(*N*,*N*-dimethylamino) ethyl methacrylate) [47]

One of the other important characteristics of these acetal-PEG-*b*-PAMA-anchored nanoparticles is the ligand installing site at the distal end of the PEG chain. The PAMA segment in the block copolymer is anchored on the surface of the GNPs, while the PEG segments extend into the aqueous exterior from the surface of the GNPs.

Table 2. Experimental characteristics of PEGylated gold nanoparticles

Mn of PEG derivatives	d_{core} (nm)	PEG chain density (chains/nm^2)	reference
5700/2800 (PEG/PAMA)	8.16	0.49	
5000 (MeO-PEG-SH)	2.8	2.86	33)
2100 (MeO-PEG-SH)	3.6	2.00	50)

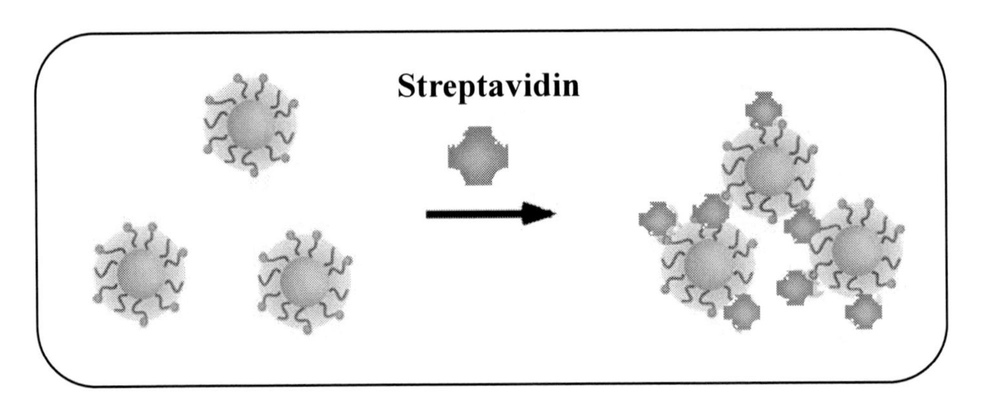

Scheme 2. Aggregation of biotinyl-PEG coated gold nanoparticles by streptavidin recognition.

To obtain aldehyde-PEG-*b*-PAMA, 1.0 g of acetal-PEG-*b*-PAMA was dissolved in acetic acid aqueous solution (10 vol %) and the mixture was stirred for 5 h at 35 ℃. After the mixture was neutralization by adding 10 mol/L NaOH aqueous solution, the solution was dialyzed against excess water. After the pH of the solution was adjusted to 9 using NaOH aqueous solution, the aldehyde-PEG-*b*-PAMA was obtained by freeze-drying. One point seven mg of biotin hydrazide was added in the 10 mL of aldehyde-PEG-*b*-PAMA aqueous solution (10 mg/mL) and stirred for 5 h. Then, 1.5 mg of sodium cyanoborohydride was added in the solution three times per 15 min. The solution was dialyzed against excess water. The α-biotinyl-PEG-*b*-PAMA was obtained by freeze-drying. The chemical structure of biotyl-PEG-*b*-PAMA was confirmed by ^{1}H-NMR (EX-400, JEOL, Japan), (400 MHz, $CDCl_3$): δ (ppm) 0.85-1.23 (d, 3H, α-CH_3), 1.81-1.87 (d, 2H, -C(COO-)(CH_3)-CH_2-), 2.52-2.57 (d, 6H, -N(CH_3)$_2$), 3.00-3.06 (d, 2H, -CH_2-CH_2-N-) 3.27 (d, 3H, -O-CH_3), 3.45-3.74 (d,

4H, -*CH*$_2$-*CH*$_2$-O-), 4.15-4.21 (d, 2H, -*CH*$_2$-CH$_2$-N-), 4.25-4.35 4.35-4.45 (d, 2H, -*CH*-NH-), 6.15-6.25 (d, 2H, -CH-*NH*-C) and 7.50-7.60 (d, 1H, -C(=O)-*NH*-NH-CH$_2$-). The biotinyl-PEG-GNPs was prepared by the same method of chapter 2.3.

Using biotin-PEG-GNPs thus prepared, molecular recognition experiment were carried out. Streptavidin, which forms an extremely stable complex with biotin (K=10^{-15}) [54-56], was used to investigate the functionality of the biotins installed on the periphery of the PEG-*b*-PAMA-anchored GNPs. Figure 5 shows a typical time course of the plasmon absorption change after the addition of 100 μg/mL streptavidin. The maximum wavelength of the absorption showed a bathochromic shift with time, indicating an aggregation of the GNPs via streptavidinbiotin interaction. At the same time, the absorption at the longer wavelength region gradually increased. Actually, the colloid solution underwent a visual color change gradually from pink to purple. On the other hand, no absorption change was observed when bovine serum albumin as a control was added to the biotinyl-PEG-GNPs solution. That is, the aggregation of biotinyl-PEG- GNPs induced by streptavidin is indeed due to the specific biotin/streptavidin interaction. A schematic representation of the molecular recognition of the biotinyl-PEG- GNPs is shown in Scheme 2.

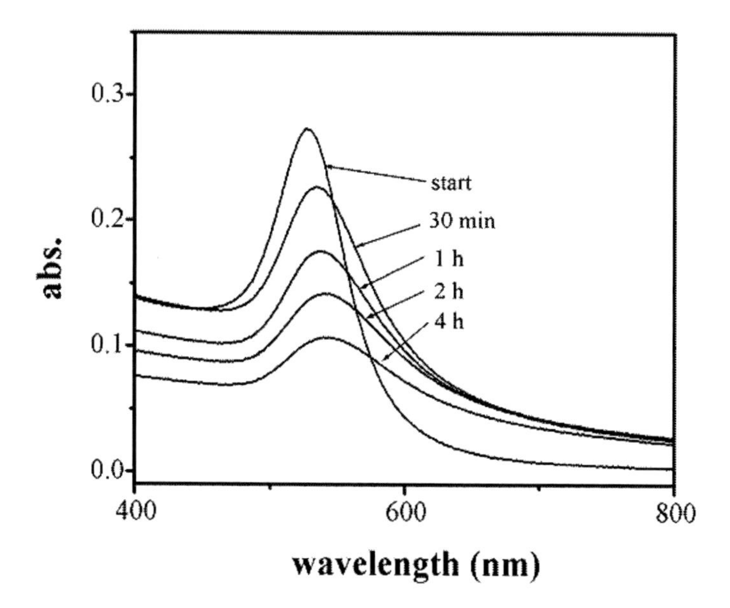

Figure 5. Time-dependent change in the absorption spectrum of R-biotinyl-PEG/PAMA-anchored gold nanoparticles after adding 100 μg/mL streptavidin. Reproduced from ref. 61 by courtesy of publishers, American Chemical Society, U.S.A.

2.5. Synthesis of Metoxy-poly(ethylene glycol)-*block*-poly(2-(*N*,*N*-dimethylamino)ethyl methacrylate) with a Variety of Polyamine Chain Length [57]

Our original oxyanion-initiated polymerization technique has been employed for the preparation of PEG-*b*-PAMA [49](chapter 2.1). This method has several advantages: fast polymerization reaction, narrow molecular weight distribution, controlling the molecular weight based on monomer/initiator ratio, and so on. In the preparation of PEG-*b*-PAMA having a short PAMA chain length, however, it is difficult to control the molecular weight because of the very fast polymerization reaction as compared to the initiation rate. In order to synthesize well-controlled oligomers, the control of the initiation reaction is crucial. Atom transfer radical polymerization (ATRP) can yield short polymer chains with controlled molecular weight and molecular-weight distribution because the propagation species is a stable copper complex, which converts to the active free form in an equilibrium manner [58,59]. If the deactivation rate constant (k_{deact}) is sufficiently larger than the activation rate constant (k_{act}), the polymerization proceeds step by step to result in the synthesis of polymers controlled from the beginning of the polymerization reaction.

Various molecular weights of PEG-*b*-PAMA were synthesized by ATRP, using methoxy-poly(ethylene glycol)-2-bromoisobutyrate (PEG-Br), copper bromide and 2,2'-bipyridine, and AMA as initiator, catalyst and monomer, respectively. The molecular weights of the block copolymer were controlled by reaction time and feed. A schematic representation of PEG-*b*-PAMA synthesis is shown in Scheme 3. The chemical structure of PEG-*b*-PAMA [49] was confirmed by ^{1}H-NMR (Avance 500 NMR Spectrometer, Bruker, Germany), (500 MHz, D_2O): δ (ppm) 0.85-1.23 (d, 3H, α-CH_3), 1.85 (d, 6H, -C(CH_3)-Br), 1.81-1.87 (d, 2H, -C(COO-)(CH_3)-CH_2-), 2.52-2.57 (d, 6H, -N(CH_3)$_2$), 3.00-3.06 (d, 2H, -CH_2-CH_2-N-) 3.27 (d, 3H, -O-CH_3), 3.45-3.74 (d, 4H, -CH_2-CH_2-O-), and 4.15-4.21 (d, 2H, -CH_2-CH_2-N-).

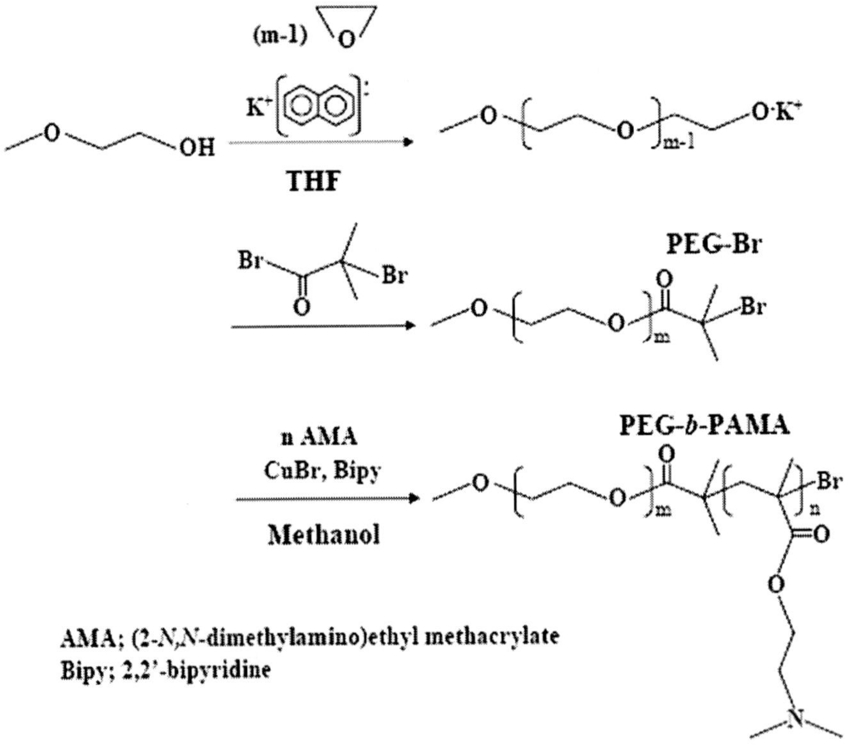

AMA; (2-*N,N*-dimethylamino)ethyl methacrylate
Bipy; 2,2'-bipyridine

Scheme 3. Schematic representation of PEG-*b*-PAMA synthesis.

The results of the synthesis and characterization of the prepared PEG-*b*-PAMAs are summarized in Table 3. The molecular weights were well controlled, with narrow molecular weight distribution factors. Especially, PEG-*b*-PAMA possessing a very short PAMA length was clearly obtained by this method. The key step of this synthesis was to add fairly large amounts of AMA monomer to the reaction system in order to increase the initiation rate. PEG-*b*-PAMAs with short PAMA chain lengths were prepared by adjusting the polymerization time. If the polymerization was quenched to be for 3 h, the degree of polymerization of the PAMA segment was adjusted to 3 (Table 3, Run 1). Thus, PEG-*b*-PAMAs with a variety of PAMA chain lengths from DP=3 to 43 (DP: degree of polymerization) were precisely synthesized. These polymers were purified through repeated precipitations and ion exchange column fractionation techniques. Using PEG-*b*-PAMA thus prepared, GNPs surface modifications were carried out.

2.6. Modification of Gold Nanoparticles Using PEG-b-PAMA(43) under Different pH Conditions [57]

In order to investigate the mechanism of GNPs surface modification with PEG-*b*-PAMA, the interaction between PEG-*b*-PAMA and commercial GNPs were investigated under several conditions. PEG-GNPs were prepared by mixing commercial GNPs with the PEG-*b*-PAMAs prepared in this study under a variety of conditions. One of the representative preparation procedures for PEG-GNPs is described as follows. Four hundred and fifty μL of commercial GNPs solution (7.1×10^{-3} wt%, pH 6.5) were added to 50 μL of PEG-*b*-PAMA(43) solution ([N]=960 μM) in 1.5 mL vial. One hundred mM phosphate buffer was used as the solvent of the block polymer for pH values below 8.5. For pH values above 8.5, 100 mM carbonate buffer was employed. The pH value was varied from 5.0 to 11.0, using these two buffers in these experiments. The solution was finally diluted to adjust the GNPs, buffer and block copolymer concentrations to be 6.4 wt%, 10 mM and 96 μM, respectively.

The protonation degree of the amino groups in the block copolymer depended on the environmental pH [49]. The effect of the state of the amino groups on GNP PEGylation was investigated using PEG-*b*-PAMA(43) and varying the environmental pH. In order to confirm the dispersion stability, DLS measurement was carried out after the PEG-*b*-PAMA modification. Figure 6 shows the particle size distribution of the PEG-GNP solutions, which was kept for 18 h at room temperature after preparation under different pH conditions in 10 mM phosphate buffer (pH 5.0-8.5). The PEG-GNPs prepared under acidic conditions increased in size after 18 h at room temperature, indicating the coagulation of the particles in the solution. On the contrary, the particle size did not increase much under alkaline conditions in the same phosphate buffer (pH<8.5). However, in phosphate buffer solution at pH 8.5, a slight but definite increase in particle size was observed. In order to further investigate the preparation conditions in the higher pH region, the buffer of the preparation experiment was changed to carbonate buffer. Figure 7 shows the particle size distribution of the PEG-GNP solutions after 18 h at room temperature in 10 mM pH 8.5-11.0 carbonate buffer. Even in carbonate buffer at pH<10, a slight increase in particle size was observed. At pH>10, on the contrary, an almost constant size (ca. 37 nm) with narrow size distribution was obtained even after 18 h. The size of the PEG-GNPs prepared at pH 10, 37 nm, is reasonable for a PEG brush composed of PEG with a molecular weight of 4.2 kDa coating the GNP (18.4 nm), assuming a single GNP for each particle.

Table 3. Synthesis of MeO-PEG-*b*-PAMA block copolymers via living radical polymerization (ATRP) of 2-(*N,N*-dimethylamino)ethyl methacrylate(AMA) initiated with MeOPEGBr coupled with CuBr and 2,2'-bipyridine[1]

Run	Code	feed (mmol/L)		Time	Yield	10^{-3} x MW of Copolymer[2]			Number of amino groups[3]
		PEGBr	AMA	h	%	Mn	Mw	Mw/Mn	
1	MeOPEG-*b*-PAMA(3)	0.10	4.0	3.0	6.1	4.7	4.8	1.02	3
2	MeOPEG-*b*-PAMA(6)	0.10	4.0	3.5	10	5.3	5.5	1.03	6
3	MeOPEG-*b*-PAMA(15)	0.10	2.0	18	54	6.6	7.2	1.09	15
4	MeOPEG-*b*-PAMA(18)	0.10	3.5	18	52	7.4	8.1	1.09	18
5	MeOPEG-*b*-PAMA(24)	0.10	4.0	18	63	8.8	9.9	1.12	24
6	MeOPEG-*b*-PAMA(36)	0.10	5.0	18	60	11.4	13.0	1.14	36
7	MeOPEG-*b*-PAMA(43)	0.10	7.0	18	62	11.7	13.3	1.14	43

[1] MeOPEGBr (Mn=4,100, Mw=4,200, Mw/Mn=1.03) : 0.1 mmol/L; CuBr : 0.1 mmol/L; 2,2'-bipyridine : 0.2 mmol/L. polymerization temperature : 25 °C.

[2] Determined by SEC data. Column :Superdex 75 10/300 GL; carrier:10 mM, pH 7.0 phosphate buffer with 140 mM NaCl; detecter: RID, flow rate: 0.7 mL/min.

[3] Determined by ^{1}H NMR data.

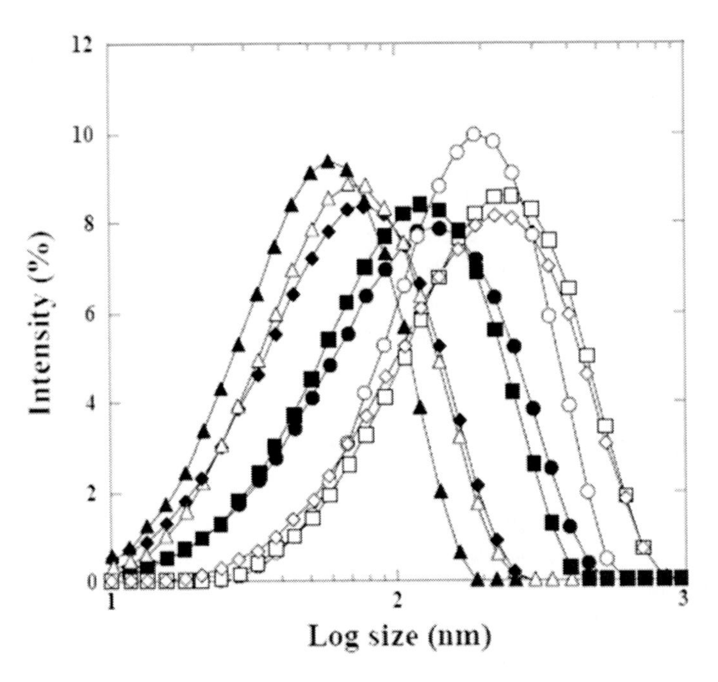

Figure 6. Particle size distribution of GNPs modified with PEG-*b*-PAMA(43) under various pH conditions in 10 mM phosphate buffer. Open circle, pH 5.0; open square, pH 5.5; open diamond, pH 6.0; closed circle, pH 6.5; closed square, pH 7.0; closed diamond, pH 7.5; open triangle, pH 8.0; closed triangle, pH 8.5.

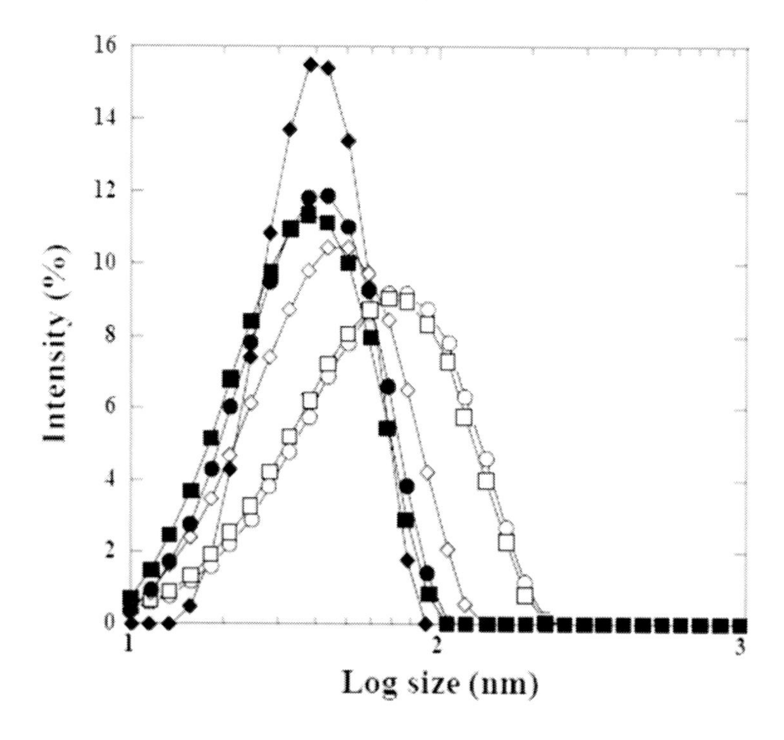

Figure 7. Particle size distribution of GNPs modified with PEG-*b*-PAMA(43) under various pH conditions in 10 mM carbonate buffer. Open circle, pH 8.5; open square, pH 9.0; open diamond, pH 9.5; closed circle, pH 10.0; closed square, pH 10.5; closed diamond, pH 11.0.

It has been reported previously that the pK_a value of the amino groups of PEG-*b*-PAMA is 7.0 [49, 60], which means that all the amino groups in the PAMA segment of the block copolymer are completely depronated at pH 10. This means that the GNPs are stabilized by the neutral PEG-*b*-PAMA, probably via the coordination of amino nitrogen atoms on the GNP surface, but not via electrostatic interaction between the protonated PAMA and the negative GNP surface, caused by the absorbed citric acid. As stated above, although the N-Au bond is not strong enough, unlike the S-Au bond, the polyvalency [61] coordination of the tertiary amino groups of PEG-*b*-PAMA might work effectively on the GNP surface.

2.7. Modification of Gold Nanoparticles Using PEG-*b*-PAMA(43) under Different Concentration [57]

Since commercial GNPs which was modified with PEG-*b*-PAMA at alkaline condition showed stable dispersion as well as the PEGylated GNPs prepared by autoreduction method, the effect of the environmental polymer concentration on GNP PEGylation was investigated using PEG-*b*-PAMA(43). PEGylated GNP was prepared by the same method as in chapter 2.6. The pH was fixed to 10.5. The concentration of the amino groups of PEG-*b*-PAMA was varied from 0.96 to 960 μM, which GNPs concentration was kept constant, 6.4 wt%.

Figure 8 shows the particle size distributions of the PEG-GNPs prepared with different polymer concentrations ([N] concentration: 0.96-960 μM). When the [N]/[GNP] ratio, which denotes the number of the tertiary amino groups to the number of GNPs, was 330, the particles size of the PEG-GNP was fairly large (75.7 nm), which indicates that the amount of polymer was not enough to stabilize the GNPs. When the [N]/[GNP] ratio was greater than 3,300, the particle sizes of the PEG-GNPs were almost equal (about 37 nm), and an [N]/[GNP] ratio of more than 3,300 was required to maintain the stability of their dispersion.

2.8. Modification of Gold Nanoparticles Using Various Molecular Weight of PEG-*b*-PAMAs [57]

The effect of the chain length of the polyamine segment was evaluated using various molecular weights of PEG-*b*-PAMAs, where the number of PAMA segments in the block copolymers was changed from 3 to 43, possessing the same PEG molecular weight (4.2 k). PEGylated GNP was prepared by the same method as in chapter 2.6. The concentration of the amino groups of PEG-*b*-PAMA and pH were fixed 96 μM and 10.5, respectively. All experiments were carried out at room temperature, and the total reaction time was fixed to 18 h. For the purification of PEG-GNPs thus prepared, the free excess polymers in the solutions was removed by repeated centrifugations (2.0 x 10^4 *g*, 4 times), and the solvent was substituted with pH 7.4 phosphate buffer saline. In order to confirm the dispersion stability of the PEG-GNPs in the protein solution, after the obtained PEG-GNPs were precipitated by centrifugation, the solvent was substituted with BSA or 95 % human serum solution by the same method as the one described above and these solutions were stored at 4 ℃ for 4 days.

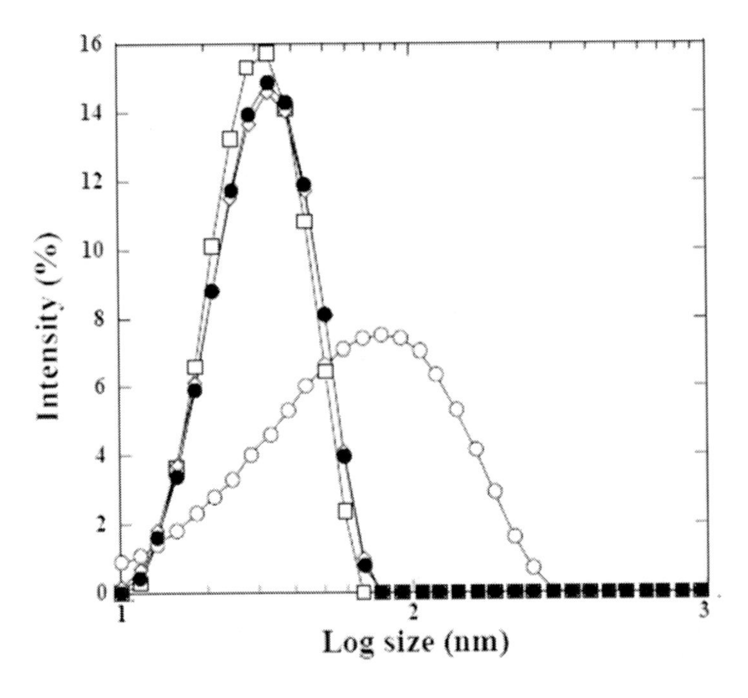

Figure 8. Particle size distribution of GNPs modified with PEG-*b*-PAMA(43) of various concentrations in 10 mM, carbonate buffer, pH 10.5 ([N] concentration, 0.96-960 μM). Open circle, 0.96 μM; open square, 9.6 μM; open diamond, 96 μM; closed circle, 960 μM.

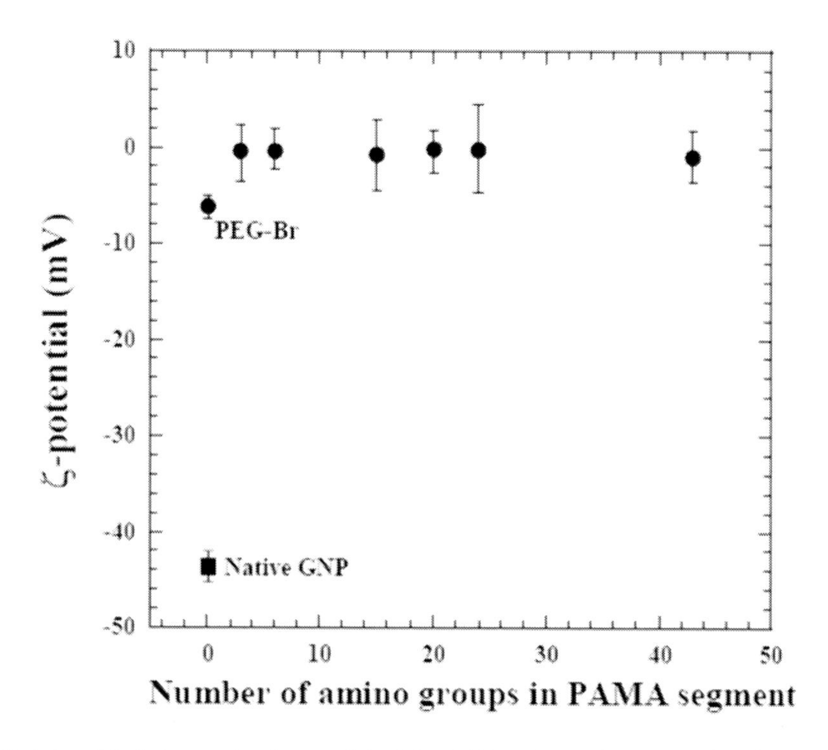

Figure 9. Relationship between the ζ-potential of the PEG-GNPs and number of amino group units in the PAMA segment of PEG-*b*-PAMA. Closed circle, PEG-GNPs(0-43); closed square, native GNPs.

Figure 9 shows the ζ-potential of the PEG-GNPs prepared from various block copolymers after the free polymer in the solution was removed and the solvent of the PEG-GNPs was changed from pH 10.5 carbonate buffer to pH 7.4 phosphate buffer saline by centrifuge purification. Since citric acid was adsorbed on the native GNPs, the ζ-potential was negative. Once the GNPs were modified by PEG-*b*-PAMA at an [N]/[GNP] ratio of 33,000, the ζ-potentials of the GNPs were almost completely shielded, regardless of the PAMA chain lengths. Actually, none of the ζ-potentials of the PEG-GNPs depended on PAMA length, and the ζ-potentials were almost 0 mV. During the immobilization of PEG-*b*-PAMAs on the GNP surface, it is not clear whether citric acid is adsorbed together with PEG-*b*-PAMAs or liberated. However, it is known that the slipping plane on the surface shifted offshore [62]. Thus, the adsorbed citric acid on the periphery of the gold surface should not affect the ζ-potential measurement in this case. The complete shielding of the surface charge indicates that PEG formed tetherlike structure when GNPs were modified by PEG-*b*-PAMA block copolymers.

As stated above, although an [N]/[GNP] ratio of 3,300 was enough to disperse the GNPs, not all the polymers in the solution interacted with the particles. When PEG-GNPs are administered *in vivo,* the free polymer in the solution must be removed. The dispersion stability of the PEG-GNPs after centrifuge purification was evaluated. Figure 10 shows the average diameter of various PEG-GNPs at an [N]/[GNP] ratio of 33,000 before and after purification. Immediately after the preparation of PEG-GNPs in the presence of excess free polymer, the PEG-GNPs showed the same average diameter regardless of the PAMA length. When the excess free polymer was removed completely from the solution by centrifugation, the particle size of the PEG-GNPs in PBS increased. The coagulation tendency was remarkable in PEG-*b*-PAMAs with long polyamine segments. When the GNPs were modified with PEG-*b*-PAMA having short AMA units (DP=3, 6), on the contrary, the PEG-GNPs showed excellent dispersion stability, which was maintained for 14 days at room temperature in pH 7.4 PBS. It is rather surprising that the purified PEG-GNPs (DP=3, 6) were stable at neutral pH even though they were prepared from deprotonated PEG-*b*-PAMA umder alkaline conditions (pH=10.5). This probably means that, once the neutral amino groups in PEG-*b*-PAMA coordinate on the GNP surface, they are not cleaved even in neutral pH. The pK_a value of the amino groups which coordinate on the gold surface might change [63]. When the GNPs were modified with PEG-*b*-PAMA having long polyamine segments, a part of the amino groups in the single polymer chain could not coordinate on the GNP surface even at high pH, due to the conformational restriction of the PAMA segments. The protonation of the free amino groups in this case may decrease the dispersion stability under physiological conditions. The dispersion stability of PEG-*b*-PAMA was further examined in the presence of BSA, as shown in Figure 10. The PEG-GNPs modified with PEG-*b*-PAMA possessing long PAMA segments aggregated further in 4.5 mg/mL BSA solution. However, the PEG-GNPs modified with PEG-*b*-PAMA(3) (PEG-GNP(3)) showed excellent dispersion even when it was stored for 4 days in BSA solution at 4 ℃.

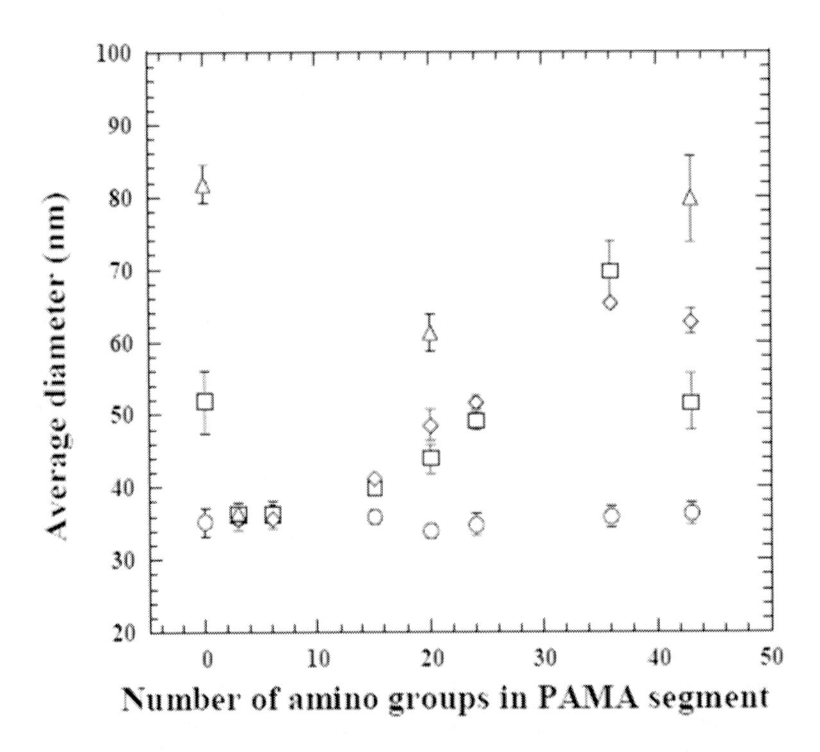

Figure 10. Average diameter of the PEG-GNPs before and after purification and in BSA solution. Open circle, before purification; open square, after purification; open diamond, after purification and stored for 14 days; open triangle, stored in BSA solution.

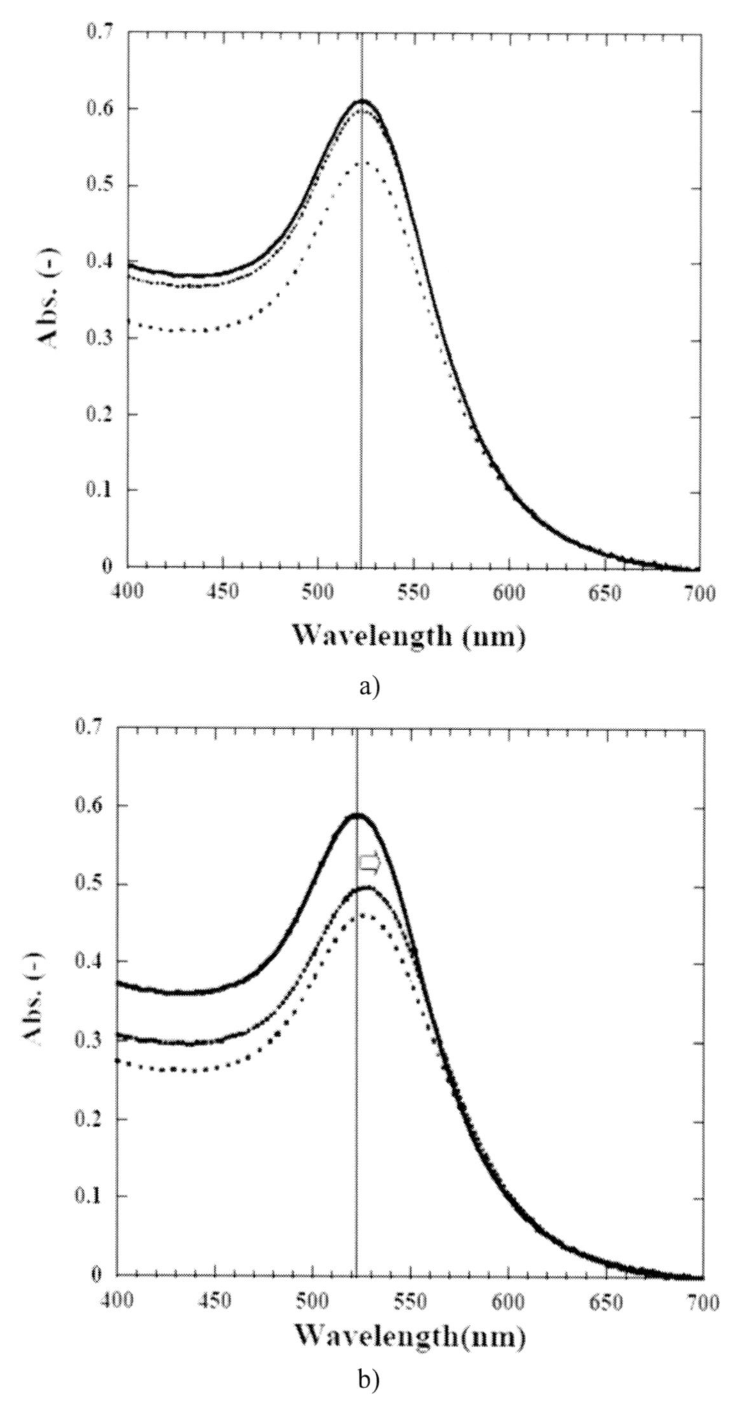

Figure 11. UV spectra of (a) PEG-GNP(3) and (b) PEG-GNP(43) before and after purification and stored for 14 days. Solid line, before purification; dashed line, after purification; dotted line, stored for 14 days after purification.

Figure 11 shows the UV spectra of GNPs modified with PEG-*b*-PAMA before and after purification. The surface plasmon peak was observed at 520.0 nm for the native GNPs. It shifted to the longer wavelength region upon modification of the GNP surface with PEG-*b*-PAMA. This means that the permittivity of the GNP surface environment was changed by the adsorption of the polymer on the GNP surface [64] and it is more proof of GNP surface modification by the block copolymer. The UV spectra of the PEG-GNPs after purification were then investigated. The UV spectra of PEG-GNP(3) after purification showed no change compared to those before purification. In the case of PEG-GNP(43), however, a slight but definite red shift was observed after purification. The red shift of the plasmon peak is known as signifying the coagulation of GNP [1]; and this is in good agreement with the DLS results. The observed bathochromic shift of the plasmon peak depended on the PAMA length of the block copolymers; *i.e.*, the shift increased with decreasing PAMA chain length. In addition to the protonation characteristics of PEG-*b*-PAMA on GNPs, as descrived above, the PEG density of the PEG-GNPs should influence on their dispersion stability. The high dispersion of the PEGylated GNPs was mainly governed by the entropic repulsion of the PEG. We assumed that the PEG density on the GNP surface depended on the PAMA chain length in the PEG-*b*-PAMA modification reaction. The PEG density of the PEG-GNPs was thus investigated by TGA.

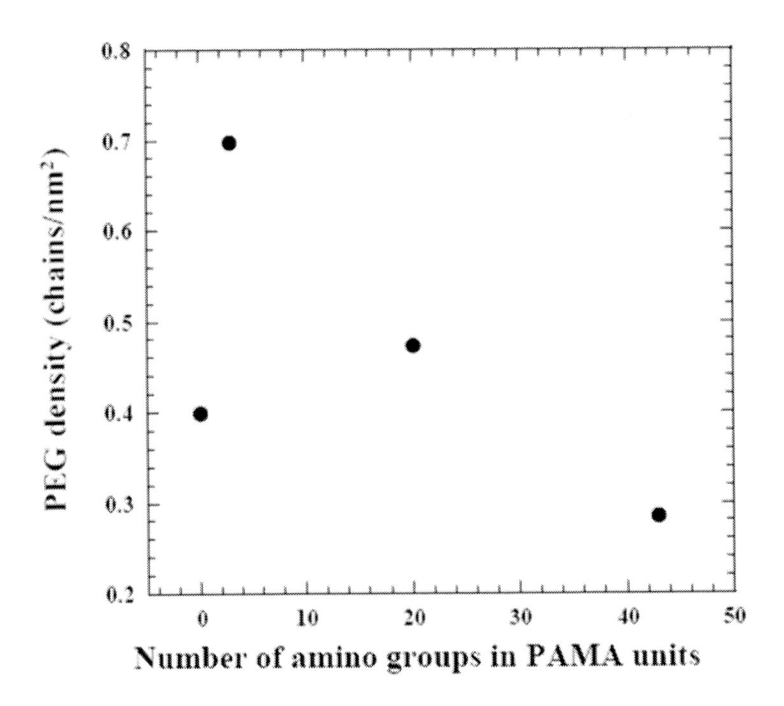

Figure 12. PEG chain density on GNPs modified with PEG-*b*-PAMA, as determined by TGA, assuming that the PAMA segments are located on the GNP surface to form a PEG brush.

2.9. Quantification of Adsorbed PEG-*b*-PAMA on GNPs [57]

After the solution of the PEG-GNPs was substituted with water through repeated centrifugations and the PEG-GNPs were completely dried *in vacuo*, thermal gravimetric analysis was carried out. Figure 12 shows the PEG brush density, as determined from the TGA data; it is a function of the number of amino groups in PEG-*b*-PAMA. As a control,

PEG-Br was used as the GNP modification agent. Since the ether oxygen in the ethylene glycol units of PEG interacted weakly with gold surface, a limited amount of PEG was adsorbed on the GNP surface. The GNPs coupled with PEG-Br were dispersed in the aqueous medium when excess PEG-Br solution was used. After removing the excess free PEG-Br from the solution by centrifugation, however, the PEG-GNPs did not redisperse completely, probably due to the insufficient amounts of PEG on the GNPs.

The amount of modified PEG-*b*-PAMA on the GNP surface increased significantly, as compared to that of PEG-Br. Among the PEG-*b*-PAMAs used in this study, the shortest PAMA (PEG-*b*-PAMA(3)) had highest density, 0.7 chains/nm^2, assuming that the PAMA segments adsorbed on the GNP surface formed a PEG brush. With increasing PAMA chain length, the brush density gradually decreased. The effect of the PAMA chain length on the PEG chain density was in good agreement with the PEG-GNP dispersion stability (*i.e.*, the dispersion stability increased with increasing PEG chain density on the GNP surface) [65-67]. Fukuda, Tsujii and co-workers reported [68] that densely packed polymer brush effectively repelled protein adsorption. They explained that the large protein can not migrate in densely packed brush layer. The colloidal stability of PEGylated nanoparticles was also interpreted in the same way [69,70], *viz.*, the PEG brushes on GNPs surface can not migrate each other when the brush density becomes high enough. When the GNPs were modified with PEG-SH, whose molecular weight was the same as that of the PEG segment in PEG-*b*-PAMA, the PEG density was 1.3 chains/nm^2 under the same modification conditions. Although this value was larger than that of PEG-GNP(3), PEG-GNP(3) showed an extremely high dispersion stability under various conditions.

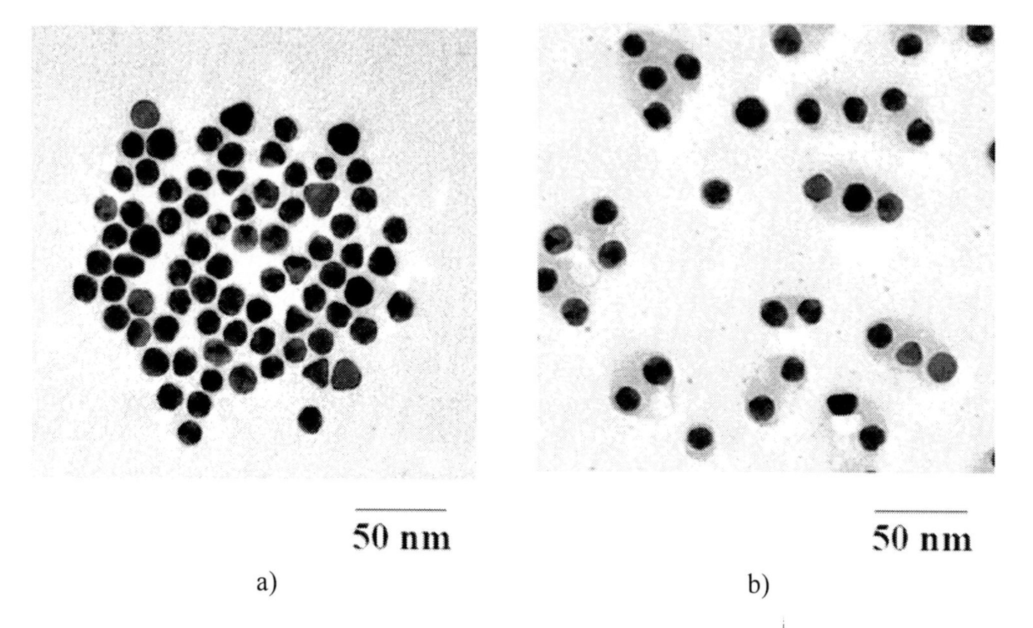

Figure 13. TEM image of PEG-GNP(3) after purifica tion (a) and PEG-GNP(3) kept in 95 % human serum for 4 days (b).

Finally, the dispersion stability of the PEG-GNPs was examined in the presence of 95 % human serum as model blood. Since direct DLS measurement of PEG-GNPs in human serum would be too difficult to conduct because of the very high protein concentration, the average diameter of the redispersed PEG-GNPs after separation of human serum proteins by centrifugation was evaluated by DLS. In the case of PEG-GNP(3), the particle size after separation of human serum proteins (ca. 43.8 nm) was almost the same as that without serum, indicating stable dispersion under high serum proteins conditions. The stable dispersion of PEG-GNP(3) in 95 % human serum was also confirmed by TEM measurement (Figure 13). On the contrary, PEG-GNP(43) in 95 % human serum precipitated on the bottom of sample tube and adsorbed on the tube surface after 4-day incubation. These results confirm the fact that PEG-GNPs(3) was completely dispersed even in 95 % human serum,, which strongly suggests that it is suitable for *in vivo* application.

Using PEG-*b*-PAMA modified GNPs thus prepared, further functionalization can be achieved. For example, PEG-*b*-PAMA and RNA-SH was able to be co-immobilized on GNPs surface (PEG/RNA-GNPs). The PEG/RNA-GNPs could thus prepared control the RNA release [71] and expected to apply for the high performance gene delivery systems.

3. Conclusion

Monodispersed GNPs were readily prepared by the simple addition of a ligand installed PEG-*b*-PAMA in $HAuCl_4$ aqueous solution at room temperature without any additional reducing reagent. The ligand-installed PEG layer locating on the outer surface of the obtained GNPs substantially contributed to sterically stabilize them even under very high salt concentrations. Furthermore, biotin as the ligand molecule installed at the end of tethered PEG chain showed specific recognition of streptavidin under physiological salt concentration indicating the high utility of these ligand-installed PEG/PAMA-gold nanoparticles as colloidal biosensor systems.

The polyvalency coordination of tertiary amino groups on the gold surface was confirmed to lead to stable modification with PEG-*b*-PAMA, and not electrostatic interaction between the protonated PAMA and the gold surface. PEG-GNPs modified with PEG-*b*-PAMA having a short PAMA chain length showed excellent dispersion stability under physiological conditions, even when kept in 95% human serum for 4 days. It was confirmed that the increased PEG chain density resulting from PEG-*b*-PAMA with a short PAMA chain length is one of the factors which improve the dispersion stability of GNPs, indicating that three amino nitrogens are enough to stabilize the PEG brush on the gold colloid surface via the polyvalency effect. The obtained PEG-GNPs may be useful in a variety of biological applications, for example, as carriers of gene delivery, model carriers of drug delivery systems, and contrast agents for *in vivo* X-ray computed tomography.

References

[1] Daniel, M. C.; Astruc, D. *Chem. Rev.* 2004, 104, 293.
[2] Brown, D. H.; Smith, W. E. *Chem. Soc. Rev.*, 1980, 9, 217.
[3] Turkevitch, J.; Stevenson, P. C.; Hiller, J. *Discuss. Faraday Soc.,* 1951, 11, 55.

[4] Frens, G. *Nature: Phys. Sci.*, 1973, 241, 20.
[5] Schmid, G. *Chem. Rev.*, 1992, 92, 1709.
[6] Brust, M.; Walker, M.; Schiffrin, D.; Whyman, R. *J. Chem. Soc., Chem. Commun.*, 1994, 801.
[7] Han, M. Y.; Quek, C. H.; Huang, W.; Chew, C. H.; Gan, L. M. *Chem. Mater.*, 1999, 11, 1144.
[8] Brown, K. R.; Walter, D. G.; Natan, M. *J. Chem. Mater.*, 2000, 12 *(2)*, 306.
[9] Weiser, H. B. *Inorg. Colloid Chem.*, 1993, 1, 21.
[10] Biggs, S.; Chow, M. K.; Grieser, F.; *J. Colloid Interface Sci.*, 1993, 160, 511.
[11] Hayat, M. A. Colloidal Gold Principles, *Method and Applications, Academic Press: New York*, 1989.
[12] Roucoux, A.; Schulz, J.; Patin, H. *Chem. Res.* 2002, 102, 3757.
[13] Ziolo, R. F.; Glannelis, E. P.; Weinstein, B. A.; OHoro, M. P.; Gamguly, B. M.; Mwhrota, V.; Russel, M. W.; Huffman, D. R. *Science*, 1992, 257, 219.
[14] Jordan, R.; West, N.; Chou, Y. M.; Nuyken, O. *Macromolecules*, 2001, 34, 1606.
[15] Selven, S. T.; Spatz, J. P.; Klock, H. A.; Moller, M. *Adv. Mater.*, 1998, 10, 132.
[16] Sayo, K.; Deki, S.; Hayashi, S. *Eur. Phys. J. D.*, 1999, 9, 429.
[17] Lee, J.; Sunder, V. C.; Heine, J. R.; Bawendi, M. G.; Jensen, K. F. *Adv. Mater.*, 2000, 12, 1102.
[18] Teichoeb, J. H.; Forrest, J. A. *Phys. Rev. Lett.*, 2003, 91, 016104-1.
[19] Raula, J.; Shan, J.; Nuoponen, M.; Niskanen, A.; Jiang, H.; Kauppinen, E. I.; Tenhu, H. *Langmuir*, 2003, 19, 3499.
[20] Corbierre, M. K.; Cammeron, N. S.; Sutton, M.; Mochrie, S. G. J.; Lurio, L. B.; Ruhm, A.; Lennox, R. B. *J. Am. Chem. Soc.*, 2001, 123, 10411.
[21] Kolb, U.; Quaiser, S. A.; Winter, M.; Reets, M. T. *Chem. Mater.*, 1996, 8, 1889.
[22] Toshima, N.; Yonezawa, T. *New J. Chem.*, 1998, 1179.
[23] Schaaf, T. G.; Whetten, R. L. *J. Phys. Chem. B*, 2000, 104, 2630.
[24] Templeton, A. C.; Wuelfing M. P.; Murray, R. W. *Acc. Chem. Res.* 2000, 33, 27.
[25] Hong, R.; Han, G.; Fernandez, J. M.; Kim, B. J.; Forbes, N. S.; Rotello, V. M. *J. Am. Chem. Soc.* 2006, 128, 1078.
[26] Barrientos, A. G.; Fuente, J. M.; Rojas, T. C.; Fernandez, A.; Penades, S. *Chem. Eur. J.* 2003, 9, 1909.
[27] Mirkin, C. A.; Letsinger, R. L.; Mucic, R. C.; Storhoff, J. J. *Nature* 1996, 382, 607.
[28] Kerman, K.; Saito, M.; Morita, Y.; Takamura, Y.; Ozsoz, M.; Tamiya, E. *Anal. Chem.* 2004, 76, 1877.
[29] Balamurugan, S.; Obubuafo, A.; Soper, S. A.; McCarley, R. L.; Spivak, D. A. *Langmuir* 2006, 22, 6446.
[30] Rosi, N. L.; Giljohann, D. A.; Thaxton, C. S.; Lytton-Jean A. K. R.; Han, M. S.; Mirkin, C. A. *Science,* 2006, 312, 1029.
[31] Shimin, R. G.; Scohch, A. G.; Braun, P. V.; *Langmuir*, 2004, 20, 5613.
[32] Kitano, H.; Kawasaki, A.; Kawasaki, H.; Morokoshi, S. *J. Coll. Int. Sci,.* 2005, 282, 340.
[33] Wuelfing W. P., Gross S. M., Miles D. T., Murray R. W. *J. Am. Chem. Soc.*, 1998, 120, 12696.
[34] Otsuka H., Akiyama Y., Nagasaki Y., Kataoka K. *J. Am. Chem. Soc.*, 2001, 123, 8226.

[35] Kim, D. K.; Park, S. J.; Lee, J. H.; Jeong, Y. Y.; Jon, S. Y. *J. Am. Chem. Soc.*, 2007, 129, 7661.
[36] Bergen, J. M.; Recum, H. A.; Goodman, T. T.; Massey, A. P.; Pun, S. H. *Macromol. Biosci.*, 2006, 6, 506.
[37] Zheng, M.; Davidson, F.; Huang, X. *J. Am. Chem. Soc.*, 2003, 125, 7790.
[38] Niidome, T.; Yamagata, M.; Okamoto, Y.; Akiyama, Y.; Takahashi, H.; Kawano, T.; Katayama, Y.; Niidome. Y. *J. Cont. Rel.*, 2006, 114, 343.
[39] Felice R. D., Selloni A. *J. Chem. Phy. B*, 2004, 120, 4906.
[40] Bearinger, J. P.; Terrettaz, S.; MicheL, R.; Tirelli, N.; Vogel, H.; Textor, M.; Hubbell, J. A. *Nature materials,* 2003, 2, 259.
[41] Cooper, E.; Leggett, G. J. *Langmuir,* 1998, 14, 4795.
[42] Castner, D. G.; Hinds, K.; Grainger, D. W. *Langmuir,* 1996, 12, 5083.
[43] Mayer, A. B. R,; Mark, J. *E. Eur. Polym. J.*, 1998, 34*(*1*)*, 103.
[44] Bronstein, L. M.; Gourkova, S. N.; Sidorov, A. Y.; Valetsky, P. M.; Hartmann, J.; Breulmann,K.; Colfen, H.; Antonietti, M. *Inorg. Chim. Acta*, 1998, 280*(*1-2*)*, 348.
[45] Harrisa L. G., Tosattib S., Wielandc M., Textorb M., Richardsa R.G. *Biomaterials*, 2004, 25, 4135.
[46] Blattler T. M., Pasche S, Textor M., Griesser H. J., *Langmuir*, 2006, 22, 5760 .
[47] Ishii, T.; Otsuka, H.; Kataoka, K.; Nagasaki, Y. *Langmuir,* 2004, 20, 561.
[48] Akiyama Y., Harada A. Nagasaki Y., Kataoka K., *Macromolecules*, 2000, 33, 5841.
[49] Kataoka K., Harada A., Wakebayashi D. Nagasaki Y. *Macromolecules*, 1999, 32, 6892.
[50] Kreibig, U.; Genzel, L. *Surf. Sci.* 1985, 156, 678-700.
[51] Collier, C. P.; Saykally, R. J.; Shiang, J. J.; Henrichs, S. E.; Heath, J. R. *Science,* 1997, 277, 1978.
[52] Heath, J. R.; Knobler, C. M.; Leff, D. V. *J. Phys. Chem. B,* 1997, 101, 189.
[53] Corvierre, M. K.; Cameron, N. S.; Lennox, R. B. *Langmuir*, 2004, 20, 2867.
[54] Green, N. M. *Adv. Protein Chem.,* 1975, 28, 85.
[55] Blankenburg, R.; Meller, P.; Ringsdorf, H.; Salesse, C. *Biochemistry,* 1989, 28, 8214.
[56] Herron, J. N.; Muller, W.; Paudler, M.;Riegler, H.; Ringsdorf, H.; Suci, P. A. *Langmuir,* 1992, 8, 1413.
[57] Miyamoto, D.; Oishi, M.; Kojima, K.; Yoshimoto, K.; Nagasaki, Y. *Langmuir*, in press.
[58] Tang, W.; Matyjaszewski, K. *Macromolecules,* 2006, 39, 4953 .
[59] Matyjaszewski, K.; Xia, J. *Chem. Rev.,* 2001, 101, 2921.
[60] Oishi, M.; Kataoka, K.; Nagasaki, Y. *Bioconjugate Chem.*, 2006, 7, 677.
[61] Mammen, M.; Choi ,S. K.; Whitesides G. M. *Angew. Chem. Int. Ed.*, 1998, 37, 2755.
[62] Koopal, L. K.; Hlady, V.; Lyklema, J. *J. Colloid Interface Sci.*, 1988, 121, 49.
[63] Nagasaki Y., Ohishi M., Nakamura T., in preparation.
[64] Nagasaki Y., Yoshinaga K., Kurosawa K., Iijima M., *Coll. Polym. Sci.*, 2007, 285, 563.
[65] Pasche, S.; Vo1ro1s, J.; Griesser, H. J.; Spencer, N. D.; Textor, M. *J. Phys. Chem. B*, 2005, 109, 17545.
[66] Paul, S. M.; Falconnet, D.; Pasche, S.; Textor, M.; Abel, A. P.; Kauffmann, E.; Liedtke, R.; Ehrat, M. *Anal. Chem.*, 2005, 77, 5831.
[67] Unsworth, L. D.; Sheardown, H.; Brash, J. L. *Langmuir,* 2005, 21, 1036.
[68] Yoshikawa, C.; Goto, A.; Tsujii, Y.; Ishizuka, N.; Nakanishi, K.; Fukuda, T. *J. Polym. Sci. Part A-Polym. Chem,*. 2007, 45, 4795.

[69] Mclean, S. C.; Lioe, H.; Meagher, L.; Craig, V. S. J.; Gee M. L. *Langmuir*, 2005, 21, 2199-2208.
[70] Romero-Cano, M. S.; Puertas, A. M.; Nieves, F. J. *J. Chem. Phys.*, 2000, 112, 8654-8659.
[71] Oishi, M.; Nakaogami, J.; Ishii, T.; Nagasaki, Y. *Chem. Lett.*, 2006,35, 1046.

In: Nanoparticles: New Research
Editor: Simone Luca Lombardi, pp. 167-171
ISBN: 978-1-60456-704-5

Chapter 5

Nanoparticles and Quantum Dots as Biomolecule Labels for Electrochemical Biosensing

Martin Pumera*
National Institute for Materials Science, 1-1 Namiki, Tsukuba, Ibaraki, Japan

Abstract

The aim of this chapter is to discuss the advantages of use of nanoparticles and quantum dots as tags for electrochemical bioassays.

Nanoscale materials bring new opportunities for electrochemical biosensing. Quantum dots are very stable (comparing to enzyme labels), they offer high sensitivity (thousands of atoms can be released from one nanoparticle) and wide variety of nanoparticles brings possibility for multiplex detection of several biomolecules.

Metallic Nanoparticles

While gold nanoparticles have been known as "colloidal gold" for centuries, their use for electrochemical detection of biomolecules were introduced in 2001 by Limoges' [1] and Wang's groups [2]. The general scheme of utilization of nanoparticles and quantum dots in biomolecule electrochemical sensing is shown in Figure 1. Probe biomolecule is immobilized on surface, then the solution containing target molecule is introduced. After the specific biorecognition event (antigen-antibody reaction or DNA hybridization), the rest of the non-complementary molecules is washed away. Consequently, the second probe molecule tagged with quantum dot or nanoparticle is introduced, interacting with target molecule and labeling it for the final step, electrochemical detection. Originally, in Wang's scheme [2], hybridization of a target oligonucleotide to magnetic bead-linked oligonucleotide probes was followed by binding of the streptavidin-coated metal nanoparticles to the captured DNA, dissolution of the nanometer-sized gold tag by HBr/Br_2 solution, and potentiometric stripping

* E-mail address: martin.pumera@gmail.com. Fax: +81-29-860-4706.

measurements of the dissolved metal tag at single-use thick-film carbon electrodes [2]. Limoges and coworkers immobilized amplified 406-base pair DNA sequence to polystyrene microwell by passive adsorption. The assay was based on the hybridization of the single-stranded target DNA with an oligonucleotide-modified Au nanoparticle probe, followed by the release of the gold metal atoms anchored on the hybrids by oxidative metal dissolution, and the indirect determination of the HBr solubilized Au^{III} ions by anodic stripping voltammetry [1]. Analogous assay using Ag nanoparticles was developed, relying on hybridization of the target DNA with the silver nanoparticle–oligonucleotide DNA probe, followed by the release of the silver metal atoms anchored on the hybrids by oxidativemetal dissolution using nitric acid and the indirect determination of the solubilized Ag^{I} ions by anodic stripping voltammetry at a carbon fiber ultramicroelectrode [3]. However, the HBr/Br_2 or HNO_3 solution is highly toxic and therefore method based on direct electrochemical detection of nanoparticle tags, which would replace the chemical oxidation agent, was developed [4, 5]. This scheme is using Cl^- chemistry where Au nanoparticles are electrooxidized at 1.2 V for short time (1-5 minutes) in 0.1 M HCl. As a result of this electrooxidation step in a highly acidic environment, $[AuCl_4]^-$ ions are produced and the biomolecules are oxidized and denatured on the electrode surface. The denaturation of the biomolecules allows an open surface for the $[AuCl_4]^-$ ions to be reduced at potential of 0.4 V (vs. Ag/AgCl) [4,5]. For DNA detection, probe DNA was immobilized on the paramagnetic beads surface via biotin-streptavidin interaction and target DNA was labeled with Au_{67} nanoparticle in ratio 1:1 preventing multiple DNA links between paramagnetic bead and nanoparticle (typical for above described Au nanoparticle-based assays), thus enhancing achievable detection limits. The hybridized paramagnetic beads were accumulated on the surface of a magnetic electrode and enabled the magnetically triggered direct electrochemical detection of gold quantum dot tracers without prior chemical dissolution of the Au quantum dots. Electrochemical magnetogenosensors for biomedical applications based on above described direct detection of gold nanoparticles was recently developed by our group [6]. The electrochemical detection of biomolecules via Au and Cl^- chemistry is gaining popularity for its non-toxicity and simplicity, and it was recently used i.e. for detection of kinase-catalyzed thiophosphorylation [7].

Biomolecule Multiplexing via Quantum Dots with Different Electrochemical Signature

Multiple target DNA/protein detection is very important feature of modern bioassays. The goal is not to detect just one biomolecule in one measurement, but several of them. This is possible to achieve using different quantum dots with different electrochemical properties (see Figure 2). Here we list some of examples of variety of nanoparticles for electrochemical detection of DNA: Core-shell Cu@Au where hybridization events between probe and target were monitored by the release of the copper metal atoms anchored on the hybrids by acidic oxidative metal dissolution and the indirect determination of the dissolved Cu^{2+} ions by anodic stripping voltammetry [8]. CdS quantum dots are another addition to the spectrum of nanoparticle tags for DNA hybridization detection [9, 10]. PbS quantum dots also offer different voltammetric detection potential for following DNA hybridization [11].

Therefore it is possible to demonstrate real hybridization assays that permit simultaneous determination of multiple DNA targets. Three encoding nanoparticles (zinc sulfide, cadmium sulfide, and lead sulfide) have been used to differentiate the signals of three DNA targets in connection with a sandwich DNA hybridization assay and stripping voltammetry of the corresponding heavy metals [12]. Multiplexing capabilities of quantum dots were also demonstrated in connection to immunoassay. Electrochemical immunoassay protocol for the simultaneous measurements of proteins, based on the use of different inorganic nanocrystal tracers was described [13]. The multiprotein electrical detection capability was coupled to the amplification attribute of electrochemical stripping transduction (to give in *fmol* detection limits). The multianalyte electrical sandwich immunoassay involved a dual binding event, based on antibodies linked to the nanocrystal tags and magnetic beads. Carbamate linkage was used for conjugating the hydroxyl-terminated nanocrystals with the secondary antibodies. Each biorecognition event provided a distinct voltammetric peak, whose position and size reflected the identity and concentration of the corresponding antigen. The concept was demonstrated for a simultaneous immunoassay of $_2$-microglobulin, IgG, bovine serum albumin, and C-reactive protein in connection with ZnS, CdS, PbS, and CuS colloidal crystals, respectively.

Soft Nanoparticles

However, not only metallic or binary nanoparticles can be used as tags for electrochemical biosensing of biomolecules. Very recently, there has been substantial research activity in development soft nanoparticle labels which are easy to synthesize and easy to detect by electrochemical means. Liu et al. developed apoferritin-templated phosphate nanoparticle labels and used them for electrochemical immunoassay [14, 15] (see Figure 3). Apoferritin is a native protein composed of 24 polypeptide subunits that interact to form a hollow cage-like structure 12.5 nm in diameter; the interior cavity of apoferritin is about 8 nm in diameter and has an interior volume that can store several thousands of metal ions in form of phosphate salt. There is wide variability of metals which can be introduced in apoferritin cavity which paves the way for highly multiplexed assays. These novel soft nanoparticles brings new possibilities for electrochemical sensing of proteins and alternatively also for DNA sensing.

Outlook

Electrochemical nanobiosensors offer without doubts an important step towards development of selective, down to few target molecules sensitive biorecognition device for medical and security applications. There is high expectation that such devices will develop toward reliable point-of-care diagnostics of cancer and other diseases, and as useful tools for molecular biosystems, intra-operation pathological testing or proteome research.

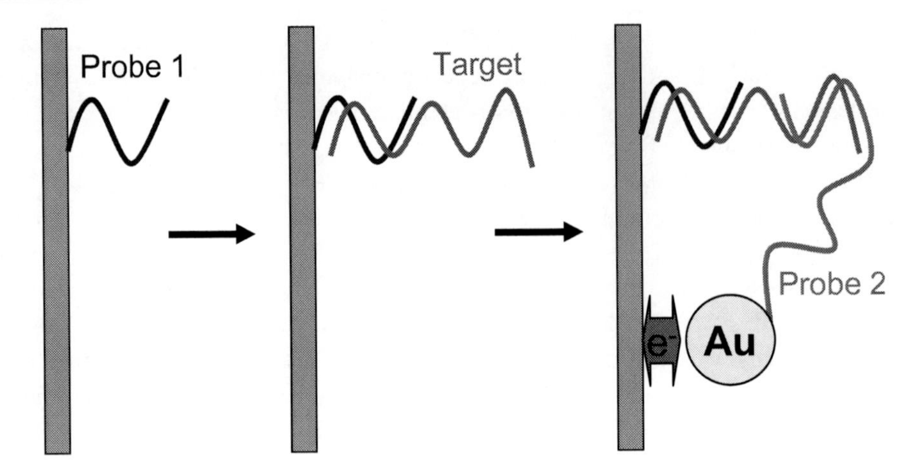

Figure 1. Schematic of sandwich bioassay employing gold nanoparticle as a electrochemical tag.

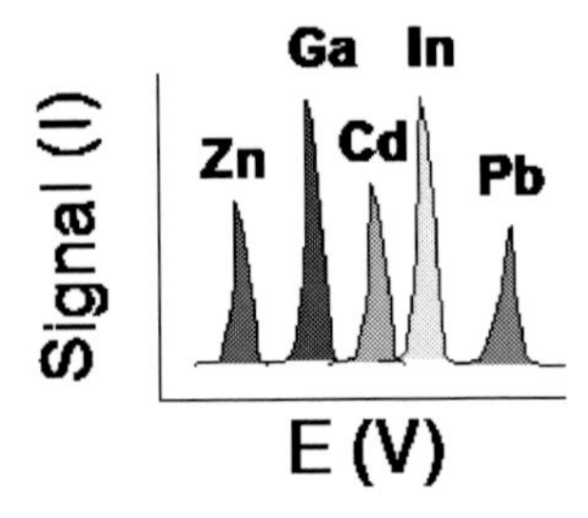

Figure 2. Schematic showing different electrochemical detection potentials of several metallic components of quantum dots.

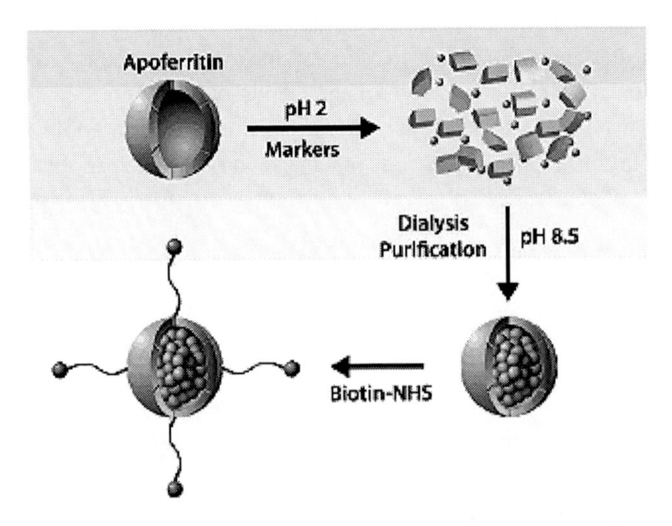

Figure 3. Preparation of biotin-functionalized appoferitin labels. Reprinted with permission from [15], G. Liu, J. Wang, H. Wu, Y. Lin, Anal. Chem. 78, 7417 (2006). Copyright (C) American Chemical Society.

References

[1] L. Authier, C. Grossiord, P. Brossier, B. Limoges, *Anal. Chem.* **73**, 4450 (2001).
[2] J. Wang, D. Xu, A.-N. Kawde, R. Polsky, *Anal. Chem.* **73**, 5576 (2001).
[3] H. Cai, Y. Xu, N. Zhu, P. He, Y. Fang, *Analyst* **127**, 803 (2002).
[4] M. Pumera, M. T. Castaneda, M. I. Pividori, R. Eritja, A. Merkoci, S. Alegret, *Langmuir* **21**, 9625 (2005).
[5] M. Pumera, M. Aldavert, C. Mills, A. Merkoci, S. Alegret, *Electrochim. Acta* **50**, 3702 (2005).
[6] M. T. Castaneda, A. Merkoci, M. Pumera, S. Alegret, *Biosens. Bioelectr.* **22**, 1961 (2007).
[7] K. Kerman, H.-B. Kraatz, *Chem. Comm.* **5019** (2007).
[8] H. Cai, N. N. Zhu, Y. Jiang, P. G. He, Y. Fang, *Biosens. Bioelectron.* **18**, 1311 (2003).
[9] J. Wang, G. Liu, R. Polsky, A. Merkoçi, *Electrochem. Commun.* **4**, 722 (2002).
[10] N. N. Zhu, A. P. Zhang, P. G. He, Y. Z. Fang, *Analyst* **128**, 260 (2003).
[11] N. N. Zhu, A. P. Zhang, Q. J. Wang, P. G. He, Y. Z. Fang, *Electroanal.* **16**, 577 (2004).
[12] J. Wang, G. Liu, A. Merkoci, *J. Am. Chem. Soc.* **125**, 3214 (2003).
[13] G. Liu, J. Wang, J., Kim, M. R. Jan, G. E. Collins, *Anal. Chem.* **76**, 7126 (2004).
[14] G. Liu, H. Wu, J. Wang, Y. Lin, *Small* **2,** 1139 (2006).
[15] G. Liu, J. Wang, H. Wu, Y. Lin, *Anal. Chem.* **78**, 7417 (2006).

In: Nanoparticles: New Research
Editor: Simone Luca Lombardi, pp. 173-200
ISBN: 978-1-60456-704-5

Chapter 6

HYBRID NANOPARTICLE BASED ON SILICA

Shaojun Guo and Erkang Wang*

State Key Laboratory of Electroanalytical Chemistry, Changchun Institute of Applied Chemistry, Chinese Academy of Sciences, Changchun, 130022, Jilin, China. Graduate School of the Chinese Academy of Sciences, Beijing, 100039, P. R. China

Abstract

Inorganic nanoparticles have received great interests due to their attractive electronic, optical, and thermal properties as well as catalytic properties and potential applications in the fields of physics, chemistry, biology, medicine, and material science and their different interdisciplinary fields. Great deals of synthetic methods such as aqueous- and organic-phase approaches have been reported to synthesize high-quality inorganic nanoparticles, but the stability of colloid nanoparticles obtained is still a great problem. In addition, most of these methods for preparing nanoparticles were based on the use of organic ligand, surfactant or polymeric templates. However, the addition of organic ligand, surfactant or polymeric templates may introduce contamination in the final product due to the formation of undesired byproduct in the reaction medium, which makes them unsuitable for further application, especially in biology. On the other hand, fluorohores such as dyes has been employed as optical probe by many groups for carrying out bioassay such as detecting target DNA, protein, etc. However, because one DNA or other probes can only be labeled with one or a few fluorophores, the fluorescence signal is too weak to be detected when the target concentration is low. Furthermore, most organic dyes suffer serious photobleaching, often resulting in irreproducible signals for ultratrace bioanalysis. Silica coating of inorganic nanomaterial or fluorohores appears as an attractive alternative for providing enhanced colloidal or photostable stability and suitable functional groups for bioconjugation. Hybrid nanoparticles based on silica has been pursued vigorously in the last few years and in this review chapter we provide a perspective on the present status of the subject. Described approaches include encapsulating inorganic nanomaterials using SiO_2 sphere, fluorophore-doped silica nanoparticles, constructing muti-functional hybrid materials using SiO_2 sphere as supporting material, and some important applications of hybrid nanoparticle based on silica. Finally, summary and outlook of these hybrid nanoparticles are also presented.

Keywords: inorganic nanoparticle, hybrid nanomaterial, quantum dot, electrochemical sensor, magnetic nanoparticle, silica

---

* E-mail address: ekwang@ciac.jl.cn; Fax: +86-431-85689711; Tel: +86-431-85262003

1. Introduction

Fabrication of functional inorganic particles with nanometer-scale dimensions has been intensively studied in the last couple of decades due to their unusual chemical and physical properties compared with their bulk materials, which enable them to be promising in application over broad areas such as electronics [1], optics [2] optoelectronics [3], nanosensor [4], information storage [5], fuel cell [6], biomedicine [7], biological labeling [8], gene delivery [9], electrocatalysis [10] and surface enhanced Raman scattering (SERS) [11, 12]. Inspired by these important applications, many groups around the world have devoted their great efforts to synthesize functional inorganic nanomaterials via diversified approaches such as chemical (bottom-up) and physical (top-down) methods. Wet-chemical approach as a simple and effective method has been employed by many scientists to synthesize functional inorganic nanoparticles with different morphologies [12-35]. For example, metal nanoparticles with different novel morphologies as stylish functional nanomaterials have been widely researched by our group [12-30] and other famous groups [31-35]; Quantum dots (QDs) as new class of fluorescent probes have been facilely synthesized via organic or aqueous phase routes [36-40]; Superparamagnetic magnetic nanoparticles [41-45], which are of great interest for researchers from a wide range of disciplines, including biotechnology/biomedicine [46], and magnetic resonance imaging [47, 48], have been prepared by a number of suitable methods. However, most of these methods for preparing nanoparticles were based on the use of organic ligand, surfactant or polymeric templates. The addition of organic ligand, surfactant or polymeric templates may introduce contamination in the final product due to the formation of undesired byproduct in the reaction medium, which makes them unsuitable for further application [49], especially in biology. Additionally, inorganic nanoparticles as friendly-environmentally materials for application in biology need to be water-soluble, colloidally stable and have a tailored surface chemistry. Currently, most syntheses for high-quality, quasi-monodisperse nanoparticles of semiconductors [50], metals [51], metal oxides [52], and magnetic material [53] involved organic solvents and coating with monolayers of hydrophobic surfactants. Such nanoparticles could not be used directly for biofunctionalization as they were water-insoluble and did not have suitable functional groups for constructing nanoparticle-biomolecule conjugates. Although many groups synthesized water-soluble nanoparticles using some polymer and surfactant as protecting agents, these protecting agents usually brought potential toxicity and side-effect for the biosystem of studying. On the other hand, fluorohores such as dyes has been employed as optical probe by many groups for carrying out bioassay such as detecting target DNA, protein, etc. However, because one DNA or other probes can only be labeled with one or a few fluorophores, the fluorescence signal is too weak to be detected when the target concentration is low. Furthermore, most organic dyes suffer serious photobleaching, often resulting in irreproducible signals for ultratrace bioanalysis. Silica coating of inorganic nanomaterial or fluorohores appears as an attractive alternative for providing enhanced colloidal or photostable stability and suitable functional groups for bioconjugation. Particularly after discovering monodisperse silica sphere by stöber et al., many excellent papers have been reported to synthesize hybrid nanoparticle based on silica because silica has very important predominance for application in biology. 1) Silica is a stabilizer which not only prevents particle coalescence but also is chemically inert and optically transparent. 2)

Silica has been reported to be good bio-compatible material. 3) Silica owns excellent surface chemistry, which enable silica coated inorganic nanoparticles or fluorophore-doped silica nanoparticles dissolvable in different solvents. 4) Hybrid nanoparticle through silica as supporting or coating material will exhibit superior multifunctional properties contributed by different inorganic nanomaterials (metal, semiconductor, fluorophore and magnetic material, etc.). Thus, hybrid nanoparticles based on silica has been pursued vigorously in the last few years and in this review chapter we provide a perspective on the present status of the subject. Described approaches include encapsulating inorganic nanomaterials using SiO_2 sphere, fluorophore-doped silica nanoparticles, constructing muti-functional hybrid materials using SiO_2 sphere as supporting material, and some important applications of hybrid nanoparticle based on silica. Finally, summary and outlook of these hybrid nanoparticles are also presented.

However, due to the explosion of publications in this field, we do not claim that this review includes all of the published work about hybrid nanoparticle based on silica. We apologize to the authors of much excellent work that due to the large activity in this field, we have unintentionally left out.

2. Encapsulating Inorganic Nanomaterials Using SiO_2 Sphere

2.1. Metal/Silica Core/Shell Nanostructure

Synthesis of metal nanomaterials such as gold and silver exhibiting particular optic, electronic, magnetic properties have received great intensity in the past decades. Particularly, the surface plasmon resonance (SPR) of metal nanomaterials (an important optical phenomenon of metal nanomaterials) can be tuned across the near-UV, visible, and near-infrared spectral range by varying the size, shape, and surrounding dielectric environment of particle. So, the SPR of silica coated metal nanomaterials can be facilely adjusted via changing the thickness of silica on the surface of metal nanomaterials. On the other hands, silica coated metal nanomaterials will commendably solve the bio-incompatible problem usually caused by protected ligands on the surface of metal nanomaterials. Most importantly, the stabilization and functionalization of metal nanomaterials with silica provides flexibility for a variety of applications, including bioassay, bioimaging and biosensor. In this part, we will discuss recent advances on encapsulating metal nanomaterials using SiO_2 sphere.

A reverse micelle and sol-gel technique has been employed to synthesize spherical Ag/SiO_2 composite particles [54, 55]. The size of the particles and the thickness of the coating can be controlled by manipulating the relative rates of the hydrolysis and condensation reactions of tetraethoxysilane (TEOS) within the microemulsion. For example, Adair et al. [54] employed the above method to facilely prepare silica coated Ag nanoparticles. The process is performed in conjunction with hydrolysis of TEOS, followed by condensation in the water nanodroplets to form a coating on the nanosized silver with reverse micelle structure. Later, Asher's group [55] reported a microemulsion method to fabricate nanocomposite silica colloidal particles containing homogeneously dispersed silver quantum dots.

Simple sol-gel technique can also be employed to fabricate silica coated metal nanoparticles. But, the silica-coating procedures reported in the literature generally involve surfaces with a significant chemical or electrostatic affinity for silica. Gold or silver metal has very little affinity for silica because it does not form a passivating oxide film in solution. So, introducing a primer layer into the surface of metal is very necessary for obtaining well-defined silica coated metal nanoparticles. (3-aminopropyl) trimethoxysilane (APTMS) and (3-mercaptopropyl) trimethoxysilane (MPTMS) as bifunctional molecules have been widely used as primers to fabricate well-defined silica coated metal nanoparticles [56-60]. For instance, Liz-Marzán et al. [56] showed that gold colloids have been homogeneously coated with silica using the silane coupling agent, APTMS, as a primer to render the gold surface vitreophilic. After the formation of a thin silica layer in aqueous solution, the particles can be transferred into ethanol for further growth using the Stöber method. The thickness of the silica layer can be completely controlled, and (after surface modification) the particles can be transferred into practically any solvent. Figure 1 shows the transmission electron microscopy (TEM) images of silica-coated gold particles with different diameters produced during the extensive growth of the silica shell. It is found that silica has homogeneously coated on the surface of gold particles. Later, Murphy and coworkers [60] employed MPTMS as a coupling agent to fabricate silica coated gold nanorods. But, in this case, it is found that for small aspect ratio nanorods, the reproducibility is poor and particle aggregation (again, preferentially through the tips) is usually observed during the sodium silicate addition step. Poly-(vinylpyrrolidone) (PVP) [61] as a functional molecule can also be employed to fabricate silica coated metal nanoparticles because PVP can provide good link points for OH group. Graf et al. [61] developed a general method to coat colloids with silica. The amphiphilic, nonionic polymer, PVP, was adsorbed to various colloidal particles such as small gold colloids, gold-shell silica-core particles, small and large silver colloids. After this functionalization, the stabilized particles could be transferred to a solution of ammonia in ethanol and directly coated with smooth and homogeneous silica shells of variable thickness by addition of tetraethoxysilane in a seeded growth process. Figure 2 shows the typical TEM images of silica coated gold nanoparticles. From the different images, gold nanoparticles have been homogeneously coated with silica. Although the above method using PVP as a coupling agent was interesting and could be applied to many particle systems, some particles such as CTAB protected gold nanoparticles or nanorods coated through PVP capping have been found to fail for these particles. CTAB-coated nanoparticles can indeed be transferred into ethanol upon functionalization with the slightly negatively charged polymer PVP, however, controlled hydrolysis and condensation of TEOS on the nanoparticle surface could not be achieved, probably because the remaining CTAB promote the formation of silica [62]. In order to solve the above problem, Pastoriza-Santos et al. [62] developed a novel silica-coating procedure that has been devised for CTAB-stabilized gold nanorods but that can be readily extended to other CTAB-stabilized nanoparticles [62]. The method comprises a combination of the polyelectrolyte layer-by-layer (LBL) technique and the hydrolysis and condensation of TEOS in a 2-propanol-water mixture and leads to homogeneous coatings with tight control on shell thickness. Figure 3 shows the effectiveness of this method for deposition of silica shells with accurate thickness control. For thin shells (Figure 3a),the silica surface is rough, probably because of the presence of polyelectrolyte chains on the nanoparticles, whereas upon subsequent additions of TEOS, the thickness of the silica shells increases (Figure 3b, c and d) and the surface quickly becomes smoother.

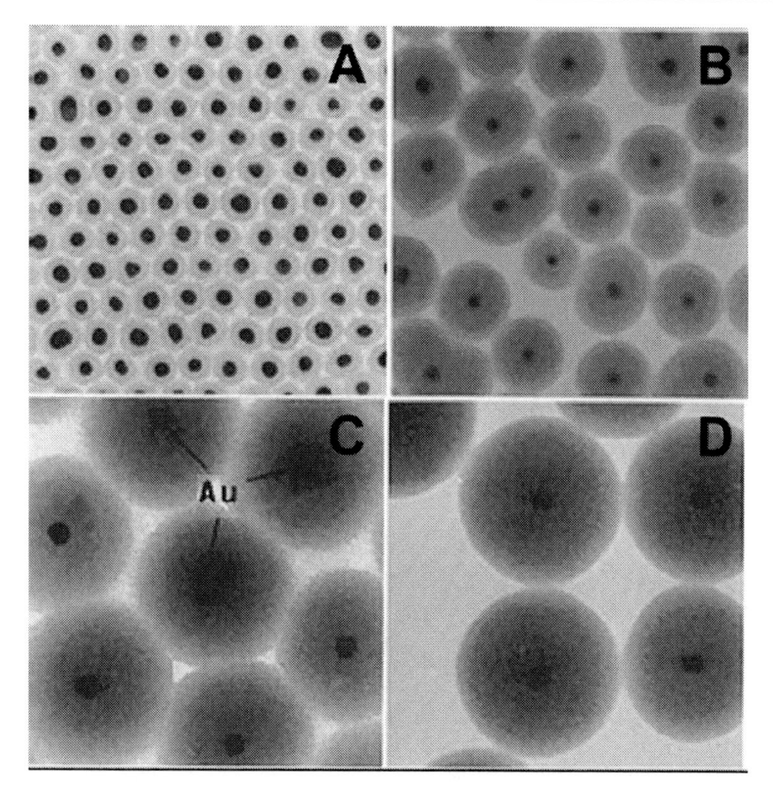

Figure 1. TEM images of silica-coated gold particles produced during the extensive growth of the silica shell around 15 nm Au particles with TEOS in 4:1 ethanol/water mixtures. The shell thicknesses are (A) 10 nm, (B) 23 nm, (C) 58 nm, and (D) 83 nm. Reprinted with permission from Ref. 56, L. M. Liz-Marzán, Langmuir 12, 4329 (1996), Copyright American Chemical Society (1996).

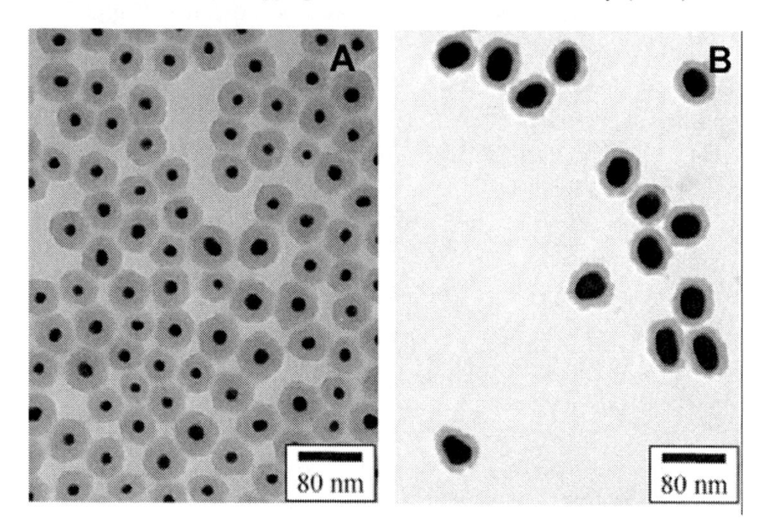

Figure 2. TEM images of gold particles coated with silica. (A) Gold particles (7nm, polydispersity 15%) with an 18nm shell (total polydispersity 9%) grown by two additions of TEOS. (B) Gold particles (20 nm radius, polydispersity 12%) with 12 nm shell (total polydispersity 9%). Reprinted with permission from Ref. 61, C. Graf, Langmuir 19, 6693 (2003), Copyright American Chemical Society (2003).

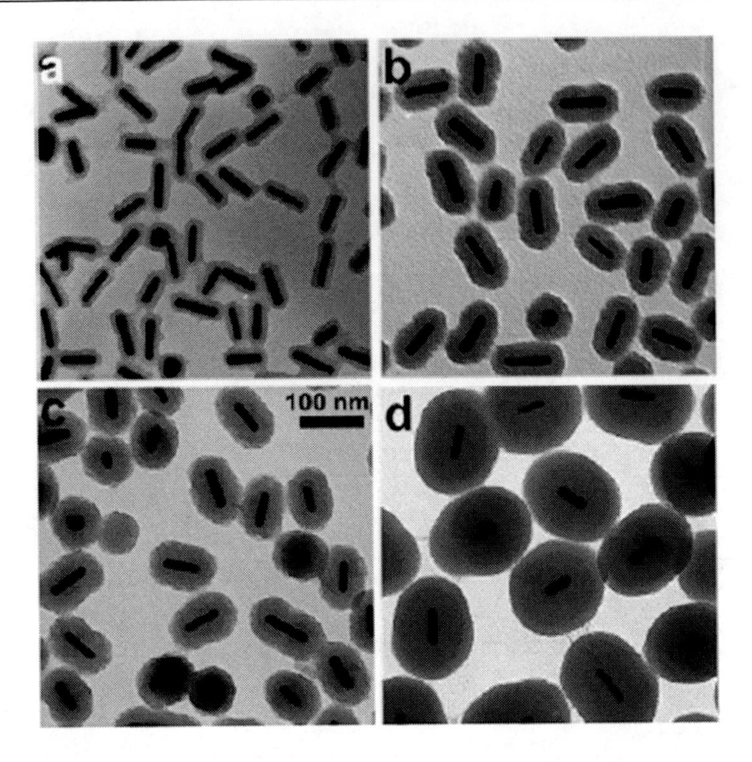

Figure 3. TEM images of silica-coated gold nanorods, with silica shell thickness increasing from a to d. The scale is the same for all images. Reprinted with permission from Ref. 62, Pastoriza-Santos, Chem. Mater. 18, 2465 (2006). Copyright American Chemical Society (2006).

Preparation of silica coated metal nanoparticles via sol-gel technique without the addition of primer have received great interests in recent years because it will greatly reduce the cost of materials and avoid the toxicity cause by the introduction of primers. Several excellent papers [63-66] have reported to fabricate the silica coated metal nanoparticles without the use of primer as coupling agent. Han et al. [63] developed a facile synthetic route to prepare monodisperse Au@SiO_2 nanoparticles with a homogeneous silica shell via well-established silica surface chemistry with readily available silane coupling agents. This simple preparation process is very fast, allowing the direct and efficient coating of the pretreated citrate-stabilized gold nanoparticles with silica, without the use of any specific, surface-coupling silane agents or large stabilizers (PVP). Typical TEM images of three Au@SiO_2 samples with different silica-shell thicknesses (~35, 75, and 90 nm) are shown in Figure 4. It is observed that monodisperse Au@SiO_2 nanoparticles with 50 nm gold cores and uniform thicknesses of silica shell were successfully prepared.

2.2. Semiconductor/Silica Core/Shell Nanostructure

Colloidal semiconductor nanocrystals have attracted extensive interest because of their variety of size- and shape-dependent optical and electrical properties and various applications such as in optoeletronic devices, photovoltaic devices, and biological fluorescence labeling. Over the past two decades, many methods for synthesizing high-quality semiconductor nanocrystals via water- or organic-phase routes have been well developed. However, semiconductor nanocrystals obtained are usually toxic, which caused by heavy metal ions effect (Cd^{2+}, Hg^{2+}). Thus, their application as in-vitro or in-vivo bioimaging agents will receive tremendous challenge, which is an agonizing problem puzzling many scientists around the word. On the other hand, although QDs have shown higher sensitivity and better photostability and chemical stability than conventional fluorophore markers [67], their photophysical behavior is usually affected by the use of solvents, ligands, environments and photoionization of QDs. In order to improve the photostability of QDs, they need to be encapsulated within a rigid matrix. Silica is an ideal choice, and it can be applied for a coating using a versatile sol-gel process. Thus, silica can act as a robust, inert layer against the degradation of optical properties and also imparts water solubility. In this part, we will discuss recent advances on encapsulating semiconductor nanomaterials (soluble in water or organic phase) using SiO_2 sphere.

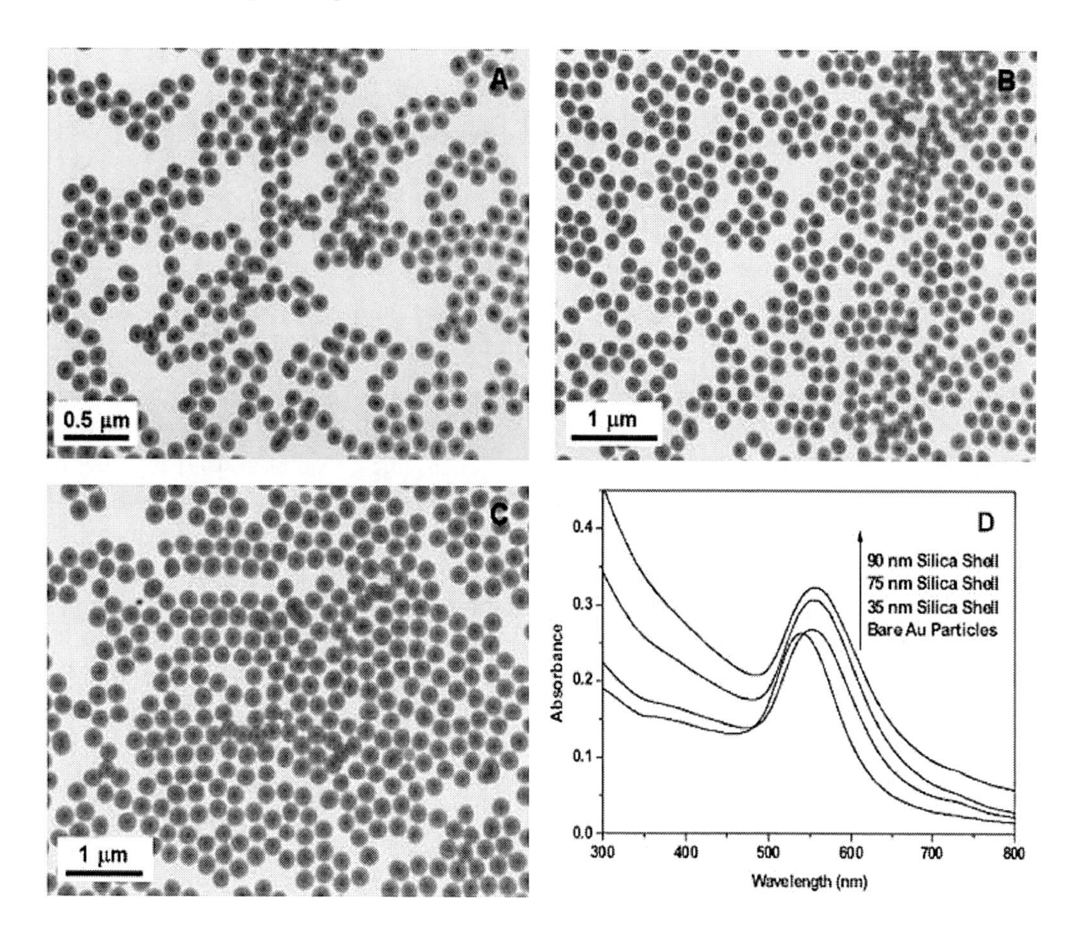

Figure 4. TEM images of Au@SiO_2 nanoparticles with silica-shell thickness of A) 35, B) 75, and C) 90 nm. D) UV-vis spectra of citrate-stabilized 50 nm gold nanoparticles with their corresponding Au@SiO_2 nanoparticles with silica-shell thickness of 35, 75, and 90 nm. Reprinted with permission from Ref. 63, M. Han, Adv. Funct. Mater. 15, 961 (2005), Copyright Wiley-VCH (2005).

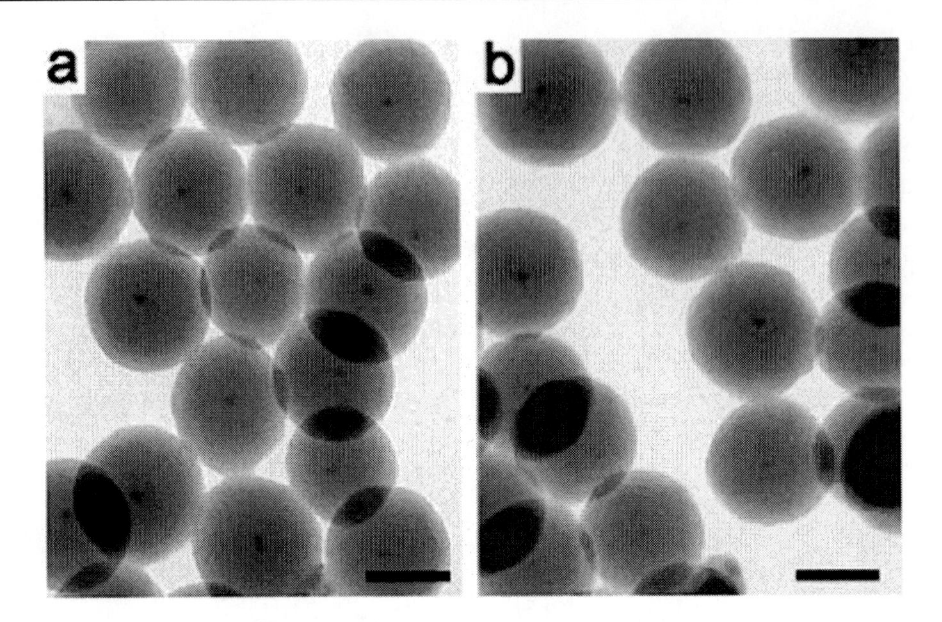

Figure 5. TEM images of core-shell-structured CdTe@SiO_2 particles prepared under different initial concentrations of CdTe QDs: a) 1.35×10^{-5} M; b) 5.40×10^{-6} M (with respect to CdTe QDs). The scale bars correspond to 50 nm. Reprinted with permission from Ref. 69, M. Gao, Adv. Mater. 17, 2354 (2005), Copyright Wiley-VCH (2005).

Semiconductor nanoparticles with water solubility coated with homogeneous silica can also be facilely prepared through the above methods for preparing silica coated metal nanoparticles. The combination of microemulsion and sol-gel technique has been developed to fabricate CdS and CdTe nanoparticles [68, 69]. For instance, Gao et al. [69] reported the preparation of CdTe@SiO_2 core/shell structured fluorescent spheres via the above method. CdTe nanocrystals stabilized by thioglycolic acid and 1-thioglycerol were first synthesized. Under optimal conditions, highly fluorescent silica spherical particles with single CdTe nanocrystal cores were obtained. Figure 5 shows the typical TEM images of two different CdTe@SiO_2 samples with different shell thicknesses. In this system, the thickness of silica can be easily controlled via changing the initial concentration of CdTe nanocrystals used. Furthermore, some primers such as PVP and MPTMS can also be used as coupling agents for water-soluble semiconductor nanocrystals to prepare silica coated fluorescent hybrid nanospheres. Zhang et al. [70] demonstrated that PVP-stabilized $NaYF_4$ nanocrystals could be directly coated with a uniform layer of silica. Kotov and coworkers [71] reported that CdTe nanowires with bristled SiO_2 coatings resembling nanoscale centipedes were produced in a modified Stöber process when mercaptosuccinic acid and MPTMS were used as a stabilizer and coupling agent, respectively. In addition, coating CdTe nanocrystals with homogeneous silica have also successfully demonstrated by Zhang et al. [72] and Weller et al. [73] via derivatived-silane as coupling agents.

Coating hydrophobic semiconductor nanoparticles with silica will have tremendous significance because high fluorescent quantum dots prepared via organic-phase route for application in biology usually renders them amenable to soluble in water. So, in the past years, a great deal of work has been reported to prepare silica coated quantum dots (organic solubility). In general, there exist two approaches to prepare silica coated hydrophobic semiconductor nanoparticles. One approach involves the use of a surface primer (a silane coupling agent) before introducing TEOS, ammonium and water into QD solution. Several excellent papers [74-83] have been mentioned to prepare silica coated QDs using MPTMS or APTMS as a coupling agent. For instance, Alivisatos et al. [77] reported that hydrophobic CdSe/ZnS core/shell nanocrystals with a core size between 2 and 5 nm are embedded in a siloxane shell through adding MPTMS as a coupling agent. Later, Nann and coworkers [81] extended the above method to directly prepare photostable, highly luminescent, water-soluble core-shell CdSe/ZnS/SiO_2 hybrid nanoparticles. To our great interest, this novel method could coat single luminescent QDs with silica shells. Figure 6 shows high-resolution TEM images of single QDs encapsulated in silica. The size of the particles is approximately 30 nm and the great majority contain single QDs at the center. In addition, the diameter of the final silica particles can be tuned from approximately 30 nm to 1 mm by means of seeded growth. More recently, Ying et al. [82] have developed a novel silica-coating method for QDs in toluene using commercially available silanes. Particularly, this silica-coating procedure can be facilely suitable for various hydrophobic inorganic nanoparticles, including Au, Ag, Fe_3O_4, and ZnS-CdSe nanoparticles. Although the success has been achieved through the above Stöber-based method and additionally introducing coupling agents, the above Stöber-based method have major disadvantages such as high requirements on purity of the reactants, the difficulty and multiplicity of the preparation steps, and the fact that it is not possible to directly coat nanoparticles with nonpolar ligands. The W/O microemulsion system in conjunction with the Stöber synthesis and silane coupling method can well solve the above problem and be used for the preparation of silica-coated semiconductor nanocrystals [84-87]. Ying et al. [84] described a simple strategy for making plain CdSe quantum dot water soluble by coating with silica. In this microemulsion system, the introducing a surfactant (Igepal CO-520) is very necessary. The surfactant interaction prior to silica coating allows the hydrophobic QDs to be encapsulated within the aqueous domains of the reverse microemulsion. Figure 7a shows typical TEM image of homogeneously dispersed 40 nm silica nanoparticles prepared by the Igepal microemulsion method. Single QDs (Figure 7b) or multiple QDs (Figure 7c, d) could be facilely encapsulated within each silica nanoparticles.

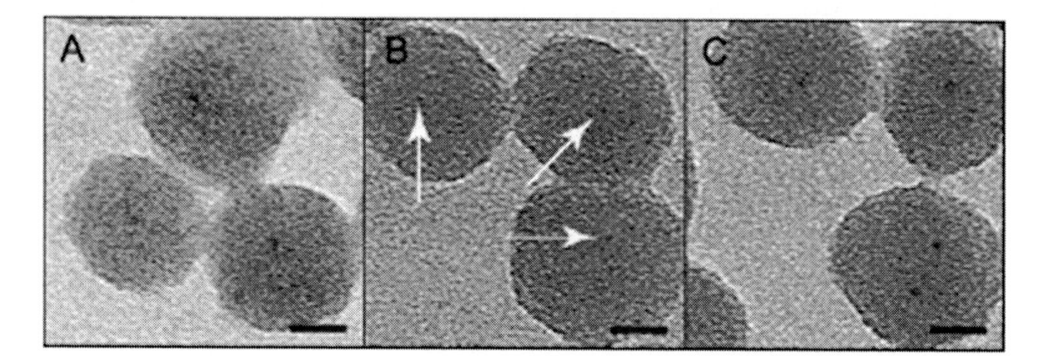

Figure 6. HRTEM images of single QDs encapsulated within a silica shell. A) QD/SiO_2 spheres of approximately 30 nm diameter. B) QDs are misarranged with respect to the electron beam. The arrows indicate the positions of the QDs. C) SiO_2 sphere with two QDs incorporated (lower right). Scalebars: 10 nm. Reprinted with permission from Ref. 81, T. Nann, Angew. Chem. Int. Ed. 43, 5393 (2004), Copyright Wiley-VCH (2004).

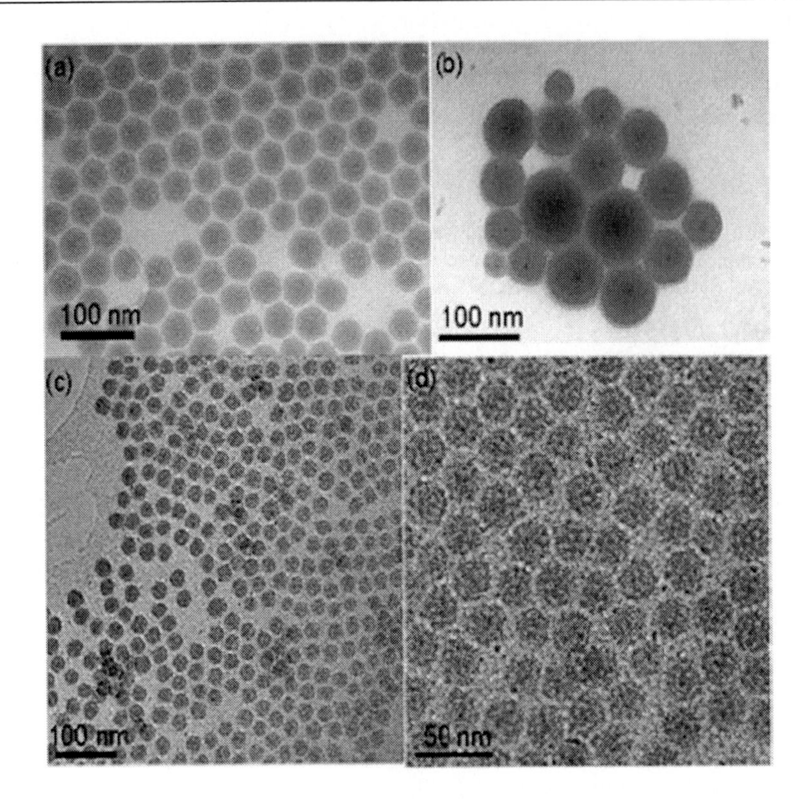

Figure 7. TEM image (a) of 40 nm SiO_2 nanoparticles (control) prepared by the Igepal microemulsion method. b) TEM image of 5 nm CdSe QDs individually encapsulated within SiO_2 nanoparticles. c, d) TEM images of multiple 5 nm CdSe QDs encapsulated within 25 nm SiO_2 nanoparticles. Unencapsulated QDs were also observed when the QD concentration was high. Reprinted with permission from Ref. 84, J. Y. Ying, Adv. Mater. 17, 1620 (2005), Copyright Wiley-VCH (2005).

Ying et al. [87] further extended this method to prepare silica coating of near-Infrared-emitting PbSe QDs. In addition, Nann and coworkers [86] found that introduction an polymer, poly (ethylene glycol) nonylphenyl ether into a microemulsion system could also be used to prepare spherical PbSe@SiO_2 core/shell nanoparticles. These core/shell nanoparticles obtained have a spherical shape with narrow size distribution and smooth surfaces. Importantly, the size of the particles and the thickness of the shells can be easily controlled by manipulating the relative rates of the hydrolysis and condensation reactions of TEOS within the microemulsion.

2.3. Magnetic Nanoparticle/Silica Core/Shell Nanostructure

Magnetic nanoparticles have been the focus of intense research for researchers from a wide range of disciplines such as magnetic fluids, catalysis, biotechnology/biomedicine, magnetic resonance imaging, data storage, and environmental remediation [88]. Particularly, synthesis of superparamagnetic magnetic nanoparticles, the size of which is below a critical value (typically around 10-20 nm) has gained more attention due to their special features and important applications in biomedicine. A number of suitable methods have been developed for the synthesis of magnetic nanoparticles of various different compositions. However, monodisperse magnetic nanoparticles, which are intensively desirable for biomedicine, are usually synthesized via organic phase route at high temperature. A great deal of organic ligands existed on the surface of magnetic nanoparticles will greatly limit the bio-application of magnetic materials themselves. On the other hand, small magnetic particles (synthesized via aqueous phase route) usually tend to form agglomerates to reduce the energy associated with the high surface area to volume ratio of the nanosized particles. Moreover, naked magnetic nanoparticles are chemically highly active, and are easily oxidized in air, resulting generally in loss of magnetism [88]. In order to solve the above problem, a great deal of methods have been developed to design novel magnetic nanoparticle/silica core/shell nanostructure. A silica shell does not only protect the magnetic cores, but can also prevent the direct contact of the magnetic core with additional agents linked to the silica surface, thus avoiding unwanted interactions. Most importantly, the protecting shells can also be used for further functionalization with other functional nanoparticles or various ligands, depending on the desired applications. In this part, we will discuss recent advances on encapsulating magnetic nanoparticles using SiO_2 sphere.

Two different approaches have been explored to generate silica coatings on the surfaces of magnetic particles. The first method relied on the well-known Stöber process, in which silica was formed in-situ on the surface of magnetic nanoparticles, particularly iron oxide particles, through the hydrolysis and condensation of a sol-gel precursor. For magnetic oxide, because the iron oxide surface has a strong affinity toward silica, no primer was required to promote the deposition and adhesion of silica. This method was originally applied to ferromagnetic rod-like nanoparticles [89], then to micrometer-sized hematite colloids by Matijevic and co-workers [90], and later extended to other magnetic nanoparticles by a number of research groups [91-100]. For instance, Xia et al. [93] developed a sol-gel approach for the coating of superparamagnetic iron oxide nanoparticles with uniform shells of amorphous silica. The thickness of silica coating could be conveniently controlled in the range of 2-100 nm by changing the concentration of the sol-gel solution. Figure 8 shows the typical TEM images of iron oxide nanoparticles whose surfaces had been coated with silica shells using different TEOS concentrations. When the thickness of silica coating was increased, the core-shell nanoparticles became more monodispersed because of a reduction in the relative size distribution. Figure 8D shows the HRTEM image of a silica-coated iron oxide nanoparticle. This image clearly indicates the single crystallinity of the iron oxide core and the amorphous nature of the silica shell. Later, Gao and coworkers [97] extended this simple method to prepare silica coated big iron oxide nanosphere. It is found that homogeneous silica can still coat on the surface of magnetic nanoparticles. Another important advance on silica coated magnetic nanoparticles was focused on the work reported by Salgueiriño-Maceira et al. [101, 102] and Luis M. Liz-Marzán et al. [103]. Particularly, Salgueiriño-Maceira et al. [101] highlighted the preparation of stable silica-coated cobalt nanoparticles through silica coating on the surface of cobalt nanoparticles, which is in-situ produced via the reduction of Co^{2+} in the presence of $NaBH_4$ and citrate. This method is simple, rapid and good reproducibility. Depending on the experimental conditions, both the diameter of the magnetic core and the thickness of the dielectric shell can be easily tuned. Figure 9 shows the typical TEM images of spherical cobalt particles coated by a uniform silica shell. It can be seen that the magnetic core varied from 32 (Figure 9a and b) to 95 (Figure 9c) nm. Carefully observing the above experimental results, it is found that it is hard to reach the good result that silica can be homogeneously coated on the surface of magnetic nanoparticles. Combining microemulsion synthesis and the stöber method probably provides a state-of-art strategy for obtaining well-defined magnetic nanoparticle/silica core/shell nanostructure because the microemulsion synthesis can be used to confine and control the coating of silica on core nanoparticles. Similar to the above metal or semiconductor coating process, the microemulsion procedure for obtaining silica coated magnetic nanoparticles still needs to introduce some surfactants into the solution. For instance, Igepal CO-520 [104-107], Triton X-100 [104] and Brij-97 [104] have been employed by several groups to prepare silica coated magnetic nanoparticles. The noticeable example is the work reported by Korgel et al. [106], who demonstrated that colloidal FePt nanocrystals, 6 nm in diameter, could be coated with silica shells in the microemulsion system. Figure 10a shows a TEM image of the oleic acid/oleylamine-capped FePt nanocrystals prior to coating with silica. The nanocrystals were relatively size-monodisperse with an average diameter of 6 nm. After coating with silica, monodisperse silica coated FePt nanoparticles can be facilely (Figure 10b and c). The SiO_2 shell thickness could be controlled from 7 to 23 nm by adjusting the TEOS concentration and the ratio of TEOS to FePt nanocrystals used during the coating step.

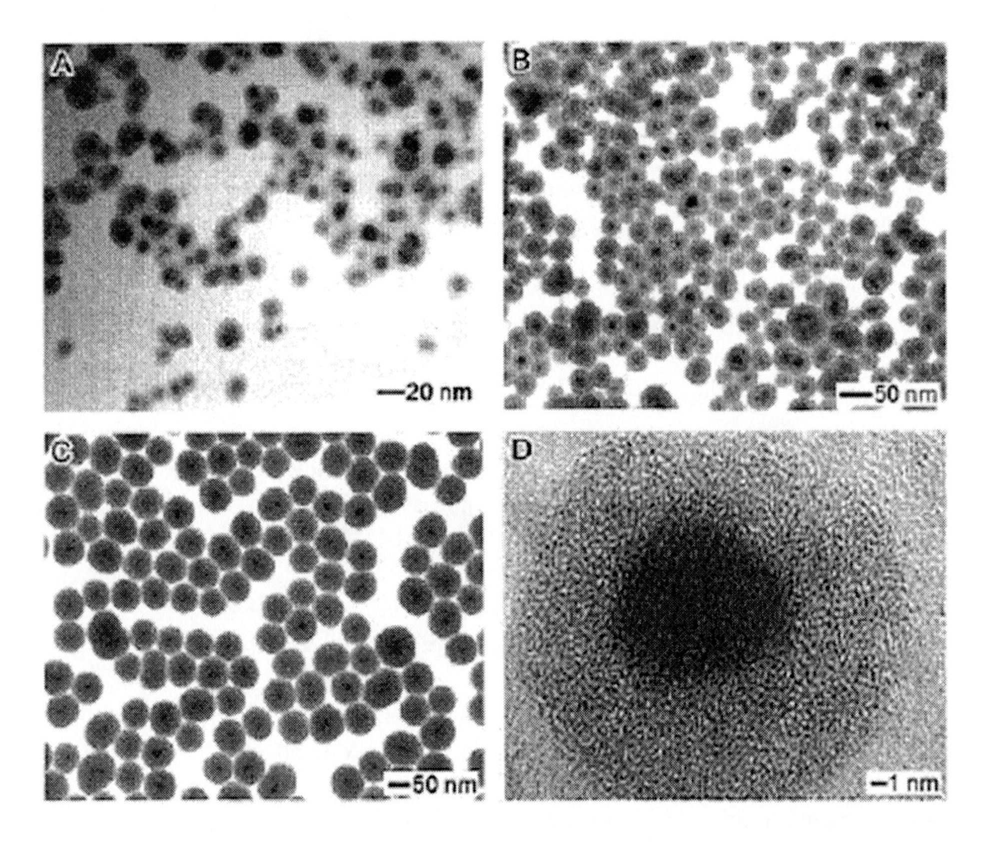

Figure 8. (A-C) TEM images of iron oxide nanoparticles whose surfaces have been coated with silica shells of various thicknesses. In this case, the thickness of silica coating could be adjusted by controlling the amount of precursor added to the solution: (A) 10, (B) 60, and (C) 1000 mg of TEOS to 20 mL of 2-propanol. (D) A HRTEM image of the iron oxide nanoparticle whose surface has been uniformly coated with 6 nm of amorphous silica shell. Reprinted with permission from Ref. 93, Y. Xia, Nano. Lett. 2, 183 (2002), Copyright American Chemical Society (2002).

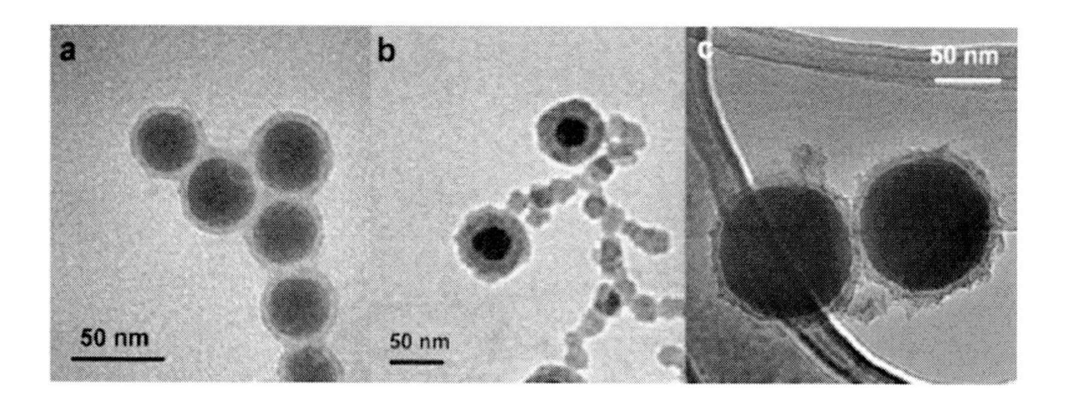

Figure 9. TEM images of silica coated cobalt nanoparticles with 32 (a and b) and 95 (c) nm cores. Reprinted with permission from Ref. 101, V. Salgueiriño-Maceira, J. Mater. Chem.16, 3593 (2006), Copyright Royal Society of Chemistry (2006).

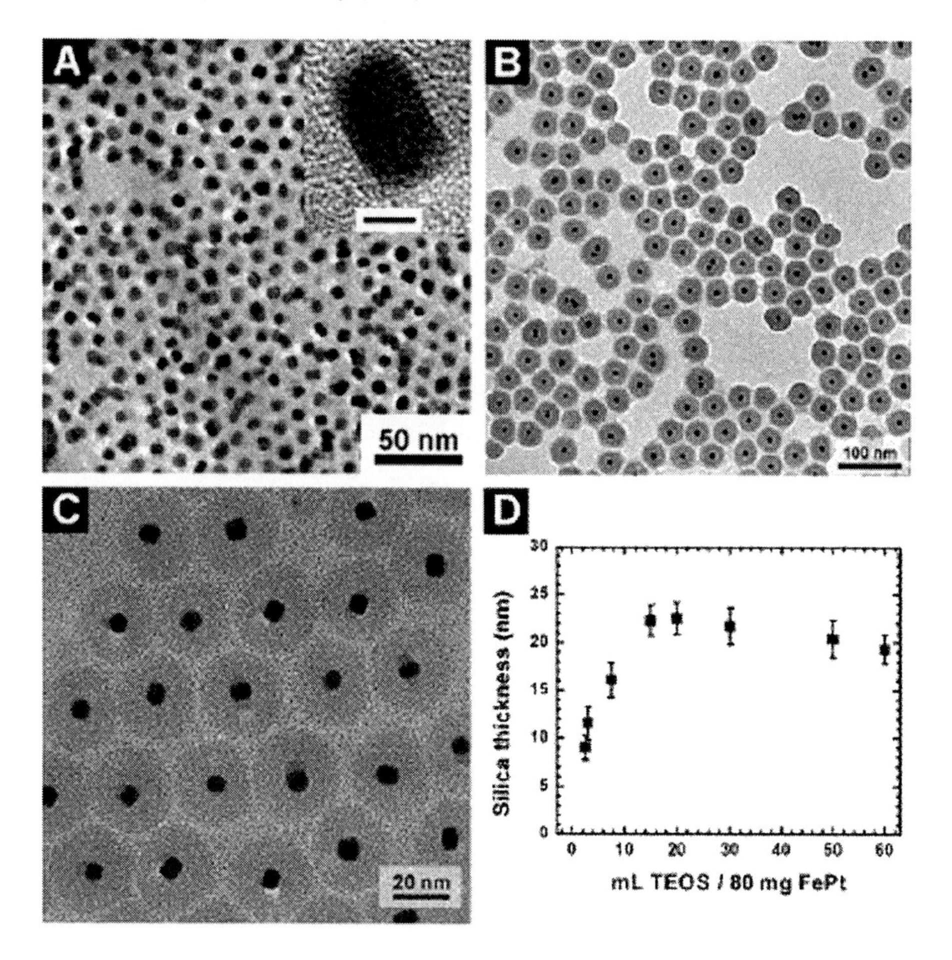

Figure 10. TEM images of FePt nanocrystals (A) prior to and (B and C) after coating with SiO_2. (The inset scale bar in part A is 3 nm.) (D) The SiO_2 shell thickness obtained as a function of TEOS concentration. Reprinted with permission from Ref. 106, B. A. Korgel, J. Phys. Chem. B 110, 11160 (2006), Copyright American Chemical Society (2006).

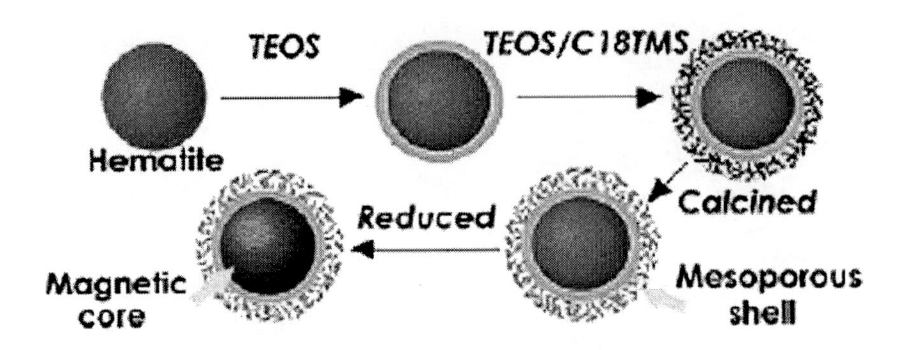

Figure 11. Illustration of synthesis of uniform magnetic nanospheres with a magnetic core/mesoporous silica shell structure. Reprinted with permission from Ref. 111, J. Shi, J. Am. Chem. Soc. 127, 8916, (2005), Copyright American Chemical Society (2005).

Herein, another important hybrid nanoparticles based on silica containing magnetic nanoparticles is mesoporous silica coated magnetic nanoparticles. This hybrid nanostructure is very important for magnetic induced catalytic application, which is caused by the properties of magnetic nanoparticles and mesoporous silica as individual component. Inspired by this, several excellent papers have mentioned to prepare this hybrid nanostructrure [108-112]. For instance, Shi et al. [111] reported a novel scheme to fabricate uniform magnetic nanospheres with a magnetic core/mesoporous silica shell structure. The synthesis strategy is presented in Figure 11. A thin and dense silica layer was first deposited on the surface of hematite particles of desired thickness in order to protect the iron oxide core from leaching into the mother system under acidic circumstances. Then, the mesoporous silica shell was formed from simultaneous sol-gel polymerization of tetraethoxysilane (TEOS) and n-octadecyltrimethoxysilane (C18TMS) followed by removal of the organic group. Finally, the hematite cores of the nanospheres were reduced in a flowing gas mixture of H_2 and N_2 to produce the final hybrid nanospheres. In addition, the thickness of could be easily tuned by changing the concentration of the TEOS/C18TMS mixture. This novel hybrid material may be applicable in targeted drug delivery, catalysis and multiphase separation in the future.

In summary, we have reviewed recent advances on preparation of silica coated inorganic nanoparticles via different coating techniques. On one hand, It should be noted that searching a general, simple and rapid method to preparation of metal (or semiconductor, magnetic nanoparticle)@silica core/shell nanoparticles is still great challenge. On the other hand, currently, the as-prepared inorganic nanoparticles generally own different surface properties such as negative or positive charges and ligands. We have to choose the best method for the as-prepared inorganic nanoparticles and render them amenable to facilely functionalize with homogeneous silica.

3. Fluorophore-Doped Silica Nanoparticles

Although silica coated semiconductor nanoparticles could be facilely obtained and used as bioprobe, an unavoidable question is the fact that the each hybrid nanoparticles own one or several semiconductor nanoparticles, which leads to low quantum yield. On the other hands, fluorohore such as dyes has been employed as optical probe by many groups for carrying out bioassay such as detecting target DNA, protein, etc. However, because one DNA or other probes can only be labeled with one or a few fluorophores, the fluorescence signal is too weak to be detected when the target concentration is low. In addition, the poor photostability of many fluorophores is anothor changlleging question. Most organic dyes suffer serious photobleaching, often resulting in irreproducible signals for ultratrace bioanalysis. Fluorophore-doped silica nanoparticles developed recently, which has a silica matrix that entraps a large number of fluorophores, can be used as alternative substitutes for solving the above existed questions because of their major advantages such as strong fluorescence signal, highly photostable performance benefiting from the silica matrix shielding effect, easy surface modifications for bioconjugation, and size uniformity and tunability.

Thus, many efforts have been directed towards synthesizing fluorophore-doped silica nanoparticles and further exploring their advanced applications [113-132]. For example, inorganic dyes such as tris(2,2'-bipyridyl)dichlororuthenium(II) hexahydrate ((Ru(bpy)) doped silica hybrid nanoparticles have been facilely obtained by several groups via the microemulsion methods [113-124]. Particularly, Tan's groups [119] studied the formation of Ru(bpy) dye-doped silica nanoparticles by ammonia-catalyzed hydrolysis of TEOS in water-in-oil microemulsion in details. It is found that the fluorescence spectra, particle size, and size distribution of Ru(bpy) dye-doped silica nanoparticles could be easily controlled via manipulating some parameters such as reactant concentrations (TEOS and ammonium hydroxide), nature of surfactant molecules, and molar ratios of water to surfactant and cosurfactant to surfactant. This optimization study of the preparation of dye-doped silica nanoparticles provides a fundamental knowledge of the synthesis and optical properties of Ru(bpy) dye-doped silica nanoparticles. But, the fluorescence intensity of inorganic dye doped silica hybrid nanoparticle is still limited due to the relatively low quantum yield of inorganic dyes. To achieve high signal amplification in ultra-sensitive bioanalysis, silica nanoparticles doped with highly fluorescent organic dyes are in great demand as organic dyes ususlly have higher quantum yield. Several important results have been reported by Tan et al. to synthesize organic dyes doped silica nanoparticles [125-132]. For instance, they have developed highly fluorescent, extremely photostable, and biocompatible organic-dye doped silica nanoparticles using a reverse microemulsion [125]. Their strategy was to link a hydrophilic dextran molecule to an organic dye molucule (hydrophobic) and use acidic conditions in the water core of a reverse microemulsion system to create a strong attractive force between the dextran linked dye and the silica matrix. In addition, the size of hybrid nanoparticles can be easily tuned by varying the microemulsion conditions. Furthermore, this method can be also extended to obtain mutil-fluorophores doped silica nanoparticles [120, 127, 129, 132]. Besides inorganic or organic dye doped silica matrix, some other inorganic molecules such as polyoxometalates (POMs) could also be doped into silica matrix to form hybrid system [133, 134]. Green et al. [133] immobilized the POMs and the enhancing agent (amine-rich species) in silica spheres, which confines the POM to the core of the particle while increasing the emission of the material. The above fluorophore doped silica nanoparticles will be probably useful in widespread fields such as bioimaging, bioassay and electrochemical detection, etc..

4. Constructing Muti-functional Hybrid Materials Using SiO_2 Sphere as Supporting Material

In the above part, we have simply reviewed the case that many scientists around the world have developed diverse methods for synthesizing stylish functional inorganic nanoparticles with particular optical, electronic and magnetic properties. With the development of nanoscience and nanotechnology, the clever combination of different nanoscale materials can probably lead to the development of multifunctional nano-assembly system with optical, electronic and magnetic properties simultaneously. Silica, as a high stable supporting material, can commendably complete the above intention due to its excellent surface chemistry. In recent years, several groups have devoted their momentous efforts to design multifunctional advanced nanoarchitectures using biocompatible SiO_2 spheres as supporting materials.

Magnetic fluorescent nanoparticles using silica as supporting material have received great attention because of their bifunctional properties [97, 135-141]. Recently, magnetic fluorescent nanoparticles with magnetic nanoparticles located at the interior of silica shell have been facilely obtained [97, 135-136]. For instance, a new class of highly fluorescent, photostable, and magnetic core/shell nanoparticles in the submicrometer size range has been synthesized by Salgueiriño-Maceira et al. [135] from a modified Stöber method combined with the layer-by-layer (LbL) assembly technique. In this synthesis system, the luminescent magnetic nanoparticles were prepared via two main steps. The first step involves controlled addition of tetraethoxysilane to a dispersion of $Fe_3O_4/\gamma\text{-}Fe_2O_3$ nanoparticles, which are thereby homogeneously incorporated as cores into monodisperse silica spheres. The second step involves the LbL assembly of polyelectrolytes and luminescent CdTe quantum dots onto the surfaces of the silica-coated magnetite/maghemite particles, which are finally covered with an outer shell of silica. Nanoparticles with such a core/shell architecture have the added benefit of providing a robust platform from silica for incorporating diverse functionalities into a single nanoparticle. In addition, silica coated magnetic fluorescent nanospheres have been well-defined demonstrated by Ying et al. [138]. For example, a novel water-soluble hybrid material consisting of QDs and magnetic nanoparticles encapsulated in a silica shell has been facilely obtained via the microemulsion method [138]. Furthermore, silica nanospheres can also be used as temples for obtaining magnetic and fluorescent nanoparticles supported on the surface of silica simultaneously [139]. Hyeon and coworkers [139] designed a muti-step strategy for obtaining multifunctional composite nanoparticle assemblies by subsequent assembly of nanoparticles of CdSe/ZnS, Au or Pd on the magnetite nanoparticle-bearing silica spheres. The general synthetic procedure for multifunctional nanoparticle/silica sphere assemblies is shown in Figure 12. First, uniformly sized silica spheres were prepared by the Stöber method and then functionalized with amino groups by treatment with APTMS. Second, the Fe_3O_4 nanoparticles were assembled on the surfaces of the amino-functionalized silica spheres. Third, functional nanoparticles of CdSe/ZnS, Au or Pd were additionally assembled on the uncovered surface of SiO_2 spheres to give multifunctional assemblies. The synthesized multifunctional silica spheres exhibited a combination of magnetism and luminescence (CdSe/ZnS), surface plasmon resonance (Au) or catalysis (Pd). Silica coated magnetic nanoparticles and dye molecules can also provide muti-functional properties [140-141] besides the above muti-functional properties provided by magnetic and fluorescent nanoparticles. Mou and coworkers [140] developed a new strategy for the synthesis of uniform magnetic/luminescent functional mesoporous silica nanoparticles with well-ordered porous structure and good aqueous dispersity through the muti-step strategy. In the above muti-functional nanostructure, the dye molecule (FITC) was employed to supply the fluorescent function of hybrid spheres.

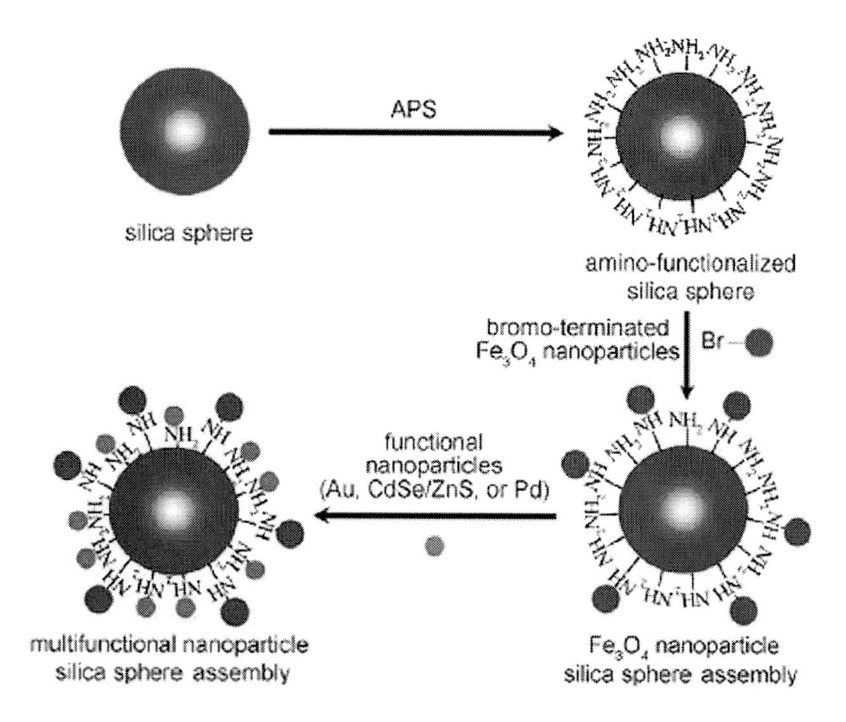

Figure 12. Synthetic procedure to obtain multifunctional nanoparticle/silica sphere assemblies. Reprinted with permission from Ref. 139, T. Hyeon, Angew. Chem. Int. Ed. 45, 4789 (2006), Copyright Wiley-VCH (2006).

Similarly, magnetic optical or magnetic catalytic hybrid nanospheres have also received great intensity in the recent years due to their applications in photothermal therapy, magnetic resonance imaging and catalysis. Several successful examples included the work reported by Salgueiriño-Maceira [142], Mirkin [143], Cho [144] and Ying [145]. For example, Salgueiriño-Maceira et al. [142] developed a method for obtaining bifunctional magnetic and optically tunable nanoparticles with a structural design involving a magnetic iron oxide core (Fe_3O_4/γ-Fe_2O_3) surrounded by a thick silica shell and further covered with an outer shell of gold. The whole preparation procedure is followed. The first step was carried out by using magnetic nanoparticles, which were coated with a silica shell using a sol-gel process based on the hydrolysis of TEOS. In this step, no primer was required because the iron oxide surface has a strong affinity for silica. The second step involved the assembly gold nanoparticles on the surface of the silica coated magnetic nanoparticles. Finally, gold shell was form on the surface of above assembling system. These obtained gold-coated magnetic silica spheres will take advantage of the strong resonance absorption gold shells show in the visible or near-infrared (NIR) range and can be controlled by using an external magnetic field, which makes them very promising in biomedical applications. Cho et al. [144] also found that magnetic and optical nanomaterials can also coated on the surface of monodisperse silica nanospheres through an advanced strategy. Figure 13a shows the synthetic approach for typical preparation process. First, 100 nm silica spheres were synthesized using the Stöber method, and then the surfaces of the particles were modified with 3-aminopropyltrimethoxysilane (Figure 13 b). Second, the Fe_3O_4 nanoparticles were then covalently attached to the amino-modified silica spheres (Figure 13c). Third, gold seed nanoparticles of 1-3 nm were attached to the residual amino groups of the silica spheres (Figure 13d). Finally, a complete 15-nm-thick gold shell with embedded Fe_3O_4 nanoparticles was formed around the silica spheres (Figure 13e). In addition, magnetic catalytic hybrid nanospheres have been well developed by Ying et al. [145], which reported on the synthesis of a nanocomposite of Pd nanoclusters supported on silica coated Fe_2O_3 nanoparticles. It is found that these bifunctional nanocomposites provided excellent catalytic reactivity and reusability in the hydrogenation of nitrobenzene.

5. Important Applications of Hybrid Nanoparticles Based on Silica

It is well known that ultimate target of nanomaterials will commendably complete their applications in many fields such as electronics, optics, optoelectronics, catalysis, sensor, biomedicine and others. Many groups have gained great benefits from studying nanoscience and nanotechnology, especially exploring the applications of different nanomaterials. Every year, excellent papers have still been published in great quantity for reporting the important applications caused by unique properties of nanomaterials. Unexceptionally, in this part, we will simply discuss recent advances on several important applications of hybrid nanoparticles based on silica.

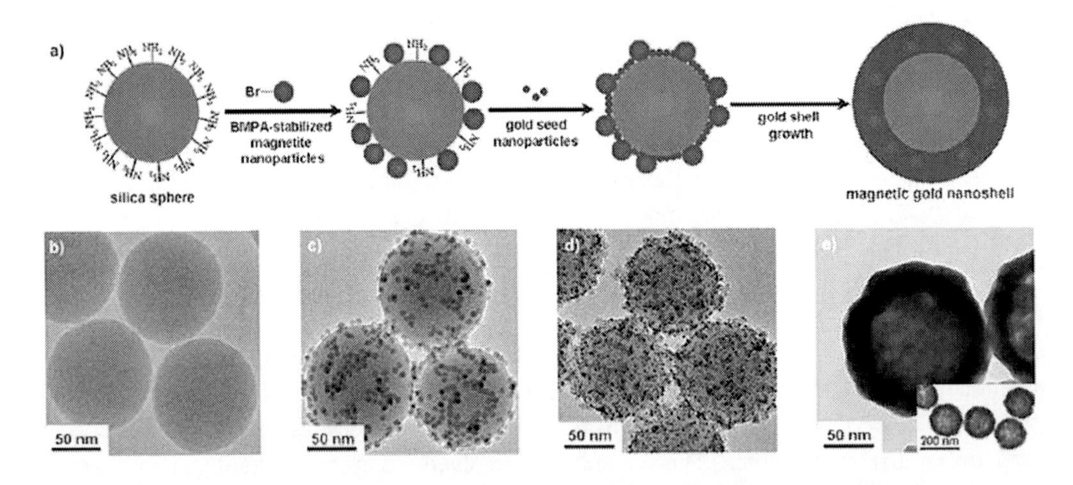

Figure 13. a) Synthesis of the magnetic gold nanoshells (Mag-GNS). TEM images of b) amino-modified silica spheres, c) silica spheres with Fe_3O_4 (magnetite) nanoparticles immobilized on their surfaces, d) silica spheres with Fe_3O_4 and gold nanoparticles immobilized on their surfaces, and e) the Mag-GNS. Reprinted with permission from Ref. 144, M. H. Cho, Angew. Chem. Int. Ed. 45, 7754 (2006), Copyright Wiley-VCH (2006).

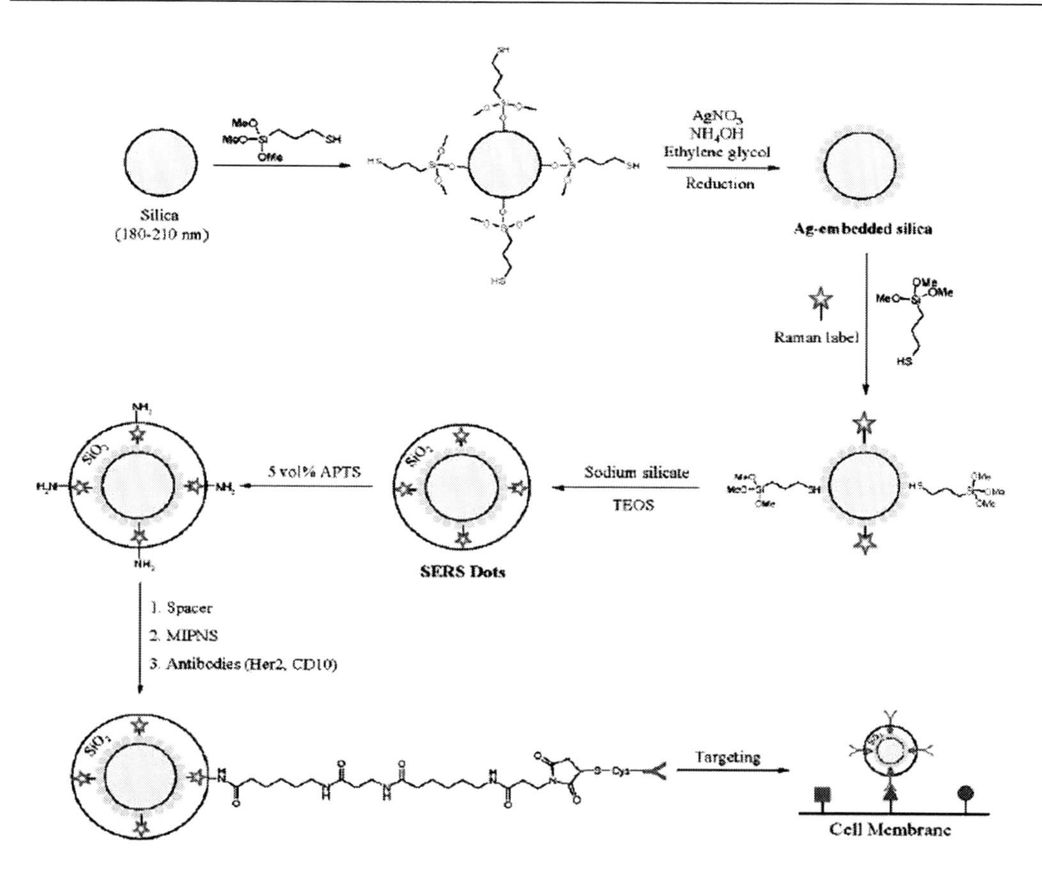

Figure 14. Schematic illustration of SERS dots synthesis and targeting of cellular cancer markers. Reprinted with permission from Ref. 147, Y. S. Lee, Anal. Chem. 78, 6967 (2006), Copyright American Chemical Society (2006).

Fluorescent or magnetic fluorescent nanoparticles based on silica have initially been used as in the field of biological and biomedical imaging [136, 137, 140, 146]. For instance, Wang et al. [136] designed a poly(N-isopropylacrylamide) (PNIPAM)-coated luminescent/magnetic silica microspheres. Further, the author demonstrated that PNIPAM-covered luminescent/magnetic microspheres could also be able to be taken up by CHO cells without inducement of any specific ligands. Ying and coworkers [137] demonstrated that Fe_2O_3/CdSe heterodimers could be coated with a thin homogeneous silica shell and used for the labeling of different live cell membranes through a simple bioconjugation method. Another important application of hybrid nanoparticles based on silica was focused on bioanalyis, biosensor and biodection [63, 66, 143, 147-148]. For instance, Ma et al. [66] designed a Au nanorod@SiO_2 nanostructure via an improved Stöber methods. These hybrid nanostructures can be easily deposited on the solid surface and further functionalized with goat anti-human-immunoglobulin G (anti-h-IgG). Finally, these anit-h-IgG modified hybrid nanostructures was successfully employed for the colorimetric detection (changes in the absorption of light) of h-IgG in a model reaction based on the specific binding affinity between the proteins. Significantly, the protein-recognition event on the Au rod@SiO_2 films could not only be monitored by changes in the absorption of light, but also was observable by the naked eye. In addition, construction of biosensors based on surface-enhanced Raman scatting (SERS) using hybrid nanoparticle based on silica is a promising direction for biodetection. For instance, Lee et al. [147] developed biocompatible, photostable, and multiplexing-compatible surface-enhanced Raman spectroscopic tagging material (SERS dots) composed of silver nanoparticle-embedded silica spheres and organic Raman labels for cellular cancer targeting in living cells. A typical preparation procedure for SERS dots synthesis and targeting of cellular cancer markers is shown in Figure 14. Finally, the obtained antibody-conjugated SERS dots were successfully demonstrated to be applied to the targeting of HER2 and CD10 on cellular membranes and exhibited good specificity for the targets. Furthermore, hybrid nanoparticles based on silica can also be used to some other important fields such as bioseparation, phototherapy and catalysis. Prominent examples include template-assisted fabrication of magnetic mesoporous silica-magnetite nanocomposite [149] and its potential for application in magnetic bioseparation (its ability to bind and elute DNA and extract RNA from bacterial cells) and preparation of magnetic gold nanoshells supported silica nanospheres for magnetic resonance imaging and photothermal therapeutic applications [150] and synthesis of magnetic nanocomposite catalysts for application in the catalytic hydrogenation of nitrobenzene with ease of recycling via magnetic separation [145].

Dye doped silica nanoparticles can be easily functionalized with different sensing probes and useful in many important fields such as biomedicine, nanobiology, and nanoelectrochemistry. Typical examples include leukemia cell identification using Ru(bpy) doped silica as an opticl probe [116, 117], parallel and high-throughput signaling of biomolecules by employing dual-dye based silica nanoparticles [120], multiplexed bacteria monitoring [127], staining probes for affymetrix genechips [128], and ultrasensitive DNA detection [130]. More recently, Ru(bpy) doped silica nanoparticles have been employed as advanced immobilization matrix for application in solid electrochemluminescence (ECL) system [113-115, 121, 123, 124]. For instance, Dong's group [113] reported a novel ECL sensor based on Ru(bpy) doped silica nanoparticles conjugated with a biopolymer chitosan membrane. Figure 15A shows cyclic voltammograms (CVs) of Ru(bpy) in the absence (dotted line) and presence (solid line) of 3×10^{-5} M tripropylamine (TPA) at the scan rate of 100 mV/s in PBS (pH 7.5). Since the TPA is quite small, it could permeate into the silica nanoparticle through the pores of the nanoparticles to react with the oxidized Ru(bpy). The presence of TPA made the oxidation current of Ru(bpy) increase clearly while the reduction current decreased. Figure 15B shows corresponding ECL intensity-time curve in the presence of 7.5×10^{-6} TPA under continuous potential scanning for ten cycles, which was no detectable change for ECL intensity. Furthermore, ECL intensity had good linearity with the TPA concentration and the linear range was wide, extending from 8.5×10^{-9} to 8.1×10^{-5} M with a remarkable detection limit (S/N=3) of 2.8 nM. Our group further extended to study the interactions between the Ru(bpy) dope silica nanoparticles and some biomacromolecules (BSA, lysozyme and ctDNA) [114]. The ECL behaviour of the RuSi nanoparticles was investigated after deposition with biomolecules through LBL self-assembly. When the surface of RuSi nanoparticles was modified using biomolecules, an ECL decrease with surface modification was observed. The more polyelectrolytes and biomolecules deposited on the RuSi nanoparticles, the more obvious ECL decrease was observed. The ECL decrease varied with different polyelectrolytes and biomolecules, and such decrease could be attributed to the changes of steric hindrance, limited diffusion of the coreactant and electrostatic interaction induced by surface modifications.

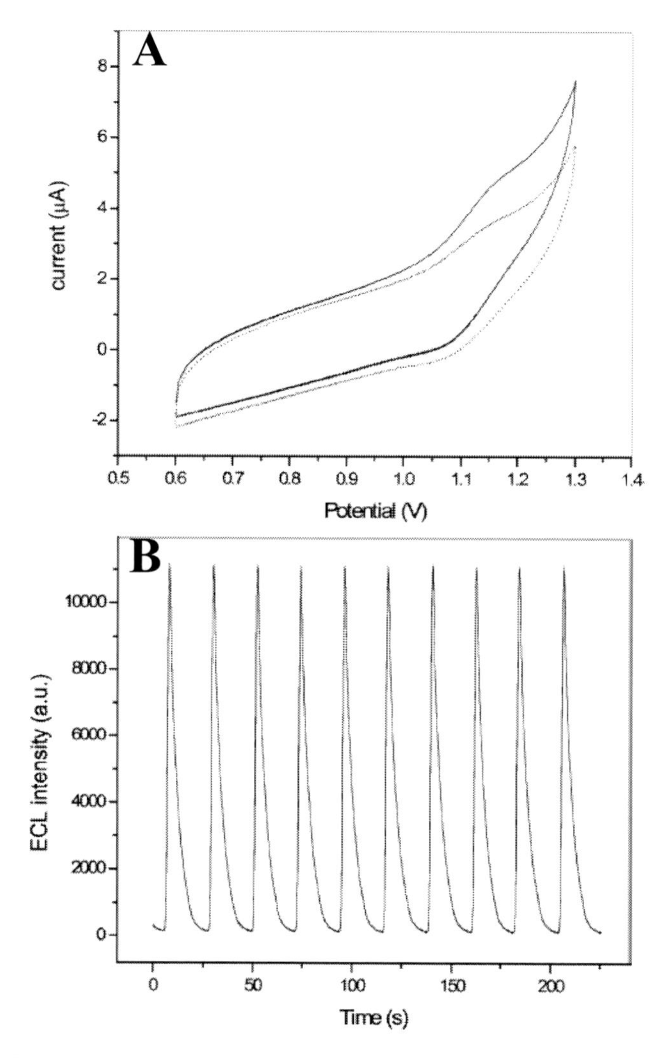

Figure 15. Cyclic voltammograms (A) of Ru(bpy) immobilized in composite film electrode in the absence (dashed line) and presence (solid line) of 3×10^{-5} M TPA in PBS (pH 7.5) at the scan rate of 100 mV/s; ECL intensity (B) of composite film electrode in PBS (pH 7.5) containing 7.5×10^{-6} M TPA under continuous CVs for 10 cycles with the scan rate of 100 mV/s. Reprinted with permission from Ref. 113, Y. S. Lee, Anal. Chem. 78, 5119 (2006), Copyright American Chemical Society (2006).

6. Summary and Outlook

The design of hybrid nanoparticle based on silica, covering a wide range of functional materials, has made substantial progress, especially over the past decade. However, synthesis of high-quality inorganic nanoparticles using disverse ligands with OH groups as protective agents is still challenges to be faced in the coming years. Although silica coated inorganic nanoparticles make great progress, silica coated inorganic nanomaterials with different morphologies (rod, wire, tube, plate, cube and other novel shapes) will a promising research field. This is because these nanomaterials with novel morphologies have been reported to exhibit more peculiar physical and chemical properties and many important applications in many fields. Furthermore, in the case of silica coated inorganic nanoparticles, it should be noted that it is difficult or impossible to achieve a fully dense and nonporous silica coating, and it is thus difficult to maintain high stability of these nanoparticles under harsh conditions. So, we must explore novel synthetic methods to ensure the stability of inorganic nanoparticles at other conditions. Thus, carbon or polymer coated inorganic nanoparticle as a good substitute developed recently, provides good opportunities for scientists engaging in coating materials.

Complex, multifunctional nanoparticle systems based on silica with designed active sites, including nanoparticles, ligands, enzymes, chiral catalysts, drugs, and other species, seem to be promising for a variety of applications. Optional combination of the above functional components will bring enormous opportunities for scientists engaging in physics, chemistry, biology, medicine, material science and their different interdisciplinary fields. On the other hand, it is noted that constructing muti-functional hybrid materials using SiO_2 sphere as supporting material usually needs complex and tedious assembly process, which usually needs long time and meantime has poor reproducible. Industrial applications of such systems can not be achieved based on current case. Recent developed bifunctional dimer nanoparticles [137,151-162] probably provide good opportunities for solving the above problem. Finally, whether industrial applications of such systems can be achieved and how far we can go in all those areas will greatly depend on our ability to synthesize hybrid nanoparticles based on silica in a simple, economical and scalable fashion.

About the Authors

Erkang Wang obtained his BS at the Univ. of Shanghai in 1952 and his PhD under Prof. J. Heyrovsky at the Czechoslovak Academy of Sciences, Prague, in 1959. He has received many national and international awards for his research, which involve electro- and bioelectrochemistry and environmental analysis. He was Visiting Prof. at many institutions and has presented over 150 seminars and 70 plenary and keynote lectures. He is the Member of Chinese Academy of Sciences (CAS) and the Third World Academy of Sciences (TWAS). His current research interests are bioelectrochemistry including biosensors and biomembranes, capillary electrophoresis, synthesis, assembly and application of micro/ nanomaterials and nanoelectrochemistry, etc.

Shaojun Guo is studying for his PhD under the supervision of Prof. Erkang Wang at Changchun Institute of Applied Chemistry, Chinese Academy of Sciences. He received her BS degree (2005) in college of chemistry from Jilin University. In 2005, he joined Prof. Wang's group in the Changchun Institute of Applied Chemistry, Chinese Academy of Sciences. His current scientific interests are focused on synthesis, design, assembly and application of functional micro/nanomaterials.

Acknowledgment

This work was supported by the National Science Foundation of China. (Nos. 20575064, 20675076 and 20735003)

References

[1] El-Sayed, M. A. *Acc. Chem. Res.* 2001, 34, 257-264.
[2] Murphy, C. J.; Jana, N. R. *Adv. Mater.* 2002, 14, 80-82.
[3] Kamat, P. V. J. Phys. Chem. B 2002, 106, 7729-7744.
[4] Zhang, J.; Liu, J.; Peng, Q.; Wang, X.; Li, Y. *Chem. Mater.* 2006, 18, 867-871.
[5] Lu, A.; Salabas, E. L.; Schüth, F. Angew. *Chem. Int. Ed.* 2007, 46, 1222-1244
[6] Srivastava, R.; Mani, P.; Hahn, N.; Strasser, P. Angew. *Chem. Int. Ed.* 2007, 46, 8988-8991.
[7] Chen, J.; Wang, D.; Xi, J.; Au, L.; Siekkinen, A.; Warsen, A.; Li, Z.-Y.; Zhang, H.; Xia, Y.; Li, X. *Nano Lett.* 2007, 7, 1318-1322.
[8] Jailswail, J. K.; Mattoussi, H.; Mauro, J. M.; Simon, S. M. *Nat. Biotechnol.* 2003, 21, 47-51.
[9] Huh, Y.; Lee, E.; Lee, J.; Jun, Y.; Kim, P.; Yun, C.; Kim, J.; Suh, J.; Cheon, *J. Adv. Mater.* 2007, 19, 3109-3112.
[10] Guo, S.; Fang, Y.; Dong, S.; Wang, E. *J. Phys. Chem.* C 2007, 111, 17104-17109.
[11] Doering, W. E.; Piotti, M. E.; Natan, M. J.; Freeman, R. G. *Adv. Mater.* 2007, 19, 3100-3108.

[12] Guo, S.; Wang, L.; Wang, E. *Chem. Commun.* 2007, 3163-3165.
[13] Guo, S.; Wang, E. *Inorg. Chem.* 2007, 46, 6740-6743.
[14] Guo, S.; Wang, E. *J. Colloid Interf. Sci.* 2007, 315, 795-799.
[15] Guo, S.; Wang, E. *Anal. Chim. Acta* 2007, 598, 181-192.
[16] Guo, S.; Wang, L.; Wang, Y.; Fang, Y.; Wang, E. *J. Colloid Interf. Sci.* 2007, 315, 363-368.
[17] Zhang, L.; Sun, X.; Song, Y.; Jiang, X.; Dong, S.; Wang, E. *Langmuir* 2006, 22, 2838-2843.
[18] Sun, X.; Dong, S.; Wang, E. *Polymer* 2004, 45, 2181-2184.
[19] Sun, X.; Jiang, X.; Dong, S.; Wang, E. *Macromol. Rapid Commun.* 2003, 24, 1024-1028.
[20] Cheng, W.; Dong, S.; Wang, E. *Chem. Mater.* 2003, 15, 2495-2501.
[21] Cheng, W.; Dong, S.; Wang, E. *Anal. Chem.* 2002, 74, 3599-3604.
[22] Cheng, W.; Dong, S.; Wang, E. *Langmuir* 2002, 18, 9947-9952.
[23] Cheng, W.; Dong, S.; Wang, E. *Langmuir* 2003, 19, 9434-9439.
[24] Cheng, W.; Wang, E. *J. Phys. Chem.* B 2004, 108, 24-26.
[25] Cheng, W.; Dong, S.; Wang, E. *J. Phys. Chem.* B 2004, 108, 19146-19154.
[26] Cheng, W.; Dong, S.; Wang, E. *Angew. Chem. Int. Ed.* 2002, 42, 449-452.
[27] Cheng, W.; Dong, S.; Wang, E. *J. Phys. Chem.* B. 2005, 109, 19213-19218.
[28] Sun, X.; Dong, S.; Wang, E. *Langmuir* 2005, 21, 4710-4712.
[29] Sun, X.; Dong, S.; Wang, E. *Angew. Chem. Int. Edit.* 2004, 43, 6360-6363.
[30] Sun, X.; Dong, S.; Wang, E. *Macromolecules* 2004, 37, 7105-7108.
[31] Xiong, Y.; McLellan, J. M.; Chen, J.; Yin, Y.; Li, Z.-Y.; Xia, Y. *J. Am. Chem. Soc.* 2005, 127, 17118-17127.
[32] Xiong, Y.; Cai, H.; Wiley, B. J.; Wang, J.; Kim, M. J.; Xia, Y. *J. Am. Chem. Soc.* 2007, 129, 3665-3675.
[33] Sun,Y.; Xia, Y. *Science* 2002, 298, 2176-2179.
[34] Tian, N.; Zhou, Z. Y.; Sun, S. G.; Ding, Y.; Wang, Z. L. *Science* 2007, 316, 732-735.
[35] Rao, C. N. R.; Vivekchand, S. R. C.; Biswas, Kanishka; Govindaraja, A.; *Dalton Trans.*, 2007, 3728-3749.
[36] Zheng, R.; Guo, S.; Dong, S. *Inorg. Chem.* 2007, 46, 6920-6923.
[37] Bao, H.; Wang, E.; Dong, S. *Small* 2006, 2, 476-480.
[38] Zheng, Y.; Gao, S.; Ying, J. Y. *Adv. Mater.* 2007, 19, 376-380.
[39] Pradhan, N.; Peng, X. *J. Am. Chem. Soc.* 2007, 129, 3339-3347.
[40] Pradhan, N.; Battaglia, D. M.; Liu,Y.; Peng, X. *Nano Lett.* 2007, 7, 312-317.
[41] Han, M.; Liu, Q.; He, J.; Song, Y.; Xu, Z.; Zhu, J. *Adv. Mater.* 2007, 19, 1096-1100.
[42] Ge, J.; Hu, Y.; Yin, Y. *Angew. Chem. Int. Ed.* 2007, 46, 7428-7431.
[43] Deng, H.; Li, X.; Peng, Q.; Wang, X.; Chen, J.; Li, Y. *Angew. Chem. Int. Ed.* 2005, 44, 2782 -2785.
[44] Ge, J.; Hu, Y.; Biasini, M.; Beyermann, W. P.; Yin, Y. *Angew. Chem. Int. Ed.* 2007, 46, 4342-4345.
[45] Chen, H. M.; Liu, R.; Li, H.; Zeng, H. *Angew. Chem. Int. Ed.* 2006, 45, 2713-2717.
[46] Gupta, A. K.; Gupta, M. *Biomaterials* 2005, 26, 3995-4021.
[47] Mornet, S.; Vasseur, S.; Grasset, F.; Verveka, P.; Goglio, G.; Demourgues, A.; Portier, J.; Pollert, E.; Duguet, E. *Prog. Solid State Chem.* 2006, 34, 237.
[48] Li, Z.; Wei, L.; Gao, M. Y.; Lei, H. *Adv. Mater.* 2005, 17, 1001-1005.

[49] Guo, S.; Fang, Y.; Dong, S.; Wang, E. *Inorg. Chem.* 2007, 46, 9537-9539.
[50] Li, J. J.; Wang, Y. A.; Guo, W.; Keay, J. C.; Mishima, T. D.; Johnson, M. B.; Peng, X. *J. Am. Chem. Soc.* 2003, 125, 12567-12575.
[51] Sun, S.; Murray, C. B.; Weller, D.; Folks, L.; Moser, A. *Science* 2000, 287, 1989-1992.
[52] Sun, S.; Zeng, H. *J. Am. Chem. Soc.* 2002, 124, 8204-8205.
[53] Park, J.; An, K. J.; Hwang, Y. S.; Park, J. G.; Noh, H. J.; Kim, J. Y.; Park, J. H.; Hwang, N. M.; Hyeon, T. *Nat. Mater.* 2004, 3, 891-895.
[54] Li, T.; Moon, J.; Morrone, A. A.; Mecholsky, J. J.; Talham, D. R.; Adair, J. H. *Langmuir* 1999, 15, 4328-4334.
[55] Wang, W.; Asher, S. A. J. *Am. Chem. Soc.* 2001, 123, 12528-12535.
[56] Liz-Marzán, L. M.; Giersig, M.; Mulvaney, P. *Langmuir* 1996, 12, 4329-4335.
[57] Ung, T.; Liz-Marzán, L. M.; Mulvaney, P. *Langmuir* 1998, 14, 3740-3748.
[58] Kobayashi, Y.; Correa-Duarte, M. A.; Liz-Marzán, L. M. *Langmuir* 2001, 17, 6375-6379.
[59] Salgueiriño-Maceira, V.; Caruso, F.; Liz-Marzán, L. M. *J. Phys. Chem.* B 2003, 107, 10990-10994.
[60] Obare, S. O.; Jana, N. R.; Murphy, C. *J. Nano. Lett.* 2001, 1, 601-603.
[61] Graf, C.; Vossen, D. L. J.; Imhof, A.; van Blaaderen, A. *Langmuir* 2003, 19, 6693-6700.
[62] Pastoriza-Santos, I.; Pérez-Juste, J.; Liz-Marzán, L. M. *Chem. Mater.* 2006, 18, 2465-2467.
[63] Liu, S.; Han, M. *Adv. Funct. Mater.* 2005, 15, 961-967.
[64] Yin, Y.; Lu, Y.; Sun, Y.; Xia, Y. *Nano. Lett.* 2002, 2, 427-430.
[65] Lu, Y.; Yin, Y.; Li, Z.; Xia, Y. *Nano. Lett.* 2002, 2, 785-788.
[66] Wang, C.; Ma, Z.; Wang, T.; Su, Z. Adv. *Funct. Mater.* 2006, 16, 1673-1678.
[67] Wu, X.; Liu, H.; Liu, J.; Haley, K. N.; Treadway, J. A.; Larson, J. P.; Ge, N. F.; Peale, F.; Bruchez, M. P. *Nat. Biotechnol.* 2003, 21, 41-46.
[68] Chang, S.; Liu, L.; Asher, S. A. *J. Am. Chem. Soc.* 1994,116, 6139-6744.
[69] Yang, Y.; Gao, M. *Adv. Mater.* 2005, 17, 2354-2357.
[70] Li, Z.; Zhang, Y. *Angew. Chem. Int. Ed.* 2006, 45, 7732-7735.
[71] Wang, Y.; Tang, Z.; Liang, X.; Liz-Marzán, L. M.; Kotov, N. A. Nano. Lett. 2004, 4, 225-231.
[72] Wolcott, A.; Gerion, D.; Visconte, M.; Sun, J.; Schwartzberg, A.; Chen, S.; Zhang, J. Z. *J. Phys. Chem.* B 2006, 110, 5779-5789.
[73] Schroedter, A.; Weller, H.; Eritja, R.; Ford, W. E.; Wessels, J. M. *Nano. Lett.* 2002, 2, 1363-1367.
[74] Rogach, A. L.; Nagesha, D.; Ostrander, J. W.; Giersig, M.; Kotov, N. A. *Chem. Mater.* 2000, 12, 2676-2685.
[75] Parak, W. J.; Gerion, D.; Zanchet, D.; Woerz, A. S.; Pellegrino, T.; Micheel, C.; Williams, S. C.; Seitz, M.; Bruehl, R. E.; Bryant, Z.; Bustamante, C.; Bertozzi, C. R.; Alivisatos, A. P. *Chem. Mater.* 2002, 14, 2113-2119.
[76] Bruchez, M. J.; Moronne, M.; Gin, P.; Weiss, S.; Alivisatos, A. P. *Science* 1998, 281, 2013-2016.
[77] Gerion, D.; Pinaud, F.; Williams, S. C.; Parak, W. J.; Zanchet, D.; Weiss, S.; Alivisatos, A. P. *J. Phys. Chem.* B 2001, 105, 8861-8871.

[78] Mokari, T.; Sertchook, H.; Aharoni, A.; Ebenstein, Y.; Avnir, D.; Banin, U. *Chem. Mater.* 2005, 17, 258-263.
[79] Chan, Y.; Zimmer, J. P.; Stroh, M.; Steckel, J. S.; Jain, R. K.; Bawendi, M. G. *Adv. Mater.* 2004, 16, 2092-2097.
[80] Sorensen, L.; Strouse, G. F.; Stiegman, A. E. *Adv. Mater.* 2006, 18, 1965-1967.
[81] Nann, T.; Mulvaney, P. *Angew. Chem. Int. Ed.* 2004, 43, 5393-5396.
[82] Jana, N. R.; Earhart, C.; Ying, J. Y. *Chem. Mater.* 2007, 19, 5074-5082.
[83] Rogach, A. L.; Kotov, N. A.; Koktysh, D. S.; Susha, A. S.; Caruso, F. Colloid Surf. A: Physicochem. *Eng. Aspects* 2002, 202, 135-144.
[84] Selvan, S. T.; Tan, T. T.; Ying, J. Y. *Adv. Mater.* 2005, 17, 1620-1625.
[85] Darbandi, M.; Thomann, R.; Nann, T. *Chem. Mater.* 2005, 17, 5720-5725.
[86] Darbandi, M.; Lu, W.; Fang, J.; Nann, T. *Langmuir* 2006, 22, 4371-4375.
[87] Tan, T. T.; Selvan, S. T.; Zhao, L.; Gao, S.; Ying, J. Y. *Chem. Mater.* 2007, 19, 3112-3117.
[88] Lu, A.; Salabas, E. L.; Schüth, F. *Angew. Chem. Int. Ed.* 2007, 46, 1222-1244.
[89] James, R. O.; Bertucci, S. J.; Oltean, *G. L. U.S. Patent* 5, 217, 804, 1993.
[90] Ohmori, M.; Matijevic, E. J. *Colloid Interface Sci.* 1992, 150, 594-598.
[91] Tang, D.; Yuan, R.; Chai, Y.; An, H. *Adv. Funct. Mater.* 2007, 17, 976-982.
[92] Correa-Duarte, M. A.; Giersig, M.; Kotov, N. A.; Liz-Marzán, L. M. *Langmuir* 1998, 14, 6430-6435.
[93] Lu, Y.; Yin, Y.; Mayers, B. T.; Xia, Y. Nano. *Lett.* 2002, 2, 183-186.
[94] Philipse, A. P.; van Bruggen, M. P. B.; Pathmamanoharan, C. *Langmuir* 1994, 10, 92-99.
[95] Aliev, F. G.; Correa-Duarte, M. A.; Mamedov, A.; Ostrander, J. W.; Giersig, M.; Liz-Marzán, L. M.; Kotov, N. A. *Adv. Mater.* 1999, 11, 1006-1010.
[96] Cannas, C.; Musinu, A.; Peddis, D.; Piccaluga, G. Chem. Mater. 2006, 18, 3835-3842.
[97] Chen, M.; Gao, L.; Yang, S.; Sun, J. *Chem. Commun.* 2007, 1272-1274.
[98] Xu, X.; Deng, C.; Gao, M.; Yu, W.; Yang, P.; Zhang, X. *Adv. Mater.* 2006, 18, 3289-3293.
[99] Deng, Y.; Qi, D.; Deng, C.; Zhang, X.; Zhao, D. 2008, 130, 28-29.
[100] Ge, J.; Hu, Y.; Zhang, T.; Yin, Y. *J. Am. Chem. Soc.* 2007, 129, 8974-8975.
[101] Salgueiriño-Maceira, V.; Correa-Duarte, M. A. *J. Mater. Chem.* 2006, 16, 3593–3597.
[102] Salgueiriño-Maceira, V.; Correa-Duarte, M. A.; Farle, M.; López-Quintela, M. A.; Sieradzki, K.; Diaz, R. *Langmuir* 2006, 22, 1455-1458.
[103] Kobayashi, Y.; Horie, M.; Konno, M.; Rodriguez-González, B.; Liz-Marzán, L. M. *J. Phys. Chem.* B 2003, 107, 7420-7425.
[104] Yi, D. K.; Lee, S. S.; Papaefthymiou, G. C.; Ying, J. Y. *Chem. Mater.* 2006, 18, 614-619.
[105] Vestal, C. R.; Zhang, Z. J. *Nano. Lett.* 2003, 3, 1739-1743.
[106] Lee, D. C.; Mikulec, F. V.; Pelaez, J. M.; Koo, B.; Korgel, B. A. *J. Phys. Chem. B* 2006, 110, 11160-11166.
[107] Santra, S.; Tapec, R.; Theodoropoulou, N.; Dobson, J.; Hebard, A.; Tan, W. *Langmuir* 2001, 17, 2900-2906.
[108] Meng, Y.; Chen, D.; Jiao, X. *J. Phys. Chem.* B 2006, 110, 15212-15217.
[109] Wu, P.; Zhu, J.; Xu, Z. Adv. *Funct. Mater.* 2004, 14, 345-351.
[110] Sen, T.; Sebastianelli, A.; Bruce, I. J. *J. Am. Chem. Soc.* 2006, 128, 7130-7131.

[111] Zhao, W.; Gu, J.; Zhang, L.; Chen, H.; Shi, J. *J. Am. Chem. Soc.* 2005, 127, 8916-8917.
[112] Lu, A. H.; Li, W. C.; Kiefer, A.; Schmidt, W.; Bill, E.; Fink, G.; Schüth, F. *J. Am. Chem. Soc.* 2004, 126, 8616-8617.
[113] Zhang, L.; Dong, S. *Anal. Chem.* 2006, 78, 5119-5123.
[114] Wei, H.; Liu, J.; Zhou, L.; Li, J.; Jiang, X.; Kang, J.; Yang, X.; Dong, S.; Wang, E. *Chem. Eur.* J. 2008, accepted.
[115] Wei, H.; Wang, E. *Chem. Lett.* 2007, 36, 210-211.
[116] Santra, S.; Zhang, P.; Wang, K.; Tapec, R.; Tan, W. *Anal. Chem.* 2001, 73, 4988-4993.
[117] Herr, J. K.; Smith, J. E.; Medley, C. D.; Shangguan, D.; Tan, W. *Anal. Chem.* 2006, 78, 2918-2924.
[118] Wang, X.; Zhou, J.; Yun, W.; Xiao, S.; Chang, Z.; He, P.; Fang, Y. *Anal. Chim. Acta* 2007, 598, 242–248.
[119] Bagwe, R. P.; Yang, C.; Hilliard, L. R.; Tan, W. *Langmuir* 2004, 20, 8336-8342.
[120] Wang, L.; Yang, C.; Tan, W. *Nano. Letter.* 2005, 5, 37-43.
[121] Chang, Z.; Zhou, J.; Zhao, K.; Zhu, N.; He, P.; Fang, Y. *Electrochim. Acta* 2006, 52, 575–580.
[122] Santra, S.; Wang, K.; Tapec, R.; Tan, W. *J. Biomed. Optics* 2001, 6, 160-166.
[123] Zhang, L.; Dong, S. *Electrochem. Commun.* 2006, 8, 1687-1691.
[124] Zhang, L.; Liu, B.; Dong, S. *J. Phys. Chem.* B 2007, 111, 10448-10452.
[125] Zhao, X.; Bagwe, R. P.; Tan, W. *Adv. Mater.* 2004, 16, 173-176.
[126] Zhou, X.; Zhou, *J. Anal. Chem.* 2004, 76, 5302-5312.
[127] Wang, L.; Zhao, W.; O'Donoghue, M. B.; Tan, W. *Bioconjugate Chem.* 2007, 18, 297-301.
[128] Wang, L.; Lofton, C.; Popp, M.; Tan, W. *Bioconjugate Chem.* 2007, 18, 610-613.
[129] Kim, S. H.; Jeyakumar, M.; Katzenellenbogen, J. A. *J. Am. Chem. Soc.* 2007, 129, 13254-13264.
[130] Zhao, X.; Tapec-Dytioco, R.; Tan, W. *J. Am. Chem. Soc.* 2003, 125, 11474-11475.
[131] Montalti, M.; Prodi, L.; Zaccheroni, N.; Battistini, G.; Marcuz, S.; Mancin, F.; Rampazzo, E.; Tonellato, U. *Langmuir* 2006, 22, 5877-5881.
[132] Wang, L.; Tan, W. *Nano. Lett.* 2006, 6, 84-88.
[133] Green, M.; Harries, J.; Wakefield, G.; Taylor, R. *J. Am. Chem. Soc.* 2005, 127, 12812-12813.
[134] Qi, W.; Li, H.; Wu, L. *Adv. Mater.* 2007, 19, 1983-1987.
[135] Salgueiriño-Maceira, V.; Correa-Duarte, M. A.; Spasova, M.; Liz-Marzán, L. M.; Farle, M. *Adv. Funct. Mater.* 2006, 16, 509–514.
[136] Guo, J.; Yang, W.; Wang, C.; He, J.; Chen, *J. Chem. Mater.* 2006, 18, 5554-5562.
[137] Selvan, S. T.; Patra, P. K.; Ang, C. Y.; Ying, J. Y. *Angew. Chem. Int. Ed.* 2007, 46, 2448 -2452.
[138] Yi, D. K.; Selvan, T.; Lee, S. S.; Papaefthymiou, G. C.; Kundaliya, D.; Ying, J. Y. *J. Am. Chem. Soc.* 2005, 127, 4990-4991.
[139] Kim, J.; Lee, J. E.; Lee, J.; Jang, Y.; Kim, S.W.; An, K.; Yu, J. H.; Hyeon, T. *Angew. Chem. Int. Ed.* 2006, 45, 4789-4793.
[140] Lin, Y. S.; Wu, S. H.; Hung, Y.; Chou, Y. H.; Chang, C.; Lin, M. L.; Tsai, C. P.; Mou, C. Y. *Chem. Mater.* 2006, 18, 5170-5172.
[141] Ma, D.; Guan, J.; Normandin, F.; Dénommeé, S.; Enright, G.; Veres, T.; Simard, B. *Chem. Mater.* 2006, 18, 1920-1927.

[142] Salgueiriño-Maceira, V.; Correa-Duarte, M. A.; Farle, M.; López-Quintela, A.; Sieradzki, K.; Diaz, R. *Chem. Mater.* 2006, 18, 2701-2706.
[143] Stoeva, S. I.; Huo, F.; Lee, J. S.; Mirkin, C. A. *J. Am. Chem. Soc.* 2005, 127, 15362-15363.
[144] Kim, J.; Park, S.; Lee, J. E.; Jin, S. M.; Lee, J. H.; Lee, I. S.; Yang, I.; Kim, J. S.; Kim, S. K.; Cho, M. H.; Hyeon, T. *Angew. Chem. Int. Ed.* 2006, 45, 7754-7758.
[145] Yi, D. K.; Lee, S. S.; Ying, J. Y. *Chem. Mater.* 2006, 18, 2459-2461.
[146] Santra, S.; Zhang, P.; Wang, K.; Tapec, R.; Tan, W. *Anal. Chem.* 2001, 73, 4988-4993.
[147] Kim, J. H.; Kim, J. S.; Choi, H.; Lee, S. M.; Jun, B. H.; Yu, K. N.; Kuk, E.; Kim, Y. K.; Jeong, D. H.; Cho, M. H.; Lee, Y. S. *Anal. Chem.* 2006, 78, 6967-6973.
[148] Zhang, L.; Liu, B.; Dong, S. *J. Phys. Chem.* B 2007, 111, 10448-10452.
[149] Sen, T.; Sebastianelli, A.; Bruce, I. J. *J. Am. Chem. Soc.* 2006, 128, 7130-7131.
[150] Kim, J.; Park, S.; Lee, J. E.; Jin, S. M.; Lee, J. H.; Lee, I. S.; Yang, Il.; Kim, J. S.; Kim, S. K.; Cho, M. H.; Hyeon, T. *Angew. Chem. Int. Ed.* 2006, 45, 7754-7758.
[151] Choi, J. S.; Jun, Y. W.; Yeon, S. I.; Kim, H. C.; Shin, J. S.; Cheon, J. *J. Am. Chem. Soc.* 2006, 128, 15982-15983.
[152] Chiang, I. C.; Chen, D. H. *Adv. Funct. Mater.* 2007, 17, 1311-1316.
[153] Yu, H.; Cen, M.; Rice, P. M.; Wang, S. X.; White, R. L.; Sun, S. *Nano Lett.* 2005, 5, 379-382.
[154] Pellegrino, T.; Fiore, A.; Carlino, E.; Giannini, C.; Cozzoli, P. D.; Ciocarella, G.; Respaud, M.; Palmirotta, L.; Cingolani, R.; Manna, L. *J. Am. Chem. Soc.* 2006, 128, 6690-6698.
[155] Xu, Z.; Hou, Y.; Sun, S. *J. Am. Chem. Soc.* 2007, 129, 8698-8699.
[156] Li, Y. Q.; Zhang, Q.; Nurmikko, A. V.; Sun, S. *Nano Lett.* 2005, 5, 1689-1692.
[157] Gu, H. W.; Zheng, R. K.; Zhang, X. X.; Xu, B. *J. Am. Chem. Soc.* 2004, 126, 5664-5665.
[158] Gu, H. W.; Yang, Z. M.; Gao, J. H.; Chang, C. K.; Xu, B. *J. Am. Chem. Soc.* 2005, 127, 34-35.
[159] Teranishi, T.; Inoue, Y.; Nakaya, M.; Oumi, Y.; Sano, T. *J. Am. Chem. Soc.* 2004, 126, 9914-9915.
[160] Shi, W.; Zeng, H.; Sahoo, Y.; Ohulchanskyy, T. Y.; Ding, Y.; Wang, Z. L.; Swihart, M.; Prasad, P. N. *Nano Lett.* 2006, 6, 875-881.
[161] Kudera, S.; Carbone, L.; Casula, M. F.; Cingolani, R.; Falqui, A.; Snoeck, E.; Parak, W. J.; Manna, L. *Nano Lett.* 2005, 5, 445-449.
[162] Kwon, K. W.; Shim, M. *J. Am. Chem. Soc.* 2005, 127, 10269-10275.

In: Nanoparticles: New Research
Editor: Simone Luca Lombardi, pp. 201-242
ISBN: 978-1-60456-704-5

Chapter 7

HEAT TRANSFER OF NANOPARTICLE SUSPENSIONS (NANOFLUIDS)

S.M. Sohel Murshed, Kai Choong Leong and Chun Yang
School of Mechanical and Aerospace Engineering
Nanyang Technological University, Republic of Singapore

Abstract

Nanofluids, a dispersion of nanoparticles in conventional heat transfer fluids, promises to offer enhanced heat transfer performance. Nanofluids are a new, innovative class of heat transfer fluids and represent a rapidly emerging field where nanoscale science and thermal engineering meet. In recent years, nanofluids have attracted great interest from researchers of multi-disciplines because of their superior thermal properties and potential applications in important fields such as microelectronics, microfluidics, transportation, and biomedical. Published research works have shown that nanofluids possess higher effective thermal conductivity and effective thermal diffusivity compared to their base fluids and the magnitudes of these properties increase remarkably with increasing nanoparticle volume fraction. Particle size and shape as well as fluid temperature are also found to influence the enhancement of thermal conductivity of the nanofluids. Despite numerous theoretical studies on model development for nanofluids, there is no widely accepted model available due to inconclusive heat transfer mechanisms of nanofluids. There are also many inconsistencies in reported experimental results and controversies in the proposed mechanisms for enhanced thermophysical properties of nanofluids. In this chapter, the current state-of-research in nanofluids including synthesis, potential applications, experimental and analytical studies on the effective thermal conductivity, thermal diffusivity, and viscosity of nanofluids are critically reviewed. Results from the authors' extensive theoretical and experimental studies on thermophysical and electrokinetic properties of nanofluids are also summarized.

1. Introduction

1.1. Development and Concept of Nanofluids

With an ever-increasing thermal load due to smaller features of microelectronic devices and more power output, cooling for maintaining desirable performance and durability of such devices is one of the most important technical issues in many high-tech industries. The conventional method to increase the cooling rate is to use extended heat transfer surfaces. However, this approach requires an undesirable increase in the size of the thermal management system. In addition, the inherently poor thermophysical properties of traditional heat transfer fluids such as water, ethylene glycol (EG) or engine oil (EO) greatly limit the cooling performance. Thus, these conventional cooling techniques are not suitable to meet the demand of these high-tech industries. There is therefore, a need to develop advanced cooling techniques and innovative heat transfer fluids with improved heat transfer performance.

It is well known that at room temperature, metals possess at least an order-of- magnitude higher thermal conductivity than fluids. For example, the thermal conductivity of copper at room temperature is about 700 times greater than that of water and about 3000 times greater than that of engine oil (Figure 1). Therefore, the thermal conductivities of fluids that contain suspended metallic or nonmetallic (oxide) particles are expected to be significantly higher than those of conventional heat transfer fluids. As thermal conductivity of a fluid plays a vital role in the development of energy-efficient heat transfer equipment, numerous theoretical and experimental studies on increasing the thermal conductivity of liquids by suspending small particles have been conducted since the treatise by Maxwell more than a century ago [1]. However, all such studies on thermal conductivity of suspensions have been confined to millimeter- or micrometer-sized particles. The major problems of such suspensions are the rapid settling of these particles, clogging of flow channels, and increased pressure drop in the fluid. Even if the fluid is kept circulating rapidly enough to prevent much settling, the micro-particles would damage the walls of the heat transfer devices (e.g. pipes and channels) and wear them thin. In contrast, nanoparticles in a liquid can maintain better suspension and thereby reduce erosion and clogging.

Over the last several decades, scientists and engineers have attempted to develop fluids, which offer better cooling or heating performance. However, it is only in 1995 that Choi and his co-workers at Argonne National Laboratory of USA [3] developed the novel concept of a "nanofluid" to meet the cooling challenges facing many high-tech industries. This new class of heat transfer fluids (nanofluids) is engineered by dispersing nanometer-sized solid particles, rods or tubes in traditional heat transfer fluids. From past investigations, nanofluids were found to exhibit significantly higher thermophysical properties, particularly thermal conductivity than those of base fluids [4-12]. Thus, nanofluids have attracted great interest from the research community due to their potential benefits and applications in numerous important fields such as microelectronics, microfluidics, transportation, manufacturing, medical, and so on.

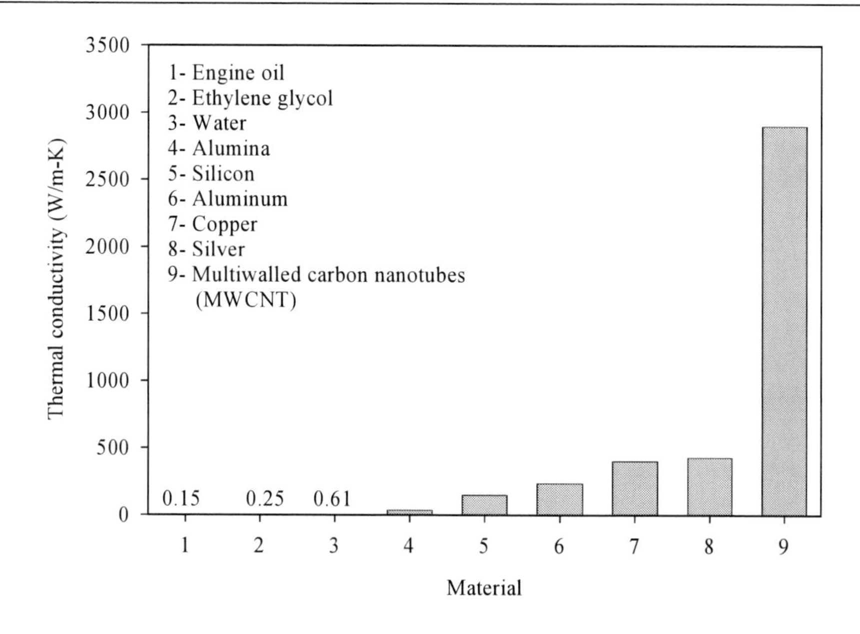

Figure 1. Thermal conductivity of typical materials (solids and liquids) at 300 K [2].

1.2. Impact and Potential Benefits of Nanofluids

The impact of nanofluids is expected to be great considering that heat transfer performance of heat exchangers or cooling devices is vital in many industries. For example, the transport industry has a strong incentive to reduce the size and weight of vehicle thermal management systems and nanofluids can increase thermal transport of coolants and lubricants. When the nanoparticles are properly dispersed, nanofluids offer numerous benefits [2] besides the anomalously high effective thermal conductivity. These benefits include:

1.2a. Improved Heat Transfer and Stability

Because heat transfer takes place at the surface of the particles, the much larger relative surface areas of nanoparticles compared to micro-particles, provide significantly improved heat transfer capabilities. In addition, particles finer than 20 nm carry 20% of their atoms on their surface, making them instantaneously available for thermal interaction [2].

1.2b. Microchannel Cooling without Clogging

Nanofluids will not only be a better medium for heat transfer in general, but they will also be ideal for microchannel applications where high heat loads are encountered. The combination of microchannels and nanofluids will provide both highly conducting fluids and a large heat transfer area. This cannot be attained with microparticles because they can easily clog the microchannels.

1.2c. Miniaturized Systems

Nanofluid technology can support the current industrial trend toward component and system miniaturization by enabling the design of smaller and lighter heat exchanger systems.

Miniaturized systems will reduce the inventory of heat transfer fluids and will result in cost savings.

The better stability of nanofluids will prevent rapid settling and reduce clogging in the walls of heat transfer devices. The high thermal conductivity of nanofluids translates into higher energy efficiency, better performance, and lower operating costs. They can reduce energy consumption for pumping heat transfer fluids. Thermal systems can be smaller and lighter. In vehicles, smaller components result in better gasoline mileage, fuel savings, lower emissions, and a cleaner environment [13].

1.3. Potential Applications of Nanofluids

With the aforementioned highly desired thermal properties and potential benefits, nanofluids are thought to have a wide range of applications, which are elaborated here.

1.3a. Nanofluids in Transportation

Because engine coolants, engine oils, automatic transmission fluids, and other synthetic high temperature fluids currently possess inherently poor heat transfer capabilities, they could benefit from the high thermal conductivity offered by nanofluids. Nanofluids would allow for smaller, lighter engines, pumps, radiators, and other components.

1.3b. In Micromechanics

Micro-electromechanical systems (MEMS) generate a lot of heat during operation. Conventional coolants do not work with MEMS because they do not have enough cooling capability. Moreover, even if solid particles are added to these coolants to enhance their thermal conductivity, they still could not work, because the particles would be too big to flow smoothly in the extremely narrow cooling channels required by MEMS. Since nanofluids can flow in microchannels without clogging, they would be suitable coolants. They could enhance cooling of MEMS under extreme heat flux conditions.

1.3c. In Electronics and Instrumentation

Demand for ultra-high-performance cooling in this area has been increasing, and conventional enhanced surface techniques have reached their limit with regards to improving heat transfer. Since nanoparticles are much smaller than the diameter of microchannels, smooth –flowing nanofluids could support this demand.

1.3d. In Medical Applications

Magnetic nanoparticles in biofluids can be used as delivery vehicles for drugs or radiation, providing new cancer treatment techniques. Nanoparticles are more adhesive to tumor cells than normal cells. Thus, magnetic nanoparticles excited by an AC magnetic field are promising for cancer therapy. The combined effect of radiation and hyperthermia is due to the heat-induced malfunction of the repair process right after radiation-induced DNA damage. In the near future, nanofluids can be used as advanced drug reduction fluids.

2. Synthesis of Nanofluids

Preparation of nanofluids is the first key step to investigate their heat transfer performance and novel characteristics. Modern fabrication technologies provide great opportunities to process materials at the nanometer scale. Nanostructured materials exhibit new or enhanced properties, which are not exhibited by the bulk solids. There are mainly two techniques for synthesizing nanofluids: the two-step process, and the direct evaporation technique or single-step process [2-4].

In the two-step process, dry nanoparticles are first produced by an inert gas condensation method and they are then dispersed into a fluid. Figure 2 shows TEM images of several types of commercially available nanoparticles. This method may result in a large degree of nanoparticle agglomeration. Thus, proper dispersion techniques and small volume fraction of nanoparticles are important to produce stable nanofluids by this technique. An advantage of the two-step process in terms of eventual commercialization of nanofluids is that the inert gas condensation technique can produce large quantities of nanopowders.

The direct evaporation technique synthesizes nanoparticles and disperses them into a fluid in a single step. As with the inert gas condensation technique, this technique involves the vaporization of a source material under vacuum conditions. An advantage of this process is that nanoparticle agglomeration is minimized. The disadvantages are that the liquid must have a very low vapor pressure and that this technique can produce very limited amounts of nanofluids. At present, most researchers used the two-step process to produce nanofluids by dispersing commercial or self-produced nanoparticles in a liquid. The optimization of thermal properties of nanofluids requires stable nanofluids, which can be ensured by proper synthesis and dispersion procedures.

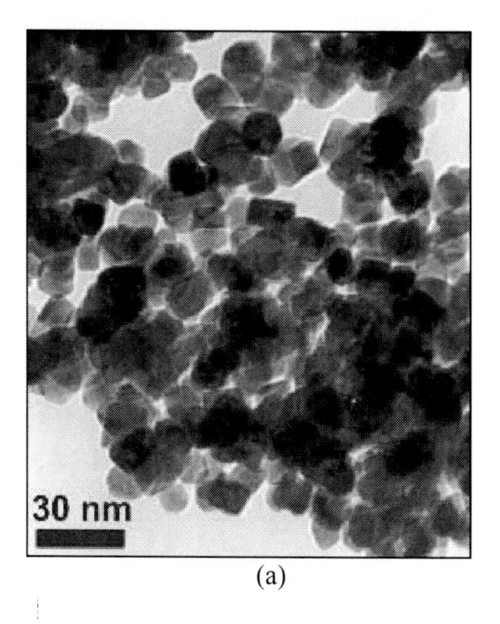

(a)

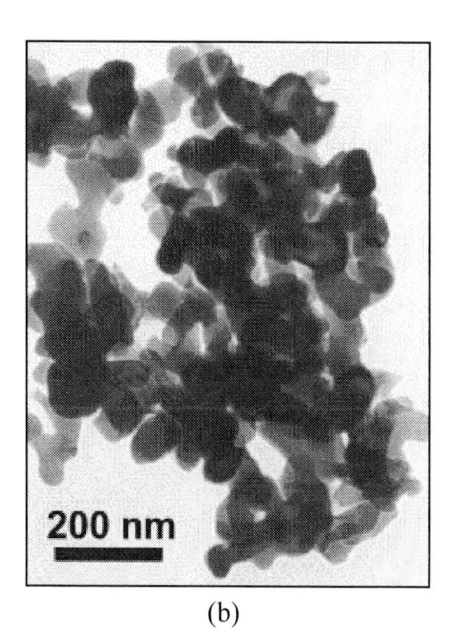

(b)

Figure 2. TEM photographs of nanoparticles from Nanostructured and Amorphous Materials, Inc., USA – (a) TiO_2 (15 nm) and (b) Al_2O_3 (80 nm).

Figure 3. Ultrasonic dismembrator probe inside sample nanofluids.

3. Preparation and Characterization of Sample Nanofluids

A nanofluid does not mean a simple mixture of liquid and solid nanoparticles. Techniques for good dispersion of nanoparticles in liquids or directly producing stable nanofluids are crucial. To prepare nanofluids by the two-step process i.e. by suspending nanoparticles into base fluids, proper mixing and stabilization of the particles are essential.

Like most of the reported nanofluids, by dispersing commercially available nanoparticles in base fluids sample nanofluids were prepared for the present study. This is the two-step process of preparing nanofluids. Several types of nanoparticles were purchased from Nanostructured and Amorphous Materials, Inc., USA. These are spherical-shape titanium dioxide (TiO_2) nanoparticles of Φ15 nm and rod-shape of Φ10 nm × L 40 nm (here Φ is the diameter and L is the length), spherical-shape alumina (Al_2O_3) nanoparticles of Φ80 nm and Φ150 nm, and spherical-shape aluminum (Al) nanoparticles of Φ80 nm. Deionized water, ethylene glycol, and engine oil were used as base fluids.

After different volumetric loading of these nanoparticles (0.1 to 5%) into base fluids, their proper mixing was ensured by 8-10 hours of homogenization using an Ultrasonic Dismembrator (Fisher Scientific Model 500) as shown in Figure 3. Ultrasonication is very important in breaking the larger clusters into smaller clusters [14]. Figure 4 depicts the procedure for the preparation and characterization of sample nanofluids. A Transmission Electron Microscope (TEM) and a Particle Size Analyzer (PSA) were then used to characterize the dispersion, clustering and morphology of nanoparticles in base fluids, which can influence heat transfer performance of nanofluids.

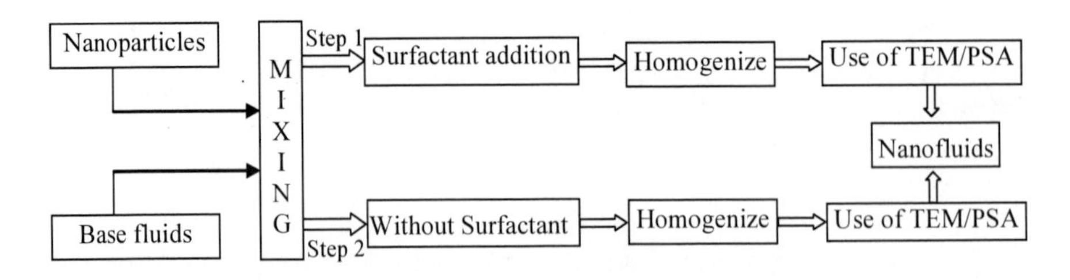

Figure 4. Process flow chart for sample preparation of nanofluids.

Primarily, the particles were neither uniformly dispersed in the base fluids nor a satisfactory stability of the nanofluids were observed. There is, therefore, a need to use surfactants to break down the particle agglomeration in suspension. Cetyl Trimethyl Ammonium Bromide (CTAB) surfactant was used to ensure better stability and proper dispersion without affecting the thermophysical properties and single-phase heat transfer performance of the nanofluids. After addition of CTAB, all types of nanoparticles were found to be well dispersed in their base liquids and formed stable suspensions. Almost no sedimentation of nanoparticles was observed within four to five days after sample preparation. Moreover, engine oil-based nanofluids were found to be stable for about few weeks. This is not surprising because engine oil is a highly viscous liquid compared to water. Based on Stokes' sedimentation law [15], the rate of particle sedimentation or Stokes settling velocity is low for small particles and highly viscous liquids. Before characterizing the thermophysical or other properties of nanofluids, proper sample preparation and its stability are very important.

Table 1. Thermophysical properties of base fluids at 300 K [16]

Base fluids	Thermal conductivity, k (W/m·K)	Thermal diffusivity, α (m^2/s)	Density, ρ (kg/m^3)	Specific heat, c_p (kJ/kg·K)	Kinematic viscosity, υ (m^2/s)
Deionized water (DIW)	0.607	14.55×10^{-8}	998	4.2	9.2×10^{-7}
Ethylene glycol (EG)	0.255	9.385×10^{-8}	1111	2.4	18.1×10^{-6}
Engine oil (EO)	0.145	8.740×10^{-8}	884	1.9	9.44×10^{-4}

Table 2. Thermophysical properties of several nanoparticles at 300 K [16, 17]

Nanoparticles	Thermal conductivity, k (W/m·K)	Thermal diffusivity, α (m^2/s)	Density, ρ (kg/m^3)	Specific heat, c_p (kJ/kg·K)
TiO_2	8.04	2.9×10^{-6}	4000	0.711
CuO	17.65	5.17×10^{-6}	6500	0.525
Al_2O_3	39	11.9×10^{-6}	3970	0.775
Al	237	97.1×10^{-6}	2700	0.877
Cu	401	117×10^{-6}	8933	0.385

4. Properties of Base Fluids and Nanoparticles

The thermophysical properties of both the nanoparticle and the base fluid are important in preparing sample nanofluids as well as determining the thermophysical properties such as thermal conductivity and viscosity of nanofluids. Therefore, the thermophysical properties of commonly used base fluids and nanoparticles are given in Tables 1 and 2.

5. Experimental Studies on Thermal Conductivity of Nanofluids

5.1. Measurement Method

Since the transient hot-wire (THW) method was first suggested by Stalhane and Pyk in 1931 to measure the absolute thermal conductivity of powders [14], many researchers have modified the method to make it more accurate. With modern electronic instrumentation and development of a proper theoretical basis, this method has become an accurate means of determining the thermal conductivity of fluids. Attempts were made by several researchers to extend the transient hot-wire method to measure the thermal conductivity of electrically conducting media. Nagasaka and Nagashima [19] performed thermal conductivity measurements of electrically conducting liquids by considering the electrical insulation layer effect on the heat transfer system in the rig. The transient hot–wire method has proved to be one of the most accurate techniques of determining the thermal conductivity of fluids [18-19]. The advantage of this method lies in its elimination of natural convection effects which will cause measurement errors. In addition, this method is very fast compared to other techniques. The conceptual design of this hot-wire apparatus is also simple when compared to other techniques. Although several studies reported the use of the steady-state technique [7, 20], temperature oscillation technique [9], and the 3ω-wire method [21] to measure the effective thermal conductivity of nanofluids, these techniques are not as accurate as the transient hot–wire method. The temperature oscillation technique measures the thermal diffusivity and the thermal conductivity is then calculated using the volumetric specific heat of the sample. Similar to the hot-wire method, the 3ω-wire method uses a metal wire suspended in a liquid. A sinusoidal current at frequency ω is passed through the metal wire and generates a heat wave at frequency 2ω, which is deduced by the voltage component at frequency 3ω. The 3ω-wire method may be suitable to measure temperature-dependent thermal conductivity.

Most researchers used the transient hot-wire method to measure the thermal conductivity of nanofluids. A schematic of the transient hot-wire apparatus used in the authors' study is shown in Figure 5. The entire experimental setup comprises several major units including experimental (hot-wire) cell, power supply, Wheatstone bridge circuit, data acquisition and control system (A/D converter and computer). Details of the transient single hot-wire setup and procedures are discussed in the paper by Murshed *et al.* [10].

Table 3. Summarized results for thermal conductivity of different types of nanofluids

Researchers	Nanofluids [Particle (size in nm) /base fluid]	Measurement technique	Observed significant enhancement of thermal conductivity at particle volume%
Eastman *et al.* [4]	Al_2O_3 (33)/water CuO (36)/water Cu (35)/oil	THW method	29% for 5 vol.% 60% for 5 vol.% 44% for 0.052 vol.%
Lee *et al.* [6]	Al_2O_3 (38)/water/EG CuO (23.6)/water/EG	THW method	CuO/EG: 22% for 4 vol.% Al_2O_3 /EG: 18% for 5 vol.%

Table 3. Continued

Researchers	Nanofluids [Particle (size in nm) /base fluid]	Measurement technique	Observed significant enhancement of thermal conductivity at particle volume%
Wang *et al.* [7]	Al_2O_3 (28)/water Al_2O_3 (28)/EG	steady-state method	12% for 3 vol.% 26% for 5 vol.%
Xuan and Li [8]	Cu (100)/water	THW method	Cu/water: 54% for 5 vol.%
Eastman *et al.* [5]	Cu (<10)/EG CuO (35)/EG	THW method	40% for 0.3 vol.% 22% for 4 vol.%
Xie *et al.* [22]	Al_2O_3 (60.4)/water/EG	THW method	Al_2O_3 / EG: 30% for 5 vol.%
Wang *et al.* [23]	Al_2O_3 (29)/EG TiO_2 (40)/EG	steady-state parallel plate method	18% for 4 vol.% 13% for 5 vol.%
Wang *et al.* [20]	CuO (50)/DIW	Quasi-steady state	17% for 0.4 vol.%
Xuan and Li [24]	Cu (10)/water	THW method	70% for 3 vol.%
Patel *et al.* [25]	Au (10-20)/toluene/ water	THW method	Au/toluene: 5.5% for 0.008 vol.% at 30 °C
Kumar *et al.* [26]	Au (4)/water/toluene	THW method	Au /water: 20% for 0.00013 vol.% at 30 °C
Hong *et al.* [11]	Fe (10)/EG	THW method	18% for 0.55 vol.%
Kwak and Kim [27]	CuO (12)/EG	THW method	6% for 1 vol.%
Li and Peterson [12]	CuO (29)/water Al_2O_3 (36)/water	steady-state method	52% for 6 vol.% at 34 °C 30% for 10 vol.% at 34 °C
Zhu *et al.* [28]	Fe_3O_4 (10)/water	THW method	38% for 4 vol.%
Hwang *et al.* [29]	CuO (35.4)/water/EG	THW method	CuO/EG: 9 % for 1 vol.%
Xuan *et al.* [30]	Cu (35.4)/water	THW method	24% for 2 vol.%
Liu *et al.* [31]	CuO (29)/EG	THW method	23% for 5 vol.%
Krishnamurthy *et al.* [32]	Al_2O_3 (20)/water	unspecified	16% for 1 vol.%
Wen and Ding [33]	TiO_2 (34)/water	THW method	6% for 0.66 vol.%
Putnam *et al.* [34]	Au (4)/ethanol	Optical beam deflection technique	1.3%$\pm$0.8% for 0.018 vol.%
Kang *et al.* [35]	Diamond (30-40)/EG Ag (8-15)/water	THW method	75% for 1.32 vol.% 11% for 0.4 vol.%

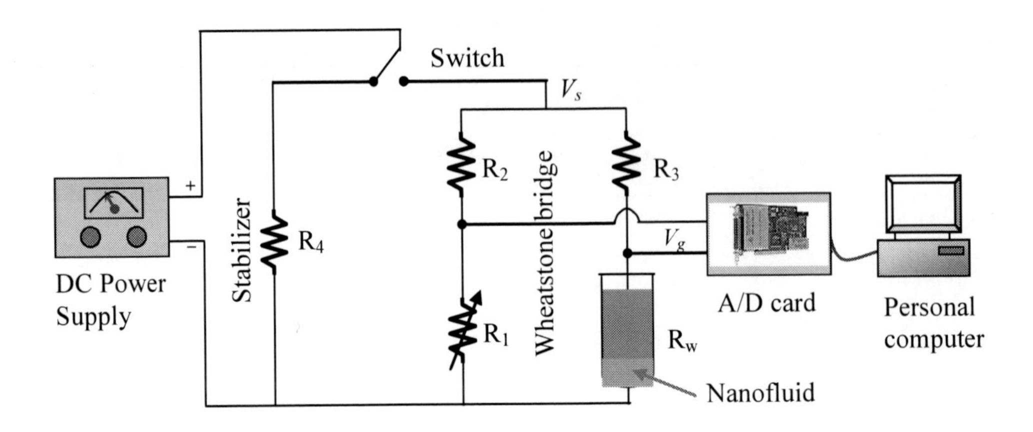

Figure 5. Schematic of transient hot-wire experimental setup [10].

5.2. Effect of Particle Volume Fraction and Base Fluids

Since Maxwell's model [1] was developed more than a century ago, it is known that the effective thermal conductivity of suspensions containing metallic or nonmetallic particles increases with the volume fraction of the solid particles. However, the enhancement of thermal conductivity of a mixture (solid/liquid) may vary significantly particularly in the case of nanofluids.

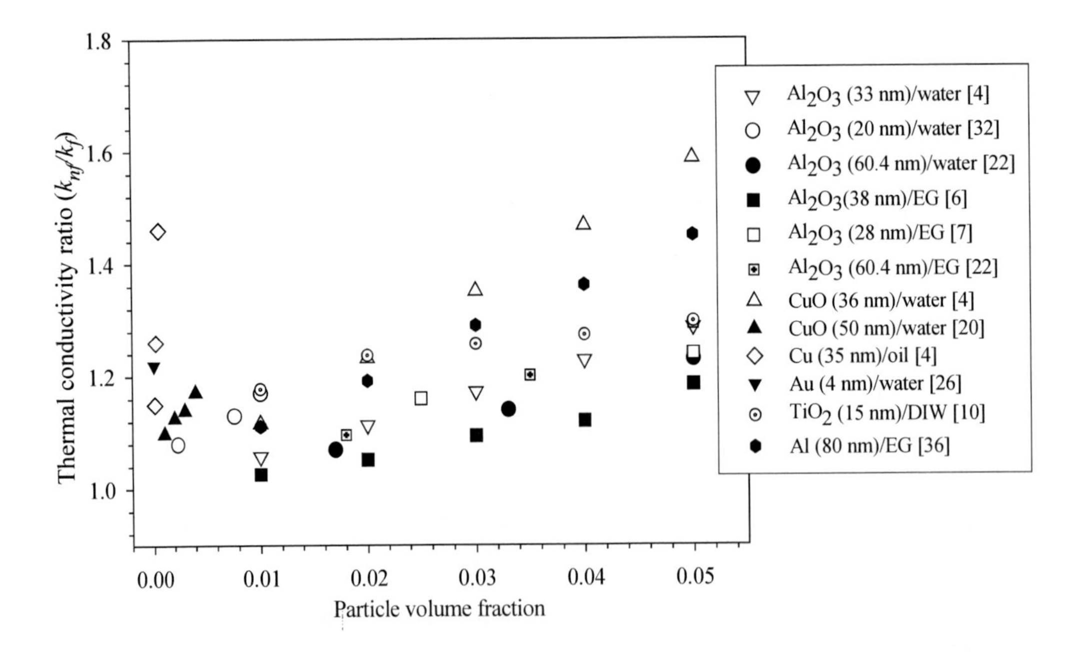

Figure 6. Comparison of experimental results of the enhanced thermal conductivity of nanofluids.

5.2.1. Results from the Literature

In the last several years, many experimental investigations on the effective thermal conductivities of nanofluids containing different concentrations, materials and sizes of nanoparticles dispersed in different base fluids have been reported. The published significant results on the thermal conductivity of nanofluids at room temperature are summarized in Table 3.

Some key results of the effective thermal conductivity of nanofluids as a function of nanoparticle volume fraction from various research groups are also shown in Figure 6.

From the reported results shown in the Table 3 and Figure 6, it is clear that nanofluids exhibit much higher thermal conductivities than their base fluids even when the concentrations of suspended nanoparticles are very low and they increase significantly with nanoparticle volume fraction. However, it is found that even for the same nanofluids different groups reported different enhancements. Several representative studies on the thermal conductivity of nanofluids are elaborated here.

The first results on the enhanced effective thermal conductivity of nanofluids were reported by Eastman *et al.* [4]. By dispersing Al_2O_3 and CuO nanoparticles in water, the reported increase in thermal conductivity were 29% and 60%, respectively for a nanoparticle volumetric loading of 5%. Surprisingly, in the case of Cu/oil-based nanofluids, the thermal conductivity was increased by about 44% by dispersing only 0.052 volume % of Cu nanoparticles in HE-200 oil. The same group [6] later showed a moderate enhancement of thermal conductivity for the same ceramic nanoparticles i.e. Al_2O_3 and CuO dispersed in water and ethylene glycol. For instance, a 20% enhancement in the thermal conductivity of ethylene glycol was observed for 4.0 volume % of CuO nanoparticles. Wang *et al.* [17] reported a surprising 17% increase in the thermal conductivity for a loading of 0.4 volume % of 50 nm sized same nanoparticles (CuO) in water. Recently, Li and Peterson [12] measured the thermal conductivity of the same ceramic nanofluids i.e. CuO (29 nm)/water and Al_2O_3 (36 nm)/water by the steady-state method. Their results were also astounding as both the CuO and Al_2O_3 nanoparticles increased the thermal conductivity of water by 52% and 22%, respectively at a volume fraction of just 6% at 34°C.

By using the steady-state parallel plate method, thermal conductivities of several types of nanofluids were measured by Wang *et al.* [21]. The Al_2O_3/EG-based nanofluids showed 18% increase in thermal conductivity at particle volume % of 4. In contrast, Xie *et al.* [20] observed about 30% increase in thermal conductivity for 5 volume % of Al_2O_3 (60.4 nm) nanoparticles in the same base fluid. Although the particle size used by Xie *et al.* was double that of the particles of Wang *et al.*, their results showed much higher thermal conductivity than those of Wang *et al.* for this nanofluid. This discrepancy could be due to different measurement methods and adjustment of pH values of the nanofluids used by Xie *et al.*

For Fe (10 nm)/EG-based nanofluids, a large increase in thermal conductivity was reported by Hong *et al.* [11]. They obtained an enhancement of 18% with just 0.55 volume % of Fe nanoparticles. They also noticed that sonication had significant effect on the thermal conductivity of nanofluids. Nonetheless, their observed enhancement for this nanofluid was even much higher than that of the Cu/EG-based nanofluids obtained by Eastman *et al.* [5]. This indicates that a suspension of high conductivity materials does not always result in an increase in the thermal conductivity of nanofluids. By using the co-precipitation method, Zhu *et al.* [26] prepared Fe_3O_4 (10 nm)/water-based nanofluids and measured their thermal conductivity values by the hot-wire method. They found a 38 % increase in the thermal conductivity for the nanoparticle volume fraction of 0.04.

Putnam *et al.* [33] measured the thermal conductivity of Au (4 nm)/ethanol-based nanofluids by using the optical beam deflection technique. For the first time, their results showed no anomalous enhancement in the thermal conductivity of nanofluids with very low particle volume fractions. Their observed maximum enhancement in thermal conductivity was 1.3% ± 0.8% for 0.018% volumetric loading of 4 nm-sized Au particles in ethanol. This result is directly in conflict with the anomalous result of Patel *et al.* [23] for the same nanofluid.

From a comparison of the reported studies, the increments of thermal conductivities are different for different types of nanofluids. The thermal conductivity of nanofluids varies with the size, material of nanoparticles as well as the base fluids. For instance, nanofluids with metallic nanoparticles were found to have a higher thermal conductivity than nanofluids with non-metallic (oxide) nanoparticles. The smaller the particle size, the larger the thermal conductivities of nanofluids. A few studies [11, 14] also reported that highly-conductive nanoparticles are not always effective in enhancing the thermal conductivity of nanofluids.

The particle size is important because shrinking it to nanoscale dimensions not only increases the surface area relative to volume but also generates some nanoscale mechanisms in the suspensions [14, 36-40]. Experimental [41] and theoretical evidence [37-38, 40] indicate that the effective thermal conductivity of nanofluids increases with decreasing particle size.

5.2.2. Results by the Authors

By using transient hot-wire method the effective thermal conductivities of several types of nanofluids are measured by the authors [10, 36]. Similar to the most studies reported in the literature, our results show a substantial increase in thermal conductivity with particle volume fraction. The choice of base fluids also influences this enhancement. From Figure 7, it can be seen that the effective thermal conductivities of TiO_2 (15 nm)/deionized water/ethylene glycol-based nanofluids increase substantially with increasing volumetric loading of TiO_2 (15 nm) nanoparticles. The thermal conductivity of TiO_2 (15 nm)/water nanofluids has a maximum enhancement of 29.7% for a particle volume fraction of 0.05 while the maximum increase of thermal conductivity of TiO_2 (15 nm)/ethylene glycol-based nanofluids is 18% for the same particle volume fraction of 0.05. Figure 7 also shows that water-based nanofluids exhibit much higher thermal conductivity compared to the ethylene glycol-based nanofluids for this nanoparticles. This is mainly due to higher value of the thermal conductivity of water compared to ethylene glycol (Table 1).

Figure 8 illustrates that the effective thermal conductivities of Al_2O_3 (80 nm)/deionized water/ethylene glycol-based nanofluids also increase significantly with increasing volumetric loading nanoparticles. By suspending 0.05 volume fraction of Al_2O_3 (80 nm) nanoparticles in deionized water and in ethylene glycol, the maximum enhancements of thermal conductivities were found to be 23% and 18%, respectively. This indicates that water is a better base fluid compared to ethylene glycol for these nanofluids.

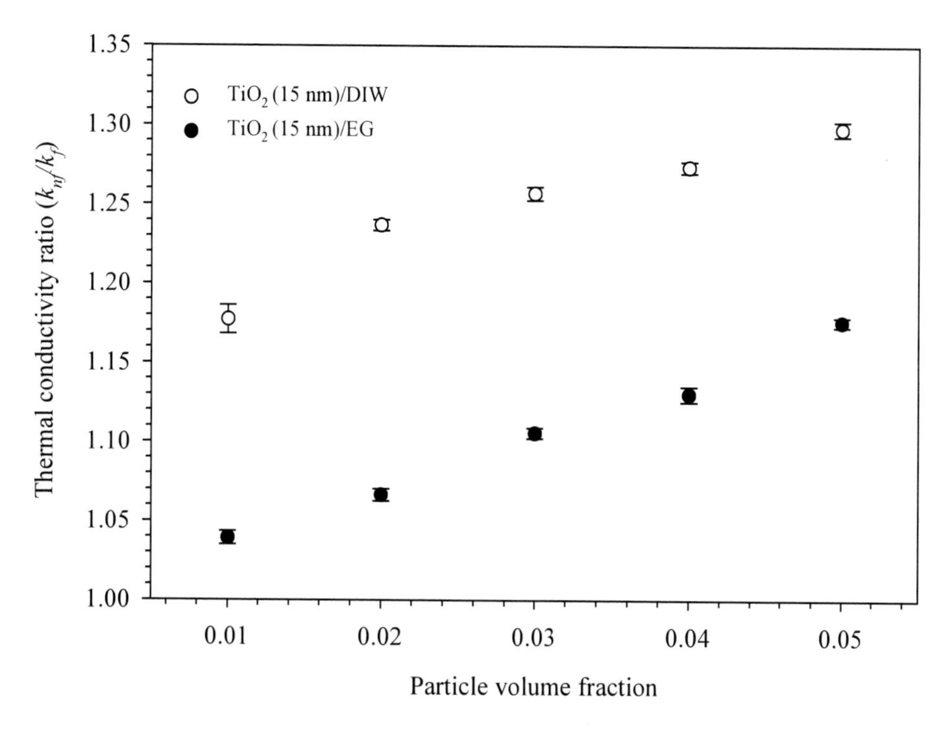

Figure 7. Enhancement of the effective thermal conductivity of TiO_2 (15 nm)/DIW/EG-based nanofluids with particle volume fraction [10, 36].

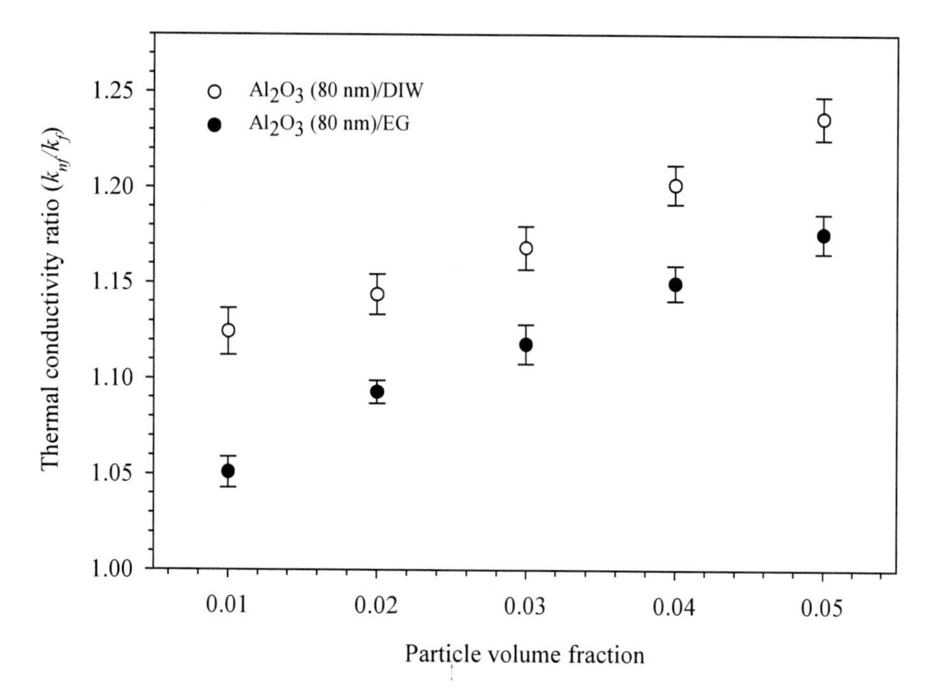

Figure 8. Results of the authors for the effective thermal conductivity enhancement of Al_2O_3 (80 nm)/DIW/EG-based nanofluids with particle volume fraction.

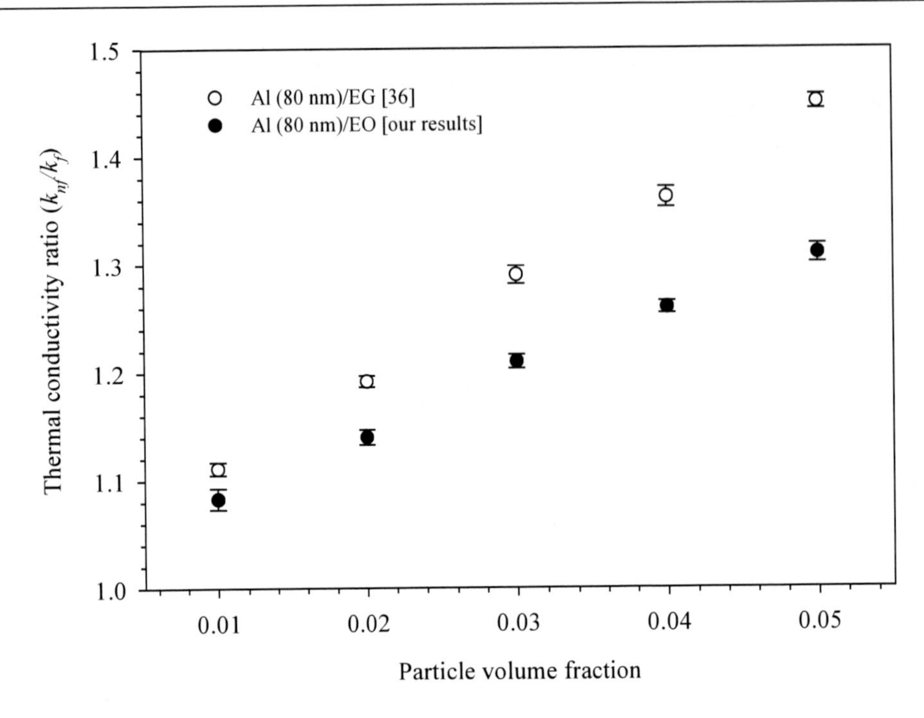

Figure 9. Enhancement of the effective thermal conductivity of Al (80 nm)/ethylene glycol /engine oil-based nanofluids with particle volume fraction.

The thermal conductivity of Al (80 nm)/ethylene glycol-based nanofluids has a maximum enhancement of 45% for a particle volume fraction of 0.05 while the maximum increase of thermal conductivity of Al /engine oil-based nanofluids is 31% for the same particle volumetric loading (i.e. 5%) as shown in Figure 9. Similar to the previous results (Figures 7 and 8), it is seen that ethylene glycol-based nanofluids show much higher thermal conductivities than those of engine oil-based nanofluids.

From the above results, it can be concluded that suspending a small volume percentage (1 to 5%) of nanoparticles in base fluids significantly increases the effective thermal conductivities of nanofluids. The thermal conductivity increases almost linearly with increasing the volumetric loading of nanoparticles. However, for lower volume fraction (< 1%) and very small-sized particles, the enhancement of thermal conductivity may not always be linear. Besides the nanoscale mechanisms such as the size of nanoparticle, nanolayer at the particle/liquid interface and Brownian motion of nanoparticles, the dissociation of surfactant may play a significant role in the enhancement of thermal conductivity. The thermophysical properties of base fluids are also found to have impact on the thermal conductivity of nanofluids. The higher the thermal conductivity of the base fluid, the larger the enhancement of thermal conductivity of nanofluids. The reason may also lie in the dispersion behavior of nanoparticles in the base fluids as well as the effect of surfactant. A small amount of CTAB surfactant was added in all sample nanofluids for a better dispersion of nanoparticles.

5.3. Effect of Particle Size and Shape

Although the thermal conductivities of nanofluids reported by different researchers were shown to be particle size dependent, very few experimental works have been reported [41] on the effect of particle size on the thermal conductivity of nanofluids. Other than results for carbon nanotubes, no experimental study of the effect of nanoparticle shape has been reported in the literature. The thermal conductivities of nanofluids containing different sizes and shapes of nanoparticles are measured by the authors and the results are presented in this section. Due to limited availability of different sizes of the same type of nanoparticles, TiO_2 nanoparticles of spherical (15 nm) and rod shape (10×40 nm) and spherical shape Al_2O_3 nanoparticles of 80 nm and 150 nm were used in our experiments.

Our results for TiO_2/deionized water-based nanofluids (Figure 10) show a nonlinear relationship between thermal conductivity and particle volume fraction at lower volumetric loading (0.005-0.02) and a linear relationship at higher volumetric loading (0.02-0.05). This nonlinear behavior of nanofluids at lower volume fractions of nanoparticles may be due to the influence of the CTAB surfactant, long time of sonication, and hydrophobic surface forces in the nanofluids. Figures 10 and 11 illustrate that for nanofluids with rod-shaped nanoparticles (TiO_2), the increase in thermal conductivity is larger than that of the nanofluids with spherical-shaped nanoparticles (TiO_2). Due to the larger shape factor and alignment, the thermal conductivity of nanofluids composed of rod-shape nanoparticles is larger than that of the nanofluids with spherical-shaped nanoparticles.

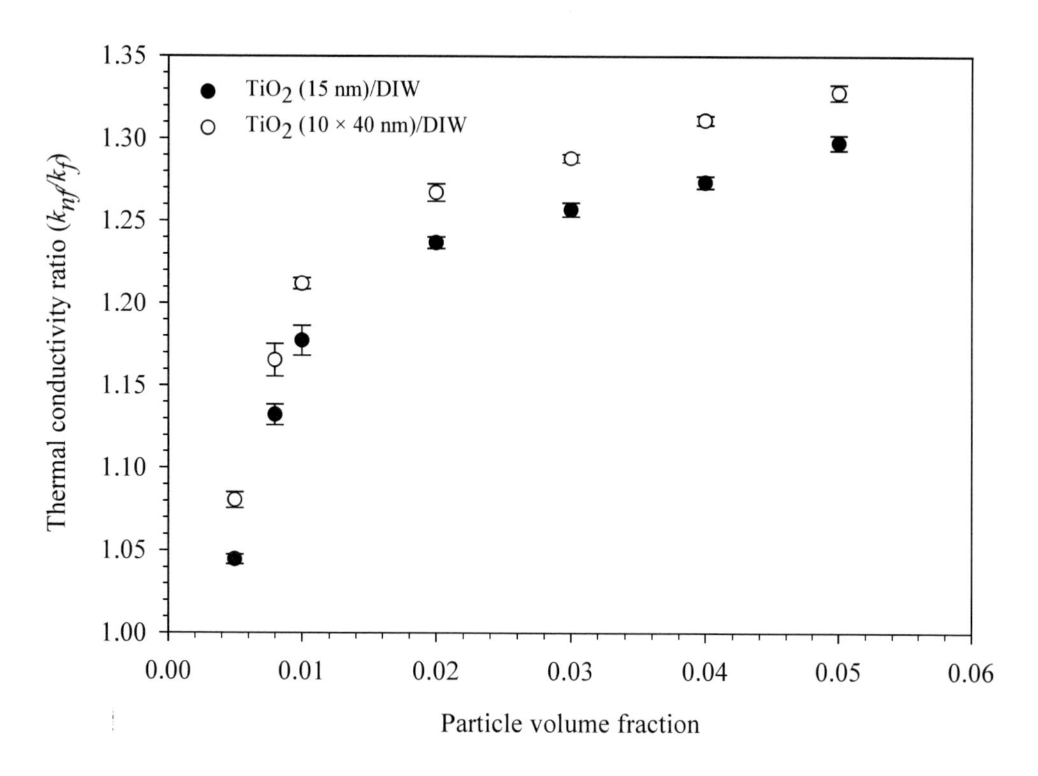

Figure 10. Enhancement of the effective thermal conductivity of TiO_2/DIW-based nanofluids [10].

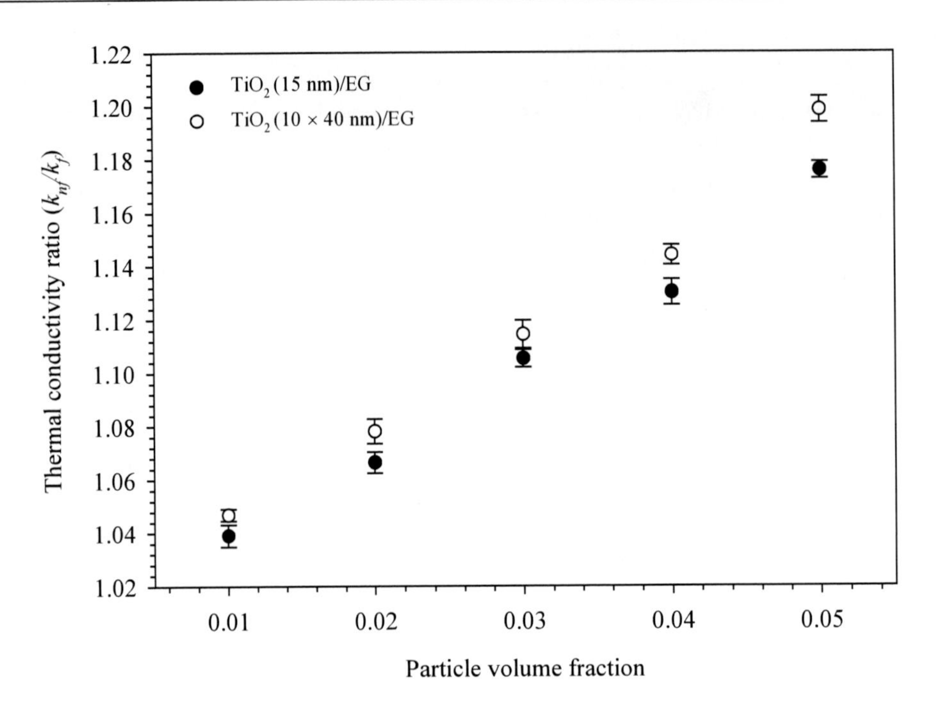

Figure 11. Results of the authors for the effective thermal conductivity enhancement of TiO_2/EG-based nanofluids.

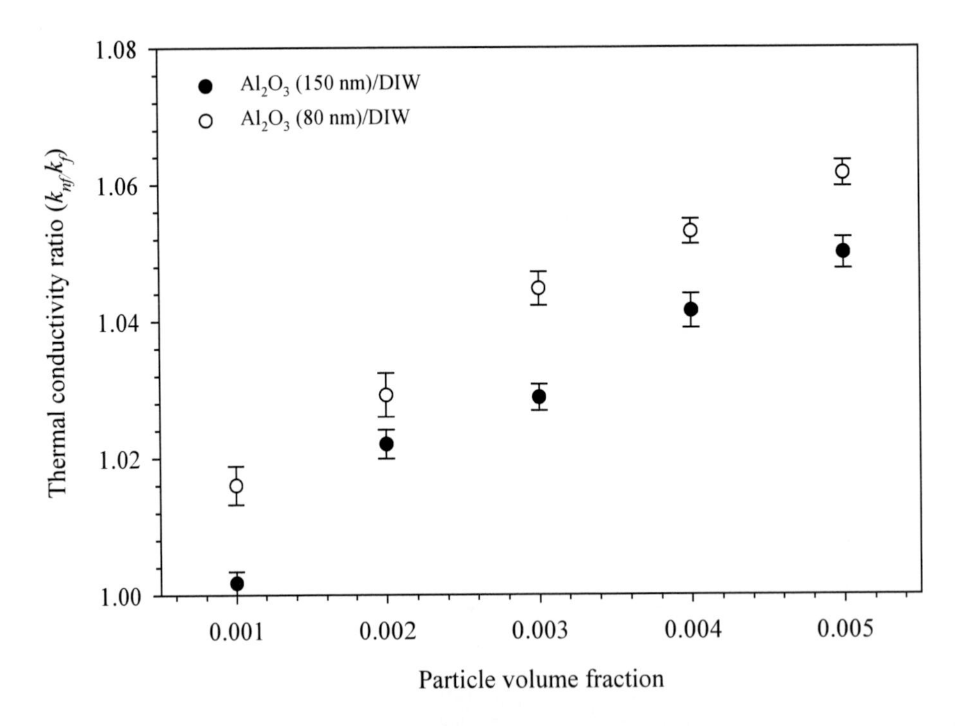

Figure 12. Results of the authors for the effective thermal conductivity enhancement of Al_2O_3/DIW-based nanofluids.

Figure 12 illustrates that for the same base fluid, smaller size (80 nm) Al_2O_3 nanoparticles show higher enhancement of thermal conductivity compared to larger-sized (150 nm) nanoparticles. Chon and Kihm [41] also found that nanofluids having smaller-sized Al_2O_3 particles have higher thermal conductivities than similar nanofluids with smaller-sized nanoparticles. The reason is that the smaller sized particles provide better dispersion and can also enhance the contribution of other nanoscale mechanisms such as Brownian motion of particles compared to larger-sized particles.

The above results demonstrate that the smaller the particle size, the higher the effective thermal conductivity of nanofluids. It is also concluded that the particle shape influences the thermal conductivity of nanofluids.

5.4. Effect of Fluid Temperature

Fluid temperature may play an important role in enhancing the effective thermal conductivity of nanofluids. Very few studies were performed to investigate the temperature effect on the effective thermal conductivity of nanofluids.

5.4.1. Results from the Literature

A summary of the published results on temperature-dependent thermal conductivity of various nanofluids is provided in Table 4 followed by brief discussion on several representative studies.

Table 4. Temperature-dependent thermal conductivity of nanofluids

Researchers	Nanofluids [Particle (size in nm)/base fluid]	Measurement technique	Observed maximum enhancement of thermal conductivity at temperature
Das *et al.* [9]	Al_2O_3(38.4)/water CuO (28.6)/water	temperature oscillation technique	for 4 vol.%: 16% at 36 °C and 25% at 51 °C for 1 vol.%: 22% at 36 °C and 30% at 51 °C
Patel *et al.* [25]	Au (10-20)/water	THW method	for 0.00026 vol.%: 5% to 21% at temperature range of 30 - 60 °C
Chon and Kihm [41]	Al_2O_3 (47)/water Al_2O_3 (150)/water	THW method	6% at 31 °C and 11% at 51 °C 3% at 31 °C and 8.5% at 51 °C [for 1 volume%]
Li and Peterson [12]	Al_2O_3 (36)/water	steady-state method	for 2 vol.%: 7% at 27.5 °C and 23% at 36 °C
Yang and Han [21]	Bi_2Te_3 (20×170) /FC72/oil	3ω wire method	for Bi_2Te_3 /FC72 at 0.8 vol.%: 8% at 10 °C and 7% at 40 °C
Venerus *et al.* [42]	Au (22) /water Al_2O_3 (30)/ Petroleum oil	Forced Rayleigh scattering	no enhancement but slight decrease with the temperature in the range of 25-75 °C

Das *et al.* [9] investigated the effect of temperature on the thermal conductivity of nanofluids containing Al_2O_3 (38.4 nm) and CuO (28.6 nm) nanoparticles in water. A temperature oscillation technique, which determines thermal conductivity through measuring the thermal diffusivity of sample medium, was used in their study. They reported a two- to four-fold increase in thermal conductivity enhancement for these nanofluids over a temperature range of 21°C to 51°C. Das *et al.* suggested that the strong temperature dependence of thermal conductivity was due to the motion of nanoparticles. Using the transient hot-wire technique, the same group [25] performed experiments for water-based nanofluids containing 10-20 nm-sized Au particles. Nanofluids with citrate stabilization showed thermal conductivity enhancement of 5-21% in the temperature range of 30 to 60°C at a very low volumetric loading of 0.00026%. Chon and Kihm [41] studied the effects of particle size and fluid temperature on the thermal conductivity of nanofluids. They observed a moderate enhancement of thermal conductivity with respect to temperature. The thermal conductivity of Al_2O_3 (47 nm)/water-based nanofluids increased by 6% to 11% when the fluid temperature was increased from 31°C to 51°C. They identified Brownian motion of the nanoparticles as the possible mechanism for increased thermal conductivity with fluid temperature. However, results obtained by Li and Peterson [12] demonstrated a surprising increase in thermal conductivity of nanofluids with a small increase of fluid temperature. For example, for Al_2O_3 (47)/water-based nanofluids at 2 volume %, an increase in fluid temperature to 36°C resulted in an enhancement of three times the thermal conductivity at 27.5°C. Yang and Han [21] measured the thermal conductivity of suspensions of Bi_2Te_3 (20×170 nm) nanorods in FC72 and in oil (hexadecane) by using the 3ω-wire method. Interestingly, they observed a slightly decrease in thermal conductivity with increasing temperature, which is in contrast to the trend observed in nanofluids containing spherical nanoparticles. The contrary trend was claimed to be due mainly to the particle aspect ratio.

Venerus *et al.* [42] used the Forced Rayleigh Scattering technique to measure the thermal conductivity of Au (22) /water and Al_2O_3 (30) /petroleum oil-based nanofluids in the range of 25 −75°C. In contrast to all other reported results, they found the level of thermal conductivity enhancement for these nanofluids to be independent of temperature.

Other than Yang and Han [21] and Venerus *et al.* [42], all published results demonstrate that the thermal conductivity of nanofluids increases significantly with the fluid temperature. However, there is a clear lack of consistency among the reported results of different research groups.

5.4.2. Results by the Authors

The effect of the temperature on the enhancement of effective thermal conductivity of several types of nanofluids was also investigated in this study. The transient hot-wire system with a fluid heating facility was used to measure the thermal conductivity of nanofluids at different temperatures ranging from 20 to 60°C. A refrigerating/heating circulator was used to the hot-wire system to generate different temperatures of nanofluids in the experiments. The measurement procedure and detailed description of the transient hot-wire technique are similar to experiments conducted at room temperature [10]. The experimental apparatus was calibrated by measuring the effective thermal conductivity of the base fluids.

Figure 13 shows that at a temperature of 60°C, the effective thermal conductivity of Al_2O_3 (80 nm)/EG based nanofluids increases by about 9% and 12% (compared to base fluid) for nanoparticle volumetric loadings of 0.5% and 1%, respectively. Whereas for the same temperature, the enhancement of effective thermal conductivity of Al (80 nm)/EO nanofluids are 20% and 37% for volumetric loadings of 1% and 3% nanoparticles in the base fluid, respectively. It is also seen from Figure 13 that dependence of temperature on thermal conductivity of nanofluids with metallic nanoparticles is more significant than that of nanofluids with nonmetallic nanoparticles.

Figures 14 and 15 compare the present results for Al_2O_3/ water based nanofluids at 1 % volumetric loading of nanoparticles with the data reported by Das *et al.* [9] and Chon and Kihm [41]. Our results are consistent with the experimental data of Das *et al.* and Chon and Kihm although the size of the Al_2O_3 nanoparticles used in our study is much larger than those of Das *et al.* and Chon and Kihm (Figure 14). For the same size (150 nm) of Al_2O_3 nanoparticles (Figure 15), our results for thermal conductivity of nanofluids were found to be higher than those reported by Chon and Kihm [41]. The observed differences could be due to difference in preparing sample nanofluids and different measurement methods used.

As shown in Figures 13 to 15, the experimental thermal conductivity values of different nanofluids increase significantly with fluid temperature. A linear increase in the effective thermal conductivity of nanofluids with temperature was also observed. This is because the high fluid temperature intensifies the Brownian motion of nanoparticles and also decreases the viscosity of the base fluid. With an intensified Brownian motion, the contribution of micro-convection in heat transport increases, thereby intensifying the enhancement in the thermal conductivity of nanofluids [7, 43].

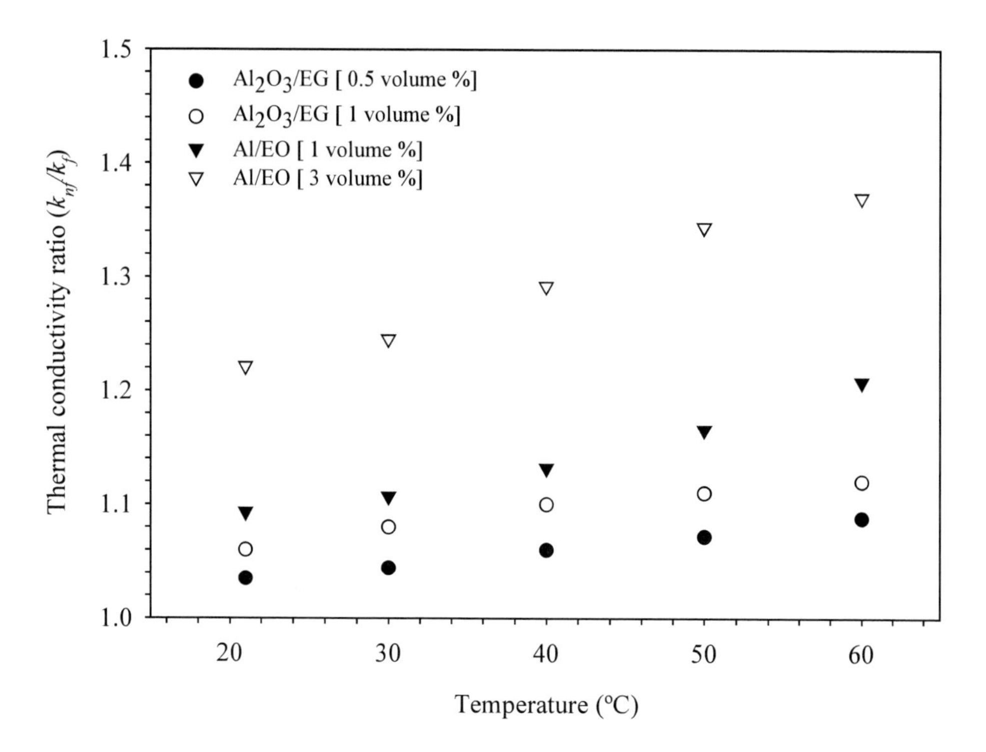

Figure 13. Thermal conductivity enhancement with temperature for Al_2O_3 (80 nm)/ethylene glycol and Al (80 nm)/engine oil-based nanofluids [36].

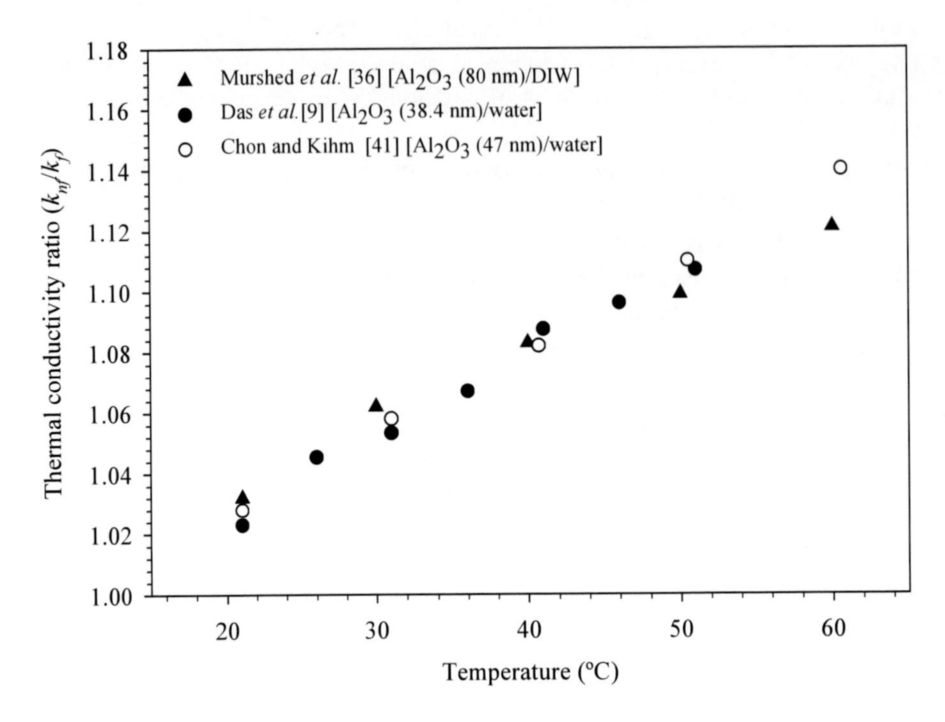

Figure 14. Thermal conductivity enhancement with temperature for Al_2O_3 (different sizes)/water-based nanofluids.

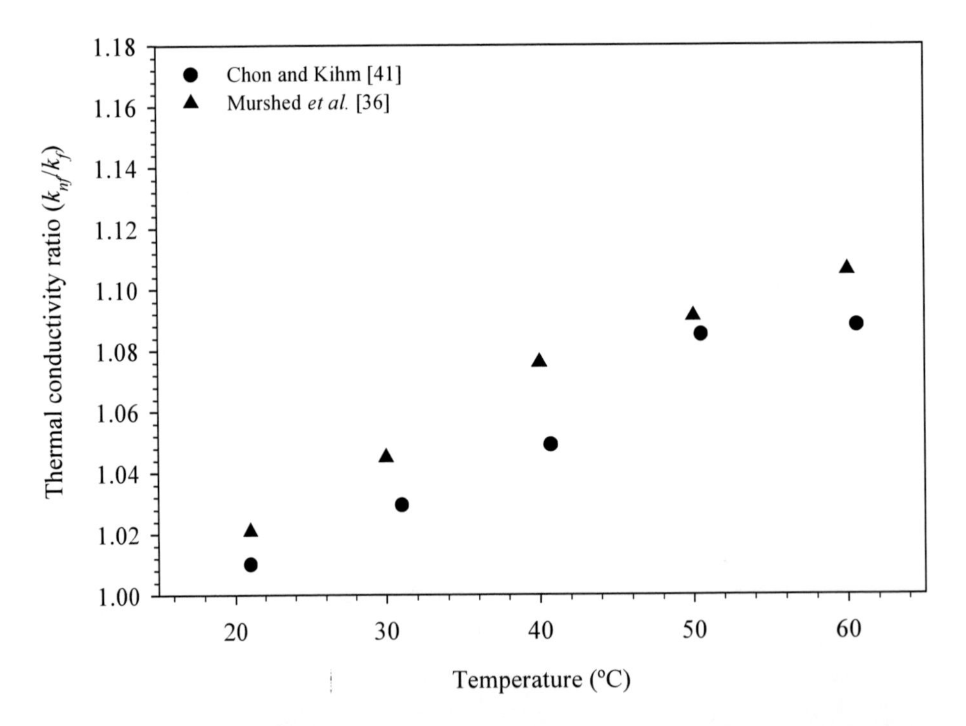

Figure 15. Thermal conductivity enhancement with temperature for Al_2O_3 (150 nm)/water based-nanofluids.

6. Theoretical Studies on Nanofluids

6.1. Mechanisms for the Enhanced Thermal Conductivity of Nanofluids

Experiments have shown that nanofluids exhibit anomalously high thermal conductivity which cannot be predicted by classical models. To explain the observed enhanced thermal conductivity of nanofluids, Wang *et al.* [7] and Keblinski *et al.* [37] proposed various mechanisms, which are not considered by classical models. Wang *et al.* [7] suggested that the microscopic motion of nanoparticles, surface properties, and the structural effects might cause enhanced thermal conductivity of nanofluids. In nanofluids, the microscopic motion of the nanoparticles due to van der Waals force, stochastic force causing Brownian motion and electrostatic force can be significant. Wang *et al.* however, showed that Brownian motion does not contribute significantly to energy transport in nanofluids. They indicated that the electric double layer and van der Waals force could have strong electrokinetic effects on the nanoparticles. The surface properties and structural effects were not confirmed as potential mechanisms in their study. Keblinski *et al.* [37] later elucidated four possible mechanisms for the anomalous increase in nanofluids heat transfer which are i) Brownian motion of the nanoparticles, ii) liquid layering at the liquid/particle interface, iii) nature of the heat transport in the nanoparticles, and iv) effect of nanoparticle clustering.

Due to the Brownian motion, particles move randomly through the liquid and may collide, thereby enabling direct solid-solid transport of heat from one to another, which can increase the effective thermal conductivity. However, by a simple analysis, Keblinski *et al.* [37] showed that the thermal diffusion of nanoparticles is much faster than Brownian diffusion and thus the contribution of Brownian motion to energy transport in nanofluids is not significant.

When the size of the nanoparticles in a nanofluid becomes less than the phonon mean-free path, phonons no longer diffuse across the nanoparticle but move ballistically without any scattering. However, it is difficult to envision how ballistic phonon transport could be more effective than a very-fast diffusion phonon transport, particularly to the extent of explaining anomalously high thermal conductivity of nanofluids. No further work or analysis has been reported on the ballistic heat transport nature of nanoparticles. Instead, the continuum approach was adopted in all reported works [20, 36, 38, 44].

Liquid layering around the particle i.e. nanolayer was offered as another mechanism responsible for higher thermal properties of nanofluids. The basic idea is that liquid molecules can form a layer around the solid particles and thereby enhance the local ordering of the atomic structure at the interface region. Hence, the atomic structure of such liquid layer is significantly more ordered than that of the bulk liquid. Given that solids, which have much ordered atomic structure, exhibit much higher thermal conductivity than liquids the liquid layer at the interface would reasonably have a higher thermal conductivity than the bulk liquid. Thus the nanolayer is considered as an important factor enhancing the thermal conductivity of nanofluids.

The effective volume of a cluster is considered much larger than the volume of the particles due to the lower packing fraction (ratio of the volume of the solid particles in the cluster to the total volume of the cluster) of the cluster. Since heat can be transferred rapidly within such clusters, the volume fraction of the highly conductive phase (cluster) is larger than the volume of solid, thus increasing its thermal conductivity [11, 37]. However, in general, clustering may exert a negative effect on heat transfer enhancement, particularly at a low volume fraction by settling small particles out of the liquid and creating a large region of "particle free" liquid with a high thermal resistance [37].

Besides these mechanisms, the effects of particle surface chemistry and particles interaction for nanometer-sized particles could be significant in enhancing the thermal conductivity of nanofluids.

6.2. Models for the Effective Thermal Conductivity

To predict the effective thermal conductivity of solid particle suspensions, several models have been developed since the treatise by Maxwell [1]. As mentioned previously, these classical models such as those attributed to Maxwell [1], Hamilton-Crosser [45], and Bruggeman [Hui *et al.* [46]] were developed for predicting the effective thermal conductivity of a continuum medium with well-dispersed solid particles. The Maxwell model [1] predicts the effective electrical or thermal conductivity of liquid-solid suspensions for low volumetric loading of spherical particles. Hamilton and Crosser [45] modified Maxwell's model to determine the effective thermal conductivity of non-spherical particles by applying a shape factor. For spherical particles, the Hamilton and Crosser (HC) model reduces to the Maxwell model. The Bruggeman model (BGM) [Hui *et al.* [46]] is another well-known model for determining the effective thermal conductivity of mixtures and composites. In this model, the mean field approach is used to analyze the interactions among the randomly distributed particles.

The classical models have been found to be unable to predict the anomalously high thermal conductivity of nanofluids. This is because these models do not include the nanoscale effects of particle size such as the interfacial layer at the particle/liquid interface, and motion of particles, which are considered as important factors for enhancing thermal conductivity of nanofluids [7, 37-40, 47-48,]. Recently, many theoretical studies have been carried out to predict the anomalously increased thermal conductivity of nanofluids. Several models have been proposed to consider various mechanisms. A detailed summary of all classical and recently developed models for the prediction of effective thermal conductivity of nanofluids is provided in review articles [49-50]. Therefore, these models are not reproduced here. Instead recently developed models are categorically briefed.

All the reported models can be categorized into two general groups, which are

(i) *static models* which assume stationary nanoparticles in the base fluid in which the thermal transport properties are predicted by conduction-based models such as those of Maxwell and Hamilton-Crosser.
(ii) *dynamic models* which are based on the premise that nanoparticles have lateral, random motion in the fluid. This motion is believed to be responsible for transporting energy directly through collision between nanoparticles or indirectly through micro liquid convection that enhances the transport of thermal energy.

Researchers in the first group [20, 35, 36, 38, 40, 47-48, 51, 52-55] use the concept of liquid/solid interfacial layer to develop models and to explain the anomalous improvement of the thermal conductivity in nanofluids. Approaches adopted in these studies can be classified under the static model category.

Other groups of researchers emphasize the contribution of dynamic part related to particle Brownian motion. Although Wang *et al.* [7] and Keblinski *et al.* [37] showed that the contribution of Brownian motion to energy transport in nanofluids is not significant, other researchers [26, 43, 56] held contrary views.

Most recently, the authors have developed an improved model to predict the effective thermal conductivity of nanofluids containing spherically shaped nanoparticles [57]. Compared to existing models, this model shows better agreement with experimental results. This was attributed to the inclusion of both static and dynamic mechanisms such as particle size, nanolayer, particle movement, particle surface chemistry and interaction potential.

7. Studies on Thermal Diffusivity and Specific Heat of Nanofluids

7.1. Thermal Diffusivity of Nanofluids

Despite the fact that thermal diffusivity is an important thermophysical property especially in convective heat transfer applications, very little work have been performed on the determination of the effective thermal diffusivity of nanofluids. Xuan and Roetzel [58] was the first to discuss the effective thermal diffusivity tensor for flowing nanofluids under both laminar and turbulent flow conditions. However, neither experimental nor theoretical result for the effective thermal diffusivity of nanofluids was provided in their paper. Wang *et al.* [59] measured the thermal conductivity and specific heat of some nanofluids and thereby calculated their effective thermal diffusivity. Their calculated results were also found to fluctuate severely with volume fraction. The accurate measurement of thermal diffusivity is much more complex for composite fluids such as nanofluids compared to solids or gases. Although there are several methods to measure the thermal diffusivity of solid, gases or liquid, no precise technique has been developed to measure the thermal diffusivity of composite fluids. Among all existing methods of measuring the effective thermal diffusivity of composite materials, the most common one is the flash method discussed by Parker *et al.* [60]. However, this method is mainly used for solid composite materials. To measure the thermal diffusivity of gases and liquids, several techniques such as the thermal–wave cavity technique [61] and the temperature oscillation technique [62] have been proposed. In the temperature oscillation technique, the thermal diffusivity is deduced by measuring and evaluating the amplitude attenuation and the phase shift between the temperature oscillations at the surface of the liquid specimen and at a specific position inside the reference specimen. The principle of the thermal–wave cavity technique consists in measuring the temperature fluctuations in a sample as a result of nonradiative deexcitation processes, which take place following the absorption of intensity-modulated radiation. The accuracy of these methods are not as good compared to the transient hot-wire method. Nevertheless, the applications of these methods are very limited due to complexities and difficulties to achieve accurate results particularly for composite fluids. They are therefore not suitable for the convenient and accurate determination of the thermal diffusivity of solid-liquid mixtures such as nanofluids.

The authors have developed a transient technique termed as the *double hot-wire (DHW)* method for the direct and precise measurement of the effective thermal diffusivity of nanofluids. Figure 16 shows the schematic of the double hot-wire experimental setup used in this study. The advantages of the double hot-wire technique are its simplicity and high measurement accuracy. Besides thermal conductivity, this technique can be used to determine the thermal diffusivity more accurately than that of the conventional hot-wire method. The theoretical basis for calculating the effective thermal diffusivity of nanofluids and detailed description of the experimental technique is provided in [63].

No investigation has been reported in the literature on the determination of the specific heat of nanofluids. Thus, the specific heat of the nanofluids is also determined from the measured thermal conductivity and thermal diffusivity. Our results for the specific heat of nanofluids are analyzed and compared with calculated values obtained from the volume fraction mixture rule (see subsection 7.2).

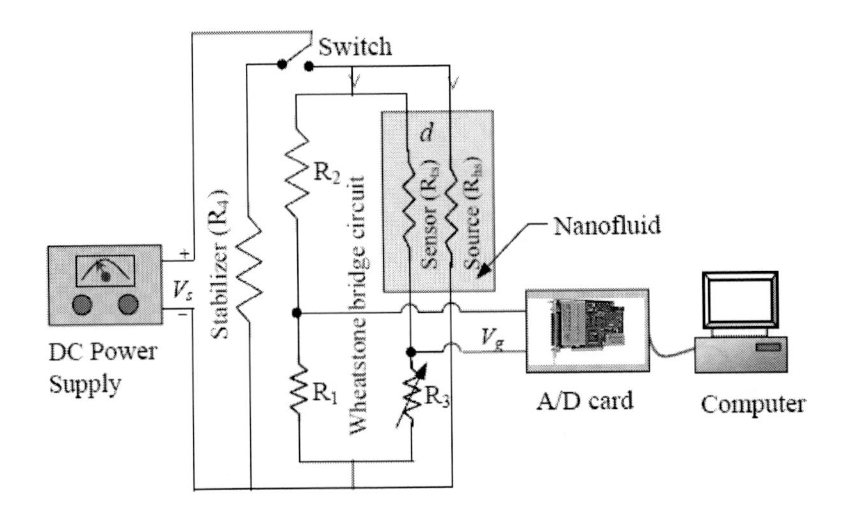

Figure 16. Schematic of the double hot-wire (DHW) experimental system [63].

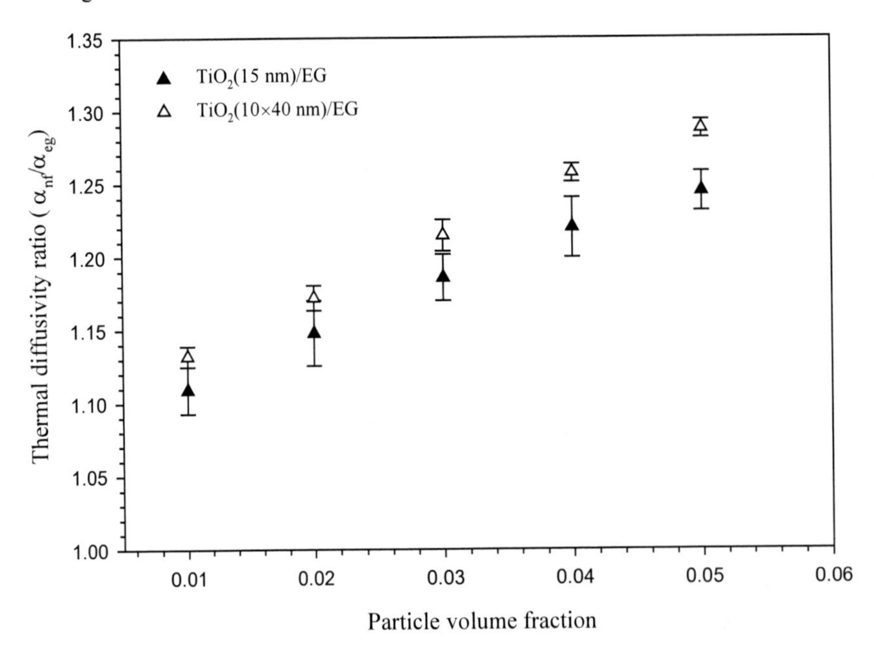

Figure 17. Thermal diffusivity of TiO_2/ethylene glycol-based nanofluids with particle volume fraction [63].

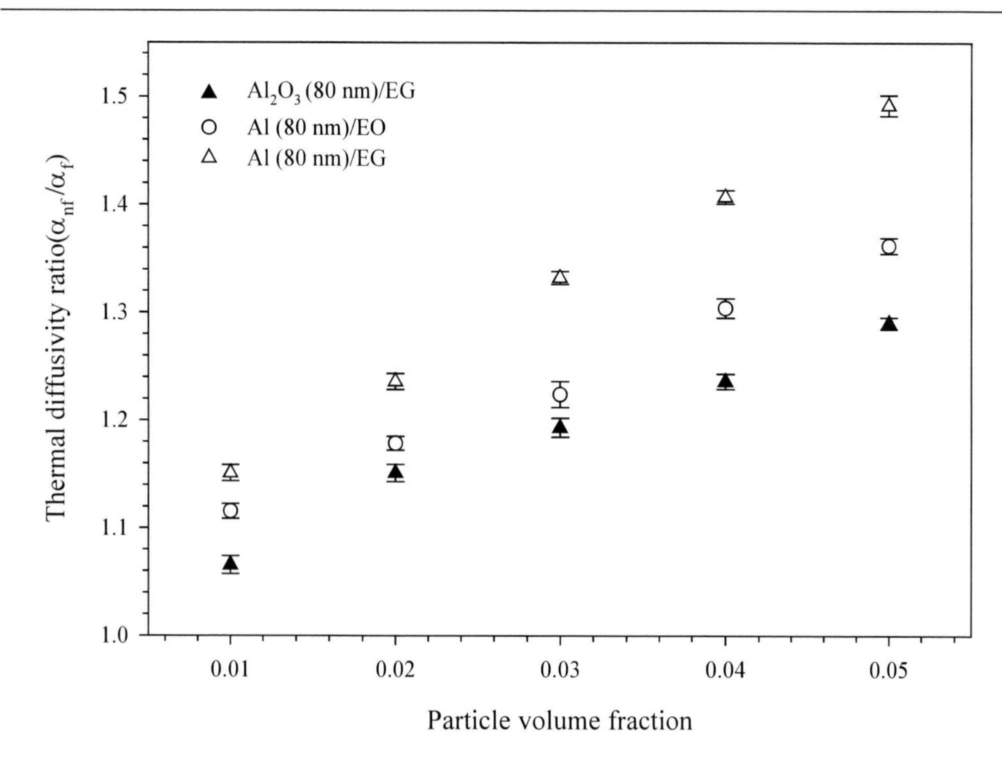

Figure 18. Effects of particle volume fraction, particle material, and base fluid on effective thermal diffusivity of nanofluids [63].

The effective thermal diffusivity of several types of nanofluids at different volume percentages (1 to 5%) of nanoparticles are measured by using double hot-wire technique. The effective thermal diffusivities of TiO_2/EG-based of nanofluids are shown in Figure 17. It is seen that for maximum 5% volumetric loading of TiO_2 nanoparticles of 15 nm and 10×40 nm in ethylene glycol, the maximum increase in effective thermal diffusivity was observed to be 25% and 29%, respectively. These results indicate that along with the particle volume fraction, particle size and shape have effect on the enhancement of thermal diffusivity of the nanofluids. This is probably because the particle size and shape affect the effective thermal conductivity of nanofluids [40-41] and the dispersion of particles in the base fluids, which eventually influence the effective thermal diffusivity of the mixture.

Figure 18 shows that nanofluids with aluminum nanoparticles in ethylene glycol and engine oil exhibit substantial enhancement of thermal diffusivity i.e. maximum 49% and 36%, respectively compared to their base fluids. The ethylene glycol-based nanofluid was found to have a higher thermal diffusivity than that of the engine oil-based nanofluids due to the higher thermal diffusivity of ethylene glycol as compared to the engine oil. It is also seen that the nanofluids with Al nanoparticles in ethylene glycol exhibit significantly higher thermal diffusivity than that of Al_2O_3 nanoparticles in same base fluid (EG). This is expected because the thermal diffusivity of Al (97.1×10^{-6} m^2/s) at 300 K is almost an order of magnitude higher than that of Al_2O_3 (11.9×10^{-6} m^2/s) at 293 K [16]. Figure 18 also demonstrates the impact of the particle material and the base fluid on the enhancement of the thermal diffusivity of nanofluids.

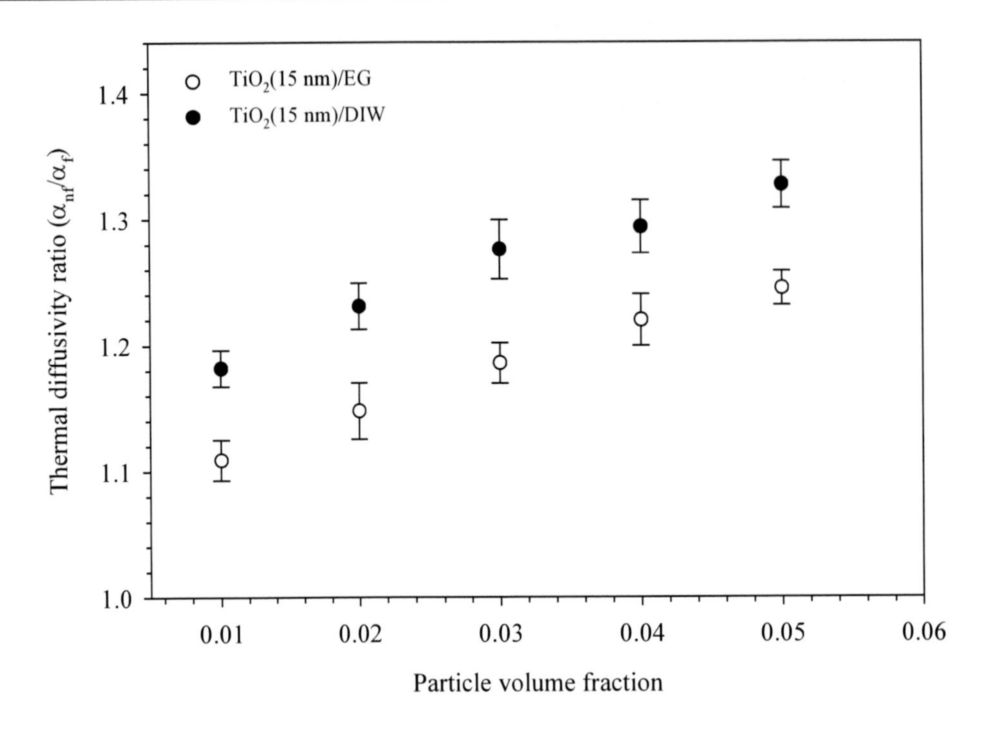

Figure 19. Thermal diffusivities of TiO_2/DIW and EG-based nanofluids with particle volume fraction [63].

Figure 19 illustrates the effect of base fluids on the effective thermal diffusivity of nanofluids. It is found that TiO_2 (15 nm)/DIW-based nanofluids have higher thermal diffusivities compared to that of ethylene glycol-based nanofluids. As mentioned previously, this is because of much higher thermal diffusivity of water compared to the ethylene glycol (Table 1).

Figures 17 to 19 show that nanofluids exhibit significant increase in thermal diffusivity with the nanoparticle volume fraction. This is mainly because the effective thermal conductivity of nanofluids was found to be substantially higher than the value for the base fluids and increases with the particle volume fraction [4-10, 22]. The dissociation of surfactant (i.e. CTAB) may play a significant role in intensifying the enhancement of thermal properties (e.g. thermal conductivity and diffusivity) of nanofluids by inducing electrostatic repulsion among the suspended nanoparticles resulting in better dispersion and stabilized suspensions. Moreover, with increased volumetric loading of nanoparticles, the specific heat of nanofluids decreases which results in an increase in the effective thermal diffusivity of the mixture (nanofluids). The calculated results of thermal diffusivity (i.e. by using the measured thermal conductivity and specific heat) of CuO (50 nm)/water-based nanofluids reported by Wang *et al.* [59] also showed significant enhancement of the thermal diffusivity of nanofluids. For example, for 0.1% volumetric loading of CuO nanoparticles in water, the enhancement of the thermal diffusivity of nanofluids was calculated to be 40%. The authors' above results also demonstrate that besides the particle volume fraction, base fluid and nanoparticle material influence the enhancement of the thermal diffusivity of nanofluids.

7.2. Determination of Specific Heat of Nanofluids

7.2.1. Theoretical Model for Specific Heat

The specific heat of sample nanofluids can be determined by using measured thermal conductivity and thermal diffusivity values and the definition of specific heat given as

$$c_{p-nf} = \frac{k_{nf}}{\alpha_{nf}\rho_{nf}} \tag{1}$$

where the density of mixture can be obtained from

$$\rho_{nf} = \phi_p \rho_p + (1-\phi_p)\rho_f \tag{2}$$

Since there is no model available to determine the effective specific heat of nanofluids, the volume fraction mixture rule can be used to formulate an expression. The effective volumetric specific heat nanofluids can be expressed as [24, 39]

$$(\rho c_p)_{eff} = [\phi_p \rho_p c_{p-p} + (1-\phi_p)\rho_f c_{p-f}] \tag{3}$$

where ϕ is the particle volume fraction and subscripts p and f stand for nanoparticle and base fluid, respectively.

From Equations (2) and (3), the effective specific heat of nanofluids can be expressed as

$$c_{p-nf} = \frac{\phi_p \rho_p c_{p-p} + (1-\phi_p)\rho_f c_{p-f}}{\phi_p \rho_p + (1-\phi_p)\rho_f} \tag{4}$$

The experimentally calculated specific heats of nanofluids will be compared with theoretically calculated results from Equation (4).

7.2.2. Results and Comparisons

Applying Equation (1), the effective specific heats of several types of nanofluids were determined from the measured thermal conductivity and thermal diffusivity obtained by double hot-wire method. As can be seen from Figures 20 to 22, the specific heats of nanofluids significantly decrease with increasing volume fraction of nanoparticles. Because of the much smaller specific heat of solid compared to that of liquid, it is obvious that the specific heat of a mixture will decrease with an increase of volumetric loading of solid particle.

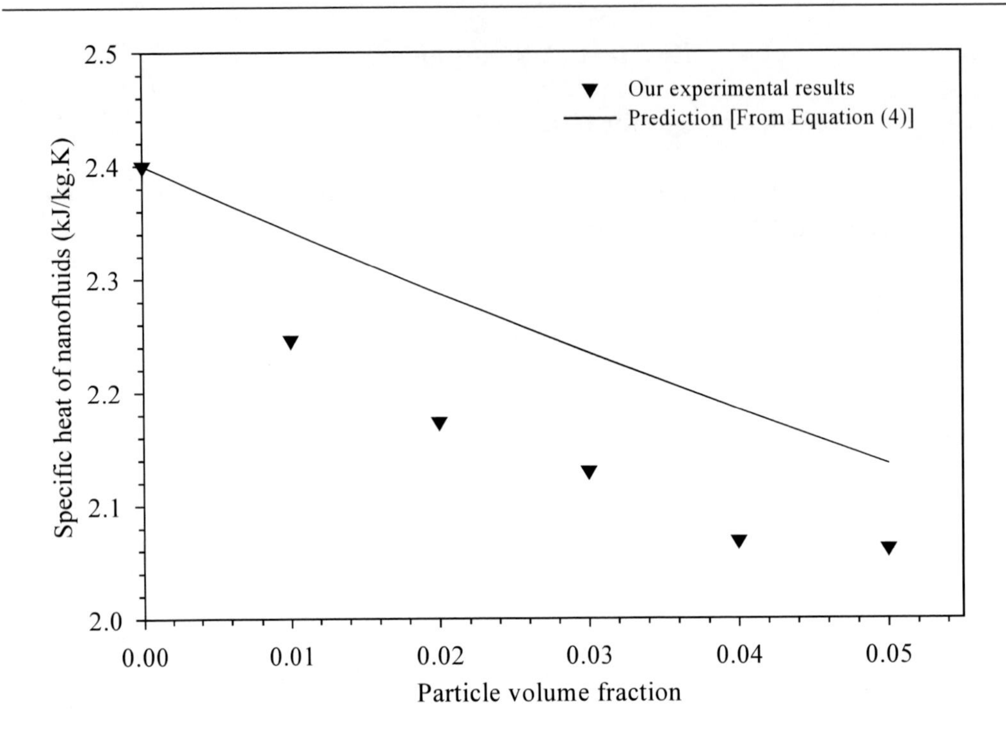

Figure 20. Our measured and theoretically calculated specific heats of TiO_2 (15 nm)/EG-based nanofluids with particle volume fraction.

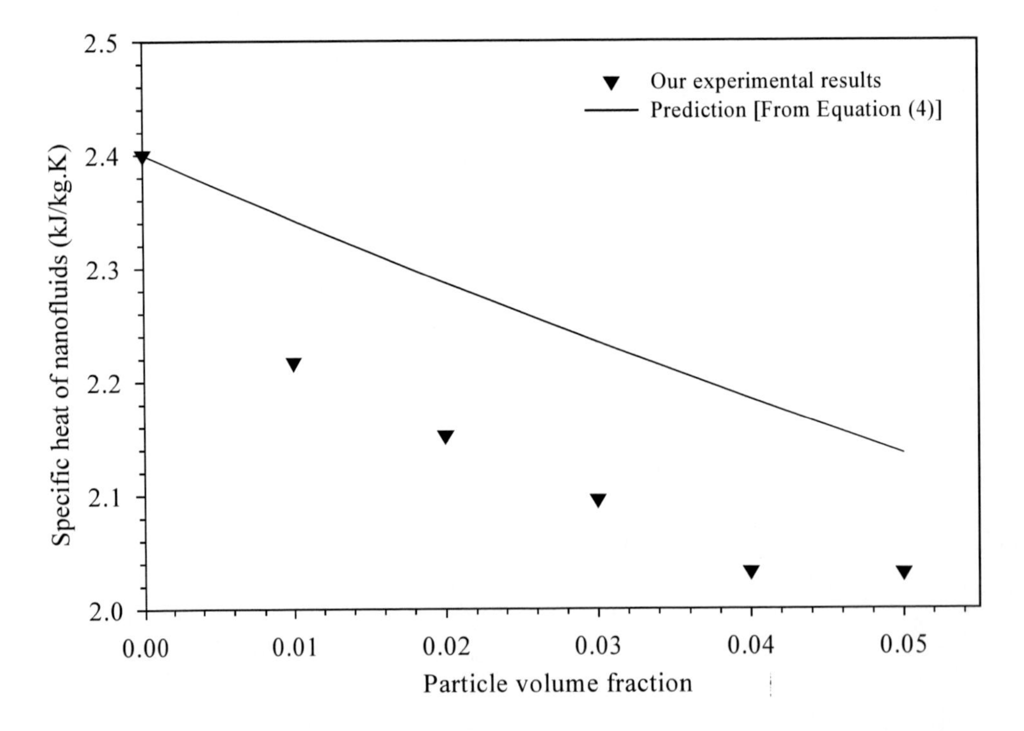

Figure 21. Measured and theoretically calculated specific heats of TiO_2 (10 × 40 nm)/EG-based nanofluids with particle volume fraction (authors' results).

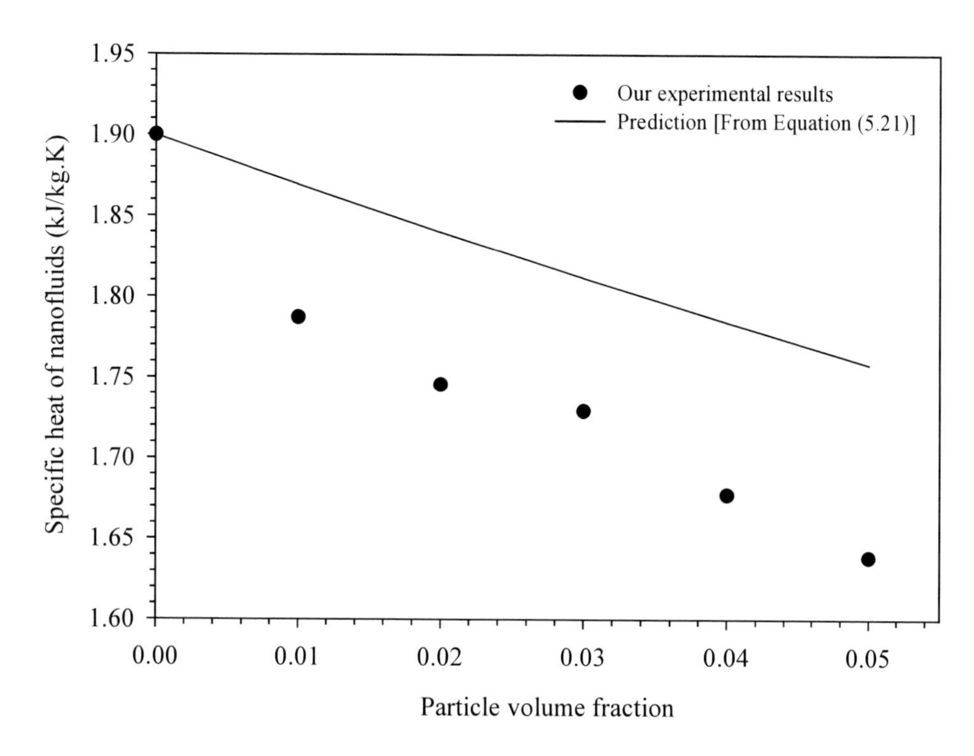

Figure 22. Measured and theoretically calculated specific heats of Al (80 nm)/EO-based nanofluids (authors' results).

Since no experimental data for the effective specific heats of nanofluids are available in the literature, no comparison of present results can be made. However, our results were compared with the calculated values obtained from the volume fraction mixture rule (Equation (4)). Figures 20 to 22 show that the effective specific heats of nanofluids determined from measured thermal conductivity and thermal diffusivity values are much lower than those calculated from Equation (4). This indicates that the volume fraction mixture rule is not suitable for the determination of the specific heats of nanofluids. The reason is unclear at the moment and requires further study.

7.3. Summary of Results

The experimental results presented in this section show that nanofluids containing a small amount of nanoparticles have much higher effective thermal diffusivity values than those of base fluids. The thermal diffusivity of nanofluids increases significantly with increasing volume fraction of nanoparticles. Moreover, the experimental results of this study demonstrate that the particle size and shape together with particle material and base fluid have effect on the effective thermal diffusivity of nanofluids. The specific heats of the nanofluids are also determined from the measured thermal conductivity and thermal diffusivity. With increasing volume fraction of nanoparticles, the specific heat of nanofluids is much lower and decreases more rapidly than the results calculated from the volume fraction mixture rule.

8. Studies on Viscosity of Nanofluids

8.1. Reported Studies in Literature

Although several articles [39, 64-65] emphasized the significance of investigating the viscosity of nanofluids, very few studies on this property were reported. It is believed that viscosity is as critical as thermal conductivity in engineering systems that employ nanofluids convective heat transfer where fluid flow is present. Pumping power is related to the pressure drop, which in turn is dependent on viscosity. In laminar flow, the pressure drop is directly proportional to the viscosity.

Masuda *et al.* [66] measured the viscosity of suspensions of dispersed ultra-fine particles in water. They found that TiO_2 (27 nm) particles at a volumetric loading of 4.3% increased the viscosity of water by 60%. Wang *et al.* [7] observed that the effective viscosity of Al_2O_3 (28 nm)/distilled water-based nanofluids was increased by about 86% for a 5% volume concentration of nanoparticles. In their case, a mechanical blending technique was used for the dispersion of Al_2O_3 nanoparticles in distilled water. They also found an increase of about 40% in viscosity of ethylene glycol at a volumetric loading of 3.5% of Al_2O_3 nanoparticle. Their results indicate that the viscosity of nanofluids depends on dispersion methods. In contrast, at 10% volume concentration of nanoparticles, the viscosities of Al_2O_3 (13 nm)/water and TiO_2 (27 nm)/water-based nanofluids used by Pak and Cho [67] were several times greater than that of water. This large discrepancy could be due to differences in dispersion techniques and differences in size of particles. Pak and Cho also used adjusted pH values and employed an electrostatic repulsion technique. As expected, the viscosity of nanofluids depends on the methods used to disperse and stabilize the nanoparticle suspension. Their viscosity results were significantly larger than the predictions from the classical theory of suspension rheology such as Einstein's model [68].

Das *et al.* [69] and Putra *et al.* [70] measured the viscosity of Al_2O_3/water and CuO/water-based nanofluids as a function of shear rate and showed Newtonian behavior of the nanofluids for a range of volume percentage between 1% and 4%. For Al_2O_3/water-based nanofluids, Das *et al.* [69] also observed an increase in viscosity with an increase of particle volume fraction.

At a given shear rate, Ding *et al.* [71] observed that the viscosity of nanofluids increased with an increasing CNT concentration and decreasing temperature. Shear thinning was observed at all concentrations. This suggests that nanofluids can offer better fluid flow performance due to the higher shear rate at the wall, which results in low viscosity.

Prasher *et al.* [72] reported results on the viscosity of alumina-based nanofluids for various shear rates, temperature, and particle volume fraction. Their results demonstrate that viscosity is independent of shear rate showing that the nanofluids are Newtonian in nature. It also is shown that with increasing nanoparticle volume fraction, the viscosity increases. On the other hand, it is found that viscosity is independent of temperature. This is contrary to the nature of liquid viscosity variation with the temperature.

For TiO_2 (34 nm)/water-based nanofluids, Wen and Ding [33] reported about 20% increase in the effective viscosity for a concentration of 2.4 weight % of particles. However, a much higher viscosity increase was observed under low share rate conditions i.e. 25-100 1/s.

All reported results show that the viscosity of nanofluids is increased anomalously and cannot be predicted by classical models such those attributed to Einstein [68], Krieger and Dougherty [73], Nielsen [74], and Batchelor [75]. No firm conclusion can be drawn from the above random data of several nanofluids. Thus, more extensive investigations need to be carried out on the effective viscosity of nanofluids.

8.2. Models for the Effective Viscosity

On top of very few experimental studies, no established model is available for the prediction of the effective viscosity of nanofluids. The Einstein model is commonly used to predict the effective viscosity of suspensions containing a low volume fraction of particles (usually < 1 volume %). The effective viscosity (η_{eff}) of dilute suspensions is commonly determined by Einstein's viscosity equation [68] expressed as

$$\frac{\eta_{eff}}{\eta_f} = 1 + 2.5\phi_p \tag{5}$$

Batchelor [75] considered interactions between particles in dilute suspension and modified the Einstein equation to

$$\frac{\eta_{eff}}{\eta_f} = 1 + 2.5\phi_p + 6.2\phi_p^2 \tag{6}$$

where ϕ_p is the volume fraction of the dispersed phase and η_f is the viscosity of the base fluid.

Since the nanofluids used in most of the studies are not dilute suspensions, the power law-based models [73, 74] are more appropriate for the prediction of the effective viscosity compared to Einstein's and Batchelor's models. For simple, hard sphere systems, the relative viscosity i.e. the ratio of the effective viscosity of suspension, η_{eff} to that of the suspending medium, η_f increases with particle volume fraction. To determine this relative viscosity, a semi-empirical equation formulated by Krieger and Dougherty (K-D) [73] expressed as

$$\frac{\eta_{eff}}{\eta_f} = (1 - \frac{\phi_p}{\phi_m})^{-[\eta]\phi_m} \tag{7}$$

is used.

A decade later, a generalized equation for the relative elastic moduli of composite materials, which is also widely used for relative viscosity, was proposed by Nielsen [74]. For low concentrations of dispersed particles, Nielsen's equation can be simplified as

$$\frac{\eta_{eff}}{\eta_f} = (1+1.5\phi_p)e^{\phi_p/(1-\phi_m)} \quad (8)$$

where ϕ_m is the maximum packing fraction and $[\eta]$ is the intrinsic viscosity ($[\eta]$ = 2.5 for hard spheres). For randomly mono-dispersed spheres, the maximum close packing fraction is approximately 0.64 [73].

8.3. Present Studies by the Authors

In this study, the effective viscosity of nanofluids was measured by a controlled rate rheometer (Contrases Low Shear 40 Model) and the results were compared with those predicted by the existing models as well as experimental data obtained from the literature.

Figure 23 shows that the viscosities of suspensions of TiO_2 and Al_2O_3 in water are independent of shear rates tested. This demonstrates that these nanofluids are Newtonian in nature. Figure 24 compares the relative viscosity data for TiO_2/water-based nanofluids from this study and those from the literature. The effective viscosity was found to increase with the particle volume fraction. For 15 nm TiO_2 nanoparticles in water, the maximum increase in viscosity was found to be 86% for a volume fraction of 0.05. For the same volume fraction of cylindrical shape (10 × 40 nm) TiO_2 nanoparticles, the viscosity of water is increased by 58%. This indicates that the shapes of the particles may influence the suspension viscosity. Through the surface charge, particle shape influences the electrostatic repulsive force. This repulsive force plays a significant role in altering the viscosity of nanofluids through the dispersion and sedimentation of nanoparticles. The particle shape affects clustering and adsorption, which may also change the viscosity of nanofluids. In fact, TiO_2 (10 × 40 nm) nanoparticles in base fluids were found to be more stable than TiO_2 (15 nm) nanoparticles. The authors' results for these nanofluids are found to be higher than those of Masuda *et al.* [66] and Pak and Cho [67]. Although both of them used the same size TiO_2 nanoparticles of 27 nm and adjusted to a high pH value (pH = 10) of their suspensions, Pak and Cho observed much higher viscosity values than those of Masuda *et al.* Figure 24 also depicts that the existing models are unable to predict the viscosity of these nanofluids.

Figure 25 presents measured viscosities of Al_2O_3 (80 nm)/water-based nanofluids with nanoparticle volume fraction and their comparisons with the models of Einstein [68], Batchelor [75], and Nielsen [74]. Al_2O_3 (80 nm) nanoparticles were found to increase the viscosity of water by nearly 82% for the maximum volumetric loading of 5%. A similar increment (86%) of the effective viscosity of Al_2O_3 (28 nm)/distilled water-based nanofluids was also observed by Wang *et al.* [7] for the same volumetric loading of 5%. In spite of their smaller particle size, Masuda *et al.'s* [66] results showed much larger increases in viscosity for the same nanofluids. For example, the viscosity of Al_2O_3 (13 nm)/water-based nanofluids at a volume % of 4.16 was about 2.68 times that of the viscosity of water in their study. Interestingly, the viscosity of the same nanofluid i.e. Al_2O_3 (13 nm)/water prepared by Pak and Cho [67] was three times higher than that of water for a 3 volume % of nanoparticles. However, a larger increase in viscosity was also observed in the present study for 150 nm sized Al_2O_3 nanoparticles. The maximum increase in viscosity was about 180% for the maximum volumetric loading of 5%. The present results for Al_2O_3/DIW-based nanofluids demonstrate the effect of particle size on the viscosity of suspensions and the larger the particle size, the higher the viscosity. Figure 25 also shows that the measured viscosities are severely under-predicted by the existing models.

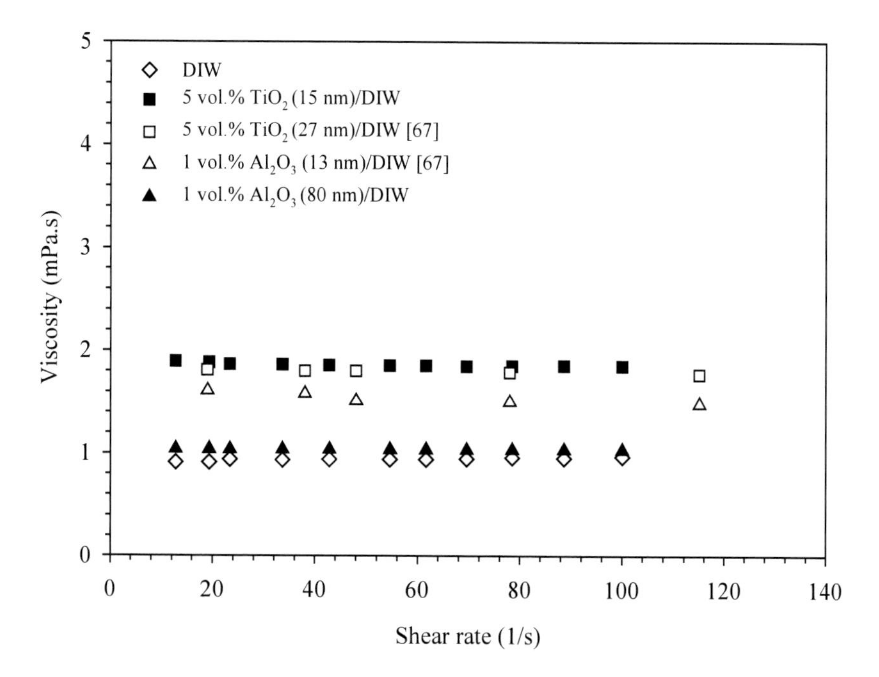

Figure 23. Comparison of authors' results and data obtained from the literature on the viscosity of nanofluids with shear rate.

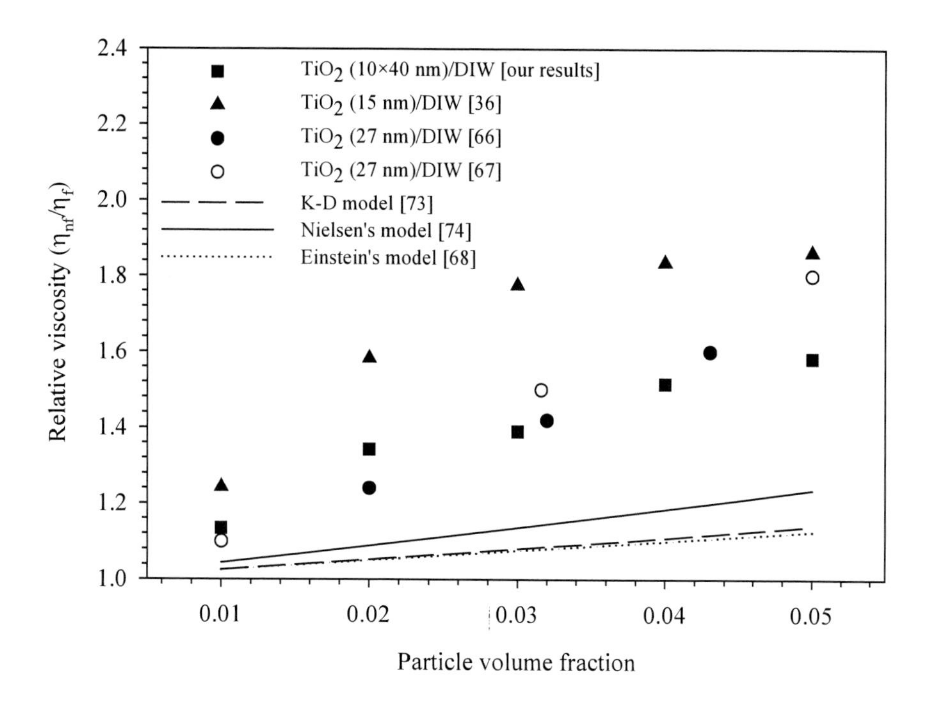

Figure 24. Experimental and predictions of relative viscosity of TiO_2/water-based nanofluids.

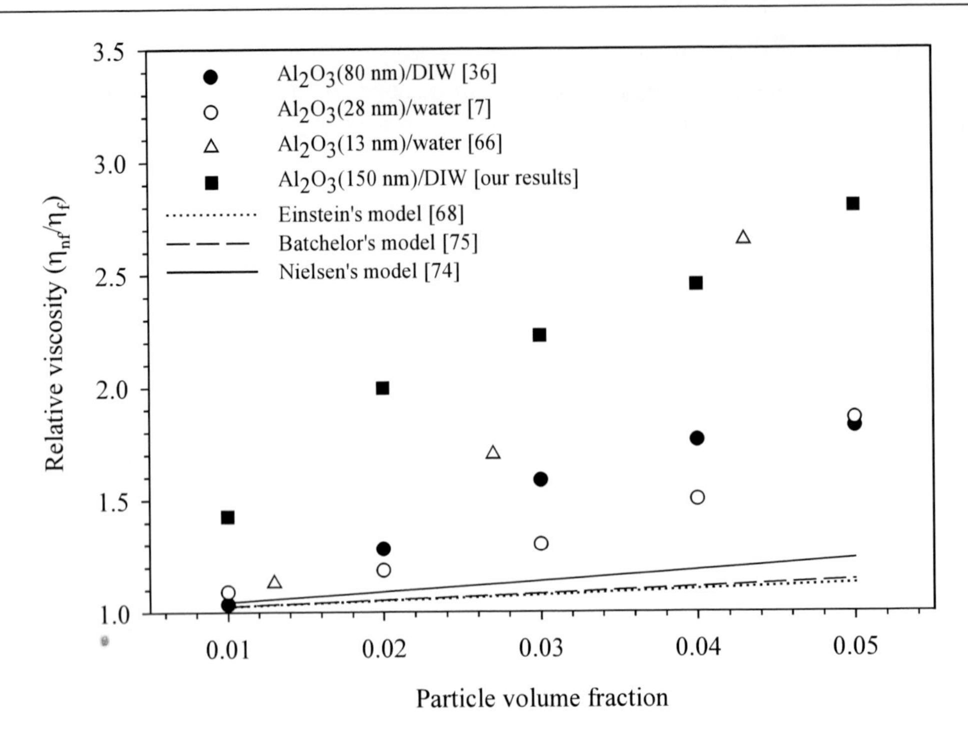

Figure 25. Experimental and predictions of relative viscosity of Al_2O_3/water-based nanofluids.

8.4. Summary

From Figures 24 and 25, it is found that the viscosity of nanofluids increases significantly with particle volume fraction. However, there are discrepancies between the present results and the reported data from the literature. The reasons for the differences could be the difference in the size of the particle clusters, differences in the dispersion techniques, and the use of surfactants. Indeed, the viscosity of nanofluids greatly depends on the methods used to disperse and stabilize the nanoparticle suspension [7]. Existing models were also found to be unable to predict the effective viscosity of nanofluids. This is probably because these models considered only particle volume fraction, whereas the nanoparticles in fluids can easily form clusters and can experience surface adsorption. Clustering and adsorption increase the hydrodynamic diameter of nanoparticles leading to the increase of relative viscosity. Besides the particle volume fraction and size, the nature of the particle surface, ionic strength of the base fluid, surfactants, pH values, inter particle potentials such as repulsive (electric double layer force) and attractive (van der Waals force) forces may play significant roles to alter the viscosity of nanofluids. However, such enhancement of viscosity may reduce the potential benefits of nanofluids.

9. Electrokinetic Phenomena of Nanofluids

9.1. Reported Studies in the Literature

Electrochemical effects certainly influence the thermal conductivity of nanofluids through stability and interparticle interactions. For example, the electrostatic repulsive force, which is described by the zeta potential, is important to avoid agglomeration and sedimentation of nanoparticles. Although some review articles [39, 76] pronounced that the surface chemistry of nanoparticles could be an important factor influencing the thermal conductivity of nanofluids, very few studies have been reported to investigate its effect on the enhanced thermal conductivity of nanofluids. Xie *et al.* [77] showed that simple acid treatment of carbon nanotubes enhanced the suspension stability of carbon nanotubes in water. It was attributed to hydrophobic-to-hydrophilic conversion of the surface nature due to the generation of a hydroxyl group. Patel *et al.* [25] reported that 4 nm gold nanoparticles with a coating of a covalent chain in toluene were about 50 times less effective for heat transport than uncoated 10-20 nm gold particles in water. This is contrary to the findings of the nanoparticle size effect on the effective thermal conductivity of nanofluids [26, 40]. Lee *et al.* [78] investigated the effect of zeta potential and particle hydrodynamic size (i.e. the mobility equivalent size) on the thermal conductivity of nanofluids. The hydrodynamic diameter of CuO nanoparticles and thermal conductivity of the nanofluids at different pH values were measured in their study. Their results demonstrated that the pH value affects the thermal performance of the nanofluids. As the solution pH value departs from the iso-electric point of particles, the colloidal particles become more stable and eventually alter the thermal conductivity of the fluid. They identified that the surface charge state is primarily responsible for the enhancement of thermal conductivity of the nanofluids. However, no further research was reported. Prasher *et al.* [79] studied the effect of aggregation on the effective thermal conductivity of nanofluids. They claimed that the observed thermal conductivity of nanofluids can be explained by the aggregation kinetics. It was also reported that the colloidal chemistry plays a significant role in deciding the thermal conductivity of nanofluids.

9.2. Studies by the Authors

The characterization of electrokinetic properties of nanofluids and their effects on thermal conductivity [80] are investigated. The effects of pH value, surfactant, and electrolyte concentration on zeta potential and thermal conductivity of TiO_2 (15 nm)/water–based nanofluids are discussed here.

9.2.1. Effects of pH value and Electrolyte Concentration on Zeta Potential

Figure 26 presents the variation of zeta potential for TiO_2 (15 nm)/water–based nanofluids with solution pH. The iso-electric point (IEP), a specific pH value at which the zeta potential is zero, was found to be between 4.9 and 5.2. This value is consistent with that reported in the literature (IEP = 5.1) for this type of nanoparticles [81]. It can also be seen from the figure that the presence of surfactant increases the magnitude of zeta potential, which indicates better stability of the nanofluids because the larger the magnitude of zeta potential, the stronger the repulsive forces. In addition, the surfactant can be adsorbed at the charged particle surface and modify the charges distribution in the electrical double layer (EDL) resulting a change in zeta potential. Due to the presence of CTAB, the IEP also decreased slightly from 5.2 to 4.9.

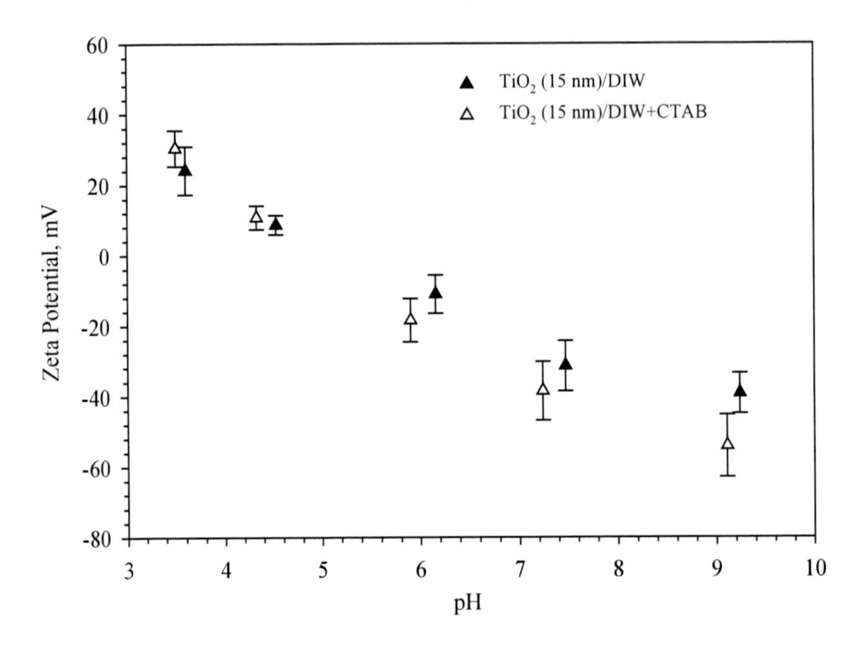

Figure 26. Effect of pH value on zeta potential of nanoparticles in deionized water [80].

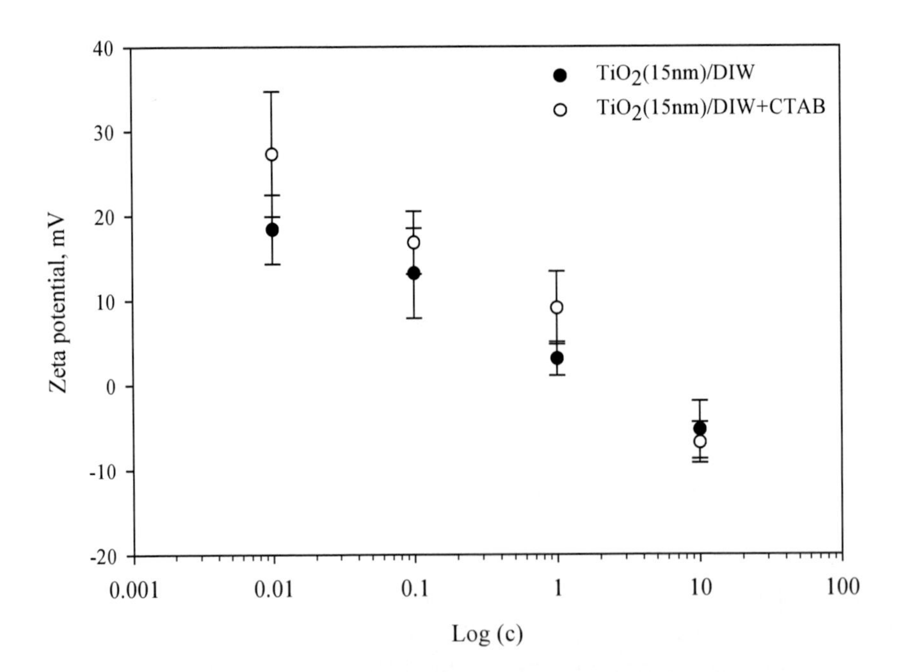

Figure 27. Effect of electrolyte concentration on zeta potential of nanoparticles [80].

The effect of electrolyte concentration (c, in mM) on the zeta potential is shown in Figure 27. Sodium chloride (NaCl) electrolyte of various concentrations was used. It is found that the ionic concentration of the suspensions has significant impact on the zeta potential. Similar to Figure 26, the addition of CTAB surfactant results in larger magnitudes of the zeta potential. The higher the concentration of the salt, the smaller the zeta potential. It is noted that the pH value at each NaCl concentration was remained the same i.e. about 6.2.

9.2.2. Effects of pH value and Electrolyte Concentration on Thermal Conductivity

Figure 28 shows that the enhanced thermal conductivity of this nanofluid decreases with an increase in pH value. For 0.2 volume % of TiO_2 nanoparticle, the thermal conductivity decreased from 5.5% to 2.55% when the pH value was increased from 3.4 to 9. A larger decrease in the enhanced thermal conductivity of Al_2O_3/water-based nanofluids was observed by Xie *et al.* [22] when the pH value was increased from 2 to 12. The trend of decreasing thermal conductivity with pH value until the IEP is reached can be explained by the DVLO theory. When the pH value is close to or equal to the IEP (zero repulsive force), the particle suspension becomes unstable and large-scale agglomerations occur which can lead to a decrease in effective thermal conductivity of nanofluids. However, when the pH value is further increased from the IEP, the effective thermal conductivity should increase gradually due to less agglomeration of particles, which was not observed in the authors' study [80]. Nevertheless, the decrease in effective thermal conductivity in this study was within 3% when the pH value changes from 3.4 to 9.

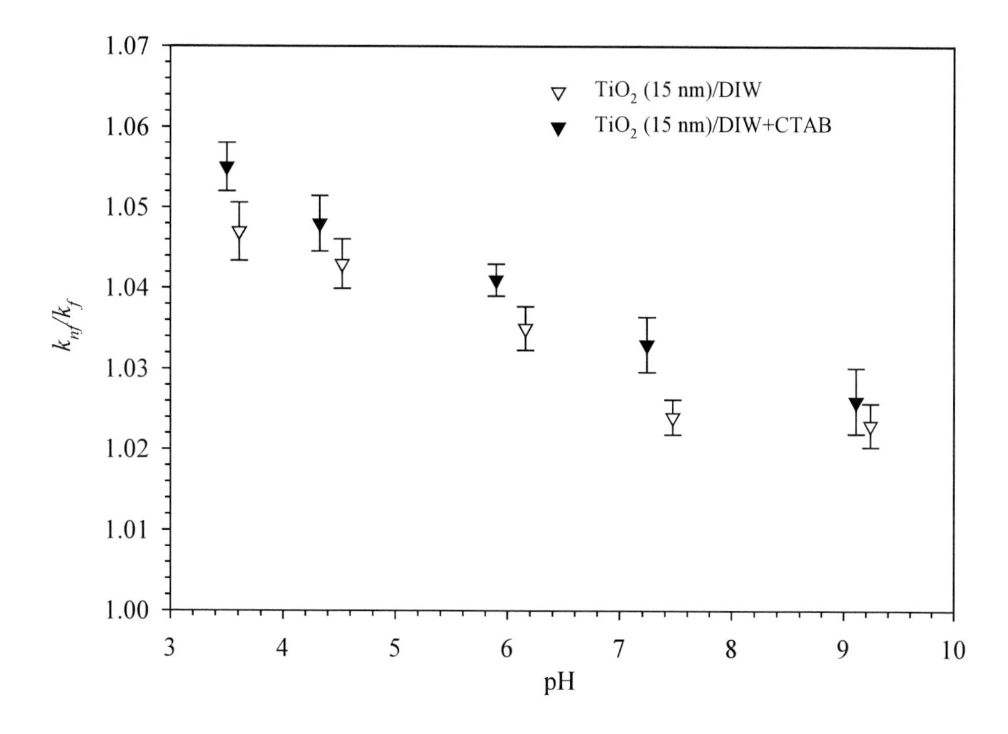

Figure 28. Effect of pH value on the effective thermal conductivity of nanofluids [8].

The influence of electrolyte concentration on the effective thermal conductivity of nanofluids is shown in Figure 29. The effective thermal conductivity was found to decrease with increasing NaCl concentration. However, the decrease of thermal conductivity was only within 1.5 to 2% when NaCl concentration was increased from 0.01 mM to 10 mM. This is consistent with the results of zeta potential (Figure 27) which decreases with increasing electrolyte concentration.

The positive effect of CTAB surfactant on the effective thermal conductivity of nanofluids can also observed from Figures 28 and 29. Surfactants promote to enhance the effective thermal conductivity of nanofluids through better particle dispersion i.e. less sedimentation/ agglomeration [82] and enhancing the interactions between components of nanofluids [83]. Such positive effect of surfactant on the enhancement of the effective thermal conductivity of Cu/ethylene glycol-based nanofluid was also reported by Yu and Choi [38]. It should be pointed out that the thermal conductivity of the base fluid is independent of pH value and electrolyte concentration. Since no other studies were reported in the literature regarding the influences of pH value and electrolyte concentration on the effective thermal conductivity for this nanofluid, no comparison of the present results can be made.

9.3. Summary

The pH value and the electrolyte concentration were found to have a significant effect on the zeta potential. The presence of a surfactant causes improved particle dispersion behavior, which resulted in better stability with less agglomeration and slightly larger effective thermal conductivity compared to nanofluids without surfactant. The enhanced thermal conductivity of this nanofluid was found to decrease with increasing pH value or electrolyte concentration.

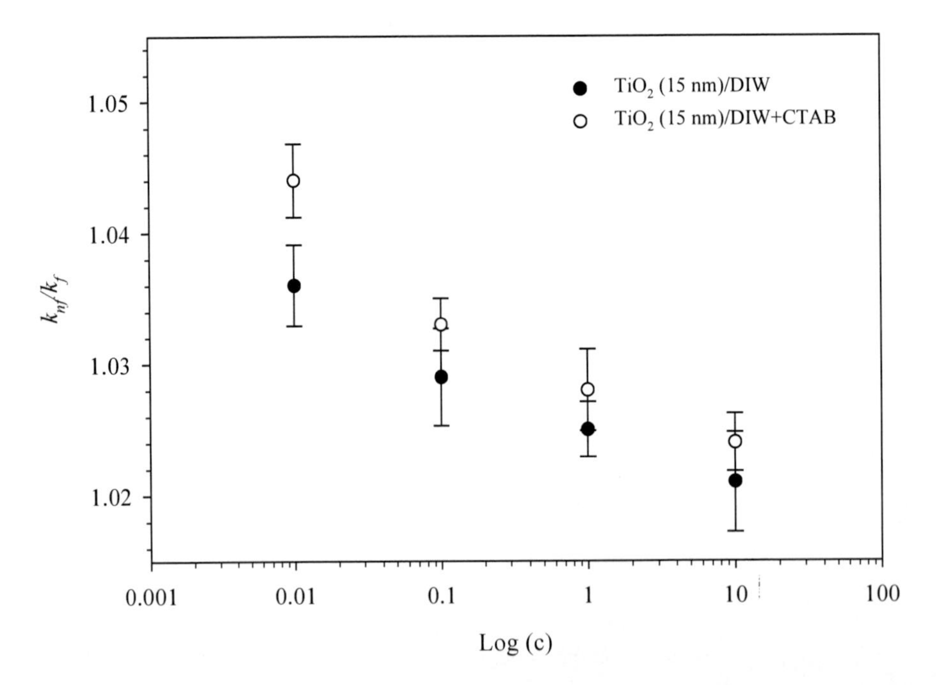

Figure 29. Effect of electrolyte concentration on the effective thermal conductivity [80].

10. Conclusions and Future Work

From the review and presented studies by the authors, it can be summarized that nanofluids exhibit enhanced effective thermal conductivity, which increases with increasing volumetric loading of nanoparticles. Review of experimental studies clearly shows the lack of consistency in the reported results from the various research groups. The effects of several important factors such as particle size and shapes, clustering of particles, temperature of the fluid, and dissociation of surfactant on the effective thermal conductivity of nanofluids have not been investigated adequately in the literature. In order to confirm the effects of these factors on the effective thermal conductivity of nanofluids, the present results are of great significance.

It is found that that the existing classical models cannot explain the observed enhanced thermal conductivity of nanofluids. On the other hand, most of the recently developed models are not based on first principles, and thus failed to explain the observed thermal conductivity of nanofluids. These models also cannot be validated with a wide range of experimental data. There is, therefore a need to develop new models, which can explicitly explain the enhanced thermal conductivity of nanofluids. Besides particle volume fraction, particles size, temperature, particles dispersions, particle interaction and clustering should be taken into account in developing model for the effective thermal conductivity of nanofluids.

Despite the importance of the effective thermal diffusivity, very little work has been reported on the determination of the effective thermal diffusivity of nanofluids. The present results show that nanofluids containing a small amount of nanoparticles have much higher effective thermal diffusivity values than those of base fluids. The thermal diffusivity of nanofluids increases significantly with increasing volume fraction of nanoparticles. The particle size and shape together with particle material and base fluid are also observed to have effect on the effective thermal diffusivity of nanofluids. With increasing volume fraction of nanoparticles, the specific heat of nanofluids is found to be much lower and decreases more rapidly than that of calculated from the volume fraction mixture rule. Effective viscosity of nanofluids is found to increase substantially with volumetric loading of nanoparticles. The existing models are also found to be unable to predict the viscosity of nanofluids. Thus, the theoretical model for the effective viscosity of nanofluids is to be developed. Furthermore, studies on characterization of electrokinetic properties and convective heat transfer of nanofluids are rather scarce. More extensive studies in these areas need to be performed for the potential applications of nanofluids in thermal management systems.

Although many studies have been performed on the heat transfer behavior of nanofluids, there has been very little agreement in the data from different research groups. Hence, proper sample preparation and repeatable and more systematic experimental studies on thermophysical properties, particularly thermal conductivity, thermal diffusivity, convective heat transfer coefficients and viscosity of nanofluids are worthwhile.

Focusing on the potential applications of nanofluids in advanced cooling techniques for more efficient cooling of electronics and microelectromechanical systems (MEMS), investigations on microchannel flow of nanofluids are of great interest.

References

[1] Maxwell, J. C. *A Treatise on Electricity and Magnetism*, Clarendon Press: Oxford, U.K., 1891.

[2] Choi, S. U. S.; Zhang, Z. G.; Keblinski, P. In *Encyclopedia of Nanoscience and Nanotechnology;* Nalwa, H. S.; Ed.; American Scientific Publishers: Los Angeles, CA, USA, 2004; Vol. 6, pp 757-773.

[3] Choi, S. U. S. In *Developments and applications of non-Newtonian flows*; Siginer, D. A.; Wang, H. P.; Eds.; ASME Publishing: New York,USA,1995; FED-Vol. 231/MD-Vol. 66, pp 99-105.

[4] Eastman, J. A.; Choi, S. U. S.; Li, S.; Thompson, L. J. In *Proceedings of the Symposium on Nanophase and Nanocomposite Materials II*; Komarneni, S.; Parker, J. C.; Wollenberger, H.; Eds.; Materials Research Society: Boston, USA, 1997; Vol. 457, pp 3-11.

[5] Eastman, J. A.; Choi, S. U. S.; Li, S.; Yu, W.; Thompson, L. J. *Appl. Phys. Lett.* 2001, 78, 718-720.

[6] Lee, S.; Choi, S. U. S.; Li, S.; Eastman, J. A. *J. Heat Transfer* 1999, 121, 280-289.

[7] Wang, X.; Xu, X.; Choi, S. U. S. *J. Thermophys. Heat Transfer* 1999, 13, 474-480.

[8] Xuan, Y.; Li, Q. *Int. J. Heat Fluid Flow* 2000, 21, 58-64.

[9] Das, S. K.; Putra, N.; Thiesen, P.; Roetzel, W. *J. Heat Transfer* 2003, 125, 567-574.

[10] Murshed, S. M. S.; Leong, K. C.; Yang, C. *Int. J. Therm. Sci.* 2005, 44, 367-373.

[11] Hong, T.; Yang, H.; Choi, C. J. *J. Appl. Phys.* 2005, 97, 064311-1-064311-4.

[12] Li, C. H.; Peterson, G. P. *J. Appl. Phys.* 2006, 99, 084314-1 – 084314-8.

[13] Choi, S. U. S.; Xu, X.; Keblinski, P.; Yu, W. In *Proceedings of 20th Symposium on Energy Engineering Science*, Illinois, USA, 2002.

[14] Hong, K. S.; Hong, T.; Yang, H. *Appl. Phys. Lett.* 2006, 88, 031901-1-031901-3.

[15] Steinour, H. H. *Ind. Eng. Chem.* 1944,36, 618-624.

[16] Kaviany, M. *Principles of Heat Transfer*, John Wiley & Sons, Inc.: New York, 2002.

[17] Bolz, R.; Tuve, G. *Handbook of Tables for Applied Engineering Science*, The Chemical Rubber Co.: Boca Raton, USA, 1973.

[18] Horrocks, J. K.; McLaughlin, E. *Proc. R. Soc. Lond. A.1963,* 273, 259-274.

[19] Nagasaka, Y.; Nagashima, A. *J. Phys. E: Sci. Instrum.* 1981, 14, 1435-1440.

[20] Wang, B.-X.; Zhou, L.-P.; Peng, X.-F. *Int. J. Heat Mass Transfer* 2003, 46 2665-2672.

[21] Yang, B.; Han, Z. H. *Appl. Phys. Lett.* 2006, 89, 083111-1–083111-3.

[22] Xie, H.; Wang, J.; Xi, T.; Liu, Y.; Ai, F.; Wu, Q. *J. Appl. Phys.* 2002, 91, 4568-4572.

[23] Wang, Y.; Fisher, T. S.; Davidson, J. L.; Jiang, L. In *Proceedings of 8th AIAA/ASME Joint Thermophysics and Heat Transfer Conference*, USA, 2002, AIAA 2002-3345.

[24] Xuan, Y.; Li, Q. *J. Heat Transfer* 2003, 125, 151-155.

[25] Patel, H. E.; Das, S. K. Sundararajan, T.; Nair, A. S.; George, B.; Pradeep, T. *Appl. Phys. Lett.* 2003, 83, 2931-2933.

[26] Kumar, D. H.; Patel, H. E.; Kumar, V. R. R.; Sundararajan, T.; Pradeep, T.; Das, S. K. *Phys. Rev. Lett.* 2004, 93, 4301-4304.

[27] Kwak, K.; Kim, C. *Kor.- Aus. Rheol. J.* 2005, 17, 35-40.

[28] Zhu, H.; Zhang, C.; Liu, S.; Tang, Y.; Yin, Y. *Appl. Phys. Lett.* 2006, 89, 023123-1–023123-3.

[29] Hwang, Y. J.; Ahn, Y. C.; Shin, H. S.; Lee, C. G.; Kim, G. T.; Park, H. S.; Lee, J. K. *Curr. Appl. Phys.* 2006, 6, 1068-1071.
[30] Xuan, Y.; Li, Q.; Zhang, X.; Fujii, M.; *J. Appl. Phys.* 2006, 100, 043507-1– 043507- 6.
[31] Liu, M. S.; Lin, M. C.-C.; Huang, I. T.; Wang, C.-C. *Chem. Eng. Technol.* 2006, 29, 72-77.
[32] Krishnamurthy, S.; Bhattacharya, P.; Phelan, P. E.; Prasher, R. S. *Nano Lett.* 2006, 6, 419– 423.
[33] Wen, D.; Ding, Y. *IEEE Trans. Nanotechnology* 2006, 5, 220-227.
[34] Putnam, S. A.; Cahill, D. G.; Braun, P. V.; Ge, Z.; Shimmin, R. G. *J. Appl. Phys.* 2006, 99, 084308-1–084308-6.
[35] Kang, H. U.; Kim, S. H.; Oh, J. M. *Exp. Heat Transfer* 2006, 19,181-191.
[36] Murshed, S. M. S.; Leong, K. C.; Yang, C. *Int. J. Therm. Sci.* 2008, 47, 560-568.
[37] Keblinski, P.; Phillpot, S. R.; Choi, S. U. S.; Eastman, J. A. *Int. J. Heat Mass Transfer* 2002, 45, 855-863.
[38] Yu, W.; Choi, S. U. S. *J. Nanoparticle Res.* 2003, 5, 167-171.
[39] Eastman, J. A.; Phillpot, S.R.; Choi, S. U. S.; Keblinski, P.; *Ann. Rev. Mater. Res.* 2004, 34, 219-246.
[40] Jang, S. P.; Choi, S. U. S. *Appl. Phys. Lett.* 2004, 84, 4316-4318.
[41] Chon, C. H.; Kihm, K. D. *J. Heat Transfer* 2005, 127, 810.
[42] Venerus, D. C.; Kabadi, M. S.; Lee, S.; Perez-Luna, V. *J. Appl. Phys.* 2006, 100, 094310-1– 094310-5.
[43] Koo, J.; Kleinstreuer, C. *J. Nanoparticle Res.* 2004, 6, 577-588.
[44] Murshed, S. M. S.; Leong, K. C.; Yang, C. *Int. J. Nanosci.* 2006, 5, 23-33.
[45] Hamilton, R. L.; Crosser, O. K. T *I&EC Fundamentals* 1962, 1, 187-191.
[46] Hui, P. M.; Zhang, X.; Markworth, A. J.; Stroud, D. *J. Mater. Sci.* 1999, 34, 5497-5503.
[47] Yu, W.; Choi, S. U. S. *J. Nanoparticle Res.* 2004, 6, 355-361.
[48] Xue, Q.-Z. *Phys. Lett. A.* 2003, 307, 313-317.
[49] Wang, X. Q.; Mujumdar, A. S., *Int. J.Therm. Sci.* 2007, 46, 1-19.
[50] Murshed, S. M. S.; Leong, K. C.; Yang, C. *Appl. Therm. Eng.* 2008 (in press, doi:10.1016/j.applthermaleng.2008.01.005).
[51] Xie, H.; Fujii, M.; Zhang, X. *Int. J. Heat Mass Transfer* 2005, 48, 2926-2932.
[52] Leong, K. C.; Yang, C.; Murshed, S. M. S. *J. Nanoparticle Res.* 2006, 8, 245-254.
[53] Yu, W.; Choi, S. U. S. *J. Nanosci. Nanotechnol.* 2005, 5, 580-586.
[54] Sabbaghzadeh, J.; Ebrahimi, S. *Int. J. Nanosci.* 2007, 6, 45-49.
[55] Ren, Y.; Xie, H.; Cai, A., *J. Phys. D: Appl. Phys.* 2005, 38, 3958-3861.
[56] Xuan, Y.; Li, Q.; Hu, W. *AIChE J.* 2003, 49, 1038-1043.
[57] Leong, K. C.; Yang, C.; Murshed, S. M. S., *Presented in Nanofluids: Fundamentals and Applications Conference*, Copper Mountain, Colorado, USA, September 2007.
[58] Xuan, Y.; Roetzel, W.; *Int. J. Heat Mass Transfer* 2000, 43, 3701-3707.
[59] Wang, B-X.; Zhou, L.-P.; Peng, X.-F. *Prog. Nat. Sci.* 2004, 14, 922-926.
[60] Parker, W. J.; Jenkins, R. J.; Butler, C. P.; Abbott, G. L. *J. Appl. Phys.* 1961, 32,1679-1684.
[61] Balderas-Lopez, J. A.; Mandelis, A., *Rev. Sci. Instrum.* 2001, 72, 2649-2652.
[62] Czarnetzki, W.; Roetzel, W., *Int. J. Thermophys.* 1995, 16, 413-422.
[63] Murshed, S. M. S.; Leong, K. C.; Yang, C. *J. Phys. D: Appl. Phys.* 2006, 39, 5316-5322.
[64] Keblinski, P.; Eastman, J. A.; Cahill, D.G. *Materials Today* 2005, 8, 36-44.

[65] Das, S. K.; Choi, S. U. S.; Patel, H. E.; *Heat Transfer Eng.* 2006, 27, 3-19.
[66] Masuda, H.; Ebata, A.; Teramae, K.; Hishinuma, N. *Netsu Bussei* 1993, 4, 227-233.
[67] Pak, B. C.; Cho, Y. I. *Exp. Heat Transfer* 1998, 11, 151-170.
[68] Einstein, A. *Investigations on the Theory of the Brownian Movement*, Dover Publications, Inc.: New York, 1956.
[69] Das, S. K.; Putra, N.; Roetzel, W. *Int. J. Heat Mass Transfer* 2003, 46, 851-862.
[70] Putra, N.; Roetzel, W.; Das, S. K. *Heat Mass Transfer* 2003, 39, 775-784.
[71] Ding, Y.; Alias, H.; Wen, D.; Williams, A. R. *Int. J. Heat Mass Transfer* 2006, 49, 240-250.
[72] Prasher, R.; Song, D.; Wang, J.; Phelan, P.E. *Appl. Phys. Lett.* 2006, 89, 133108-1–133108-3.
[73] Krieger, I. M.; Dougherty, T. J. *Trans. Soc. Rheology* 1959, 3, 137-152.
[74] Nielsen, L. E. *J. Appl. Phys.* 1970, 41, 4626-4627.
[75] Batchelor, G. K. *J. Fluid Mech.* 1977, 83, 97-117.
[76] Kabelac, S.; Kuhnke, J. F. In *Proceedings of International Heat Transfer Conference*, Sydney, Australia, August 2006.
[77] Xie, H.; Lee, H.; Youn, W.; Choi, M. *J. Appl. Phys.* 2003, 94, 4967-4971.
[78] Lee, D.; Kim, J. W.; Kim, B. G. *J. Phys. Chem. B* 2006, 110, 4323-4328.
[79] Prasher, R.; Phelan, P. E.; Bhattacharya, P. *Nano Lett.* 2006, 6, 1529-1534.
[80] Murshed, S. M. S.; Leong, K. C.; Yang, C. *J. Nanosci. Nanotechnol.* 2008, 8, 1-6..
[81] Kosmulski, M.; *J. Colloid Interface Sci.* 2006, 298, 730-741.
[82] Kim, S. H.; Choi, S. R.; Kim, D. *J. Heat Transfer* 2007, 129, 298-307.
[83] Assael, M. J.; Metaxa, I. N.; Kakosimos, K.; Constantinou, D. *Int. J. Thermophys.* 2006, 27, 999-1017.

In: Nanoparticles: New Research
Editor: Simone Luca Lombardi, pp. 243-275
ISBN: 978-1-60456-704-5

Chapter 8

RECENT DEVELOPMENTS IN THE EFFECTIVE THERMAL CONDUCTIVITY OF NANOPARTICLE SUSPENSIONS (NANOFLUIDS) RESEARCH

Calvin H. Li
Department of Mechanical, Industrial, and Manufacturing Engineering,
University of Toledo, Toledo, Ohio 43606, USA
G.P. Peterson
Department of Mechanical Engineering, University of Colorado
Boulder, Colorado 80309, USA

Abstract

Since the first report in 1995 that indicated that the addition of nanometer copper metal particles to traditional heat transfer liquids, such as water and Ethylene Glycol, could result in an enhanced effective thermal conductivity which was considerably higher than that predicted by traditional mean-field theory, tremendous experimental studies have been conducted. Most of these support this conclusion, but there is some variation in the degree of the level of the enhancement. These investigations include experiments on metal nanoparticle/liquid suspensions, metal oxide nanoparticle/liquid suspensions, liquid nanoparticle/liquid suspensions, and carbon nanotube/liquid suspensions, and cover a wide range of nanoparticle shapes and volume fractions. These investigations include variations in the size of the nanoparticle, the type of nanoparticle material, the type of base fluid, the effect of surfactants, the effect of pH value of the suspension, the temperature of the suspension, and the ultrasonic vibrating time. Moreover, a number of experimental techniques and methods have been adopted to measure the effective thermal conductivity of different types of suspensions, including steady-state cut-bar methods, transient hot-wire methods and other less well known methods.

The theoretical exploration of the fundamental mechanisms for enhanced effective thermal conductivity of nanoparticle suspensions has resulted in a number of theoretical models. There are basically two different mechanisms proposed for these variations, the Brownian motion effect and the agglomeration effect. In addition to these, several other alternatives, such as the effect of the adsorption layer of the fluid molecules on the surface of the nanoparticles and the thermophoresis of the nanoparticles inside the suspension have been proposed.

The Brownian motion mechanism in nanoparticle suspensions is a relatively new concept, which has grown in acceptance with the addition of additional experimental evidence. While the agglomeration mechanism is not new, it assumes that the agglomeration of the nanoparticles replaces the micron size and/or millimeter size particles in the mean-field theory. Both major mechanisms are supported by the comparisons between predictions of models and reported experimental results. Moreover, there is almost equal support for both from numerical simulation studies and experimental observations.

An extensive review of the available experimental reports and theoretical developments indicates that a majority of the experiments have confirmed the enhanced effective thermal conductivities of nanoparticle suspensions and indicated that these are greater than the values predicted from mean-field theory models. In addition, the mean-field theory can not explain why the effective thermal conductivity of nanoparticle suspensions could have this unusually high enhancement, or why it increases with increases in the bulk temperature, and with decreases in the nanoparticle size. Finally, the evolution process of newly developed Brownian motion models and agglomeration models is discussed in detail.

As a result, what is clear from the available research is that it is necessary to illuminate why effective thermal conductivities of nanoparticle suspensions are higher than the predictions of mean-field theory models. Extensive study on the contributions of the mechanisms of Brownian motion, agglomeration and correlation of both is critically needed. This review will help the development of a more robust understanding of the new phenomenon and the further application of effective thermal conductivity of nanoparticle suspensions in the future.

1. Introduction

The term nanofluids is used to describe in general terms 1-100 nm diameter nanoparticle/liquid suspensions, which are used to achieve enhanced thermal conductivity for heat transfer and thermal management purposes. The concept was first proposed in 1995 when it was observed that by dispersing metallic nanoparticles into traditional heat transfer fluids, the effective thermal conductivity of copper nanoparticle/water suspensions could be dramatically enhanced and the pumping power significantly decreased [1, 2]. Based on the theoretical model of Hamilton and Crosser [3], it was estimated that the Cu nanoparticle with a sphericity of 0.3 could dramatically enhance the effective thermal conductivity of nanoparticle suspensions by as much as 3.5 times that of pure water, at a volume fraction of 20 %, and Masuda et al.'s experimental result of the effective thermal conductivity of γ Al_2O_3 nanoparticle/Distilled water suspensions indicated that a 30 % enhancement could be achieved at a nanoparticle volume fraction of 4.3 % [4].

Adding high thermal conductivity particles into a low thermal conductivity matrix to achieve increased thermal conductivities while retaining the other desirable physical properties of the matrix is far from a novel idea. This kind of work has been extensively studied for many centuries. Due to the limitations of technology, however, the size of the dispersed particles has been restricted to the order of millimeter and microns, such as the experimental results reported by Hamilton and Crosser [3] and by Meredith and Tobias [5]. Previous theoretical research can be traced even earlier than the first experimental investigations, and many successful mean-field theory models have been developed and have been used to accurately predict the effects of the thermal conductivity of solid/solid mixtures, solid/liquid mixtures, and liquid/liquid mixtures. A number of these models have been used to predict the effective thermal conductivities of nanoparticle suspensions, but deviations of experimental data from the predictions of those mean-field theory models were reported, and meanwhile this lack of correlation between the measured and predicted values required further study of the fundamental mechanisms of the effective thermal conductivity of nanoparticle suspensions.

2. The Mean-Field Theory and Models

The mean-field theory has a long history which can be traced back as far as the 18th century. The first mean-field model capable of predicting the effective thermal conductivity of solid particles and fluid mixtures was initially developed by Maxwell [6] in 1873 who studied conduction through heterogeneous media. In this initial work, the potentials at the interface were assumed to be continuous and equal, thus the current passing through the interface of the two media would be the same. Examples of two and three different physical properties with interfaces that were spherical were conducted, and the mathematical treatment was applied to introduce the potential expanded in solid body harmonics and the surface harmonics. By analyzing the algebraic sign of the constants in the harmonics, it became clear that if the conduction of one of the materials were higher than the rest, the material with the higher conduction would tend to equalize the potential throughout the material. Otherwise, the current was prevented from reaching the inner of the lower conduction materials. By examining the case of n spheres of radius a_1 and resistance R_1, placed in a medium of resistance material, R_2, where all the spheres and media were contained in a sphere of radius a_2, and assuming no interaction between the spheres or medium, the effective resistance of the sphere of radius a_2 with a volume fraction of ϕ could be calculated as follows:

$$R = \frac{2R_1 + R_2 + \phi(R_1 - R_2)}{2R_1 + R_2 - 2\phi(R_1 - R_2)} R_2 \tag{1}$$

This equation serves as the original formula, which was subsequently modified by other researchers and has been extensively used to predict the behavior of these types of systems, with the limitation that the volume fraction, ϕ, should be very small in order to avoid the interaction between spheres and that the magnitude of $\frac{R_1 - R_2}{2R_1 + R_2}$ should not be large.

Based on Maxwell's work, a number of models were developed, and only the models which were used to compare the predictions and experimental data in the nanoparticle suspension study will be introduced in this chapter.

2.1. Models with Consideration of the Shape Influence of Dispersed Particle

In 1924, Fricke [7] developed the mean-field theory of Maxwell's by including the effect of particle shape to the effective thermal conductivity. For other particle geometries, such as ellipsoid or spheroid particles, the geometrical arrangement of the suspended particles was introduced. For suspensions of non-spherical particles, the effective thermal conductivity could be calculated using a form somewhat analogous to the equation for a suspension with spherical particles as Maxwell's equation,

$$(k_e / k_1 - 1)/(k_e / k_1 + x) = \phi(k_2 / k_1 - 1)/(k_2 / k_1 + x) \tag{2}$$

where k_e, k_1, and k_2 are the specific conductivities of the suspension, the suspending medium, and the suspended spheroids, ϕ is the volume concentration of the suspended spheroids. The influence of the geometric factors of the suspended particles was considered through a parameter, x, which is a function of the ratio k_2/k_1 and the ratio a/b of the axis of symmetry of the spheroids to the other axis. Fricke compared the result of this equation with experimental data for the conductivity of blood (k_2=0, a/b=1/4.25, x=1.05), which showed excellent agreement for concentrations from 10 % to 90 %. For a constant volume concentration, the conductivity of a suspension was shown to be independent of the size of the suspended particles and also nearly independent of the form of the particles when the difference between the conductivities of the dispersed particles and the continuous matrix was not very large, especially with suspensions of prolated spheroids that are less conducting than the suspending medium. This was a pure mathematical treatment, which gave the hints to calculate the effective thermal conductivity of suspensions in the later theoretical study.

Following this lead, a significant theoretical work was performed and resulted in the development of equations to calculate effective thermal conductivity. R. L. Hamilton and O. K. Crosser (1962) [3] proposed an equation to predict the effective thermal conductivity of heterogeneous systems consisting of a continuous matrix with dispersed particles within it. This was based on the work of Maxwell and Frick and introduced a new constant n, which was dependent on the shape of the particle, and independent of the ratio of the thermal conductivities of the dispersed particles and the continuous matrix. When the particle was a sphere, n would not demonstrate any dependence at all. Thus, in the development of this equation, the effect of the particle shape was referred to the spherical shape, and a shape factor was added into the equation, which was based on the experiments of spherical particles and non-spherical particle mixtures. A coefficient ψ was then introduced and defined as the ratio of the surface area of a sphere to the surface area of a non-spherical particle having the same volume as the spherical particle. For spherical particles, $n=3/\psi$; for prolate ellipsoids, $n=3/\psi^2$; and for oblate ellipsoids $n=3/\psi^{1.5}$. The equation for the effective thermal conductivity was

$$k_e = k_1 \left[\frac{k_2 + (n-1)k_1 - (n-1)\phi(k_1 - k_2)}{k_2 + (n-1)k_1 + \phi(k_1 - k_2)} \right] \tag{3}$$

With this equation, it was found that if the continuous matrix had a higher conductivity than that of the dispersed particles, the particle shape would have little effect on the effective thermal conductivity of the mixture. However, if the dispersed particles had a higher conductivity, there was a strong shape effect. When using Eq. 3 with n in terms of sphericity and the conductivity of the dispersed particles at least 100 times larger than the conductivity of the continuous matrix, equation 3 agreed with the experimental results of silastic rubber and variously shaped particles of aluminum and balsa at selected volume compositions at 95 ^oF, in which the diameters of particles were in the magnitude of a millimeter. Finally, it was also pointed out that the agglomeration might change the shape, size, and orientation of particles, which would make the prediction of the equation deviate from the experimental data.

Another significant theoretical development on the shape effect was accomplished one year later after Hamilton and Crosser's model, J. B. Keller (1963) [8] considered the extreme situation in which the spheres and circular cylinders imbedded in a continuous matrix of conductivity k_1 have perfect conduction with a dense cubic array, and nonconducting cylinders in a medium with square array. For the perfectly conducting spherical particles, it was determined that the maximum value of the fractional volume is π/6. With this maximum value of fractional volume, the equation to calculate the effective conductivity was derived as follows:

$$k_e / k_1 = -(\pi / 2)\log[(\pi / 6) - \phi] + \cdots (\pi / 6) - \phi << 1 \tag{4}$$

When ϕ=0.5161, the result yielded k_e/k_1=7.65, which agreed with the measured result of k_e/k_1=7.6 in the experiment of Merdith and Tobias [5].

For the perfectly conducting circular cylinders, the maximum value of the fractional volume is π/4. With the same procedure, the equation for medium containing perfectly conducting circular cylinders was obtained as Eq. 5 for the square array of cylinders, which agreed with the numerical results obtained by J. B. Keller [8].:

$$k_e / k_1 = -(\pi^{\frac{3}{2}} / 2)\log[(\pi / 4) - \phi]^{\frac{1}{2}} + \cdots (\pi / 4) - \phi << 1 \tag{5}$$

This investigation also considered a situation in which the cylinders contained in the continuous matrix were non-conducting, and there was a resistance of each gap between two adjacent cylinders, which was $R=1/k_e c$. Here, c was the distance between the axes of the two adjacent cylinders. It was found that the equation of effective thermal conductivity for nonconducting cylinders and the equation for perfectly conducting cylinders were the reciprocal of each other, and had the following relationship:

$$\frac{k_e^{\infty}}{k_1} = \frac{k_1}{k_e^0} \tag{6}$$

where k_e^{∞}, k_e^0 were the effective thermal conductivities of perfectly conducting cylinders and of nonconducting cylinders. Here, k_1 was the conductivity of a continuous medium.

A. Rocha and A. Acrivos (1973) [9, 10] made the assumption that no significant interactions existed in the dilute suspension as previous researchers had, and got a general expression from Fourier's law for the heat conduction in statistical homogeneous suspension as Eq. 7. Their principal contribution is that the effective thermal conductivity of suspensions was studied not only with spherical particles but with arbitrary shaped particles as well.

$$q + k_1 \frac{\partial T}{\partial x_i} = (k_1 - k_2) c \int_{S_p} T n_i dS \tag{7}$$

where q was the heat flux, k_1 was the thermal conductivity of continuous matrix, k_2 was the thermal conductivity of dispersed particle, c was the number of dispersed particles per unit volume, S_p was the surface of a particle and n_i was the normal on the surface. The effective thermal conductivity of the suspension with spherical particles was then derived as

$$\frac{k_e}{k_1} = \{1 + 3[(\alpha - 1)/(\alpha + 2)]\phi\}\delta \tag{8}$$

where α was the ratio of thermal conductivity of dispersed particles to that of the continuous matrix, ϕ was the volume fraction of particles, and δ was the Kronecker delta.

When the particles were slender or ellipsoids, the effective thermal conductivity was not isotropic, but had different values along different particle axes direction in the suspension. Finally, the effective thermal conductivity equations for suspensions with arbitrary shaped particles, axially symmetric shaped particles, and slender particles were obtained under the steady state and non-interaction condition. The general equation was

$$\frac{k_e}{k_1} = \delta_{ij} + 2[(\alpha - 1)/(\alpha + 1)]\phi \times \left\{\delta_{ij} + v_2^{-1}\sum_{m=1}^{\infty}(2\pi)^m[(\alpha - 1)/(\alpha + 1)]^m \int_\Omega\right\} \tag{9}$$

It was pointed out that the effective thermal conductivity of the suspension would still be isotropic if the function was truncated to the first order of (α-1)(α+1), and reduce to only a function of volume fraction of dispersed particles.

I. C. Kim and S. Torquato (1993) [11] applied computer simulation to calculate the effective thermal conductivity of aligned spheroidal particle suspensions with first passenger time algorithm. The dispersed particle had a wide range of aspect ratio from 0.1 to 10 and the ratio of the thermal conductivities of dispersed particle and the continuous matrix was from zero to infinite. This method was mainly concerned with the Brownian motion of dispersed particles and was a totally different approach to calculating the effective thermal conductivity of suspensions from the previously dominant thought, which held that the derivative of the Maxwell's equation should be conducted to the high order of volume ratio to include the far and near range interactions between two or more particles. This approach was from the viewpoint of motion instead of the statistic force field to consider the mechanism of the enhancement of effective thermal conductivity of suspension.

2.2. Models with Consideration of Interaction between Particles

Due to the nature of the atoms and molecules, when the dispersed particles are close to each other, their interaction would influence the effective thermal conductivity of the suspension, too.

G. K. Batchelor [12] reviewed the study made on the transport properties of heterogeneous mixtures that were composites of two phases that separately were homogeneous, and pointed out that the use of probability theory should be very important and essential for the study on the bulk properties of these two-phase dispersed systems. Batchelor started from the Fourier theory that $F=K\bullet G$, where F was the flux, K was the bulk transport coefficient characteristic of the medium, and G was the gradient of intensity. Then a table of the physical phenomena that represented the problem of determining an effective transport coefficient was given, and the theoretical analysis was conducted based on the common principles of these phenomena.

This table discussed the applicability of the equations and theories to all the volume fractions and gave the upper and lower limits of effective thermal conductivity for arbitrary volume fractions found by Z. Hashin and S. Shtrikman [12] for the effective magnetic permeability of multiphase materials. Then, it discussed the calculations on the mean flux F in two-phase statistic homogeneous mixtures, in which the local heat flux always had a linear relation to the local intensity gradient. Here, a new parameter was introduced: the dipole strength S, which was a measure of the result when the matrix material was replaced by the particle material, and a statistical property of microstructure of the mixture representing the interaction between particles.

Table 1. Effective transport coefficient for two-phase disperse systems [12]

Cases of two-phase disperse systems with random structure in which the problem is to determine an effective transport coefficient

Nature of the two uniform media of which the two-phase disperse system is composed	Quantity represented by F	Quantity represented by G	Transport coefficient (or scalar components in case of isotropic structure)	Local differential equation satisfied in each phase (in steady state)
(1) Thermal conductor	Heat flux	Temperature gradient	Thermal conductivity	$\mathbf{F}=\mathbf{K}\cdot\mathbf{G}$ $\nabla\cdot\mathbf{F}=0$
(2) Electrical conductor	Electric current	Electric field	Electrical conductivity	
(3) Electrical insulator	Electric displacement	Electric field	Dielectric constant	
(4) Dia- or para-magnetic material	Magnetic induction	Magnetic field	Magnetic permeability	
(5) Small fluid or rigid particles sedimenting through Newtonian incompressible fluid (a "fluidized bed")	Flux of particle number relative to zero-volume-flux axes	Gravitational force on particles in unit volume of mixture (minus the buoyancy force)	Mobility	$\nabla p=\mu\nabla^2\mathbf{u}$ $\nabla\cdot\mathbf{u}=0$ where $\mathbf{u}$ = velocity p = pressure μ = viscosity
(6) "Porous medium" consisting of a fixed array of small rigid particles through which incompressible Newtonian fluid is passing	Force on particles in unit volume of mixture (= pressure gradient calculated from pressure drop between distant parallel planes)	Flux of fluid volume relative to particles	Permeability (Darcy constant) divided by μ	
(7) Small fluid or rigid particles suspended in incompressible Newtonian fluid	Deviatoric stress	Rate of strain	Shear viscosity	As immediately above; or $\mathbf{F}=2\mu\mathbf{G}$, $\nabla\cdot\mathbf{F}=0$ G = trace $\mathbf{G}=0$
(8) Elastic inclusions embedded in an elastic matrix	Stress	Strain	Lamé constants (or rigidity and bulk moduli)	$\mathbf{F}=2\mu\mathbf{G}+\lambda G I$ $\nabla\cdot\mathbf{F}=0$ where μ,λ = local Lamé constants

Three kinds of interacting conditions of particles that dilute suspensions were discussed: (1) non-interaction between particles, (2) first order of interaction between particles, and (3) strong interaction in small volume fraction suspensions. The corresponding expression for the effective thermal conductivity of the first condition was given as

$$k_e = k_1\left\{1+\left[2(\alpha-1)/(\alpha+2)\right]\phi+\left[3(\alpha-1)^2/(\alpha+2)^2\right]\phi^2\right\} \tag{10}$$

where k_e was the effective thermal conductivity, k_1 was the thermal conductivity of continuous matrix, α was the ratio of the thermal conductivity of particle material to that of the matrix material, and ϕ was the volume fraction of particles.

For the last two conditions, a broader analysis considered the situation in which there was a bulk movement in the suspensions. This was a more complicated situation which included the interactions between particles plus the influence of other external forces.

D. J. Jefferey (1973) [13] suggested that the probability $P(r|o)$ for one sphere only in the distance $O(r)$ of the reference sphere was $O(\phi)$ and the probability $P(r|o)$ for two spheres in the distance $O(r)$ was $O(\phi^2)$. The choice for the probability $P(r|o)$ satisfied

$$\begin{aligned} P(r|o) &= 0, r \leq 2a \\ P(r|o) &= n, r >> a \end{aligned} \tag{11}$$

where a is the radius of the particle and r is the distance of two particles.

This could be interpreted as following. The first condition was that the spheres did not overlap and the second condition was that all members of spheres were far away from the reference sphere. When solving the problems involving two spheres, the twin spherical expansions were applied. After obtaining the expressions for the heat flux and temperature within and out of the spheres, the author took the interactions between pairs of spheres into consideration through a second order of the volume ratio between particles and fluid. This equation depends on the way pairs of spheres are distributed with respect to each other. As to the convergence, this work was done on the basis of Batchelor's method in 1972 [12] for the average interactions between spheres to obtain bulk properties of suspensions. The equation developed is shown as Eq. 12.

$$k_e / k_1 = 1 + 3\beta\phi + 3\beta\phi^2\left(\beta + \sum_{p=6}^{\infty}\frac{B_p - 2A_p}{(p-3)2^{p-3}}\right) \tag{12}$$

where $\beta=(\alpha-1)(\alpha-2)$ and $\alpha=k_2/k_1$, volume fraction $\phi = 4/3\pi a^3 n$ and a is the radius of the particle, P is the probability density for a second sphere being at location r to the center of the reference sphere and A_p and B_p are known coefficients.

R. H. Davis (1986) [14], fully aware of the difficulty of the integral convergence encountered by early researchers, considered a situation that inside the composite material, the undisturbed temperature gradient varied inversely with the square of the distance from the heated body. Thus, the correction to the dipole strength of the reference sphere varied as $(a/r)^5$ instead of $(a/r)^3$ due to the presence of a second sphere, and only convergent integrals were encountered. Finally, Davis proposed an equation (Eq. 13) based on Green's theorem to calculate the rate of heat transfer from the heated body to the composite material, which was similar to Jeffrey's model [13], shown as below.

$$k_e / k_1 = 1 + \frac{3(\alpha - 1)}{[\alpha + 2 - (\alpha - 1)\phi]} \{\phi + f(\alpha)\phi^2 + 0(\phi^3)\} \tag{13}$$

here $f(\alpha) = \sum_{p=6}^{\infty} \frac{B_p - 3A_p}{(p-3)2^{p-3}}$ was used to represent the decaying temperature field in the process of developing this equation, while early researchers had considered it a composite material of infinite extent on which an undisturbed linear temperature field was imposed. The $O(\phi^2)$ in the equation contributing to the effective thermal conductivity was not given entirely by the solution for two spheres placed in the undisturbed temperature field, but the actual environment of these two spheres was the undisturbed temperature plus an $O(\phi)$ term from all of the remaining spheres. It was also pointed out that contributions of $O(\phi^3)$ and higher orders were even more difficult to obtain and not generally needed for the dilute suspension cases. Here, the maximum volume ratio is 0.62 in order to prevent the spheres from touching one another.

J. F. Brady et al. (1988) [15] presented a method to calculate hydrodynamic interactions among particles in infinite suspensions with infinite small particle Reynolds number. The numerical calculation applied to the spatially periodic suspension problem was the Stokesian-dynamics method.

Based on this work, R. T. Bonnecaze and J. F. Brady [16, 17], in order to avoid the convergent problem in integration over the infinite extent of matrix, developed a method for determining the individual dipoles for a finite number of particles, which included the particles interactions in both long range and short range fields. The individual particle potentials and the potential gradient were related to the sum of particle moments to generate a "potential" matrix. Then, by inverting this potential matrix, a capacitance matrix was formed. The potential invert produced the many-body reflections of the particle moments that were included in its formulation, and the potential invert included the far field interactions of the particles. Plus, the exact two-body interactions added into the potential invert, the accurate capacitance matrix was obtained. With this capacitance matrix, the dipole strength of particles could be calculated directly under the given external potential field. The equation developed was

$$k_e = k_1 - n(S / G^E) \tag{14}$$

where S was the dipole strength and G^E was the potential gradient applied.

This method was applied to the infinite suspensions by using the O'Brien's method [16] and periodic boundary conditions which reduced the system to a finite number of linear equations. With this method, the effective thermal conductivities of the suspensions with cubic arrays of spheres which had infinite thermal conductivity, 10 times the thermal conductivity of the matrix's, and 0.01 times the thermal conductivity of the matrix's, were calculated.

S. Y. Lu and S. T. Kim 1990 [18] studied the relationship between the microstructure of the dispersed particles and the effective thermal conductivity of the anisotropic composite, consisting of homogeneous matrix with spheroidal particles dispersed within it. The systems with fixed aligned dispersed particles were the only ones considered, and only the variation of the particles' positions influenced the effective thermal conductivity in this study. The dipole strength and renormalization approach in the integration were used to calculate the dipole strength as early researchers had done. The model that included the assumption that dispersed particles would take non-overlapping positions with equal probability was calculated to show the flaw of the absence of potential energy existing between particles. Then, the hard-sphere fluid model was adopted, in which the pair distribution function was a function only of the distance between two axes. This model showed a better agreement with the former work and the experimental data.

$$k_e = \widehat{I} + \left(-\frac{M}{V_2 r_1}\right)\phi + \left(\frac{-\Xi + \Omega}{V_2 r}\right)\phi^2 \quad (15)$$

where $\widehat{I}$ was the unit tensor, V_2 was the volume of dispersed particles, M was the second-order tensor for one-body contribution, Ξ was the second-order tensor for two-body contribution, and Ω was the renormalization quantity.

S. Y. Lu and H. C. Lin (1996) [19] developed an expression for the effective conductivity tensor to calculate the effective conductivity in the form of the viral expansion of the dispersed particle volume fraction ϕ which was truncated at the $O(\phi^2)$ term. The particle shape discussed was ellipsoid and its mathematical expression was

$$\frac{x^2}{a^2} + \frac{y^2}{b^2} + \frac{z^2}{c^2} = 1, a \geq b \geq c \quad (16)$$

In the study, pair interaction, the second-order tensor for one-body contribution M, the second-order tensor for two-body contribution Ξ, and the renormalization quantity Ω for the two aligned spheroids, were evaluated by solving a boundary-value problem involving two aligned spheroids with a boundary collocation scheme. This boundary collocation scheme was derived based on the pre-limit that the governing equation was the Laplace equation and the system was a two-particle system. The temperature fields outside the particles and inside two particles were expressed, first, with the general solution for the Laplace equation, then forced to satisfy the boundary conditions at the particle surfaces, and finally the relevant thermal dipole strength. The series ratio inferred from the viral expansion of the equivalent inclusion estimate served as a reasonably good validity test parameter for the second-order virial approximation. The magnitude of this series ratio was also a reflection of intensity of thermal interactions within the system. Two models for the pair distribution function were studied; one was the well-stirred mode and the other one was the hard spheroid model. The equivalent inclusion estimate method of Hatta and Taya was applied here and derived as

$$\frac{k_e}{k_1} = 1 + \frac{\alpha - 1}{(\alpha - 1)h_i + 1}\phi + \frac{(\alpha - 1)^2 h_i}{[(\alpha - 1)h_i + 1]^2}\phi^2 + \frac{(\alpha - 1)^3 h_i}{[(\alpha - 1)h_i + 1]^3}\phi^3 + \ldots \quad (17)$$

where $\alpha = k_2/k_1$, and the equation would converge with the condition $\frac{\alpha - 1}{(\alpha - 1)h_i + 1} < 1$.

2.3. Models with the Consideration of Interfacial Resistance

Based on the experimental results of spherical nickel dispersions in a sodium borosilicate glass matrix and carbon fiber-reinforced glass ceramic, D. P. H. Hasselman and L. F. Johnson (1987) [20] pointed out that the thermal barrier resistance at the interface between the dispersed particle and the continuous matrix would have an important influence on the effective thermal conductivity of the composite, which the previous study did not take into consideration in the development of the equation or in the calculation of the effective thermal conductivity. The idea that potential was continuous and the current was the same at the two sides of interface between dispersed particle and continuous matrix, was still adopted in this study, just as Maxwell had proposed.

$$k_1 \left(\frac{\partial T_1}{\partial r} \right) = k_2 \left(\frac{\partial T_2}{\partial r} \right) \tag{18}$$

$$T_1 - T_2 = -\frac{k_2}{h_c} \frac{\partial T_2}{\partial r} \tag{19}$$

where T_1 was the temperature in the continuous matrix out of the dispersed particle, T_2 was the temperature inside of the dispersed particle, k_1 was the thermal conductivity of the continuous matrix, k_2 was the thermal conductivity of the dispersed particle, r was the radius of the interfacial surface, and h_c was the interfacial thermal barrier resistance. If the interfacial thermal barrier resistance was zero, then $T_1 = T_2$ at the interface, which agreed with the solutions of the Maxwell equations [6].

Physically the interfacial thermal barrier resistance h_c was not zero, so the effective thermal conductivity of the composite was affected by the cumulative effect on T_1, and the radius of the large sphere, b, which included n dispersed spheres of radius a. When the large sphere was large enough, $b \to \infty$, the equation for the effective thermal conductivity was

$$k_e = k_1 \frac{\left[2\left(\frac{k_2}{k_1} - \frac{k_2}{ah_c} - 1 \right)\phi + \frac{k_2}{k_1} + \frac{2k_2}{ah_c} + 2 \right]}{\left[\left(1 - \frac{k_2}{k_1} + \frac{k_2}{ah_c} \right)\phi + \frac{k_2}{k_1} + \frac{2k_2}{ah_c} + 2 \right]} \tag{20}$$

The equations for two kinds of particles, circular cylinders and flat plate, were conducted as:

$$k_e = k_1 \frac{\left[\left(\frac{k_2}{k_1} - \frac{k_2}{ah_c} - 1\right)\phi + \frac{k_2}{k_1} + \frac{2k_2}{ah_c} + 1\right]}{\left[\left(1 - \frac{k_2}{k_1} + \frac{k_2}{ah_c}\right)\phi + \frac{k_2}{k_1} + \frac{2k_2}{ah_c} + 1\right]} \tag{21}$$

$$k_e = \frac{k_2}{\left[\left(1 - \frac{k_2}{k_1} + \frac{2k_2}{ah_c}\right)\phi + \frac{k_2}{k_1}\right]} \tag{22}$$

respectively, and they all agreed with the previous reported solutions when the interfacial thermal barrier resistance was zero. The equation (20) was the most widely used equation, beside the Hamilton-Crosser model [3], in the later nanoparticle suspension effective thermal conductivity study.

The original contribution of work is that it did not only consider the influence of the interfacial resistance to the effective thermal conductivity, but included the particle size in the model as well.

2.4. Models of Liquid-Liquid Mixture

In 1961, Meredith and Tobias [5] found that the Maxwell's equation [6], the Fricke equation [7] and the Bruggeman equation [5] all failed to predict the experimental results of effective thermal conductivity of the liquid-liquid emulsions when the volume fraction is greater than 20 %, by comparing their experimental data with the predictions of the models. Hence they developed a new expression to predict the effective conductivity of the water-oil emulsion suspensions which have both a wide range of size distribution of dispersed spherical oil particles in a water matrix and a very narrow size distribution of dispersed spherical water particles in an oil matrix, which is shown as Eq. 23 below,

$$\frac{k_e}{k_1} = \left(\frac{2 + 2X\phi}{2 - X\phi}\right)\left(\frac{2 + (2X - 1)\phi}{2 - (X + 1)\phi}\right) \tag{23}$$

where $X = \frac{k_2 / k_1 - 1}{k_2 / k_1 + 2}$. Here, an assumption was made that spherical shape liquid particles were dispersed in the other liquid matrix.

L. G. Leal [21] analyzed the effective conductivity of a dilute suspension in a simple shear flow and studied the role of the mechanical and thermal properties of the particles in contributing to the bulk thermal transport characteristics of the suspension. The bulk heat flux of the suspension was calculated by the averaging of the microscale heat flux. The local temperature field was considered under the assumption that the suspension was experiencing a simple shear flow with a linear bulk temperature field across the shear gradient in which the Reynolds number and the Peclet number were very small. In this investigation, it was found that the effective conductivity was largely determined by the thermal conductivity of the suspended particles and the shear flow disturbance contribution. Further, it was pointed out that the particle thermal conductivity might have either a positive or a negative contribution, according to the conductivity ratio between the particle and base fluid, while the flow always contributed positively. The study gave people another perspective of the enhancement of the effective thermal conductivity of suspensions, stating that not only the physical and thermal properties of the suspended particles would change the effective thermal conductivity of suspension, but their motion could also result in a considerable change to the effective thermal conductivity as well. However, because this work lacked information on the effect of the particle size, shape, volume fraction, and other factors, the resulting expression was not capable of providing good predictions on the experimental data.

$$k_e = k_1[1+\phi\{\frac{3(k_2-k_1)}{k_2+2k_1}+(\frac{1.176(k_2-k_1)^2}{(k_2+2k_1)^2}+\frac{2\mu_1+5\mu_2}{\mu_1+\mu_2}(0.12\frac{2\mu_1+5\mu_2}{\mu_1+\mu_2}$$
$$-0.028\frac{k_2-k_1}{k_2+2k_1}))Pe_1^{3/2}+0(Pe_1^2)\}] \tag{24}$$

Here μ is the viscosity, and subscript 1 represents the basefluid; 2 represents the particle if the particle was formed by another fluid.

All of the above discussed models are well constructed and have been successfully used to predict the effective thermal conductivity of many kinds of mixtures. However, the application to effective thermal conductivity of nanoparticle suspensions was not quite well. One problem is that the predictions of those models are significantly lower than the experimental results of nanoparticle suspension effective thermal conductivities at room temperature; Second, those models could not predict the elevation of nanoparticle suspension effective thermal conductivity with an increase of temperature; Third, only few of those models considered the nanoparticle size and could not predict that the effective thermal conductivity would increase with a decrease of nanoparticle size. Hence, a new theory, which could well describe the new phenomena of the effective thermal conductivity variation of nanoparticle suspensions, is critical needed.

3. Development of Nanoparticle Suspension Study and Two Major Possible Mechanisms: Brownian Motion and Agglomeration

In the following several years after the first proposal that indicated that adding nanoparticle into the traditional heat transfer fluid media could obtain much higher effective thermal conductivity than that of the base fluids, Choi's prediction was confirmed by several experimental reports that nanoparticle suspensions could obtain unusually high effective thermal conductivity compared to that of base fluids [22-24]. In those first experimental reports, both a transient hot-wire method and a parallel plates method were employed to measure the effective thermal conductivity of suspensions consisting of Cu, CuO, and Al_2O_3 nanoparticles in Distilled (DI) water, Ethylene glycol and engineer oil [22-24].

3.1. Experimental Reports on Effective Thermal Conductivity of Nanoparticle Suspensions

Eastman et al. [22] reported that a 36 nm diameter CuO nanoparticle/DI water suspension could obtain 60 % enhancement on thermal conductivity at a volume fraction of 5 % over that of DI water. And a 33 nm diameter Al_2O_3 nanoparticle/DI water suspension could achieve around 30 % enhancement on thermal conductivity at 5 % volume fraction over that of DI water. Interestingly, the 18 nm diameter Cu nanoparticle/HE-200 oil suspension could have as high as 43 % enhancement over the thermal conductivity of HE-200 oil at a volume fraction as small as around 0.05 %, which is two orders less in volume fraction than that of CuO and Al_2O_3 nanoparticle/DI water suspensions at the same magnitude of enhancement in thermal conductivity [4]. However, in another later experimental report by the same group of researchers [23], the enhancements on thermal conductivity were much lower than previously reported. The 23.6 nm diameter CuO nanoparticle/DI water suspension had an enhancement above 20% at the volume fraction of 4%, and the 38.4 nm diameter Al_2O_3 nanoparticle/DI water suspension had an enhancement around 12 % at the volume fraction of 4 %, and an enhancement around 20 % at the volume fraction of 5 %. The Hamilton and Crosser model was used to compare the predictions and the experimental results. It was illustrated that this model could well predict the effective thermal conductivity of Al_2O_3 nanoparticle suspensions, while predictions of the CuO nanoparticle suspension model were much smaller.

The new experiment of the Cu nanoparticle suspensions showed that, with the Cu nanoparticles of diameter less than 10 nm, the Cu/ethylene glycol suspension could have an enhancement above 40 % at the volume fraction of 0.3 %, which is much less than the 43 % enhancement achieved at around 0.05 % volume fraction of 18 nm Cu/HE-200 oil suspension [4].

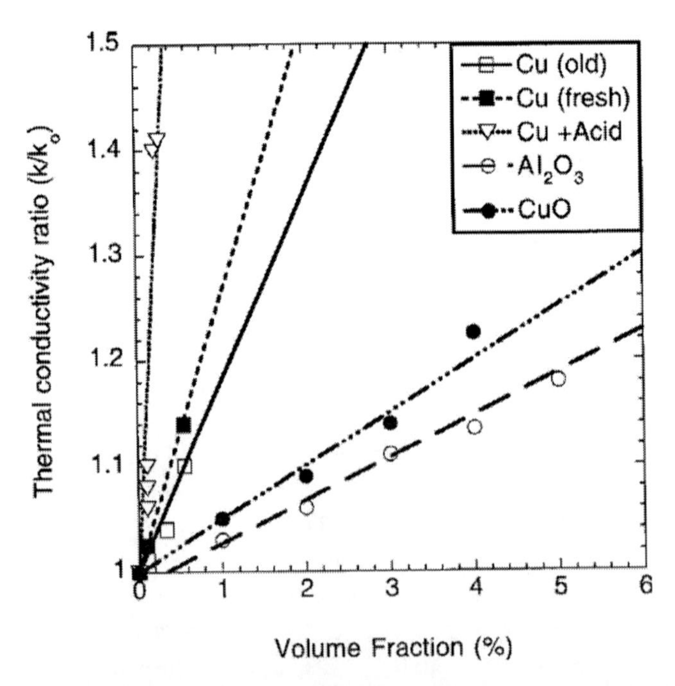

Figure 1. Effective thermal conductivities of metallic and oxide nanoparticle suspensions [22].

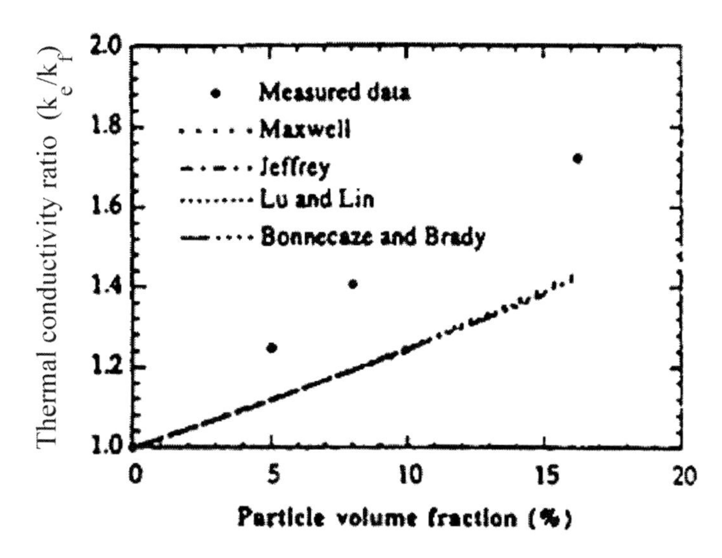

Figure 2. Measured thermal conductivities of Al_2O_3/ethylene glycol mixtures vs. effective thermal conductivities calculated from theories [24].

In 1999, Wang and Xu et al., [24] employed a steady-state parallel-pate method to measure the effective thermal conductivities of suspensions with 20 nm diameter Al_2O_3 and CuO nanoparticles in DI water, ethylene glycol, engine oil, and vacuum pump fluid. The experimental results were compared with a series of mean-field models, such as Maxwell's model [6], Jeffrey's model [13], Davis' model [14], Lu and Lin's model [19], Bonnecase and Brady's model [16, 17], and Hamilton-Crosser's model [3]. The comparisons between the experimental results and predictions on Al_2O_3 nanoparticle/ethylene glycol suspensions pointed out that those available models all failed to give predictions that were large enough to match the experimental results of thermal conductivity enhancements. Specifically, it was pointed out that the microscopic motion and the particle structure could be the reasons which induced the unusually high effective thermal conductivity of nanoparticle suspensions. This report is the first theoretical analysis of the two extensively debated mechanisms of enhanced thermal conductivity, Brownian motion effect and agglomeration effect, on which researchers have been focusing in the later investigations.

Since 2001, there has been an explosive growth in interest in this new heat transfer liquid media, both experimentally and theoretically [25-83]. The experimental data dramatically enriched the database of all kinds of effective thermal conductivity tests of nanofluids. Therefore, it is possible to understand the behavior of this new heat transfer fluid media and to find out the relationships between the effective thermal conductivity and many factors, such as the nanoparticle shape, nanoparticle material, nanoparticle volume fraction, nanoparticle size, base fluid material, and bulk temperature. In 2002, Xie et al., [25, 26] measured the effective thermal conductivity of suspensions with a range of diameter (5 nm, 25 nm, 58 nm, 101 nm, 122 nm and 124 nm) and two crystalline phase Al_2O_3 nanoparticles (α crystalline and γ crystalline) in pump oil, DI water at different pH value, and ethylene glycol, as the base fluids. There were several important discoveries in this report. First, it illustrated that the thermal conductivity enhancement of Al_2O_3 nanoparticle suspensions decreased with an increase of pH value in the DI water base fluid. Second, it showed that the crystalline phase of Al_2O_3 did not have an influence on the effective thermal conductivity of suspensions. Third, the nanoparticle diameter had an influence on the effective thermal conductivity and the best enhancement of effective thermal conductivity happened to be that of 25 nm diameter Al_2O_3 nanoparticle suspension. And last, the enhancement of suspension effective thermal conductivity increased with a decrease of base fluid thermal conductivity.

The same year, another two experimental reports brought one more parameter, bulk temperature, which could influence the effective thermal conductivity of nanoparticle suspensions [27, 28]. The first experimental report introduced a temperature oscillation technique to obtain the effective thermal conductivity through measuring the thermal diffusivity of the nanoparticle suspension [27].

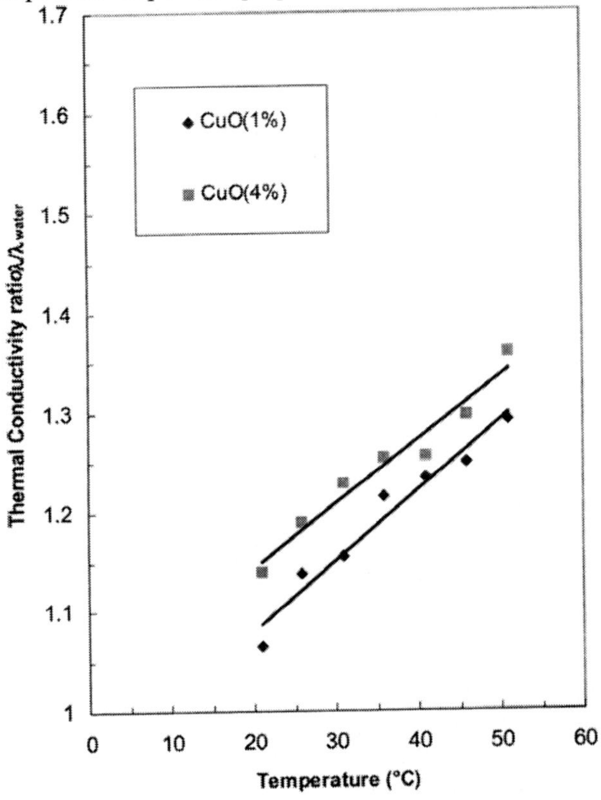

Figure 3. Temperature dependence of thermal conductivity enhancement for Al_2O_3/water nanoparticle suspensions [27].

It was demonstrated that the effective thermal conductivities of both 38.4 nm diameter Al_2O_3/DI water and 28.6 nm diameter CuO/DI water nanoparticle suspensions would increase with an increase of bulk temperature [27]. Additionally, it pointed out that the Hamilton-Crosser model could not predict the elevated effective thermal conductivity of nanoparticle suspensions with the increase of temperature, although the effective thermal conductivity of Al_2O_3/DI water nanoparticle suspension generally agreed with the prediction of the Hamilton-Crosser model at room temperature. The second report confirmed this temperature dependence relation by investigating nanoparticle suspensions with 60-80 nm diameter Ag nanoparticle, 10-20 nm diameter Au nanoparticle and 3-4 nm diameter Au nanoparticle covered with a monolayer of Octadecanethiol in water and toluene base fluids. The 3-4 nm diameter Au nanoparticle covered with a monolayer of Octadecanethiol in toluene base fluid at 0.0079 % volume fraction showed a thermal conductivity enhancement of 4.5 % at 30 ° C and 6.3 % at 60 ° C. The 10-20 nm diameter Au nanoparticle/water suspensions at 0.00026 % volume fraction had a thermal conductivity enhancement of 4.6 % at 30 ° C and 8.3 % at 60 ° C. The 60-80 nm diameter Ag nanoparticle/water at 0.001 % volume fraction had a thermal conductivity enhancement of 3.2 % at 30 ° C and 4.5 % at 60 ° C [28].

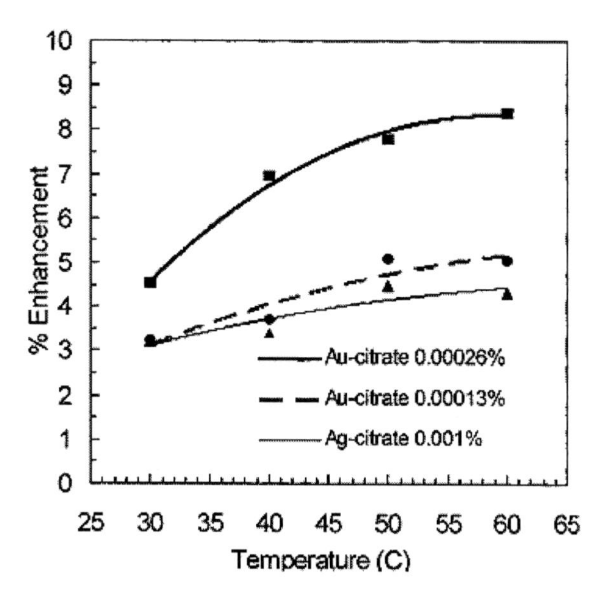

Figure 4. Percentage of enhancement in thermal conductivity vs. the temperature of Au-citrate and Ag-citrate with reference to the conductivity of blank water (with 5 nM trisodium citrate) at each temperature [28].

In 2006, a visual experiment of enhanced mass transport in nanoparticle suspensions was reported, which indirectly proved the Brownian motion effect toward the effective thermal conductivity of the nanoparticle suspension [29]. It was observed that the same dye in water and in nanoparticle suspensions showed different transport abilities. The distances dye transported in nanoparticle suspensions were much greater than the distance dye transported in pure water. And there is an optimal volume fraction, 0.5 %, in which the dye transported best. As a result, it was explained that nanoparticle Brownian motion could create a velocity disturbance field and this disturbance field enhanced the transport ability of dye inside nanoparticle suspensions.

A new experimental study by Putnam et al. [30] introduced an optical beam deflection technique to measure the thermal diffusivity of C_{60}-C_{70} fulleren nanoparticle/tolune suspensions, and alkanethiolate-protected Au nanoparticle/ethanol and toluene suspensions. It was stated that in the dilute volume fraction (0.01 % to 1 %), the enhancement was not as great as it was previously reported by [28]. However, there was another interesting discovery in this report that the mixture of 2 % volume fraction water in ethanol could have an effective thermal conductivity enhancement of 6 %, two folds higher than the prediction of 3 % by mean-field model.

When researchers focused on the Brownian motion effect of the nanoparticle, some experimental reports brought the attention to other factors, such as the agglomeration effect [31-39].

A cryogenic transmission electron microscope technique was employed to determine the morphology of nanoparticles and agglomerations in the suspensions [33]. In this method, the suspension specimen, which could not be studied under a conventional transmission electron microscope, was frozen with liquid nitrogen below the devitrification temperature to reserve the original structure of the nanoparticles/agglomerations inside the suspension. Then, it was fractured into two parts in the middle, which gave a cross-section surface. On this surface, it had the original size distribution and morphology information of nanoparticles/ agglomerations in the suspension. Then, the carbon and platinum ions beams were sprayed on the surface at an angle of 45 ^{o} and formed a thin layer membrane of 1.5 ~ 2.0 nm thickness. This thin layer membrane could duplicate the information of size distribution and morphology of nanoparticles/agglomerations on the surface, and be studied under a transmission electron microscope. The picture of 25 nm diameter nanoparticle in water is shown below.

Hong and Yang et al. [34, 35] experimentally investigated the dependence of effective thermal conductivity of 10 nm diameter Fe nanoparticle suspensions on the Fe nanoparticle cluster size. It was reported that the effective thermal conductivity of nanoparticle suspensions was not necessarily dependent on the thermal conductivity of nanoparticles by comparing this experimental results with that of Cu nanoparticle suspensions reported previously [22]. It was also shown in another paper [35] that the effective thermal conductivity of Fe nanoparticle suspensions would decrease with an increase of the time after ultrasonication was stopped, and the size of the Fe nanoparticle cluster increased with an increase in the time after ultrasonication was stopped. These experimental results clearly demonstrated that, first, the mean-field theory could not be applied to predict the effective thermal conductivity of nanoparticle suspensions because the prediction of effective thermal conductivity by mean-field theory is always dependent on the thermal conductivity of nanoparticles; and second, the effective thermal conductivity of Fe nanoparticle suspensions decreased with an increase in the Fe nanoparticle cluster size, which confirmed the previous conclusion that the smaller the nanoparticle size is, the better the effective thermal conductivity will be. But here, it also attributed the better enhancement of Fe nanoparticle suspensions over that of Cu nanoparticle suspensions as a result of clustering of Fe nanoparticles. While another experimental report by Liu and Wang et al. [36] showed that the effective thermal conductivity of Cu/DI water nanoparticle suspensions decreased with an increase of time after the measurement started. Moreover, it was found that Cu nanoparticle would agglomerate into 50-100 nm diameter spherical and square shaped clusters. The ratio between the effective thermal conductivity of the nanoparticle suspension and the thermal conductivity of base fluid, DI water, would decrease quickly in the first ten minutes and then kept a relatively stable value which was slightly higher than unity due to the severe agglomeration of the Cu nanoparticles.

Peterson, Li and Kim [40-43] noticed the possible contribution of nanoparticle clustering and tried to experimentally verify this by measuring the effective thermal conductivities of the CuO nanoparticle, the Al_2O_3 nanoparticle, and the single-walled carbon nanotube and multiwalled carbon nanotube suspensions. The relationships between the effective thermal conductivity of metal oxide nanoparticles and carbon nanotube suspensions and volume fraction, temperature, and base fluid were confirmed. It was pointed out that for the carbon nanotube suspensions, the effective thermal conductivity might be well predicted with the extended mean-field theory by choosing the correct shape factors [40]. The work [40] corresponded to the other experimental work by Choi et al. [31], Xie and Choi et al. [32], and Liu et al. [36], which all reported that both single-walled and multi-walled carbon nanotube suspensions could enhance the effective thermal conductivity much greater than the predictions of mean-field models.

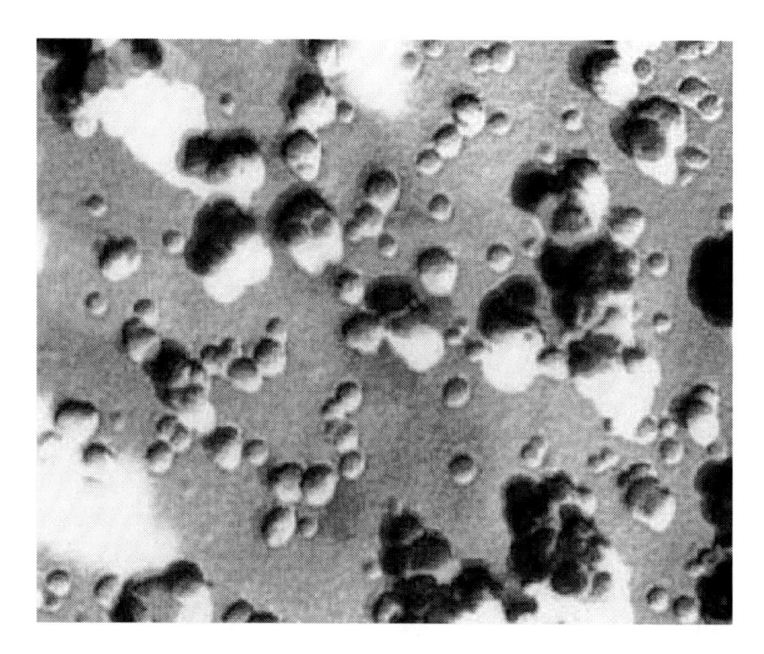

Figure 5. Cry-TEM pictures of SiO_2 nanoparticles/agglomerations in suspension [33].

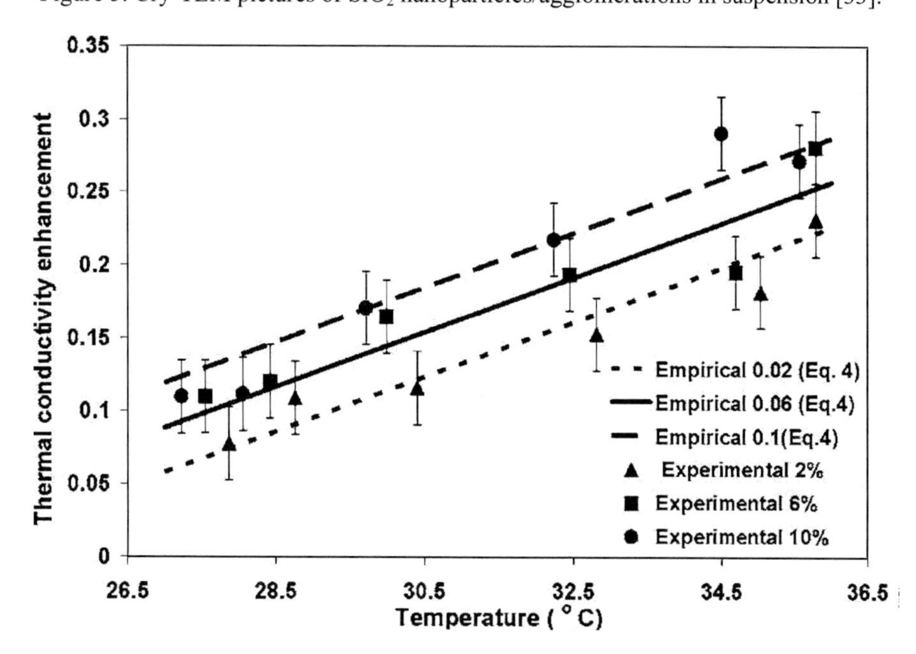

Figure 6. Thermal conductivity enhancement of Al_2O_3/DI water suspensions vs. temperature [40].

The Al_2O_3 nanoparticle suspensions were further studied by measuring the effective thermal conductivities of different diameter Al_2O_3 nanoparticle suspensions, and it once more confirmed that nanoparticle size would influence the enhancement of the effective thermal conductivity [41, 42]. Li and Peterson et al. [43] also conducted a comparing experiment of steady state cut bar method and transient hot wire method to respond to the questioning on the validity of employing the transient hot wire method to measure the effective thermal conductivity of nanoparticle suspensions. The experiment well illustrated that the transient hot wire method was fully capable of measuring the enhancement of the effective thermal conductivity with very good accuracy.

Zhu and Yin et al. [44] measured the effective thermal conductivity of 9.8 nm diameter Fe_3O_4 nanoparticle/DI water suspensions, and found that the effective thermal conductivity of Fe_3O_4 nanoparticle/DI water suspensions was higher than that of other metal oxide nanoparticle suspensions, such as Al_2O_3, CuO, and TiO_2, and significantly higher than the predictions of Maxwell's model [6], the renovated Maxwell's model [45], and Das' Brownian motion model [46]. It is interesting to note in this report that the cluster size of the Fe_3O_4 nanoparticle agglomerations increased with an increase of volume fraction. For example, at a volume fraction of 0.5 %, only few loose clusters were found with a size of 40 nm, at volume fraction of 1 %, the cluster size increased to around 60 nm, and at volume fraction of 3 %, the cluster size dramatically increased to several hundreds nanometers. The alignment of Fe_3O_4 nanoparticles was observed to be random in all the samples.

Chopkar and Kumar et al. [47] introduced another new experimental method of a modified thermal comparator to measure the effective thermal conductivities of 15 nm diameter Al_2Cu and 18 nm diameter Ag_2Al nanoparticles in DI water and ethylene glycol base fluids. It was once again confirmed that the effective thermal conductivity of nanoparticle suspensions increased with an increase of volume fraction and the enhancements of this experiment were much greater than the predictions of the Maxwell model [6] and the Hamilton-Crosser model [3]. A new kind of needle-like 30 nm CuO nanoparticle/DI water suspensions were measured by Jwo et al. [48]. This work reported that for the needle-like shape CuO /DI water nanoparticle suspension, the enhancement of the effective thermal conductivity could be even greater than that of spherical shape CuO/DI water nanoparticle suspensions. At the volume fraction of 0.8 %, the enhancement of effective thermal conductivity could be as high as 5.9 % at 5 ^{o}C to 38.2 % at 40 ^{o} C. Even at the volume fraction of 0.2 %, the enhancement of effective thermal conductivity could be 14.7 % at 40° C.

Kim et al. [49] experimentally studied the size dependence of the effective thermal conductivity of suspensions with 38 nm diameter Al_2O_3 nanoparticle, 10 nm, 30 nm and 60 nm diameter ZnO nanoparticles, and 10 nm, 34 nm, and 70 nm diameter TiO nanoparticles in DI water and ethylene glycol base fluids. The results of the 38 nm diameter Al_2O_3 nanoparticle in both DI water and ethylene glycol suspensions confirmed the previous reported data by Lee et al. [23], and the results of both ZnO nanoparticle suspensions and TiO nanoparticle suspensions demonstrated that the smaller size could induce greater enhancement in effective thermal conductivity. The experimental data were compared with both a mean-field model, the Hamilton-Crosser model [3], and a Brownian motion model, Jang-Choi's model [50, 51]. There was no satisfactory agreement achieved. This report also presented the concern of the elevation of the effective thermal conductivity by a laser beam shooting on the ZnO nanoparticle suspensions.

Except for the measurement of effective thermal conductivities of the solid/liquid suspensions, liquid/liquid suspensions were also measured. Yang and Han [52, 53] adopted a 3-ω method to measure the effective thermal conductivity of water in FC72 nanoemulsion suspensions, and found a remarkable enhancement of 126 % at a volume fraction of 12 %. This work extended the selection of nanoparticle suspensions from solid/liquid matches towards the liquid/liquid matches.

3.2. The Simulation Reports on the Effective Thermal Conductivity of Nanoparticle Suspensions

In 2004, Bhattacharya and Phelan et al. [54] adopted a Brownian dynamics simulation technique with the equilibrium Green-Kubo method to compute the effective thermal conductivity of nanoparticle suspension. This simulation work confirmed the experimental reports [4] that Al_2O_3/ethylene glycol nanoparticle suspension and Cu/ethylene glycol nanoparticle suspension had much higher effective thermal conductivities compared to the predictions of the Hamilton-Crosser model. Another simulation work of Lattice Boltzmann simulation agreed with the results of the Brownian dynamics simulation [55], and discussed the role of the irregular motion of the nanoparticle and its contribution toward the effective thermal conductivity of nanoparticle suspension.

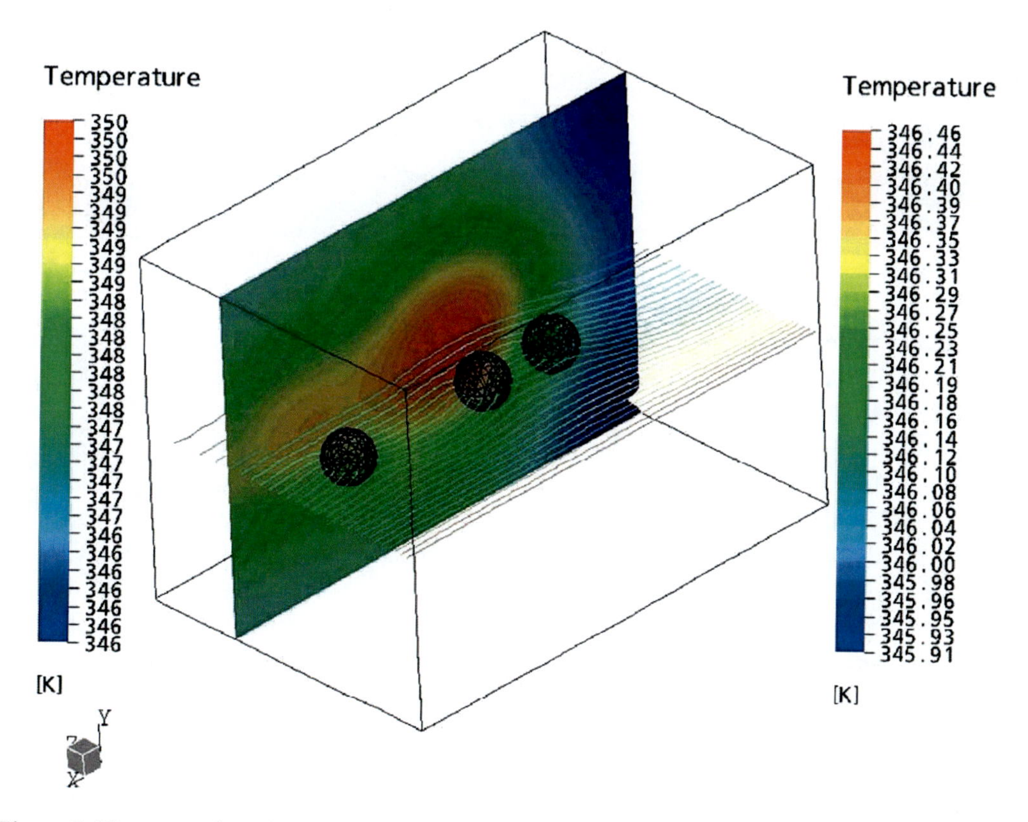

Figure 7. The comparison for simulated temperature field on vertical plane (left, the temperature scale of perpendicular contour; right, the temperature scale of horizontal contour) [56].

Li and Peterson [56] demonstrated in a simulation that the Brownian motion of nanoparticle could significantly change the temperature field in the suspensions and smooth the temperature gradient not only along the moving direction of the Brownian motion, but the temperature gradient perpendicular to the moving direction of the Brownian motion through the base fluid microconvection introduced by the nanoparticle Brownian motion. A comparison of one single nanoparticle case and a two-adjacent nanoparticles case showed that the hydrodynamic interaction between nanoparticles would dramatically enhance the influence on the temperature gradient and field compared to the effect of two single non-interacting nanoparticles. This simulation work demonstrated that the microscale hydrodynamic interaction of Brownian motion could explain the reason that macroscale effective thermal conductivity would increase with an increase of temperature and an increase of volume fraction. Moreover, Keblinski and Thomin [57] studied the hydrodynamic field around a nanoparticle with molecular dynamics simulation, which was conducting Brownian motion. The simulation showed that time-integrated response of the Brownian motion velocity field around a nanoparticle should be the same as that when the nanoparticle was having a constant velocity, which might imply that the convective heat transfer of a nanoparticle conducting Brownian motion could be the same as that when it was having a constant velocity.

As researchers had focused on adopting the Brownian motion effect to explain the effective thermal conductivity of nanoparticle suspensions, some criticizing voices on the Brownian motion effect have been reported, too. In 2006, Evans and Keblinski et al. [58] employed molecular dynamics simulation to study the Brownian motion effect of nanoparticles with 296 atoms of each in fluid. The interaction potential between solid atoms was FENE potential, $U_{FENE} = -5.625\varepsilon \cdot \ln[1-(r/1.5\sigma)^2]$, and the interaction potential between base fluid molecules was the standard Lennard-Jones potential, $U_{LJ} = 4\varepsilon[(\sigma/r)^{12} - (\sigma/r)^6]$, where ε is the unit of energy, σ is the unit of length, and r is the distance between a pair of atoms or molecules. The interaction potential between base fluid molecule and the solid atom on the surface of nanoparticle is still a Lennard-Jones potential, but the cutoff distance was set to be 1.5σ, which led to attractive forces between base fluid molecule and the solid atom. Moreover, the unit of energy between the base fluid molecule and solid atom was set to be three values, $\varepsilon_{SF}=0.25\varepsilon$, $\varepsilon_{SF}=1.25\varepsilon$, and $\varepsilon_{SF}=2.25\varepsilon$ for different wetting conditions between the nanoparticles and the base fluids. It concluded that only minor differences of thermal conductivity compared to that of the pure fluid was observed in this simulation, and the thermal conductivity increases were rather small and could not correspond to spectacular increases in experiments.

Eapen et al. [59, 60] employed molecular dynamics simulation to study the mechanisms of heat transfer in nanoparticle suspensions. The microscale heat flux is separated into the self-correlations and cross-correlations between the kinetic, potential, and collision modes. It showed in a demonstrating picture that the total effective thermal conductivity of Pt/Xenon nanoparticle suspension would be the sum of the contributions for all correlations, and much higher than the prediction of the Maxwell mean-field model [6]. Moreover, it was pointed out later that the Brownian motion models gave too high predictions on effective thermal conductivity of nanoparticle suspensions over the corresponding experimental results, and the enhanced effective thermal conductivity would be the reason of chain-like agglomeration of nanoparticles [60]. This simulation work was echoed by another molecular dynamics simulation report by Vladkov and Barrat [39], which stated that the enhanced thermal conductivity was caused by the collective effects of confinement, nanoparticle mass, Brownian motion, nanoparticle agglomeration and percolation.

However, the Brownian motion effect was also assured by a molecular dynamics simulation work by Sarkar and Selvam [61] that the effective thermal conductivity was much higher than the prediction of Hamilton-Crosser model. An equilibrium molecular dynamics simulation was conducted with the Green-Kubo method to study the effective thermal conductivity of different volume fraction Cu/argon nanoparticle suspensions. The mean square displacement analysis showed that both the enhancement of effective thermal conductivity and the enhancement of diffusivity had the similar mechanism, which was caused by the base fluid molecule microconvection induced by the nanoparticle Brownian motion.

3.3. Brownian Motion Models of Effective Thermal Conductivity of Nanoparticle Suspensions

Based on the previous experimental and simulating investigations on effective thermal conductivity of nanoparticle suspensions, there were several Brownian motion models proposed [46, 50, 51, 62,-65]. The first model took consideration of nanoparticle size, volume fraction, and bulk temperature, and based its velocity on the Stokes-Einstein formula [46], shown as in Eq. 25. This model believes that nanoparticles would carry thermal energy from hot place to cold place by mass diffusion, and enhance the effective thermal conductivity of the suspension as represented by the parameter, *c,* in the equation.

$$\frac{k_e}{k_1} = 1 + c \cdot \overline{u}_p \frac{\eta r_m}{k_1(1-\phi)r_p} \tag{25}$$

where k_e is the effective thermal conductivity of the nanoparticle suspension, k_l is the thermal conductivity of the base fluid, c is a constant, $\overline{u}_p = \frac{2k_b T}{(\pi \eta d_p^2)}$ is the average nanoparticle velocity, k_b is Boltzmann's constant, η is the base fluid dynamic viscosity, d_p=$2r_p$ is the diameter of nanoparticle, r_m is the radius of base fluid molecule, and ϕ is the volume fraction of nanoparticle. Moreover, the constant is estimated as $c = \frac{nlc_v}{3}$, where n is the number concentration of nanoparticle, l is the mean free path of the nanoparticle and c_v is the specific heat of nanoparticle.

The second Brownian motion model included one more factor, the adsorption layer of base fluid molecules by Jang and Choi [50], as show in Eq. 26. The first two parts on the right side of the equation are the contributions of the thermal conductivities of the nanoparticle and base fluid, and the third part is the contribution of Brownian motion and adsorption layer of base fluid molecules.

$$k_e = k_1(1-\phi) + k_2\phi + \phi h\,\delta_T \tag{26}$$

where k_e is the effective thermal conductivity, k_1 is the thermal conductivity of base fluid, ϕ is the volume fraction of nanoparticle, k_2 is the thermal conductivity of the nanoparticle, h is the convective heat transfer coefficient, δ_T is the thickness of the thermal boundary layer. The third part on the right side of the equation was further developed as $3C_1 \dfrac{d_f}{d_p} k_1 \mathrm{Re}_{dp}^2 \Pr\phi$, where $h = \dfrac{k_1}{d_p}\mathrm{Re}_{dp}^2 \Pr^2$ is the heat transfer coefficient for flow past nanoparticles, $\delta_T = \dfrac{\delta}{\Pr} = \dfrac{3d_f}{\Pr}$, and $\mathrm{Re}_{dp} = \dfrac{Cd_p}{\upsilon}$. Here $C = \dfrac{D}{l_f}$ is the Brownian motion velocity of nanoparticle, D is the nanoparticle diffusion coefficient, l_f is the mean free path of a base fluid molecule, and υ is the kinematic viscosity of base fluid.

Jang and Choi [51] also tried to clarify some doubts of their Brownian motion model later and explained the assumptions made during the building up of the model that the nanoparticles would carry thermal energy by the nanoparticle mass diffusion through different temperature regions. This effort was made to ensure the Brownian motion effect to the effective thermal conductivity, whose intensity was decided by the bulk temperature, nanoparticle size, and base fluid properties.

The third Brownian motion model [62] included the base fluid effect that was carried by the Brownian motion of nanoparticle, instead of the adsorption layer [50, 51] shown in Eq. 27. It argued that it was the base fluid flow carried by the nanoparticle, instead of the nanoparticle's mass movement, that helped to transfer the thermal energy in the suspension.

$$k_e = 5\times10^4 \beta\alpha_p\rho_f c_f \sqrt{\frac{kT}{\rho_p D}}\, f(T,\alpha_p,etc.) \tag{27}$$

where ρ_f and ρ_p are the density of the base fluid and the nanoparticle, c_f is the specific heat of the base fluid, k is the Boltzmann constant, T is bulk temperature, D is the diameter of nanoparticle, α_p=NV_d/V, N is the total number of nanoparticles in a cell, V_d is the volume of a nanoparticle, V is the volume a nanoparticle sweeping in a certain time period, β is the fraction of the base fluid volume travelling with nanoparticles, and $f(T,\alpha_p,\text{etc})$ is a factorial function of temperature, T, α_p, and the physical properties of the base fluid and nanoparticle materials, which depends on the properties of the intervening base fluid and the nanoparticle interactions.

This model attributed the temperature dependence of effective thermal conductivity to the long impact range of the interparticle potential, and illustrated that this temperature dependence was not due to the thermophoresis or osmophoresis of nanoparticles [62]. The thermophoresis influence on effective thermal conductivity was represented with Eq. 28, and the osmophoresis influence was represented with Eq. 29, as below. It was clearly demonstrated through the comparison of these three influences (Eq. 27-29) that the impact of the Brownian motion is much more significant than that of thermophoresis or osmophoresis effect, and the thermophoresis and osmophoresis motion effects were independent of nanoparticle size.

$$k_{thermoph} = \frac{1}{6\pi}\frac{\alpha_p k}{\mu_f}\frac{3k_1}{k_2 + 2k_1}(1\times 10^5 \rho_f c_f)\nabla T \tag{28}$$

$$k_{osmoph} = \frac{1}{3\pi}\frac{a_p^2 k}{\mu_f}\frac{3k_1}{k_2 + 2k_1}(1\times 10^5 \rho_f c_f)\nabla T \tag{29}$$

where α_p is the same as in Eq. 27, k is the Boltzmann constant, μ_f is the dynamic viscosity, k_1 and k_2 are the thermal conductivities of base fluid and nanoparticle, ρ_f and c_f are density and specific heat of base fluid, and ∇T is the temperature gradient.

Still another Brownian motion, the forth one, was proposed with an order-of-magnitude analysis of various possible mechanisms [63, 64]. This model combined the Maxwell-Garnett mean-field model [20] and a Brownian motion convection model together, and would reduce to only the Maxwell-Garnett model [20] once the nanoparticle size is large enough. Due to the Maxwell-Garnett mean-field model [20], the interfacial thermal boundary resistance was considered in this model, too. In this report, the heat transport in nanoparticle suspensions was compared with the particle-to-fluid heat transfer in fluidized beds in the way that convective currents from various particle interactions in fluidized bed enhanced the heat transfer. Henceforth, a convective heat transfer model in the Stokes regime was adopted for the heat transfer enhancement of Brownian motion effect of the nanoparticles. The model was presented as Eq. 30.

$$\frac{k_e}{k_1} = (1 + A\cdot \mathrm{Re}^m \,\mathrm{Pr}^{0.333}\,\phi)[\frac{(1+2\alpha)+2\phi(1-\alpha)}{(1+2\alpha)-\phi(1-\alpha)}] \tag{30}$$

where k_e is the effective thermal conductivity of the nanoparticle suspension, k_1 is the thermal conductivity of base fluid, Re is the Reynolds number based on the nanoparticle Brownian motion velocity, u, and diameter, d, Pr is the Prandtl number of the base fluid, ϕ is nanoparticle volume fraction, $\alpha=2R_{\mathrm{interfacial}}k_1/d$, $R_{interfacial}$ is the interfacial thermal resistance, and A and m are constants.

In 2005, Chon and Kihm et al. [66] measured the effective thermal conductivities of Al_2O_3/DI water suspensions with diameter size at 11 nm, 47 nm, and 150 nm and bulk temperature ranging from 21 $^\circ$ C to 71 $^\circ$ C. This report clearly illustrated that for the same nanoparticle material, Al_2O_3, and the same base fluid, DI water, the effective thermal conductivity of nanoparticle suspension increased with an increase of bulk temperature, and increased with a decrease of nanoparticle diameter. Based on the experimental results, an empirical equation was generated via a linear Buckingham-Pi theorem regression, Eq. 31. With this equation, it confirmed the importance of the Brownian motion on the effective thermal conductivity, and concluded that the elevation of the effective thermal conductivity at high temperature was a primary result of the intensity increase of nanoparticles Brownian motion. This finding was also supported by another experimental report later [67].

$$\frac{k_e}{k_1} = 1 + 64.7 \cdot \phi^{0.746} \left(\frac{d_f}{d_p}\right)^{0.369} \left(\frac{k_2}{k_1}\right)^{0.7476} \times \mathrm{Pr}^{0.9955}\ \mathrm{Re}^{1.2321} \quad (31)$$

where ϕ is the volume fraction of nanoparticles, d_f and d_p are the diameter of the base fluid molecule and nanoparticle, k_2 and k_1 are the thermal conductivity of the nanoparticle and base fluid, *Pr* is the Prandtl number of the base fluid, and *Re* is the Reynolds number based on nanoparticle Brownian motion velocity and diameter.

3.4. Related Research on the Agglomeration Mechanisms

Wang and Li et al. [68, 69] measured the effective thermal conductivity of the 25 nm diameter SiO_2 nanoparticle/DI water suspensions and demonstrated the clustering of SiO_2 nanoparticles inside the suspension with TEM pictures. It was proposed that the microscope motion should be the microconvection introduced by the nanoparticle Brownian motion and the structure should be the agglomeration of nanoparticles. The microconvection and agglomeration could be the major mechanisms for the enhanced thermal conductivity of nanoparticle suspensions, as well as the adsorption fluid layer on nanoparticles. In the same year, Keblinski et al. [70] broke further down the proposed mechanisms of the microscope motion and the nanoparticle structure, and reported the theoretical analysis results on the Brownian motion effect, liquid layering effect, ballistic transport inside of nanoparticles effect, and the nanoparticle clustering effect. In this report, the role of Brownian motion was argued to produce the nanoparticle agglomeration. It was concluded that the agglomeration produced by the nanoparticle Brownian motion with the assistance of liquid layering effect and ballistic transport inside nanoparticles served as a fast lane for heat flux, and the agglomeration of nanoparticles was the mechanism of the unusual high thermal conductivity of nanoparticle suspension [69]. In 2003, Wang et al. [71], considering the fact of nanoparticles clustering inside the suspensions, further developed a fractal model to predict the effective thermal conductivity of nanoparticle suspensions, which tried to include the size effect, the fractal of nanoparticle agglomerations and the adsorption layer of the base fluid.

Because the agglomerations of nanoparticles are relatively loose compared to the same size of solid particles, the base fluid is filled in the void of the cluster to connect the nanoparticles together as a whole body. If the agglomeration of nanoparticles could serve as a fastline for the heat flux insider the suspensions, the adsorbed base fluid inside the cluster should play an important role. The effect of the adsorption layer was extensively studied with molecular dynamics simulation by Xue and Keblinski et al. [72-75]. A simple Lennard-Jones interaction potential was applied to the liquid, solid and the interface of the liquid-solid in the simulation. And the binding energy parameter of the liquid-solid interaction was set as proportional to the binding energy parameter of liquid-liquid interaction, $\varepsilon_{ls}=\alpha\varepsilon_{ll}$, where α was chosen from 0.2 to 5 to represent the non-wetting, weakly-wetting, and the wetting interaction of the liquid-solid interface. It was concluded that the adsorption layer could not play an important role in the effective thermal conductivity even in the limit of nanometer size particles because the non-wetting/weakly-wetting liquid-solid interface showed a high thermal resistance, as well as that the wetting liquid-solid interface had a much smaller but still large thermal resistance, which would be equal to that of liquid-liquid situation over a distance of around 1 nm [72-74]. It was also found that the interfacial resistance of weakly coupled between carbon nanotube and octane liquid was relatively larger and the adsorbed liquid layer did not result in an observable effect in the heat transfer [73]. Despite the possible negative effect from simulation results, Xue and Choi [74] considered the possible positive effect of adsorbed base fluid layer and tried to modify the Hamilton-Crosser model to include the base fluid interfacial layer. Xie and Zhang et al. [76] also proposed that the adsorbed base fluid layer could contribute to the effective thermal conductivity of nanoparticle suspension in a theoretical analysis. Prakash and Giannelis [77] tried to incorporate the interfacial resistance effect of the adsorbed base fluid layer with the Brownian motion of the nanoparticles together, to explain the enhanced effective thermal conductivity of nanoparticle suspension.

Vadas [78] used both the DuPhlag solution and the Fourier solution to interpret the experimental data reported previously, and showed that the solid-fluid interfacial heat transfer mechanism should be accounted to reach identical results. Lee [79] introduced the electrical double layer concept and calculated the thermal conductivity and thickness of interfacial layer. It was concluded that the interfacial layer could enhance the overall effective thermal conductivity. Evans and Keblinski [80] studied the effective thermal conductivity of water molecules with different orientational and translational orders induced by an electric field. It was found that the electric-field induced crystallization and translational order of water molecules could bring a three fold increase in thermal conductivity compared to that of the base water. This nonequilibrium molecular dynamics simulation confirmed the possibility of the contribution from the adsorbed base fluid layer, but also pointed out that it might not be possible to have the crystallinity at room temperature in nanoparticle suspensions. Tillman and Hill [81] analyzed the adsorbed base fluid layer with mathematical approach and disagreed any possibility of contribution from the adsorbed layer to the effective thermal conductivity.

Based on the experimental reports on agglomeration of nanoparticles inside nanoparticle suspensions, and the reported experimental relationship between the effective thermal conductivity of nanoparticle suspensions and the agglomeration sizes, Prasher et al. [82, 83] also tried to modify the mean-field theory by including the fractal clusters, and to use the modified model to predict the contribution of agglomeration to the effective thermal conductivity enhancement. It was concluded in this effort that the morphology of nanoparticle clusters might be very important when using mean-field theory to predict the effective thermal conductivity.

4. Discussion and Conclusion

From all of the experimental reports on the effective thermal conductivity of nanoparticle suspensions, it is for certain that the effective thermal conductivity of nanoparticle suspensions will increase with an increase of volume fraction, bulk temperature, and a decrease of nanoparticle size. The stable nanoparticle suspensions could be used in many areas in which the thermal control and heat transfer are critical issues, such as the automobile industry, manufacturing industry, microelectronic industry, and many other cutting-edge scientific research fields. In order to guide the discovery of high effective thermal conductivity nanoparticle suspensions to satisfy the urgent need of better heat transfer liquid media for the rapidly growing heat flux cooling demands, many models were proposed based on the Brownian motion effect and the agglomeration, as well as other possible mechanisms. However, the general acceptance of any one of the models has not yet been reached and how those mechanisms play a role in the effective thermal conductivity enhancement is still not clear. Before the new heat transfer media, nanoparticle suspensions, could be massively used in the daily cooling of devices and other applications, there is still a long way to go in the development of a fundamental understanding of the microscale thermophysical transport phenomena.

References

[1] Choi, S. U. S., Enhancing thermal conductivity of fluids with nanoparticles. In *''Developments and Applications of Non-Newtonian Flows''* (D.A. Siginer and H.P. Wang, eds.). American Society of Mechanical Engineers, New York, 1995.

[2] S. P. Lee and U. S. Choi, Application of metallic nanoparticle suspensions in advanced cooling sytems, Recent advances in solids/structures and application of metallic materials, eds. Y. Kwon, D. Davis, and H. Chung, *The American Society of Mechanical Engineers,* New York, PVP-Vol 342/MD-Vol 72, pp. 227-234, 1996.

[3] Hamilton, R. L., and Crosser, O. K., *Thermal conductivity of heterogeneous Two component Systems, I&EC Fundamentals,* Vol. 1 (3), pp. 187–191, 1962.

[4] Eastman, J. A., Choi, S. U. S., Li, S. Thompson, L. J., and Lee, S., Enhanced thermal conductivity through the development of nanofluids. In *''Material Research Society Symposium Proceedings''* (S. Komarnei, J.C. Parker and H.J. Wollenberger, eds.), Vol. 457, pp. 9–10, Warrendale, PA, 1997.

[5] Meredith, R. E., and Tobias, C. W., *Conductivities in emulsions.* J. Electrochem. Soc, Vol. 108(3), pp. 286–290, 1961.

[6] Maxwell, J. C., *A Treatise on Electricity and Magnetism,* 3rd edn. Oxford University Press, Oxford, 1873.

[7] Fricke, H., *A mathematical treatment of the electric conductivity and capacity of disperse systems. Physical Review,* Vol. 24, pp. 575–587, 1924.

[8] Keller, J. B., Conductivity of a medium containing a dense array of perfectly conducting spheres or cylinders or nonconducting cylinders. *Journal of Applied Physics,* Vol. 34(4), pp. 991–993, 1963 .

[9] Rocha, A. and Acrivos, A., On the effective thermal conductivity of dilute dispersions: highly conducting inclusions of arbitrary shape. *Quarterly journal of mechanics and applied mathematics,* Vol. 26(4), pp. 441-455, 1973.

[10] Rocha, A. and Acrivos, A. On the effective thermal conductivity of dilute dispersions. *Quarterly journal of mechanics and applied mathematics,* Vol. 26(2), pp. 217-233, 1973.

[11] Kim, I. C. and Torquato, S. Effective conductivity of composites containing spheroidal inclusions: comparison of simulations with theory. *Journal of applied physics,* Vol. 74(3), pp. 1844-1854, 1993.

[12] Batchelor, G. K., Transport properties of two-phase materials with random structure. *Annual review of fluid mechanics,* Vol. 6, pp. 227-255, 1974.

[13] Jeffrey, D. J., Conduction through a random suspension of spheres. *Proceedings of the royal society of London,* series a (mathematical and physical sciences), Vol. 335(1602), pp. 355-367, 1973.

[14] Davis, R. H., The effective thermal conductivity of a composite material with spherical inclusions. *International journal of thermophysics,* Vol. 7(3), pp. 609-620, 1986.

[15] Brady, J. F., Phillips, R. J., Lester, J. C., and Bossis, G. Dynamic simulation of hydrodynamically interacting suspensions. *Journal of fluid mechanics,* Vol. 195, pp. 257-280, 1988.

[16] Bonnecaze, R. T. and Brady, J. F. A method for determining the effective conductivity of dispersions of particles. *Proceedings of the royal society of London: Mathematical and physical sciences,* Vol. 430(1879), pp. 285-313, 1990.

[17] Bonnecaze, R. T. and Brady, J. F. The effective conductivity of random suspensions of spherical particles. *Proceedings of the royal society of London, series a* (mathematical and physical sciences), Vol. 432, pp. 445-465, 1991.

[18] Lu, S. Y. and Kim, S. T. Effective thermal conductivity of composites containing spheroidal inclusions. *AIChE journal,* Vol. 36(6), pp. 927-938, 1990.

[19] Lu, S. Y. and Lin, H. C. Effective conductivity of composites containing aligned spheroidal inclusions of finite conductivity. *Journal of applied physics,* Vol. 79(9), pp. 6761- 6769, 1996.

[20] Hasselman, D. P. H., and Johnson, L. F., Effective thermal conductivity of composites with interfacial thermal barrier resistance. *Journal of composite materials,* Vol. 21, pp. 508-515, 1987.

[21] Leal, L., G. On the effective conductivity of a dilute suspension of spherical drops in the limit of low particle peclet number. *Chemical engineering communication,* Vol. 1, pp. 21-31, 1973.

[22] Eastman J. A., Choi S. U. S., Li S., Anomalously increased effective thermal conductivities of ethylene glycol-based nanofluids containing copper nanoparticles, *Applied Physics Letters, V*ol. 78 (6), pp. 718-720, 2001 .

[23] Lee, S., Choi, S. U. S., Li, S., and Eastman, J. A. Measuring thermal conductivity of fluids containing oxide nanoparticles. *Transactions of the ASME,* Vol. 121, pp. 280-289, 1999.

[24] Wang, X. and Xu, X. Thermal conductivity of nanoparticle-fluid mixture. *Journal of thermophysics heat transfer,* Vol. 13(4), pp. 474-480, 1999.

[25] Xie, H., Wang, J., Xi, T., Liu, Y., and Ai, F. Dependence of the thermal conductivity of nanoparticle-fluid mixture on the base fluid. *Journal of materials science letters,* Vol. 21, pp. 1469-1471, 2002.

[26] Xie, H., Wang, J., Xi, T., Liu, Y., and Ai, F. Thermal conductivity enhancement of suspensions containing nanosized alumina particles. *Journal of applied physics,* Vol. 91(7), pp. 4568-4572 , 2002.

[27] Das, S. K., Putra, N., Thiesen, P. and Roetzel, W. Temperature dependence of thermal conductivity enhancement for nanofluids. *Journal of heat transfer,* Vol. 125, 567-574, 2003.

[28] Patel, H. E., Das, S. K., and Sundararajan, T. Thermal conductivities of naked and monolayer protected metal nanoparticle based nanofluids: Manifestation of anomalous

enhancement and chemical effects. *Applied physics letters,* Vol. 83(14), pp. 2931-2933, 2003.

[29] Krishnamurthy, S., Bhattacharya P., Phelan P. E., and Prasher R., Enhanced mass transport in nanofluids, *Nano letters,* Vol. 6, pp. 419-423, 2006.

[30] Putnam, S. A., Cahill, D. A., and Braun, P. V., Thermal conductivity of nanoparticle suspensions, *Journal of Applied Physics,* Vol. 99 (8), pp. 084308, 2006.

[31] Choi, S. U. S., Zhang, Z. G., Yu, W., Lockwood, E. A., Grulke, E. A., Anomalous thermal conductivity enhancement in nanotube suspensions, *Applied Physics Letters,* Vol. 79, pp. 2252-2254, 2001.

[32] Xie, H., Lee, H., Youn, W., and Choi, M. Nanofluids containing multiwalled carbon nanotubes and their enhanced thermal conductivities. *Journal of applied physics,* Vol. 94(8), pp. 4967-4971, 2003.

[33] Peterson, G. P., and Li, C. H., An Overview of Heat and Mass Transfer in Fluids with Nano-particle Suspensions" , *Advances in Heat Transfer,* Vol. 39, J. P. Hartnett and T. F. Irvine (eds.), Academic Press, New York, NY, pp. 261-392, 2006.

[34] Hong T. K., Yang H. S., Choi C. J., Study of the enhanced thermal conductivity of Fe nanofluids, *Journal of Applied Physics,* Vol. 97 (6), pp. 064311, 2005.

[35] Hong T. S., Hong T., K, and Yang H. S., Thermal conductivity of Fe nanofluids depending on the cluster size of nanoparticles, *Applied Physics Letters,* Vol. 88, pp. 031901, 2006.

[36] Liu M. S., Lin M. C. C., Tsai C. Y., and Wang, C., Enhancement of thermal conductivity with Cu for nanofluids using chemical reduction method, *International Journal of Heat and Mass Transfer,* Vol. 49 (17-18), pp. 3028-3033, 2006.

[37] Jana S., Salehi-Khojin A., Zhong W. H., Enhancement of fluid thermal conductivity by the addition of single and hybrid nano-additives, *Thermochimica Acta,* Vol. 462 (1-2), pp. 45-55, 2007.

[38] Wensel, J., Wright, B., Thomas, D., Douglas, W., Mannhalter, B., Cross, W., Hong, H., Kellar, J., Smith, P., and Roy, W., Enhanced thermal conductivity by aggregation in heat transfer nanofluids containing metal oxide nanoparticles and carbon nanotubes, *Applied Physics Letters,* Vol. 92, pp. 023110, 2008.

[39] Vldakov, M., and Barrat, J., Modeling transient absorption and thermal conductivity in a simple nanofluid, *Nano Letters,* Vol. 6, pp. 1224-1228, 2006.

[40] Kim, B. H., and Peterson, G. P., Effect of morphology of carbon nanotubes on thermal conductivity enhancement of nanofluids, *Journal of Thermophysics and Heat Transfer,* Vol. 21, pp. 451-459, 2007.

[41] Li, C. H. and Peterson, G. P., "Experimental investigation of temperature and volume fraction variations on the effective thermal conductivity of nanoparticle suspensions (nanofluids)", *Journal of applied physics,* Vol. 99(8), pp. 084314, 2006.

[42] Li, H. C. and Peterson, G. P., "The effect of particle size on the effective thermal conductivity of Al_2O_3-water nanofluids", *Journal of Applied Physics,* Vol. 101(4), pp. 044312, 2007.

[43] Li, H. C., Williams, W., Hu, L., Buonqiorno, J. and Peterson, G. P., "Transient and steady-state experimental comparison study of effective thermal conductivity of Al_2O_3/water nanofluid", *Journal of Heat Transfer,* Vol. 130(4), 2008, in print.

[44] Zhu H. T., Zhang C. Y., Liu, S. Q., Tang Y. M., and Yin, Y. S., Effects of nanoparticle clustering and alignment on the thermal conductivities of Fe_3O_4 aqueous nanofluids, *Applied Physics Letters,* Vol. 89, pp. 023123, 2006.

[45] Yu W., Choi S. U. S., The role of interfacial layers in the enhanced thermal conductivity of nanofluids: A renovated Maxwell model, *Journal of Nanoparticle Research,* Vol. 5 (1-2), pp. 167-171, 2003.

[46] Kumar, H. D., Patel, H. E., Kumar, V. R. R., Sundararajan, T., Pradeep, T., and Das, S. K., Model for heat conduction in nanofluids, *Physical Review Letters,* Vol. 93, pp. 144301, 2004.

[47] Chopkar M., Kumar S., Bhandari D. R., Development and characterization of Al_2Cu and Ag_2Al nanoparticle dispersed water and ethylene glycol based nanofluid, *Materials Science and Engineering B-Solid State Materials for Advanced Technology,* Vol. 139 (2-3), pp. 141-148, 2007.

[48] Jwo C. S., Chang H., Teng T. P., A study on the effects of temperature and volume fraction on thermal conductivity of copper oxide nanofluid, *Journal of Nanoscience and Nanotechnology,* Vol. 7 (6), pp. 2161-2166 Sp. Iss. SI, 2007.

[49] Kim S. H., Choi, S. R., Kim D., Thermal conductivity of metal-oxide nanofluids: Particle size dependence and effect of laser irradiation, *Journal of Heat Transfer-Transactions of The ASME,* Vol. 129 (3), pp. 298-307, 2007.

[50] Jang, S. P., and Choi, S. U. S., Role of Brownian motion in the enhanced thermal conductivity of nanofluids, *Applied Physical Letters,* Vol. 84, pp. 4316-4318, 2004.

[51] Jang S. P., Choi S. U. S., Effects of various parameters on nanofluid thermal conductivity, *Journal of Heat Transfer-Transactions of The ASME,* Vol. 129 (5), pp. 617-623, 2007.

[52] Yang, B. and Han, Z. H., Temperature-dependent thermal conductivity of nanorod-based nanofluids, *Applied Physics Letters,* Vol. 89, pp.083111, 2006.

[53] Han, Z. H., and Yang, B., Thermophysical characteristics of water-in--FC72 nanoemulsion fluids, *Applied Physics Letters,* Vol. 92, pp. 013118, 2008.

[54] Bhattacharya P., Phelan P. E., and Prasher R. Brownian dynamics simulation to determine the effective thermal conductivity of nanofluids, *Journal of Applied Physics,* Vol. 95, pp. 6492-6494, 2004.

[55] Xuan, Y., and Yao, Z., Lattice Boltzmann model for nanofluids, *Heat Mass Transfer,* Vol. 41, pp. 199-205, 2005.

[56] Li, H. C. and Peterson, G. P., "Mixing effect on the enhancement of thermal conductivity of nanoparticle suspensions (nanofluids)", *Int. J. Heat and Mass Transfer,* Vol. 50, pp. 4668-4677, 2007.

[57] Keblinski, P., and Thomin, J., Hydrodynamic field around a Brownian particle, *Physical Review E,* Vol. 73, pp. 010502, 2006.

[58] Evans W., Fish J., Keblinski P., Role of Brownian motion hydrodynamics on nanofluid thermal conductivity, *Applied Physics Letters,* Vol. 88 (9), pp. 093116, 2006.

[59] Eapen J., Li J., Yip S., Mechanism of thermal transport in dilute nanocolloids, *Physical Review Letters,* Vol. 98 (2), pp. 028302, 2007.

[60] Eapen J., Williams W. C., Buongiorno J., Mean-field versus microconvection effects in nanofluid thermal conduction, *Physical Review Letters,* Vol. 99 (9), pp. 095901, 2007.

[61] Sarkara S., Selvam R. P., Molecular dynamics simulation of effective thermal conductivity and study of enhanced thermal transport mechanism in nanofluids, *Journal of Applied Physics,* Vol. 102 (7), pp. 074302, 2007.

[62] Koo J., Kleinstreuer C., A new thermal conductivity model for nanofluids, *Journal of nanoparticle research,* Vol. 6, pp. 577-588, 2004.

[63] Koo J., Kleinstreuer C., Impact analysis of nanoparticle motion mechanisms on the thermal conductivity of nanofluids, *International Communications in Heat and Mass Transfer,* Vol. 32 (9), pp. 1111-1118, 2005.

[64] Prasher R., Bhattacharya P., Phelan P. E., Thermal conductivity of nanoscale colloidal solutions (nanofluids), *Physical Review Letters,* Vol. 94, pp. 025901, 2005.

[65] Prasher R., Bhattacharya P., Phelan P. E., Brownian-motion-based convective-conductive model for the effective thermal conductivity of nanofluids, *Journal of Heat Transfer-Transactions of The ASME,* Vol. 128 (6), pp. 588-595, 2006.

[66] Chon C. H., Kihm K. D., Lee S. P., Empirical correlation finding the role of temperature and particle size for nanofluid (Al2O3) thermal conductivity enhancement, *Applied Physics Letters,* Vol. 87 (15), pp. 153107, 2005.

[67] Chon, C. H., and Kihm, K. D., Thermal conductivity enhancement of nanofluids by Brownian motion, *Journal of Heat Transfer,* Vol. 127, pp. 810, 2005.

[68] Wang, B.X., Li, H., Peng, X.F., Research on the Heat-Conduction Enhancement for Liquid with Nano-Particle Suspensions, *Journal of Thermal Science,* Vol. 11(3), pp. 214-219, 2002.

[69] Wang, B.X., Li, H., Peng, X.F., Effect of Surface Adsorption on Heat Transfer Enhancement for Liquid with Nano-particle Suspensions, *Journal of Engineering Thermophysics,* Vol. 24(4), pp. 664-666, 2003.

[70] Keblinski, P., Phillpot, S. R., Choi, S. U. S., and Eastman, J. A. Mechanisms of heat flow in suspensions of nano-sized particles (nanofluids). *International journal of heat and mass transfer,* Vol. 45, pp. 855-863, 2002.

[71] Wang B. X., Zhou L. P., Peng X. F., A fractal model for predicting the effective thermal conductivity of liquid with suspension of nanoparticles, *International Journal of Heat and Mass Transfer,* Vol. 46 (14), pp. 2665-2672, 2003 .

[72] Xue, L., Leblinski, P., Phillpot, S. R., Choi, S. U. S., and Eastman, J. A. Two regimes of thermal resistance at a liquid-solid interface. *Journal of chemical physics,* Vol. 118(1), pp. 337-339, 2003.

[73] Shenogin, S., Xue, L., Ozisik, R., and Keblinski, P. Role of thermal boundary resistance on the heat flow in carbon-nanotube composites. *Journal of applied physics,* Vol. 95(12), pp. 8136-8144, 2004.

[74] Xue, l., Keblinski, P., Phillpot, S. R., Choi, S. U. S., and Eastman, J. A., Effect of liquid layering at the liquid-solid interface on thermal transport, *International Journal of Heat and Mass Transfer,* Vol. 47, pp. 4277-4284, 2004.

[75] Evans, W., Fish, J., and Keblinski, P., Thermal conductivity of ordered molecular water, *The Journal of Chemical Physics,* Vol. 126, pp. 154504, 2007.

[76] Xie H. Q., Fujii M., Zhang X., Effect of interfacial nanolayer on the effective thermal conductivity of nanoparticle-fluid mixture, *International Journal of Heat and Mass Transfer,* Vol. 48 (14), pp. 2926-2932, 2005.

[77] Prakash M., Giannelis E. P., Mechanism of heat transport in nanofluids, *Journal of Computer-Aided Materials Design,* Vol. 14 (1), pp. 109-117, 2007.

[78] Vadasz P., Heat conduction in nanofluid suspensions, *Journal of Heat Transfer-Transactions of The ASME, Vol.* 128 (5), pp. 465-477, 2006.

[79] Lee, D., Thermophysical properties of interfacial layer in nanofluids, *Langmuir,* Vol. 23, pp. 6011-6018, 2007.

[80] Evans, W., Fish, J., and Keblinski, P., Thermal conductivity of ordered molecular water, *The Journal of Chemical Physics,* Vol. 126, pp. 154504, 2007.

[81] Tillman P., Hill J. M., Determination of nanolayer thickness for a nanofluid, International Communications In *Heat and Mass Transfer,* Vol. 34 (4), pp. 399-407, 2007.

[82] Prasher R., Phelan P. E., Bhattacharya P., Effect of aggregation kinetics on the thermal conductivity of nanoscale colloidal solutions (nanofluid), *Nano Letters,* Vol. 6 (7), pp. 1529-1534, 2006.

[83] Prasher R., Evans, W., Meakin, P., Phelan P. E., and Keblinski, P., Effect of aggregation on thermal conduction in colloidal nanofluids, *Applied Physics Letters,* Vol. 89, pp. 143119, 2006.

In: Nanoparticles: New Research
Editor: Simone Luca Lombardi, pp. 277-306
ISBN 978-1-60456-704-5

Chapter 9

Enhancement and Temperature Variation in the Thermal Conductivity of Nanofluids

Pei Tillman, Miccal T. Matthews and James M. Hill
School of Mathematics and Applied Statistics,.
University of Wollongong, Wollongong, Australia

Abstract

One of the most important findings for nanofluids is the discovery of a significant enhancement and a strong temperature effect in their thermal conductivity. Research confirms that the standard rules for mixtures is not capable of describing the nanoparticle's contribution to the thermal conductivity, and some researchers believe that the Brownian motion is the key mechanism affecting the variation of thermal conductivity with temperature. In this paper, the authors review existing experimental results and explore different approaches to estimating the effective thermal conductivity of nanofluids from recent theoretical models. Focus is primarily on an explanation of the temperature-dependence on the thermal conductivity of nanofluids. It is found that the mobility of nanoparticles is theoretically conjectured as having a key role in determining the temperature effect on the thermal conductivity of a nanofluid. The limitations of existing models and future research directions are discussed.

1. Introduction

The development of thermal energy-efficient heat transfer fluids is required in many industrial and commercial applications. Since solid materials typically have a thermal conductivity several orders of magnitude larger than commonly-used liquids, many heat transfer fluids containing millimeter or micrometer-sized particles were developed for improved heat exchange systems. However, these so-called "charged" fluids had several drawbacks, most commonly due to particle settling and cogging of machine components. With the emergence of nanotechnology, a new class of heat transfer fluids was developed by suspending nanoparticles and carbon nanotubes in these liquids. At present, three types of nanofluids containing single- and multi-wall carbon nanotubes, vapor grown carbon fibres, amorphous carbon and copper, CuO and Al_2O_3 nanoparticles have been developed, manufactured, and tested [1]. These are:

1. heat transfer nanofluids or nanocoolants based on three base fluids: water, water/ethylene glycol mixtures, water/antifreeze mixtures,

2. nanolubricant fluids: commercial 15W-40 oil, BP Amoco DS-1666 Durecene oil (synthetic poly-α-olefin oil), military specification-based fluids,

3. nanolubricant grease, based on fluids and greases: BP 1666 Durecene oil, military specification-based greases.

Typical applications of these types of nanofluids are vehicle radiator coolants, air conditioning systems, cutting fluids, quenching fluids and high-stress contact gears for rotary machines. The enhanced thermal conductivity of the nanofluids tends to remove unwanted heat from the engines while providing lubrication and temperature control. The anticipated benefits of nanofluids are lower operation costs, improved thermal transfer properties, minimal clogging, suspension stability and long-term performance life.

The thermal conductivities of nanofluids are significantly higher than those of the base fluid, even when the loaded nanoparticle concentration is less than 1 % by volume. The main reason for this is that the suspended nanoparticles remarkably increase the thermal conductivity of the fluid, primarily as a result of the high surface-to-volume ratio of the nanoparticles (for example, the surface-to-volume ratio of a 10 nm particle is 1,000 times larger than that of a 10 μm particle). In nanoparticle-fluid mixtures, other effects such as the microscopic motion of particles, particle structure and surface properties may cause additional heat transfer mechanisms, which are not well understood at present. Traditional multiphase effective medium theories such as those of Maxwell [2] and Hamilton and Crosser [3] cannot account for the thermal conductivity enhancement achieved in nanofluids. The effective medium theories predict no change in the conductivity enhancement with temperature, since they are only dependent on the volume fraction of the solid particles and the thermal conductivity of the pure components. The model of Hamilton and Crosser [3] is dependent on the volume fraction and the sphericity of the particles, which is defined as the ratio of the surface area of a particle with a perfectly spherical shape to that of a non-spherical particle with the same volume. These, and other effective medium theories, predict no effect of particle size and only a weak effect of particle conductivity.

Various attempts have been made in the past ten years to theorize this phenomenon by taking into account the effects of particle radius and shape, concentration, properties of the nanolayer surrounding the nanoparticle, nanoparticle aggregation, nanoparticle motion, ballistic phenomena, dynamic viscosity of the liquid and the temperature of the medium. However, many models do not have any temperature-dependence and cannot be treated as a general theoretical framework. The purpose of this article is to present a review of the progress of research in the important aspect of the temperature-dependence on the thermal conductivity enhancement in nanofluids. Our review of the temperature effect on the thermal conductivity behavior of materials begins with single-phase systems, such as solids and liquids, then multiphase systems, in particular nanofluids. We examine the physical reasoning behind the models adopted, which correlate the thermal conductivity of nanofluids as a function of temperature, particle size and volume fraction. Limitations of the existing models and some future research directions are discussed.

2. Thermal Conductivity of Liquids

Accurate knowledge of the thermal transport properties of pure liquids is essential for the development of a theory of the transport properties of nanofluids. In heat conduction, heat is transferred due to an energy exchange between heat carriers in the collision process. It is believed that liquids mainly conduct heat through the vibration of single molecules, while other methods of transfer such as molecular convection and radiation impart a small contribution to the overall heat transfer mechanism. Based on statistical mechanics (in particular kinetic gas theory) and correlated with experimental data, Avsec and Oblak [4] extended the Chung-Lee-Starling model (CLS) [5, 6] and developed equations to calculate the transport properties of pure fluids.

For most simple liquids (except water) the thermal conductivity is generally characterized as decreasing linearly with increasing temperature. The thermal conductivity of liquids can generally be expressed as a function of temperature by

$$k_f = A_1 + A_2T + A_3T^2 \quad (\mathrm{Wm^{-1}K^{-1}}), \tag{1}$$

where the subscript f represents the fluid, the temperature T is in Kelvin (K) and A_1, A_2 and A_3 are constants in units of $\mathrm{Wm^{-1}K^{-1}}$, $\mathrm{Wm^{-1}K^{-2}}$ and $\mathrm{Wm^{-1}K^{-3}}$, respectively. For water, $A_1 = -0.76760$, $A_2 = 7.5390 \times 10^{-3}$, $A_3 = -9.8250 \times 10^{-6}$; for toluene ($C_7H_8$), $A_1 = 0.22050$, $A_2 = -3.000 \times 10^{-4}$, $A_3 = 0$; for the hexadecane oil ($C_{16}H_{34}$), $A_1 = 0.1963$, $A_2 = -1.9\times10^{-4}$, $A_3 = 0$ [7]. Figure 1 displays the variation of the thermal conductivities of water, toluene, hexadecane oil, and pure perfluoro-n-hexane (FC72) with temperature, as measured by Yang and Han [8].

3. Thermal Conductivity of Solids

The thermal conductivity of solids (metals and semiconductors) has a more complex behavior than those of liquids. Starting from $T = 0$ K, the thermal conductivity rises to a maximum k_{max} in the range 10 to 50 K in the case of metals, and then decreases rapidly, finally leveling off at about 300 K for typical metals. The maximum value generally occurs at a higher temperature for non-metals, since this behavior is highly dependent on the purity of the material, and the maximum thermal conductivity may change by an order-of-magnitude with a small change in the impurity level. It is found that the temperature T_{max} corresponding to the thermal conductivity peak k_{max} can be significantly increased by the impurity level of the material [9], hence a lower T_{max} corresponds to a material with a higher purity. Thus, it is difficult to treat the low-temperature thermal conductivity as a true property of the material.

Heat conduction in pure metals actually results from the random motion of free electrons in the system, which carry thermal energy from one location to another. In dielectric solid materials (such as insulators and semiconductors) heat is conducted via the vibration of atoms, and the atoms are bonded to each other through interatomic force interactions.

Each atom in a solid vibrates around its equilibrium position, and the atoms near the hot side will have large vibrational amplitudes, which will affect the atoms on the other side of the system. The vibration of any one atom can cause the vibration of the entire system by creating lattice waves. A phonon is defined as a quasiparticle with an energy equal to that of all of the vibration modes of the lattice [10]. Heat in a dielectric crystal is conducted by such a phonon gas, similar to that in a box of gas molecules. The collision of phonons is due to the interaction of phonon waves. In impure metals or disordered alloys, the electron mean free path is reduced by collisions with impurities, and the phonon contribution may be comparable with the electronic contribution. In semiconductors at low temperatures where they are not electrical conductors, heat is transported only by phonon thermal conduction.

A phonon gas model of heat conduction is adopted in insulators or semiconductors in the lower temperature range. When the temperature is above the Debye temperature, defined by

$$T_\theta = \frac{\hbar v_s}{k_B}\left(\frac{6\pi^2 N}{V}\right)^{\frac{1}{3}}, \tag{2}$$

where V is the volume of the specimen, k_B is the Boltzmann constant (1.38×10^{-23} JK^{-1}), N is the number of atoms in the specimen and v_s is the speed of sound, semiconductors start to generate current and phonon thermal conduction is accompanied by a significant fraction of electron thermal conduction. Therefore, the thermal conductivity of semiconductors is generally evaluated by use of the additive law

$$k = k_e + k_{ph}, \tag{3}$$

where k_e and k_{ph} are the thermal conductivities due to electron and phonon heat transport, respectively. The unified temperature-dependence of thermal conductivity for metals and insulators or semiconductors is deduced from the well-known Debye formula for thermal conductivity as [9]

$$k_e = \tfrac{1}{3} C_v \ell_e u_e B_1 B_2 = \tfrac{1}{3} C_v \tau_e u_e^2 B_1 B_2, \tag{4}$$

where C_v is the heat capacity per unit volume, u_e is the velocity of the electrons, ℓ_e is the mean free path of the electrons, τ_e is the relaxation time of the electrons, B_1 is a function of the electron scattering by the sample boundary and B_2 is a function of the electron scattering by defects and other electrons. An identical expression for k_{ph} may also be applied, with an obvious change in notation.

The electron/phonon mean free path ℓ is determined by two processes; geometrical scattering and scattering by other electrons or phonons [11]. When the mean free path becomes comparable with the size of the nanoparticles at low temperature, the size effect dominates which causes a reduction of the thermal conductivity of the solid.

Kittel [11] explains the temperature-dependence of the phonon thermal conductivity from the theory of anharmonic coupling. If the force between atoms was purely harmonic, there would be no mechanism for collisions between different phonons, and the mean free path is solely based on collisions between a phonon and a crystal boundary or lattice imperfections. There are situations where these effects are dominant. At temperatures higher than the Debye temperature T_θ, the determining factor for the thermal conductivity is phonon-phonon scattering. With anharmonic lattice interactions, there is a coupling between different phonons which limits the value of the mean free path. This theory predicts that ℓ is

Table 1. Thermal conduction mechanism in solids.

Temperature range	Electron/phonon scattering by boundary (B_1)	Electron/phonon scattering by defect (B_2)	Electron/phonon scattering by electron/phonon
$0 < T < 10\text{K}$	$\surd$	-	-
$10 < T < T_{ph}$	-	$\surd$	$\surd$
$T > T_{ph}$	-	-	$\surd$

proportional to $1/T$ at high temperatures, in agreement with many experiments [11]. At low temperatures, the only temperature-dependent term on the right hand side of Eq. (3) is C_v, the heat capacity per unit volume, which varies as T^3 [11].

Avsec and Oblak [4] express the phonon contribution to the thermal conductivity of an isotropic solid as the integral of the product of the phonon mean free path ℓ, heat capacity per unit volume of a single phonon mode C_v and phonon density of states F over the acentric factor ω via

$$k_{ph} = \frac{Nv_g}{3V}\int_0^{\omega_{max}} \ell(\omega)C_v(\omega)F(\omega)d\omega, \tag{5}$$

where ω_{max} is the phonon spectrum width, v_g is an average phonon group velocity, and the ratio N/V is the number of atoms per volume. For the low and high temperature region where the temperature is lower or higher than the Debye temperature T_θ, they used the following solution for the thermal conductivity of an isotropic solid

$$k_{ph} = \begin{cases} k_0 \exp\left(-\dfrac{T_\theta}{T}\right), & \text{if} \quad T < T_\theta, \\ \dfrac{BM\Omega_a^{1/3}k_B^3T_\theta^3}{(2\pi)^3\hbar^3\gamma_G^2 T}, & \text{if} \quad T > T_\theta, \end{cases} \tag{6}$$

where B is a dimensionless constant, Ω_a is atomic volume, M is the molecular mass and γ_G is the Grüneisen constant.

Okhotin *et al.* [9] introduce a characteristic temperature T_{ph}, defined as a certain temperature above which electron scattering by defects of any kind exert practically no effect on the thermal conductivity. They outlined the thermal conduction mechanisms in different temperature ranges, as summarized in Table 1. The Dulong-Petit law is a chemical law which offers fairly good predictions for the specific heat capacity of solids with relatively simple crystal structures at high temperatures. It states that the dimensionless heat capacity $C_v/(Nk_B)$ is equal to 3, where N is the number of identical oscillators. Above T_{ph}, scattering by defects of any kind exerts practically no effect on the thermal conductivity because the product of the heat capacity ($C_v = 3k_BN$ by the law of Dulong and Petit [11]) and the terms allowing for election/phonon scattering by boundaries and defects (B_1 and B_2) disappear. Their ideal gas model for each temperature-dependant factor in Eq. (4) is listed in Table 2. Table 2 represents Okhotin *et al*'s complete theory; from the temperature ranges given in Table 1, the expressions may be simplified [9].

Table 2. Temperature-dependent factors in Okhotin *et al.*'s model [9].

	Phonon gas model	Electron gas model
C_v	$9R\left(\frac{T}{T_\theta}\right)^3 \int_0^{\frac{T_\theta}{T}} \frac{x^4 e^x}{(e^x-1)^2}dx$	$9R\left(\frac{T}{T_\theta}\right)^3 \int_0^{\frac{T_\theta}{T}} \frac{x^4 e^x}{(e^x-1)^2}dx$
B_1	$1-\exp(-aT^b)$, $a=(4\sim 50)\times 10^{-3}$, $b=1.5\sim 3.25$	$1+a_1T^{-b_1}\exp\left[-c_1\left(\frac{T}{T_{ph}}\right)^3\right]$. The values of a_1, b_1 & c_1 for some materials can be found in [9]
B_2	$\exp(a\pm bT+cT^2)$, $a=0.2\sim 0.7$, $b=(5\sim 10)\times 10^{-3}$, $c=(2.5\sim 10)\times 10^{-4}$	$\left\{\exp\left[a_2+b_2\left(\frac{T}{T_{ph}}\right)+c_2\left(\frac{T}{T_{ph}}\right)^2\right]\right\}^{-1}$. The values of a_2, b_2 & c_2 for some materials can be found in [9]
τ	$\frac{\left(\frac{k_BT}{\hbar}+\frac{k_BT_{max}}{\hbar}\right)^{-1}}{\exp\left(\frac{T_{max}}{T}\right)-1}$	$\frac{\left(\frac{k_BT}{\hbar}+\frac{k_BT_{max}}{\hbar}\right)^{-1}}{\exp\left(\frac{T_{max}}{T}\right)+1}$
u	$\sqrt{\frac{v_s(6\pi^2)^{\frac{1}{3}}}{r_c m_{ph}}}$	$\frac{k_BTm_e}{\pi\hbar^3\rho v_s^2}\left[\frac{3}{4m_e}\left(\frac{\hbar}{r_c}\right)^2\left(\frac{T}{T_{ph}}\right)^{-n}\right]^2$ For most metals $n=0.3$.

A reduction in material dimensionality can cause a change in the phonon dispersion, phonon density of states, phonon speed and phonon relaxation time. These factors may lead to significant changes in the magnitude and temperature variation of the thermal conductivity. For example, a carbon nanotube is considered a rolled-up graphene sheet, and the high thermal conductivity of a nanotube is believed to be associated with the large phonon mean free paths in the system [12], which is limited by scattering from sample boundaries, point defects and phonon-phonon umklapp scattering processes [13]. This is obvious from Eq. 4, which shows that $k \sim C_v u\ell$. Figure 2 shows the temperature-dependence on the thermal conductivity of some metals, semiconductors [14,15], a (10,10) carbon nanotube calculated from molecule dynamics simulation [12] and a multiwall carbon nanotube (MWCNT) bundle [13]. The calculation for the (10,10) carbon nanotube by Berber *et al.* [12] suggests that at room temperature the thermal conductivity of the nanotube is 6,600 $\mathrm{Wm^{-1}K^{-1}}$, exceeding the thermal conductivity of 3,320 $\mathrm{Wm^{-1}K^{-1}}$ for near-isotropic pure diamond. Kim *et al.*'s [13] study indicates that the thermal conductivity of a MWCNT bundle decreases with the diameter of the MWCNT bundle, with a thermal conductivity of about 3,000 $Wm^{-1}K^{-1}$ at room temperature.

Studies of the thermal conductivity of nanofluids are normally performed in a temperature range of 278 K to 323 K, although data over a much larger temperature range is valid for the pure components of the nanofluid. Figure 3 shows the variation of the thermal conductivity of nanoparticle materials normally used in nanofluids (Al_2O_3, Au and CuO) with temperature [7, 15]. It is obvious that the thermal conductivity of the nanoparticles decreases with increasing temperature in the temperature range 200 K to 500 K.

4. Experimental Data of Nanofluids

An important finding concerning nanofluids was the discovery of a significant enhancement and a strong temperature effect on their thermal conductivity. Experimental investigations on the effective thermal conductivity of nanofluids have examined a wide variety of different solvent-particle systems.

For a 1 %-loaded Al_2O_3/water nanofluid at 21 °C there is a 2 % enhancement, while at 51 °C there is a 10.8 % enhancement; for a 4 %-loading the enhancement goes from 9.4 % to 24.3 %. For a CuO/water nanofluid the results were an enhancement from 6.5 % to 29 % and 14 % to 36 % corresponding to a 1 %-load and 4 %-loading, respectively [16]. A similar increase in the thermal conductivity of CuO and Al_2O_3 nanofluids was also observed by Li and Peterson [17]. Patel *et al.* [18] observe a 5 % $\sim$ 21 % increase in thermal conductivity for a temperature range of 30-60 °C at a loading of 0.00026 % Au in water. There have been several other experimental studies on various types of nanofluids [8, 19–24], and the maximum enhancements obtained are summarized in Table 3.

Kim *et al.* [24] investigate the effect of the morphology of carbon nanotubes (CNT) on the effective thermal conductivity of suspensions as a function of the volume fraction, aspect ratio (defined as the ratio of the length to the diameter of the nanotubes) and temperature. The measured data was classified into three different groups: CNT A (functionalized SWCNT/H_2O), B (SWCNT/H_2O) and C (MWCNT/H_2O). Their findings indicate that the degree of thermal conductivity enhancement increases with the aspect ratio of the carbon nanotube, volume fraction and temperature. Wen and Ding's [23] finding concerning the effect of tempearture on the thermal conductivity of carbon nanotube-nanofluids is similar to Kim *et al.*'s results [24]; the effective thermal conductivity of nanofluids increases with temperature, but they discovered a nonlinear dependence on temperature and nanotube concentration.

In contrast to Kim *et al.* [24] and Wen and Ding's [23] findings concerning non-spherical nanoparticle suspensions, Yang and Han [8] experimentally found a thermal conductivity enhancement for Bi_2Te_3 nanorods in FC72 and in oil nanofluids which decreases with increasing temperature, and proposed the relation $k_{eff} \propto T^{-a}$, where $a = 1.68 \sim 1.95$. They analyzed the thermal diffusion coefficient of the nanorods and the thermal conductivity predicted by the effective medium theory versus the particle aspect ratio. They found that the diffusive conduction benefits from the increased aspect ratio of the nanoparticles. Although the long fibres are less mobile than spherical particles in a nanolfluid, and hence the effects of Brownian motion are reduced, they can promote heat

Table 3. Experimental studies of the temperature effect on the thermal conductivity of nanofluids. Only the maximum enhancements obtained are shown.

Author	Nanofluid	Temp. (°C)	Enhancement
Das *et al.* [16]	CuO/H_2O	51	36 % at $\phi = 4$ %
Patel *et al.* [18]	Au thiolate/H_2O	60	21 % at $\phi = 0.00026$ %
Kumar *et al.* [19]	Au/H_2O	50	9 % at $\phi = 0.00026$ %
Chon *et al.* [20]	Al_2O_3/H_2O	71	29 % at $\phi = 4$ %
Zhang *et al.* [21]	Al_2O_3/H_2O	10, 30, 50	20 % at $\phi = 15$ %
Zhang *et al.* [22]	CNT/H_2O	23	40 % at $\phi = 0.9$ %
Li *et al.* [17]	CuO/H_2O (distilled)	34	52 % at $\phi = 6$ %
Yang & Han [8]	Bi_2Te_3/FC72	3	7.7 % at $\phi = 0.8$ %
Wen *et al.* [23]	$MWCNT/H_2O$	45	31 % at $\phi = 0.84$ %
Kim *et al.* [24]	$MWCNT/H_2O$	44	37 % at $\phi = 1$ %

transfer by providing a rapid and longer heat flow path. Their analysis suggests that the conduction-based mechanism should be primarily responsible for the thermal conductivity enhancement in nanofluids containing nanorods.

5. Present Models to Account for the Temperature Effect

From the classic effective medium theories the definition for the thermal conductivity of a two-phase mixture is

$$\frac{k_{eff}}{k_f} = 1 + \frac{n(k_p - k_f)\phi}{k_p + (n-1)\,k_f - (k_p - k_f)\phi}, \tag{7}$$

where n is an empirical constant for mixtures containing particles of arbitrary shape. For Maxwell's formula $n = 3$, while for that of Hamilton and Crosser we have $n = 3/\Phi$ where Φ is the sphericity defined as the ratio of the surface area of a sphere, with a volume equal to that of a particle, to the surface area of the particle.

The Hamilton and Crosser model yields excellent results for particles larger than 13 nm, but for smaller particles the theory breaks down with a deviation of more than 100 % compared to experimental results. This is primarily because the theory possesses no dependence on particle size and surface phenomena.

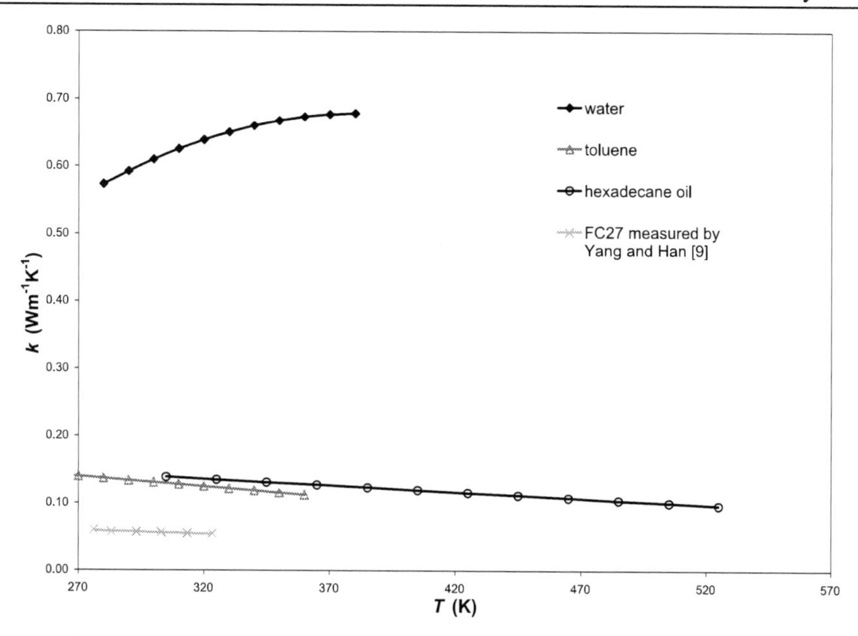

Figure 1. Thermal conductivity of pure fluids as a function of temperature [7, 8].

It is important to note that the bulk thermal conductivity of metals and ceramics such as Au, Al_2O_3 and CuO and fluids such as toluene and oil decrease with increasing temperature at 0 °C to 100 °C, but water does not (see Fig. 1). If we use a linear function $k_p(T) = A_p - B_pT$ to represent the thermal conductivity of a nanoparticle and $k_f(T) = A_f + B_fT$ for that of the base fluid, Maxwell's formula implies that the effective thermal conductivity of nanofluids is a monotonically decreasing function with respect to T for volume fraction $\phi \leq 10$ %. This behavior is in contrast to most of the experimental findings shown in Figs. 4 and 5. As a result, it is clear that the effective thermal conductivity of nanoparticle suspensions is not a simple function of the combined thermal properties of the base fluid and the nanoparticle. Maxwell's rule of mixtures is not adequate to describe the nanoparticle contribution to the thermal conductivity.

Various approaches have been developed in the past ten years based on the phenomenology and morphology of the agglomerates of observed nanoparticles. Several theoretical descriptions have been proposed that incorporate a variety of fundamental mechanisms as explanations of the thermally-driven particle flow in liquids. Many researchers believe that the Brownian motion of the nanoparticles constitutes a key mechanism for the thermal conductivity enhancement with increasing temperature and decreasing nanoparticle size. The effects of Brownian motion on small particles are more pronounced at high temperature, as has been observed experimentally [16–21]. Based on the theory of Brownian motion of particles suspended in a fluid, many researchers propose a microscopic convective-current model which takes into account the dependence of temperature on the effective thermal conductivity of nanofluids, along with the traditional effective medium theories. The motion of particles inside a fluid is modelled as low Reynolds number flow past a sphere. In the postulated micro-convection model, four parameters are used: the Reynolds number $Re = u_pd_p/\nu$ which represents the mobility of the nanoparticles, the

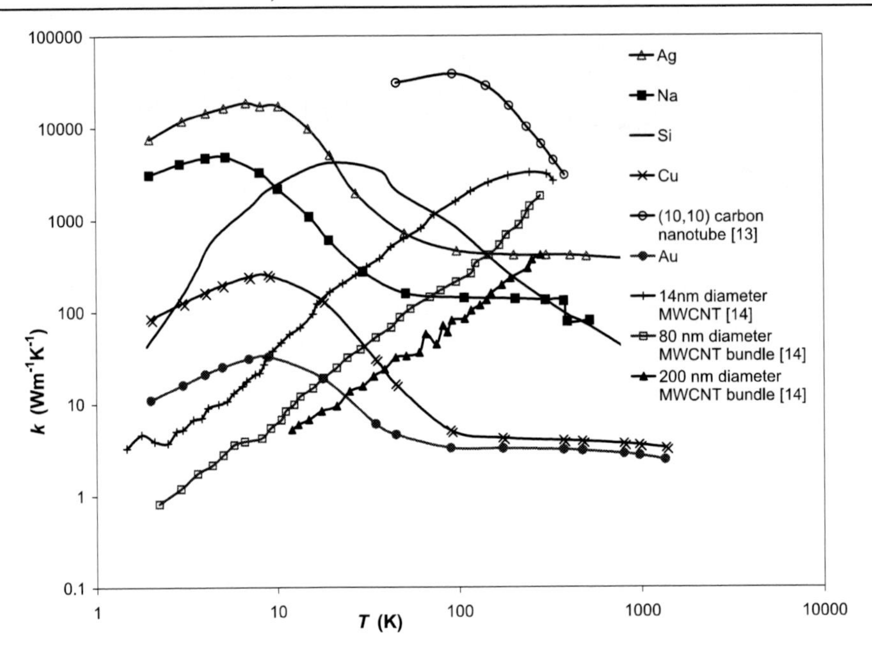

Figure 2. Thermal conductivity of some metals and semiconductors [12–15].

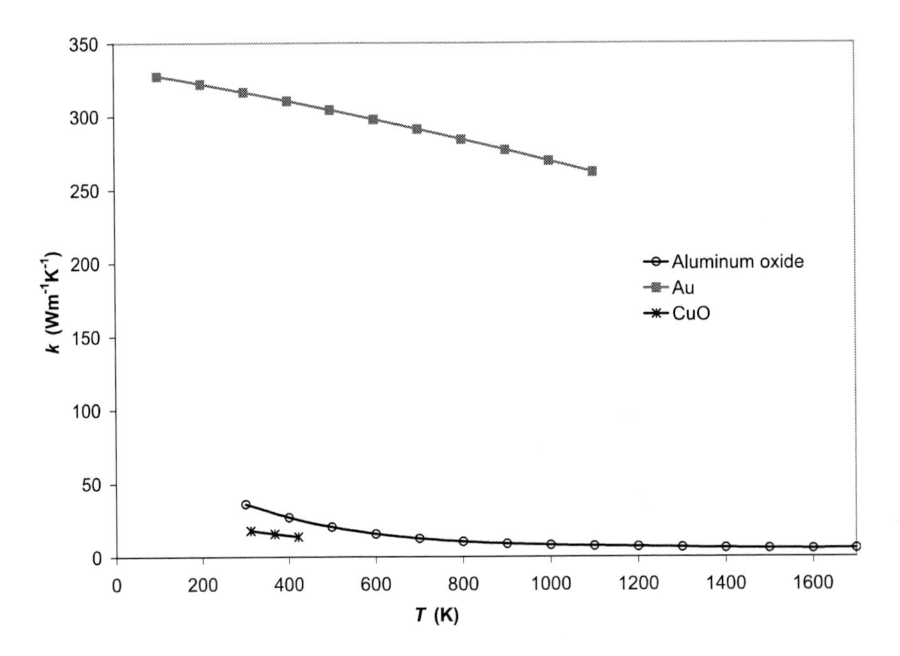

Figure 3. Thermal conductivity of Au, Al_2O_3 and CuO [7, 15].

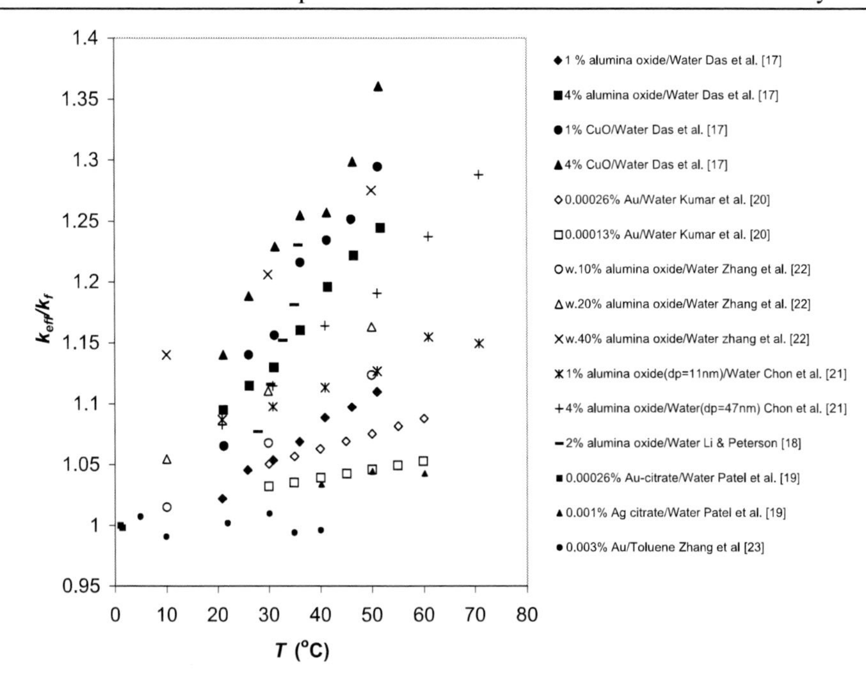

Figure 4. Measured thermal conductivity of spherical particle suspensions versus temperature.

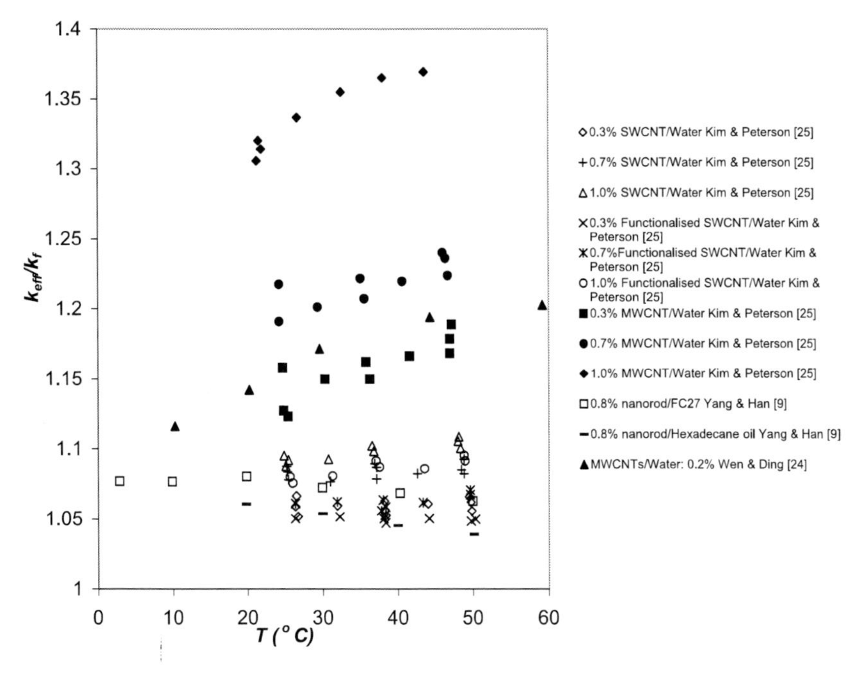

Figure 5. Measured thermal conductivity of non-spherical particle suspensions versus temperature [8, 24].

Prandtl number $Pr = \nu/\alpha_f = \mu C_{p,f}/k_f$, Nusselt number $Nu = hd_p/k_f$, and Péclet number $Pe = u_p d_p/\alpha_f = RePr$. The classical solution for flow past a sphere is given by [25]

$$Nu = 2.0 + 0.5RePr + O(Re^2Pr^2) = 2.0 + 0.5Pe + O(Pe^2). \quad (8)$$

Table 4. Overview of development of models including temperature effects for evaluation of the thermal conductivity of nanofluids.

Model type	Formula
Analytical Wang *et al.* [36]	Nanosphere (1999) $k_{Brownian} = k_r + k_t$ $k_r = \left[1.176\left(\frac{k_p - k_f}{k_p + 2k_f}\right)^2 + 5\left(0.6 - 0.028\frac{k_p - k_f}{k_p + 2k_f}\right)\right]\phi k_f Pe_f^{1.5}$ $k_t = \left(0.0556Pe_t + 0.1649Pe_t^2 - 0.0391Pe_t^3 + 0.0034Pe_t^4\right)k_f$ $Pe_f = \frac{r_p^2\gamma\rho_f C_{p,f}}{k_f}$ $Pe_t = \frac{u_p\rho_f C_{p,f} L\phi^{0.75}}{k_f}$ $L = r_p\left(\frac{4\pi}{3\phi}\right)^{\frac{1}{3}}$
Analytical Xuan *et al.* [42]	Nanocluster (2003) $k_{eff} = k_{static}(\text{Maxwell}) + \frac{1}{2}\rho_p c_p \phi\sqrt{\frac{k_B T}{3\pi r_c \mu}}$
Analytical Jang *et al.* [27,28]	Nanosphere (2004) $k_{eff} = (1-\phi)k_f + \phi k_p + \frac{3c_1\phi d_f k_f Re_p^2 Pr^2}{d_p}$ $Re_p = \frac{u_p d_p}{\nu}$ $u_p = \frac{k_B T}{3\pi\mu d_p \ell}$ $C_1 = 18 \times 10^6$
Analytical Kumar *et al.* [19]	Nanosphere (2004) $k_{eff} = k_f + \frac{c u_p \phi r_f}{(1-\phi) r_p}$ $u_p = \frac{2k_B T}{\pi\mu d_p^2}$ $c \approx \frac{N\ell\hat{C}_v}{3}$

Table 4. Continued

Model type	Formula
Analytical Ren *et al.* [29]	Nanosphere (2005) $k_{eff} = k_f \left[1 + F(Pe) + 3\Theta\phi_T + \frac{3\Theta^2\phi_T^2}{1 - \Theta\phi_T}\right]$ $F(Pe) = 0.0556Pe + 0.1649Pe^2 - 0.0391Pe^3 + 0.0034Pe^4$ $Pe = \frac{u_p L \phi_T^{0.75}}{\alpha_f}$ $L = (r_p + d)\sqrt[3]{\frac{4\pi}{3\phi_T}}$ $\Theta = \frac{\beta_{lf}\left[\left(1 + \frac{d}{r_p}\right)^3 - \beta_{pl}\beta_{fl}\right]}{\left(1 + \frac{d}{r_p}\right)^3 + 2\beta_{lf}\beta_{pl}}$ $\beta_{ij} = \frac{k_i - k_j}{k_i + 2k_j}$ $u_p = \sqrt{\frac{3k_BT}{m_c}}$ $\phi_T = \phi\left(1 + \frac{d_l}{r_p^3}\right)$
Analytical Xu *et al.* [37]	Nanoparticle cluster (2006) $k_{eff} = k_{static}$ (Hamilton & Crosser) $+ k_{Brownian}$ $k_{Brownian} = \frac{ck_f Nu\,(2 - D_f)\,d_f\left[\left(\frac{d_{p,max}}{d_{p,min}}\right)^{1-D_f} - 1\right]^2}{Prd_{p,min}\,(1 - D_f)\left[\left(\frac{d_{p,max}}{d_{p,min}}\right)^{2-D_f} - 1\right]}$ $D_f = 2 - \frac{\ln(\phi)}{\ln\left(\frac{d_{p,min}}{d_{p,max}}\right)}$ $c = O(10^2)$
Analytic Sabbaghzadeh *et al.* [30]	Cylindrical nanoparticles (2007) $k_{eff} = k_f[1 - \phi(1 + M)] + \phi(k_p + k_l M) + \frac{k_f d_f \phi\,(1 + M)}{2Pr_f(r_p + d)}\left(0.35 + 0.56Re^{0.52}\right)Pr_f^{0.3}$ $Re = \frac{u_p L}{\nu_f}$ $u_p = \sqrt{\frac{3k_BT}{m_p}}$ $L = \sqrt[3]{\frac{\pi r_p^2 \ell M^2}{\phi}}$ $M = \left(\frac{d}{r_p} + 1\right)^2 - 1$

Table 4. Continued

Model type	Formula
Semi-empirical Shukla & Dhir [38]	Nanosphere (2005) $k_{keff} = k_{static}(\text{Maxwell}) + \frac{c\phi(T-T_0)}{\mu r_p^4}$ c, T_0 are experimentally determined constants.
Semi-empirical Prasher [31]	Nanosphere (2005) $k_{eff} = k_f\left(1 + ARe^m Pr^{0.333}\phi\right)\left[\frac{(1+2a)+2\phi(1-a)}{(1+2a)-\phi(1-a)}\right]$ $a = \frac{2R_b k_f}{d_p}$ $u_p = \sqrt{\frac{3k_B T}{m_p}}$ $A = 40,000$, $m = 2.5 \pm 15\,\%$
Semi-empirical Patel *et al.* [26]	Nanosphere (2005) $k_{eff} = k_f + \frac{k_p d_f \phi}{d_p(1-\phi)}\left(1 + \frac{c u_p d_p}{\alpha_f}\right)$ $u_p = \frac{2k_B T}{\pi \mu d_p^2}$ $c = 25,000$
Analytic and Correlation Koo *et al.* [39]	Nanosphere (2004) $k_{eff} = k_{static}(\text{Maxwell}) + k_{Brownian}$ $k_{Brownian} = (5 \times 10^4)\,\beta_1 \phi \rho C_v \sqrt{\frac{k_B T}{\rho d_p}} \times [(-6.04\phi + 0.4705)T + (1722.3\phi - 134.63)]$ $u_p = \sqrt{\frac{18k_B T}{\pi \rho d_p^3}}$ For Au-citrate, Ag-citrate & CuO: $\beta_1 = 0.0137(100\phi)^{-0.8229}$, $\phi < 1\,\%$ For CuO: $\beta_1 = 0.0011(100\phi)^{-0.7272}$, $\phi > 1\,\%$ For Al_2O_3: $\beta_1 = 0.0017(100\phi)^{-0.0841}$, $\phi > 1\,\%$
Correlation Chon *et al.* [20]	Nanosphere(2005) $\frac{k_{eff}}{k_f} = 1 + c\left(\frac{1}{d_p}\right)^{0.369}\left(\frac{T^{1.2321}}{10^{\frac{2.4642B}{T-C}}}\right)$ For water: $B = 247.8$, $C = 140$ c is an empirical constant

Table 4. Continued

Model type	Formula
Correlation	Nanosphere (2006)
Li & Peterson [17]	For Al_2O_3/water:
	$\frac{k_{eff} - k_f}{k_f} = 0.7644481464\phi + 0.018688867T - 0.462147175$
	For CuO/water:
	$\frac{k_{eff} - k_f}{k_f} = 3.761088\phi + 0.0017924T - 0.30734$

In the analysis of the Stokes flow regime, there are two approaches to evaluate the average velocity of the nanoparticles, u_p. One approach is estimated from the Stokes-Einstein diffusion theory, assuming that a nanoparticle freely moves over a distance equal to the mean-free path of a base fluid molecule ℓ_f. Thus $u_p = D_o/\ell_f = k_B T/(6\pi\mu d_p \ell_f K)$, where D_o is the particle thermal diffusion coefficient and K is a shape factor ($K = 1/2$ for spheres). Another approach is the root-mean-square velocity from the kinetic theory of gases, $u_p = \sqrt{3k_B T/m_p}$, where m_p is the mass of a nanoparticle. Finally, a direct expression for the effective thermal conductivity of nanofluids, k_{eff} as a function of T, is obtained via the mathematical relation $u_p = u_p(T)$. Table 4 gives a chronological account of some of the prediction methods developed over the past decade concerning the temperature-dependence on the thermal conductivity enhancement. These existing theoretical models can be categorized into three groups: analytical; semi-empirical; correlation.

5.1. 1D Parallel Process-Based Heat Flow Paths

One group of researchers [19, 26–31] begins their analysis by considering one-directional heat transfer over the base fluid and the solid. Fourier's law of heat conduction, and the heat transfer micro-convection from the particles is applied to the liquid and the solid. The effective thermal conductivity of the nanofluid is subsequently derived by substituting relevant parameters from thermal statistical theory and Stokes flow theory. These sort of models involve process-based techniques. This technique is quite useful for examining what physical processes cause heat flow in a nanofluid medium.

In Jang and Choi's model [27,28] based on kinetics, Kapitza resistance and convection, the whole process of energy transport is assumed to consist of three additive modes; (a) collisions between base fluid molecules; (b) thermal diffusion in nanoparticles in fluids; (c) collision of nanoparticles with each other due to Brownian motion; and (d) thermal interaction of nanoparticles with the base fluid (nanoconvection). In the first mode, the energy carriers (the base fluid molecules) travel freely only over the mean free path ℓ_f before they collide, and the net energy flux q_1 across a plane at z is given by

$$q = -\frac{1}{3}\ell_f C_{v,f} u_f (1-\phi)\frac{dT}{dz} = -k_f(1-\phi)\frac{dT}{dz}, \quad (9)$$

where $C_{v,f}$, u_f and T are the heat capacity per unit volume, mean speed and temperature of the base fluid molecules, respectively.

In the second mode, the net energy transport is carried by electrons or phonons in the suspended nanoparticles, so that the energy flux is given by

$$q = -\frac{1}{3}\ell_{nano}C_{v,nano}u_{nano}\phi\frac{dT}{dz} = -k_{nano}\phi\frac{dT}{dz}, \quad (10)$$

where k_{nano}, ℓ_{nano} and u_{nano} are the thermal conductivity of the suspended nanoparticles, the mean free path and the mean speed of the electrons or phonons, respectively. The Kapitza resistance is considered in order to evaluate the thermal conductivity of the suspended nanoparticle and the effect of nanoparticle size, and is given by

$$k_{nano} = \beta k_p, \quad (11)$$

where $\beta \sim O(10^{-2})$ is a constant related to the Kapitza resistance per unit area.

The third mode concerns the collision of nanoparticles with each other via Brownian motion. The translational speed of the of the nanoparticle may be calculated from kinetic theory, and from an order-of-magnitude analysis it may be deduced that this mode is smaller than the other three modes and may be neglected.

The last mode of the energy transport is the thermal interactions of the randomly moving nanoparticles with the base fluid molecules by thermally-induced fluctuations. They postulate that the Brownian motion of the nanoparticles in nanofluids produces convection-like effects at the nanoscale. Hence the Brownian-motion-induced nanoconvection energy flux for the last mode may be defined by

$$q = -h\delta_T\phi\frac{dT}{dz}, \quad (12)$$

where δ_T is the thickness of the thermal boundary layer, estimated from $\delta_T \sim \delta/Pr \sim d_f/Pr$, where δ is the the hydrodynamic boundary layer thickness. For typical nanofluids, the Reynolds and Prandtl numbers are of the order of 1 and 10, respectively, and the Nusselt number for the flow past a sphere, Eq. (7), is simplified to $Nu \sim (RePr)^2$. The heat transfer coefficient for the flow past nanoparticles is then defined by

$$h \sim \frac{k_f}{d_p}(RePr)^2. \quad (13)$$

The expression for the thermal conductivity of nanofluids k_{eff} is derived as a function of k_p, k_f, ϕ, d_p, d_f, Re and Pr, as shown in Table 4. As the particle size is decreased or the temperature is increased, the random motions become larger in magnitude and convection-like effects dominate. This model is able to predict a particle size- and temperature-dependent thermal conductivity.

Sabbaghzadeh *et al.* [30] follow the same approach as Jang and Choi's model [27,28] and in addition consider the effects of liquid layering and the nanolayer structure [32] for the case of a cylindrical nanoparticle suspension. They assume a linear distribution of the thermal conductivity of the nanolayer and a fixed nanolayer thickness. The heat transfer coefficient h is given by Fand [33] for a flow direction perpendicular to the cylinder axes where $10^{-1} < Re < 10^5$ as

$$h = \frac{k_f}{2(r_p + d)}(0.35 + 0.56Re^{0.52})Pr^{0.3}, \quad (14)$$

where d is the nanolayer thickness, $Re = u_p L/\nu$, L is the specific length of the cylindrical shape calculated from $L = \sqrt[3]{\pi r_p^2 l (r_p + d)^2/(r_p^2 \phi)}$ and l is the length of the cylinder. This process links k_{eff} with temperature through the Reynolds number. Their model successfully shows that the effective thermal conductivity increases with decreasing nanoparticle diameter if the nanolayer thickness increases, and increases with the volume fraction of the nanoparticle. However, they didn't plot the variation of the effective thermal conductivity with temperature. We found that the contribution from the thermal interaction of dynamic complex nanoparticles with the base fluid molecules is $O(10^{-5})$ for their model, which is very small compared to the other processes. As a result, the relationship between k_{eff} and temperature T is insignificant.

Kumar *et al.* [19] divide the heat transfer in nanofluids into two portions only, namely by the fluid and the particles. From the one-dimensional heat flux of the parallel model, they found

$$q = -k_p A_p \left(\frac{dT}{dx}\right)_p - k_f A_f \left(\frac{dT}{dx}\right)_f. \tag{15}$$

Assuming that the nanofluid system is in thermodynamic equilibrium such that

$$\left(\frac{dT}{dx}\right)_p = \left(\frac{dT}{dx}\right)_f = \left(\frac{dT}{dx}\right), \tag{16}$$

and that the ratio of the heat transfer areas A_p/A_f is proportional to the total surface area of the nanoparticles and the liquid species per unit volume of the suspension, the effective thermal conductivity of the nanofluid is expressed as

$$q = -k_{eff} A_m \frac{dT}{dx}, \tag{17}$$

where

$$k_{eff} = k_f \left[1 + \frac{k_p \phi r_f}{k_f (1-\phi) r_p}\right]. \tag{18}$$

Kumar *et al.* [19] also postulated that the effective thermal conductivity of the particle k_p is directly proportional to its mean velocity according to kinetic theory, $k_p = c u_p$, where c is a constant. The velocity of the particles u_p is taken to be that due to Brownian motion at a given temperature of the suspension and may be calculated using the Stokes-Einstein formula $u_p = 2k_B T/(\pi \mu d_p^2)$. After substituting $k_p = c u_p$ into Eq. (17), a comprehensive model is derived which explains the enhancement of the thermal conductivity of a nanofluid with respect to variations in particle size, particle volume fraction and temperature. However, the assumption $k_p = c u_p$ results in a zero thermal conductivity of a nanoparticle when the mean velocity of the particle is zero - this is clearly incorrect, since the thermal conductivity of the particle should be equal to its own property value.

To circumvent this error, Patel *et al.* [26] improved Kumar *et al*'s model [19] by introducing the concept of micro-convection. They assumed that heat flows through the nanofluid via conduction through both the solid and liquid and advection from the Brownian motion of the particles. The motion of particles inside the liquid is modelled as flow past a sphere, and from the 1D parallel model, the overall heat transfer is written as

$$q = q_p + q_f + q_{adv} = -k_p A_p \left(\frac{dT}{dx}\right)_p - k_f A_f \left(\frac{dT}{dx}\right)_f + \frac{1}{3} h A_p \Delta T. \tag{19}$$

As the size of the the particles is of the order of 10^{-7} to 10^{-8} m and the Brownian motion velocity is also very low, Pe^2 is negligible compared to Pe. As the conduction through the liquid and the particle conduction is already separately accounted for in q_f and q_p, Patel *et al.* [26] simplified the classical solution of the Nusselt number for the flow past a sphere Eq. (8) via

$$Nu = 0.5Pe. \tag{20}$$

They postulate that the thermal conductivity of the fluid is $k_f = ck_p$, where c is constant, which circumvents the zero thermal conductivity of the nanoparticles at $T = 0$ K in Kumar *et al*'s [19] model. The Nusselt number is then defined as

$$Nu = \frac{hd_p}{ck_p}. \tag{21}$$

The new semi-empirical model of the effective thermal conductivity is obtained by deriving h from Eqs. (20) and (21) and substituting all relevant parameters for the situation of micro-heat convection of flow over a sphere, yielding

$$\frac{k_{eff} - k_f}{k_f} = \frac{k_p d_f \phi}{k_f d_p (1-\phi)} \left(1 + \frac{c u_p d_p}{\alpha_f}\right), \tag{22}$$

where α_f is the thermal diffusivity of the fluid and the constant c is determined experimentally. The Brownian velocity of the particles u_p, which causes the convection, is calculated the same way as in Kumar *et al.* [19], that is from the Stokes-Einstein formula using the particle diameter d_p as the characteristic length. In their work, $c = 25,000$ gives accurate predictions for a wide range of experiments.

Ren *et al.* [29] presented a model that considered four heat transport modes; heat flux through the nanoparticle; heat flux through the nanolayer; heat flux through caused by micro-convection; and heat flux through the base fluid - $q = q_p + q_l + q_c + q_f$. They assumed a linear linear function for the thermal conductivity of the nanolayer between the values for the nanoparticle and the base fluid, the thickness of which is a variable. Expressions for q_f, q_p and q_l were based on classical theories, while they derived an expression for the heat transfer from micro-convection. In their model q_c is described as a function of the modified Péclet number $Pe = u_p L/\alpha_f \phi_T^{0.75}$ via

$$q_c = -k_f \nabla T F(Pe), \tag{23}$$

where the function F is given by

$$F(Pe) = 0.0556Pe + 0.1649Pe^2 - 0.0391Pe^3 + 0.0034Pe^4. \tag{24}$$

The specific length L is calculated from $L = (r_p + d_l)\sqrt[3]{4\pi/(3\phi_T)}$, and the total volume fraction of the original nanoparticle and the nanolayer is given by $\phi_T = \phi(1 + d_l/r_p)^3$. The particle velocity is estimated from the kinetic theory of a monatomic gas molecule, $u_p = \sqrt{3k_B T/m_c}$, where m_c is the mass of the complex nanoparticle. The results fit well for ceramic particle suspensions. The model reveals that the enhancement in the effective thermal conductivity increases with the decreasing nanoparticle radius and temperature, due primarily to the contribution of the particle motion.

Prasher *et al.* [31] analyzed convection in the liquid where a sphere is imbedded in a semi-infinite medium. They considered the effects of possible mechanisms for the thermal energy transfer in nanofluids, and an order-of-magnitude analysis demonstrated that local convection caused by the Brownian motion of the nanoparticles is the only important mechanism to explain the thermal conductivity enhancement. The heat transfer coefficient h is calculated from Eq. (8) as

$$h = \frac{k_f}{r_p}\left(1 + \frac{RePr}{4}\right). \tag{25}$$

If a sphere is imbedded in a semi-infinite medium of thermal conductivity k_m, then Nu based on the radius of the particle ($r_p = d_p/2$) is given to be 1, that is $h = k_m/r_p$ according to [34]. Substituting this h into the left side of Eq. (25) gives the new effective thermal conductivity k_m of the fluid due to the micro-convection caused by the movement of a single sphere as

$$k_m = k_f\left(1 + \frac{RePr}{4}\right). \tag{26}$$

The root-mean-square velocity of a Brownian particle from the kinetic theory of gases is used as a particle velocity for modelling the thermal conductivity of nanoparticles, $u_p = \sqrt{3k_BT/(m_p)}$. Substituting the estimated k_m due to the convection contribution into the Maxwell-Garnett model, the theory is extended to predict the thermal conductivity k_{eff} of a nanofluid which includes the conduction contribution of the particles, the thermal interfacial resistance and the Brownian convection contribution from a single nanoparticle, and is given by

$$\frac{k}{k_f} = \left(1 + \frac{RePr}{4}\right)\left[\frac{(1+2a) + 2\phi(1-a)}{(1+2a) - \phi(1-a)}\right], \tag{27}$$

where $a = 2R_bk_f/d_p$ and R_b is the interfacial resistance between the fluid and a nanoparticle. However, their analysis showed that even at very small values of ϕ the flow field due to two particles will interact, and Eq. (27) needs to be modified to reflect this. Heat transport in a nanofluid systems is similar to the particle-to-fluid heat transfer in fluidized beds. They followed the findings of Brodkey *et al.* [35] for the Nu-correlation for the particle-to-fluid heat transfer in fluidized beds, and proposed a general correlation for h due to the presence of multiple nanoparticles. With two introduced empirical coefficients C and m, this modification leads to

$$\frac{k}{k_f} = (1 + CRe^m Pr^{0.333}\phi)\left[\frac{(1+2a) + 2\phi(1-a)}{(1+2a) - \phi(1-a)}\right]. \tag{28}$$

They found that $m = 2.5 \pm 15$ % is the best value for water-based nanofluids, and $C = 40,000$ matched experimental data well.

5.2. Models of the Additive Type

In a quiescent suspension, nanoparticles move randomly and carry a relatively large volume of the surrounding fluid with them, which may result in a heat flux between hot and cold regions. A group of researchers [36–39] propose that the effective thermal conductivity of nanofluids k_{eff} is composed of the particle's conventional static part k_{static} and a Brownian motion part $k_{Brownian}$, that is

$$k_{eff} = k_{static} + k_{Brownian}. \tag{29}$$

The stationary effective thermal conductivity models assume no bulk motion of the fluid, and k_{static} is predicted from traditional effective medium theories such as those of Maxwell [2] and Hamilton and Crosser (HC) [3]. In a nanoparticle-fluid mixture, microscopic forces can be significant; forces such as the Van der Waals force, the electrostatic force resulting from the electric double layer at the particle surface, the stochastic force that gives rise to the Brownian motion of particles, and of course hydrodynamic forces. The motion of the particles and the fluid are induced and affected by the collective effect of these forces. At present, these forces have not been calculated accurately, and only $k_{Brownian}$ due to the convective effect from the particle Brownian motion have been studied in detail.

Wang *et al.* [36] propose that $k_{Brownian}$ consists of the rotational and translational motions of a spherical particle k_r and k_t, respectively. They can be estimated from provided functions of k_p, k_f, ϕ and the modified Peclet numbers Pe_f and Pe_t. Pe_f is defined as $Pe_f = r_p^2 \gamma \rho_f C_{p,f}/k_f$, where γ is the velocity gradient calculated from the mean Brownian motion velocity and the average distance between particles, and ρ_f and $C_{p,f}$ are the base fluid density and specific heat, respectively. Pe_t is defined as $Pe_t = (u_p L \rho_f C_{p,f}/k_f)\phi^{3/4}$, where $L = (r_p/\phi^{1/3})(4\pi/3)^{1/3}$. Their calculation reveals that the thermal conductivity increase by Brownian motion is less than 0.5 % up to a volume fraction of 10 % for an Al_2O_3-liquid mixture. They conclude that Brownian motion does not contribute significantly to the energy transport in nanofluids. Thus the variation of the thermal conductivity of nanofluids with temperature is insignificant based on their model.

Xu *et al.* [37] presented a new model by taking into account the fractal distribution of nanoparticle sizes and the heat convection between the nanoparticles and the fluid due to the Brownian motion of the nanoparticles. Combined with the HC model [3] for stationary nanoparticles, they proposed a heat transfer model by micro-convection of all the particles in the fluid. The heat flux due to the Brownian motion of the nanoparticles is expressed as an integral of the heat transfer by a single particle assuming thermal equilibrium, that is

$$\begin{aligned} q_c &= -\int_{d_{p,min}}^{d_{p,max}} hA_p(T_p - T_f)dN \\ &= -\frac{\Delta T}{\delta_T}\int_{d_{p,min}}^{d_{p,max}} hA_p\delta_T dN, \end{aligned} \tag{30}$$

where $d_{p,min}$ and $d_{p,max}$ are the minimum and maximum diameters of the nanoparticle, respectively, h_p is the heat transferred by convection from a single nanoparticle moving in the liquid, A_p is the surface area of a particle and the thickness of the thermal boundary layer is $\delta_T \sim \delta/Pr = 3d_f/Pr$. The fractal power law for the size distribution of the nanoparticles in nanofluids is introduced in the system via

$$-dN = D_f d_{p,max}^{D_f} d_p^{-(D_f+1)} d(d_p) > 0, \tag{31}$$

where D_f is the fractal dimension determined by $D_f = 2 - \ln\phi/\ln(d_{p,min}/d_{p,max})$. Since $Re \ll 1$ Tomotika *et al.*'s [25] low-Reynolds number expression for the Nusselt number

$Nu = 2.0 + 0.5RePr + O(Re^2Pr^2) \approx 2.0$ is applied to a single particle, which implies $h = k_f Nu/d_p$. Substituting expressions for h and dN into Eq. (30), the equivalent thermal conductivity contributed by heat convection is approximated by

$$k_{Brownian} = \frac{q_c}{-A\frac{\Delta T}{\delta_T}} = \frac{\int_{d_{p,min}}^{d_{p,max}} hA_p\delta_T dN}{\int_{d_{p,min}}^{d_{p,max}} A_p dN}, \tag{32}$$

where A is the total area of all the nanoparticles. By introducing an empirical constant c for the thickness of the thermal boundary layer such that $\delta_T = c\delta/Pr$ and a viscosity given by $\mu = \mu_o 10^{B/(T-C)}$ into the Prandtl number $Pr = \mu c_p/k_f$, the model shows that the thermal conductivity of a nanofluid due to heat convection by Brownian motion of the nanoparticles is proportional to the temperature and the reciprocal of the nanoparticle size d_p^{-1}, via

$$\frac{k_{Brownian}}{k_f} = c\frac{d_f k_f Nu\,(2 - D_f)\left[\left(\frac{d_{p,max}}{d_{p,min}}\right)^{1-D_f} - 1\right]^2}{10^{\frac{B}{T-C}} AC_p\,(1 - D_f)\,d_{p,min}\left[\left(\frac{d_{p,max}}{d_{p,min}}\right)^{2-D_f} - 1\right]}. \tag{33}$$

It is found that c is around $O(10^2)$. Their model suggests that the maximum value of the contribution from the heat convection due to the nanoparticle's Brownian motion is about 21 % and 2.1 % for nanoparticles with an average diameter of 5 and 60 nm, respectively. There is a critical concentration of nanoparticles at which $k_{Brownian}$ reaches a maximum value, below which $k_{Brownian}$ increases with concentration and above which $k_{Brownian}$ decreases with concentration. It is found that this critical concentration is 12.6 % for all nanoparticles. It is believed that at higher concentrations, the nanoparticles may easily agglomerate to form clusters, and the strength of the Brownian motion is thus weakened, leading to the decrease in heat convection.

Shukla and Dhir [38] describe the Brownian motion of the nanoparticles using Langevin dynamics. They derived an explicit expression for the thermal conductivity contribution $k_{Brownian}$ for N spherical Brownian particles suspended in a system occupying volume V at an equilibrium temperature T using the Green-Kubo relation

$$k_{Brownian} = \frac{V}{k_B T^2}\int_0^\infty \langle J_x(\tau)J_x(0)\rangle d\tau = C\frac{\phi\,(T - T_0)}{\mu r_p^4}, \tag{34}$$

where C and T_0 are constants to fit the individual solid/liquid combination. T_0 represents a reference temperature, below which the contribution of Brownian motion to the effective thermal conductivity is negligible. In their work $C = 7.0 \times 10^{-36}$ and $T_0 = 294$ K for an alumina/water nanofluid, and $C = 1.5 \times 10^{-33}$ and $T_0 = 273$ K for an gold/water nanofluid. Reasonable agreement with experimental data is obtained for the variation of the effective thermal conductivity enhancement with temperature

Koo and Kleinstreuer [39] propose a model where k_{static} is calculated from Maxwell theory. To derive $k_{Brownian}$, they consider two nanoparticles with translational time-averaged Brownian velocities u_p in two different temperature fields of extent l, where l is the average distance a particle travels in one direction without changing its direction due to the Brownian motion of the other particles. The relationship between the translational time-average speed u_p and temperature T is described by the real-gas conduction law

$$u_p = \sqrt{\frac{18 k_B T}{\pi \rho d_p^3}}, \tag{35}$$

where ρ is the particle density and d_p is its diameter. They introduce a probability p for a particle to travel along any direction, and assume that each of the two particle cells are in thermal equilibrium at temperatures T_1 and T_2. The energy the particles carry across the interface from multiple particles moving to neighboring cells is deduced to be

$$q = \frac{\Delta Q}{A \Delta t} \approx -p\phi\rho C_v u_p d_p \nabla T = -k_{Brownian} \nabla T. \tag{36}$$

They argue that each Brownian particle generates a velocity field in the surrounding fluid described by Stokes flow past a sphere. The ability of a large volume of fluid dragged by the nanoparticle to carry a substantial amount of heat is represented by the fraction of liquid volume β_1 which travels with a particle and a factorial function $f(T, \phi)$. Together with Eq. (36), and the expected value of the probability for particles to move in one direction as $p = 0.197$, the model is described by

$$k_{Brownian} = \left(5 \times 10^4\right) \beta_1 \phi \rho C_p f(T, \phi) \sqrt{\frac{k_B T}{\rho D}}. \tag{37}$$

Both β_1 and $f(T, \phi)$ are determined from experimental data. The function $f(T, \phi)$ is given by

$$f(T, \phi) = (-6.04\phi + 0.4705)T + (1722.3\phi - 134.63), \tag{38}$$

while β_1 has the form

$$\beta_1 = A\phi^{-B}, \tag{39}$$

where$A,\ B > 0$ are experimentally determined constants.

5.3. Models of the Correlation Type

The last group of models considered are of the correlation type [17,20], which are fitted to experimental data using classical linear regression techniques. Chon *et al.* [20] provide an empirical correlation to describe the effect of individual physical parameters on the thermal conductivity of a nanofluid. The thermal conductivity of a nanofluid is derived as a function of Pr and Re, which are dependent on temperature, and is given by

$$\frac{k_{eff}}{k_f} = 1 + 64.7\phi^{0.7460} \left(\frac{d_f}{d_p}\right)^{0.3690} \left(\frac{k_p}{k_f}\right)^{0.7476} Pr^{0.9955} Re^{1.2321}, \tag{40}$$

where the temperature-dependence of the viscosity $\mu = \mu_o 10^{B/(T-C)}$ is included in the estimation of Pr and Re, and the Brownian velocity of the nanoparticles is based on Einstein's diffusion theory by using the mean free path of liquid molecules as a characteristic length. Thus, the empirical correlation of k_{eff} can explicitly show the effect of nanofluid temperature and size as

$$\frac{k_{eff}}{k_f} = 1 + c\left(\frac{Pr\,(T)^{0.9955}\,T^{1.2321}}{d_p^{0.369}k_f\,(T)^{0.7476}\,\mu\,(T)^2}\right) = 1 + c\left(\frac{1}{d_p}\right)^{0.369}\left(\frac{T^{1.2321}}{10^{\frac{2.4642B}{T-C}}}\right), \quad (41)$$

where c is an empirical coefficient. The model is fitted to the thermal conductivity measurements from three different batches of fairly mono-dispersed Al_2O_3 samples. The most important finding of their work is the effect of nanoparticle mobility, or u_p, is a dominant function of temperature for a given particle size, and that Re dominates the temperature dependence whereas Pr shows a slightly decreasing temperature dependence.

Li and Peterson [17] applied a two-factor linear regression analysis in the measured effective thermal conductivity of CuO and Al_2O_3 nanoparticle suspensions to examine the effects of variations in temperature and volume. For the Al_2O_3/water nanofluid they found

$$\frac{k_{eff} - k_f}{k_f} = 0.7644481464\phi + 0.018688867T - 0.462147175,\ R^2 = 0.9171, \quad (42)$$

where the temperature T is measured in °C. For the CuO/water nanofluid they found

$$\frac{k_{eff} - k_f}{k_f} = 3.761088\phi + 0.0017924T - 0.30734,\ R^2 = 0.9078. \quad (43)$$

It may be noted that the coefficient for the volume fraction is much larger than that for temperature for both correlations, indicating the relative importance of these two parameters.

6. Conclusions

Various models and mechanisms that depend on an extension of classical theory involving the movement of the particles has been developed in an effort to explain the enhancement and temperature-dependence on the thermal conductivity of nanofluids. The measurements show that in a nanofluid containing spherical nanoparticles or MWCNT bundles with a high aspect ratio, the thermal conductivity enhancement increases rapidly by two-to-four orders of magnitude with increasing temperature over a temperature range of 20 °C to 50 °C. In contrast, the thermal conductivity of a nanorod nanofluid as examined by Yang and Han [8] decreases with increasing temperature. For a low volume fraction of SWCNT suspensions [24], the thermal conductivity enhancement was found to be nearly insensitive to temperature.

As shown in Sections 2 and 3, the thermal conductivity of liquids such as toluene and oil usually decrease with increasing temperature, while the thermal conductivities of nanoparticles, both metallic and metallic oxide, decrease with temperature. If we simply use a linear function $k_f = A_f - B_fT$ or $k_p = A_p - B_pT$ to represent temperature-dependence of the thermal conductivity of the nanoparticles and the base fluid over the temperature range of 5 °C to 50 °C and substitute it into any traditional effective medium theory, such as those of Maxwell [2] and Hamilton and Crosser [3] and calculate the weighted average volume, the effective thermal conductivity of a nanofluid always appears as a decreasing function of temperature, even for an aqueous suspension when the volume fraction of the nanoparticles ϕ is less than 10 %. Although $k_f = A_f + B_f T$ for water and a small loading of nanoparticles (< 1 % volume fraction), k_{eff} of an aqueous suspension still appears as a decreasing function of temperature. As a result, it is clear that the effective thermal conductivity of nanofluids is not a simple function of the combined properties of the base fluid and the nanoparticles. This analysis indicates that the traditional effective medium theories cannot explain the phenomena of thermal conductivity variation with temperature.

Many authors believe that Brownian motion is the key mechanism which affect the variation of the thermal conductivity with temperature. The existing models are able to show this phenomena, that is, the higher the temperature and the smaller the average size of the nanoparticles in a fluid, the higher the velocity of the nanoparticle's Brownian motion. Xu *et al.* [37] found that the contribution from the heat convection due to a nanoparticle's Brownian motion is between 2.1 % and 21 % for nanoparticles with an average diameters of 5 and 60 nm, respectively. The intensive Brownian particle motion causes a high heat flux along with micro-convention currents, thus it improves the heat transfer performance.

A nanoparticle's Brownian velocity u_p is considered to be the single dominating function describing the temperature-dependence of the thermal conductivity of a nanofluid, dominating over other parameters such as k_f and Pr. It has been noticed that models lack consistency in the formulation of the nanoparticle Brownian velocity, u_p. Some authors estimate u_p from the kinetic theory of dilute gases, which is $u_p \propto \sqrt{T}$ - the validity of this estimate for dense liquids is questionable. Others use the Stokes-Einstein formula to estimate $u_p \propto T/d_p$. However, in the postulated micro-convection model, to account for the heat transfer between the particles and the fluid the Reynolds number, defined as $Re = u_p d_p/\nu$, becomes independent on the particle diameter based on u_p estimated from the Stokes-Einstein formula. However, Re is linearly dependent on the particle diameter in the correlation for the flow past a solid sphere.

Some researchers believe that the interaction of the flow field due to the movement of multiple nanoparticles is important even at small values of ϕ [31], but others believe that it is insignificant in dilute cases where $\phi < 1$ % due to retardation effect on the inter-particle potential [39], and can be neglected.

In the postulated micro-convection model to account for the heat transfer between particles and fluid, based on the theory of Tomotika *et al.* [25], the Nusselt number can be expressed as $Nu = 2.0 + 0.5RePr + O(Re^2Pr^2)$. We have noted that each investigator applies it in their own model in different ways. Some assume $Re \ll 1$ and use $Nu = 2$ [37], others $Nu = 0.5RePr$ with the assumption $Pe^2 \ll Pe$ [26], and even others assume the Reynolds number and the Prandtl number are of the order 1 and 10 and use $Nu \sim Re^2Pr^2$ [27,28]. The validity of these simplified Nusselt number correlations is questionable.

Several authors have argued that the large thermal conductivity increase is due to the hydrodynamic effects of the large volume of fluid dragged by the Brownian motion of the nanoparticles [36,40,41]. The calculation by Wang *et al.* [36] shows that up to a volume fraction of 10 % the thermal conductivity increase by Brownian motion is less than 0.5 % for an Al_2O_3-liquid mixture. They concluded that Brownian motion does not contribute significantly to the energy transport in a nanoparticle-fluid mixture. Evans *et al.* [40] uses a kinetic theory-based analysis of heat flow in nanofluids. Their results demonstrate that the hydrodynamics effects associated with Brownian motion have a only minor effect on the effective thermal conductivity of nanofluids. Their analysis is also supported by the results of molecular dynamics simulations of heat flow in a model of a nanofluid. Keblinski *et al.* [41] compares the time scale of a particles motion with that of heat diffusion in the liquid. The comparison shows that the thermal diffusion is much faster than the Brownian diffusion, even in the limit of extremely small particles. Their analysis demonstrates that the movement of nanoparticle due to Brownian motion is too slow to transport a significant amount of heat through a nanofluid, a conclusion supported by the results of their molecular dynamic simulations.

As mentioned above, several arguments presented in these models are neither consistent nor convincing. It is obvious that success of these models has been very limited, and no concrete conclusions have been reached. Due to the complexities in the convective currents arising from a system comprising multiple nanoparticles, most existing extended models for the prediction of the effective thermal conductivity of a nanofluid are semi-empirical in nature. Based on Yang and Han's analysis [8], the thermal conductivity enhancement of a nanofluid containing a moderate aspect ratio is attributed to the combined effects of the particles Brownian motion and diffusive heat conduction. According to Sabbaghzade *et al.*'s [30] model for cylindrical nanoparticles, we found that the contribution from the heat convection is about $0.004\ \%$ for a volume fraction of $\phi = 1\ \%$ of carbon nanotubes (diameter = 25 nm, length = 50 μm). The nanoparticles with a cylindrical shape are generally bulky and less mobile than spherical nanoparticles. A more comprehensive study is needed to explore the effects of particle shape on the Brownian motion in nanofluids.

Despite the theoretical effort on the subject, a generally accepted theory for the temperature-dependence of k_{eff} has yet to be established. A unified theoretical model is needed for the temperature-dependence of the thermal conductivity of nanofluids which matches well with experimental results. Accounting for Brownian motion is the predominant mechanism for spherical nanoparticles, while the diffusive heat conduction mechanism will become dominant as the particle aspect ratio increases. To improve the accuracy and versatility of the temperature-dependency of the effective thermal conductivity of nanofluids, more experimental data sets, in particular for non-spherical particle suspensions, are required.

Acknowledgments

The authors would like to acknowledge the support of the Discovery Project scheme of the Australian Research Council. JMH is grateful to the Australian Research Council for provision of an Australian Professorial Fellowship.

Nomenclature

a	constant
A	surface area, m^2/constant
B	constant
c	constant

Nomenclature. Continued

C	constant
C_p	specific heat capacity at constant pressure, $\text{Jg}^{-1}\text{K}^{-1}$
C_v	specific heat capacity at constant volume, $\text{Jm}^{-3}\text{K}^{-1}$
d	diameter/thickness, m
D_f	fractal dimension
D_0	nanoparticle thermal diffusion coefficient, m^2/s
F	defined function
$F(\omega)$	phonon density of states
h	convective heat transfer coefficient, $\text{Wm}^{-2}\text{K}^{-1}$
$\hbar$	Planck's constant, 1.05459×10^{-34} Js
J_x	heat current vector in x-direction
k	thermal conductivity of material, $\text{Wm}^{-1}\text{K}^{-1}$
k_B	Boltzmann constant, 1.38×10^{-23} JK^{-1}
K	shape factor
ℓ	mean free path, m
L	length/specific length, m
m	mass, g
M	molecular mass
n	constant
N	number of particles or atoms
Pe	Péclet number
Pr	Prandtl number
q	heat flux
r	radius, m
R_b	thermal resistance, Km^2W^{-1}
Re	Reynolds number
T	temperature, K
T_θ	Debye temperature, K
u	velocity, m/s
v_s	speed of sound, m/s
V	volume, m^3
Greek and symbols	
α	thermal diffusion coefficient of fluid, m^2/s
β	constant related to the Kapitza resistance
γ	velocity gradient
γ_G	Grüneisen constant
δ	hydrodynamic boundary layer thickness, m
δ_T	thermal boundary layer thickness, m
ν	kinematic viscosity of fluid, m^2/s
μ	dynamic viscosity of fluid, Pa.s

Nomenclature. Continued

ω	angular frequency, s^{-1}
Ω_a	atomic volume
ϕ	volume fraction of loaded particles
ϕ_T	volume fraction of loaded particles and nanolayer
Φ	sphericity of particles
ρ	density of material, g/cm^3
τ	relaxation time, s
$\hbar$	Planck's constant, 1.05459×10^{-34} Js
Subscripts	
adv	advective component
c	cluster/convestion
$Brownian$	Brownian motion component
e	electron
eff	effective thermal conductivity
f	base fluid
l	nanolayer
max	maximum
min	maximum
$nano$	nanolayer
p	nanoparticle
ph	phonon
r	rotational component
s	sound
$static$	static component
t	translational component

References

[1] Marquis, E. D. S & Chibante, L. P. F. Improving the heat transfer of nanofluids and nanolubricants with carbon nanotubes. *The Journal of the Minerals, Metals & Materials Society*. 2005, vol. 57, pp. 32-43.

[2] Maxwell, J. C. A *Treatise on Electricity and Magnetism*. 3rd ed.; Oxford: Clarendon Press, 1892, vol. 1, pp. 440-435.

[3] Hamilton, R. L. & Crosser, O. K. Thermal conductivity of heterogeneous two-component systems. *Industrial & Engineering Chemistry Fundamentals*. 1962, vol. 1, pp. 187-191.

[4] Avsec, J. & Oblak, M. The calculation of thermal conductivity, viscosity and thermodynamic properties for nanofluids on the basis of statistical nanomechanics. *International Journal of Heat and Mass Transfer*. 2007, vol. 50, pp. 4331-4341.

[5] Chung, T., Ajlan, M., Lee, L. L. & Starling, K. E. Generalized multiparameter correlation for nonpolar and polar fluid transport properties. *Industrial & Engineering Chemistry Research.* 1988, vol. 27, pp. 671-679.

[6] Chung, T., Lee, L. L. & Starling, K. E. Applications of kinetic gas theories and multiparameter correlation for prediction of dilute gas viscosity and thermal conductivity. *Industrial & Engineering Chemistry Fundamentals.* 1984, vol. 23, pp. 8-13.

[7] Lide, D. R. & Kehiaian, H. V. CRC *Handbook of thermophysical and thermochemical data.* ISBN-0849301971. Boca Raton: CRC Press: USA, FL, 1994; pp. 417-419.

[8] Yang, B. & Han, Z. H. Temperature-dependent thermal conductivity of nanorod-based nanofluids. *Applied Physics Letters*. 2006, vol. 89, 083111.

[9] Okhotin, A. S., Zhmakin, L. I. & Ivamyuk, A. P. The temperature dependence of thermal conductivity of some chemical elements. *Experimental Thermal and Fluid Science*. 1991, vol. 4, pp. 289-300.

[10] Chen, G. *Nanoscale energy transport and conversion;* ISBN-10:019515942X; Oxford University Press; USA, 2005, pp18-22.

[11] Kittel, C. *Introduction to Solid State Physics,* Eighth edition; ISBN 0-471-41526-X; John Wiley & Sons, Inc: USA, NJ, 2005, pp121-128.

[12] Berber, S., Kwon, Y. K. & Tománek, D. Unusually high thermal conductivity of carbon nanotubes. *Physics Review Letters.* 2000, vol. 84, pp. 4613-4616.

[13] Kim, P., Shi, L., Majumdar, A. & McEuen, P. L. Thermal transport measurement of individual multiwalled nanotubes. *Physics Review Letters.* 2001, vol. 87, 215502.

[14] Ho, G.Y.; Power, R.W.; Liley, P.E. Thermal conductivity of the elements: A Comprehensive Review. *J. Phys. Chem. Res. Data.* 1974, vol.4, suppl.1, 340.

[15] Shackelford, J.F.; Alexander, W. *CRC Material Science and Engineering Handbook*; ISBN-0849326966; Boca Raton: CRC Press: USA, FL, 2001:252.

[16] Das, S. K., Putra, N., Thiesen, P. & Roetzel, W. Temperature dependence of thermal conductivity enhancement for nanofluids, *Journal of Heat Transfer.* 2003, vol. 125, pp. 567-574.

[17] Li, C. H. & Peterson, G. P. Experimental investigation of temperature and volume fraction variations on the effective thermal conductivity of nanoparticle suspensions (nanofluids). *Journal of Applied Physics.* 2006, vol. 99, 084314.

[18] Patel, H. E., Das, S. K., Sundararajan, T., Nair, A. S., George, B. & Pradeep, T. Thermal conductivities of naked and monolayer protected metal nanoparticle based nanofluids: Manifestation of anomalous enhancement and chemical effects. *Applied Physics Letters*. 2003, vol. 83, pp. 2931-2933.

[19] Kumar, D. H., Patel, H. E., Kumar, V. R. R., Sundararajan, T., Pradeep, T. & Das, S. K. Model for heat conduction in nanofluids. *Physics Review Letters*. 2004, vol. 93, 144301.

[20] Chon, C. H., Kihm, K. D., Lee, S. P. & Choi, S. U. S. Empirical correction finding the role of temperature and particle size for nanofluid (Al_2O_3) thermal conductivity enhancement. *Applied Physics Letters*. 2005, vol. 87, 153107.

[21] Zhang, X., Gu, H. & Fujii, M. Experimental study on the effective thermal conductivity and thermal diffusivity of nanofluids. *International Journal of Thermophysics*. 2006, vol. 27, pp. 569-580.

[22] Zhang, X., Gu, H. & Fujii, M. Effective thermal conductivity and thermal diffusivity of nanofluids containing spherical and cylindrical nanoparticles. *Journal of Applied Physics*. 2006, vol. 100, 044325.

[23] Wen, D. & Ding, Y. Effective thermal conductivity of aqueous suspensions of carbon nanotubes (carbon nanotube nanofluids). *Journal of Thermophysics and Heat Transfer*. 2004, vol. 18, pp. 481-485.

[24] Kim, B. H. & Peterson, G. P. Effect of morphology of carbon nanotubes on thermal conductivity enhancement of nanofluids. *Journal of Thermophysics and Heat Transfer*. 2007, vol. 21, pp. 451-459.

[25] White, F.M. *Viscous fluid flow*; ISBN 0-07-069710-8; McGraw-Hill Inc.: USA, 1974; pp.212-214.

[26] Patel, H. E., Sundararajan, T., Pradeep, T., Dasgupta, A., Dasgupta, N. & Das, S. K. A micro-convection model for thermal conductivity of nanofluids. Pramana - *Journal of Physics*. 2005, vol. 65, pp. 863-869.

[27] Jang, S. P. & Choi, S. U. S. Role of Brownian motion in the enhanced thermal conductivity of nanofluids. *Applied Physics Letters*. 2004, vol. 84, pp. 4316-4318.

[28] Jang, S. P. & Choi, S. U. S. Effects of various parameters on nanofluid thermal conductivity. *Journal of Heat Transfer*. 2007, vol. 129, pp. 617-623.

[29] Ren, Y., Xie, H. & Cai, A. Effective thermal conductivity of nanofluids containing spherical nanoparticles. *Journal of Physics D: Applied Physics*. 2005, vol. 38, pp. 3958-3961.

[30] Sabbaghzadeh, J. & Ebrahimi, S. Effective thermal conductivity of nanofluids containing cylindrical nanoparticles. *International Journal of Nanoscience*. 2007, vol. 6, pp. 45-49.

[31] Prasher, R. Thermal conductivity of nanoscale colloidal solutions (Nanofluids), *Physics Review Letters*. 2005, vol. 94, 025901.

[32] Tillman, P., & Hill, J. M. Determination of nanolayer thickness for a nanofluid. *International Communications on Heat and Mass Transfer*. 2007, vol. 34, pp. 399-407.

[33] Fand, R. M. Heat transfer by forced convection from a cyclinder to water in crossflow. *International Journal of Heat and Mass Transfer.* 1965, vol. 8, pp. 995-1010.

[34] Incropera, F.P.; DeWitt, D.P. *Fundamentals of heat and mass transfer*, Second edition; ISBN 0-471-51728-3; John Wiley & Sons Inc.; NY, 1990, pp357-409.

[35] Brodkey, R.S.; Kim, D.S.; Sidner, W. Fluid to particle heat transfer in a fluidized bed and to a single particle. *Int. J. Heat Mass transfer* 1991, vol.34, 2327-2337.

[36] Wang, X., Xu, X. & Choi, S. U. S. Thermal conductivity of nanoparticle-fluid mixture. *Journal of Thermophysics and Heat Transfer.* 1999, vol. 13, pp. 474-480.

[37] Xu, J., Yu, B., Zou, M. & Xu, P. A new model for heat conduction of nanofluids based on fractal distributions of nanoparticles. *Journal of Physics D: Applied Physics.* 2006, vol. 39, pp. 4486-4490.

[38] Shukla, R. K. & Dhir, V. K. Study of the effective thermal conductivity of nanofluids. *Proceedings of IMECE2005, 2005 ASME International Mechanical Engineering Congress and Exposition*. IMECE05-80281.

[39] Koo, J. & Kleinstreuer, C. A new thermal conductivity model for nanofluids. *Journal of Nanoparticle Research*. 2004, vol. 6, pp. 577-588.

[40] Evans, W., Fish, J. & Keblinski, P. Role of Brownian motion hydrodynamics on nanofluid thermal conductivity. *Applied Physics Letters.* 2006, vol. 88, 093116.

[41] Keblinski, P., Phillpot, S. R., Choi, S. U. S. & Eastman, J. A. Mechanisms of heat flow in suspensions of nano-sized particles (nanofluids). International *Journal of Heat and Mass Transfer*. 2002, vol. 45, pp. 855-863.

[42] Xuan, Y., Li, Q. & Wu, W. Aggregation structure and thermal conductivity of nanofluids. *AIChE Journal of Thermodynamics*. 2003, vol. 49, pp. 1038-1043.

In: Nanoparticles: New Research
Editor: Simone Luca Lombardi, pp. 307-333

ISBN: 978-1-60456-704-5

Chapter 10

NANOTECHNOLOGY: OBTAINING OF NANOPARTICLES AND NANOCOMPOSITES AND THEIR USE IN FOOD AND DRUG PACKAGING

A. Edwin Moncada[*]

BRASKEM S.A. III Polo Petroquímico, Via Oeste Lote 5 - Passo Raso.
Braskem. S.A. Triunfo, RS, Brasil

Abstract

This work aims to review the immense influence that nanotechnology has having along the last years in all areas of knowledge, as will as, the method of obtaining of nanoparticles, trough of clays modification and Sol-gel synthesis and the used of the nanoparticles obtained in the formation of nanocomposites with polyolefin, also is discussed the interaction nanoparticle-polymer, and the used of a new compatibilizer agent in the properties of nanocomposites. Is presented one of its main applications and the main capital investment from the governments is the packaging industry of food and pharmaceutical products. This is due, among many other factors, to the possibility of using nano-sensors that when combined with the package communicate to the consumer/producer, as well an in control situations, the characteristics of package material such as: microbiological quality, storage conditions (humidity, temperature, light, etc), time-temperature treatment and shelf-life, among others. Finally is shows some trends for the application of nanoparticles.

Introduction

The development of materials has been fundamental to modify the behavior and human culture throughout the time. It is possible to describe the historical evolution periods of human race through advance, manipulation and knowledge of chemistry and physics of materials. That is, how its development can be categorized in different ages: Stone age, Bronze age, Iron age, Steel age (industrial revolution), Polymers age, and Silicone and Silicon

[*] E-mail address: edwin.moncada@braskem.com.br; moncada.edwin@gmail.com. Phone: 55.51.3721.8251, Fax:55.51.3457.1084

age (the revolution of telecommunication). For this reason, materials have been so important to man and the researches in this area present advance in a rate, every time, more accelerated.

This century has been characterized for getting huge advances in every area of research. In the last years have had a growth phenomenon that has centered the attention of scientist all over the world, leading the government to make huge investments in this area, we are talking about: nanotechnologies. These are characterized to allow the manipulation of the matter structure to a smaller scale, in the order of nanometers (nm, thousandth of micrometers, millionth of mm or thousand-millionth of meters), generating materials and structures with different characteristics from those, which are normally used.

The nanotechnologies, today, are like an innovative and radical jump, that will affect the whole economic sectors. The spread effect with extensive dissemination of nanotechnologies is principally due to; the substantial changes caused in the field of material that are used in the whole manufacturing and service sectors, conditioning many time its development. By means of these technologies it is possible to modify the properties of known materials much more radically that what has been achieved until now, as well as creating completely new materials. On the other side, nanotechnologies allow to adjust materials in a scale better reduced than the actual manufacture of microprocessors.

Everything looks like to indicate that we are now in the initial stages of an expansive wave equivalent to that initiated in the first years of 70´s in correlation with the introduction of microprocessors.

The nanotechnology word was presented in 1974, by the researcher Nori Tanoguchi of the Tokio University, who showed the differences between microelectronic and the new engineering field in a submicrometric scale. [1] According to the National Science Foundation (NSF) and the National Nanotechnology Initiative (NNI) of EUA, nanoscience and nanotechnology are defined like the understanding, control, and exploration of materials and systems whose structures and components show physical, chemical and biologist properties and phenomena that are, significantly, new and modified due to their nanometric scale (1-100 nm).

Some questions emerge, such as: Is Nanotechnology a radical technology that involves the construction of macroscopic products through direct manipulation of software? Or are we near to a decade of virtual utopian technology where the materials can be constructed through molecular manufacture? These questions were asked on the Nottingham University Symposium (August 26th 2005), where they were extensively discussed.[2,3] One important point was that: the concept of molecular manufacture was established (enunciated originally by Drexler in 1981 [4]) in which, later on, it was proposed as a molecular nanotechnology (MNT) by Drexler and others.[5,6,7,8]

In the figure 1, it is possible to see the evolution of nanoscience and nanotechnology through the time. The indexed annual publications (where the word nano is on the title of the article) (Web of Science) and international patents found in Internet (Derwent). In the graph, it is possible to see an exponential growth in publications (since 1993, with 955 articles on indexed magazines) and international patents (initiating on 1999, with 296 registered patents). In the year 2006, 26967 international publications were found in indexed magazines and 5369 concessions of international patents.

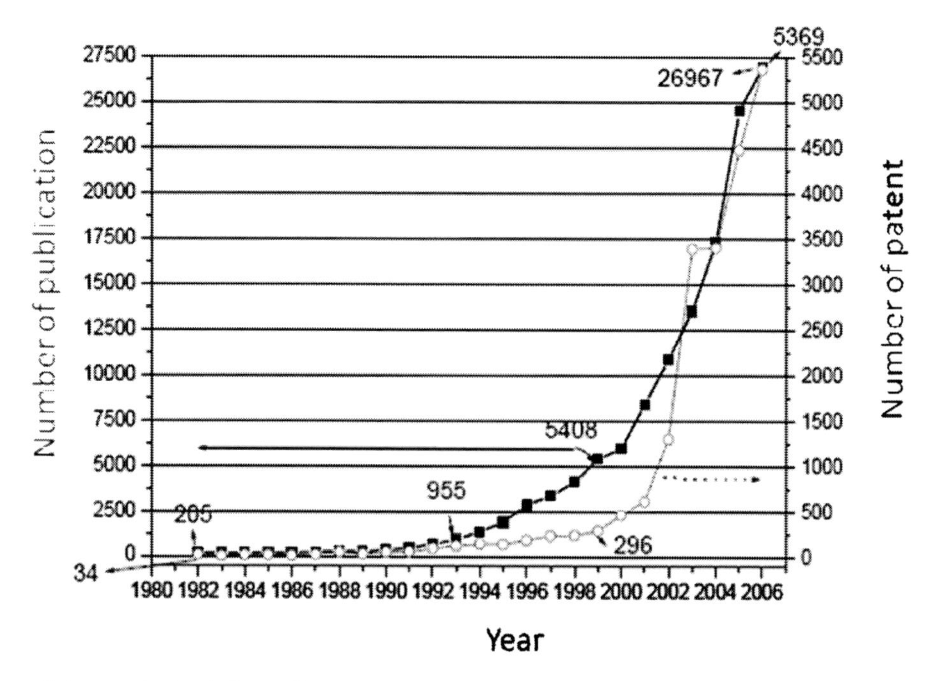

Source: Scioence citation index and Derwent, 2007.

Figure 1. Numbers of scientific publication per year, where the word nano is found on the title of the published work and the international patents of the last 15 years. The search was done on dice base "web of science" and "Derwent" of Thomson Corporation on April 2007 [9].

What said before gives an idea of the great increase in researches, measured by the quantity of articles, industrial applications, quantity of patents and how it will be seen through this chapter, the tendency for the following years continue increasing the amount of publication and patents. Besides the governments invest every time more money to finance, specifically, nanoscience and nanotechnology.

Inside of investigations in this scale of physical greatness, Public Universities, Institutes of Governmental Investigation and some of their departments in several countries in the world, it was used around US$ 4 billions in the year 2005, in the construction of "facilities" and in the establishment of the research and development (P&D) lines. On the other hand companies of P&D are investing more than nine times of used resource in other field of research. In 2005, 1331 bound companies to 76 industries, second to LUX Research Inc/EUA, invested US$ 3,2 billions in nanotechnology (NT) and sold approximately US$ 32 billions in products that used some way of NT. Or CEO, of "General Electric", Jeffrey Immelt fixed nanotechnology (NT), personalized medicine and renewable energy, like the most priority of P&D. [10] According to the published article on February 2006 in the magazine "Science", or amount that is being estimated to 2015, it is of a global market for Nanotechnology (NT) and Nanoscience (NC), around of US$1,0 trillion (NEL, 2006). To May 16th 2007, the products stocktaking to the consumer, organized by the "Woodrow Wilson International Center for Scholars", it already was of 475 produced products, or in production lines, coming from nanotechnology.

Industrial Applications of Nanotechnology

The applications to produce nanostructure materials and nanotechnologies are being developed with extreme speed and a simple list of limited number of applications only gives a small idea of its potentialities. Since long time, there are techniques that allow working in a nanostructural level, like some areas of condensed matter, science of colloid and the growth of superficial coating among others. The great development, in molecular biology fields and biotechnologies since 80´s, has increased its expansion towards every kind of materials: metallic, non-metallic, plastics and compound materials, and through them towards the most diverse fields: scientific, technological and industrials.

There are a lot of areas in science that has been, are and will be implicated in nanotechnologies, such as: medicine, biology, pharmacology and materials with all theirs applications in the engineering sectors like: civil, construction, electronic, mechanics, chemistry and foods engineering among others.

Some examples of innumerable applications of nanotechnology, already present at the market or near to its commercialization are as follows. [11]

- Sensors: They are destined to detect the presence of specific compounds in different environments; closed and opened or aromas that characterize the quality of beverages and food products (electronic nose, intelligent packaging).
- Photovoltaic systems of high efficiency for the conversion of the solar energy.
- New materials with an elevated relation between resistance and mass to airspace, biomedical and transport applications.
- Packaging of food products with better barrier characteristics for gases penetration and capacity to indicate the state of conservation (intelligent packaging).
- Diagnosis techniques based on a system named "lab-on-a-chip" to make the clinical and genetics analysis with minimum sample and on real time.
- Cosmetics, especially to protect against the solar radiation.
- Materials for the filtration and hydrocarbons catalysis and other substances.
- Superficial coatings with resistance to corrosion scratched and notably improved ragged.
- Cut tools with highest tenacity and reduced fragility.
- More functional and lighter screen videos based on electronics of polymers.
- New prosthesis and implant to placement in vivo.
- Work techniques of pieces to micromechanics and microelectronic in a 100 nm scale.

The nanostructured materials can be gotten in two ways: one named "top down" in which the nanostructures are sculptured on a block of material and the other called "bottom up" where the nanostructured materials are gotten through nanoparticles. The "top down" techniques have similarity with the actual technique of productions of electronic microprocessors. "Bottom up" techniques are based on similar processes used in technology of materials and can get powders, compact objects or thin coatings with great changed properties with regard to the same materials gotten by conventional technologies.

Nanoparticles refers to particles where at least one of its space dimensions is in nanometric scale (smaller than 100 nm) and it can be found nanospheres, nanotubes, nanosheet, nanofiber, etc. A possible works listed where the word nanoparticles appears, would be too extensive to be named, but between the more used are those coming from clays, inorganic oxides and concerning ones to the carbon.

It will be presented in the following pages, related concepts to nanotechnology and nanoparticles, some methodologies of obtaining nanoparticles, specifically by means of clays treatments and Sol-gel synthesis method, besides, the possibility to form sensors with these particles and their use in the industry of food and pharmaceutical products, specifically in food and drug packaging.

Nanoparticles Obtain by Clays

Clay is a mineral coming from the decomposition of rocks originated in a natural process during thousands of years. Physically it is considered a colloid of small particle and flat surface. Chemically it is a hydrated alumina silicate which general formula is $Al_2O_32SiO_22H_2O$. It is, also, called aluminosilicatos or philosilicates due to their structure in sheets. The main chemical component of clay is the silicate minerals (SiO4) depending on the structural organization and cations presented by the family of generated clay. The most commonly utilized clays for obtaining nanoparticles are those coming from aluminosilicates 2:1. These structures are called TOT due to their tetrahedral-octahedral-tetrahedral layers. The thickness of the sheet is 0.96 nm. A scheme of the structure of these compounds (aluminosilicates) is presented in the figure 2.

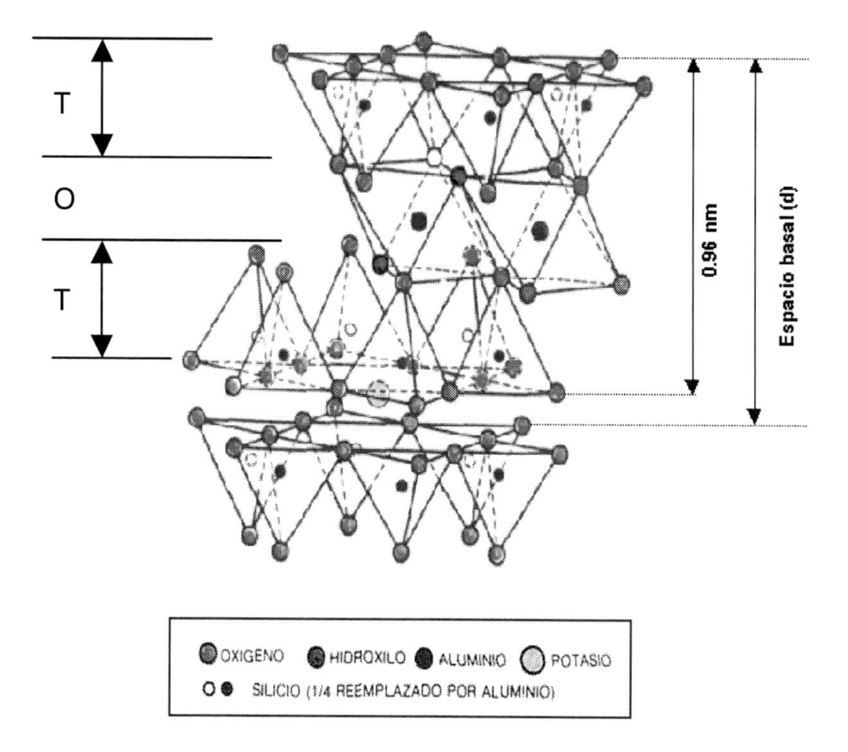

Figure 2. Structure in philosilicates layer 2:1 [12].

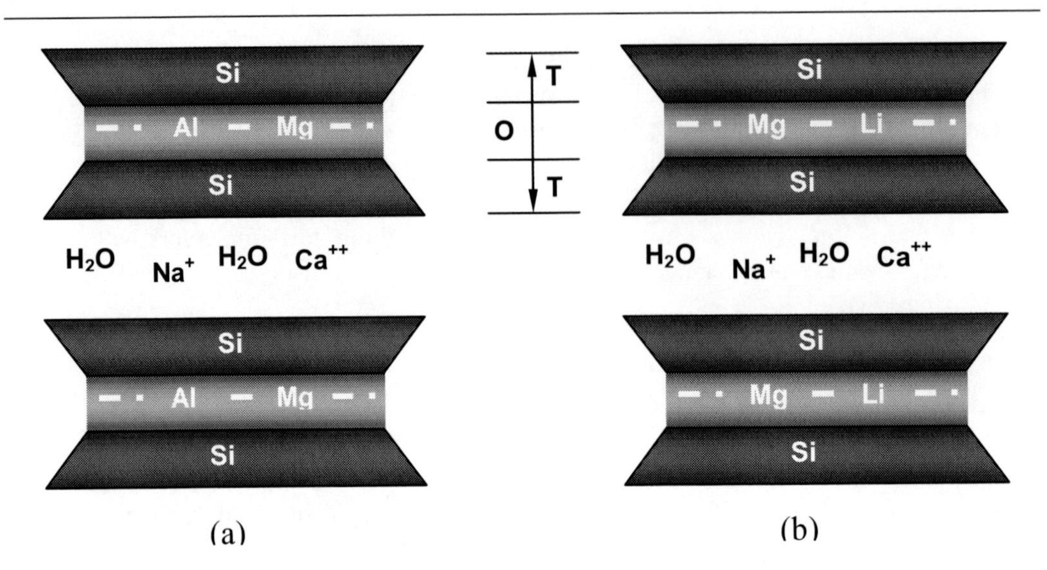

Figure 3. Schematic representation of (a) Montmorillonite clays (b) Hectorite clays.

But if in the octahedral layer, the magnesium atoms predominate and they present isomorphic substitutions for lithium, this clay is named Hectorite (HTA). The differences between these two important clays are shown in the figure 3, where it is appreciated the characteristic ions of each one of the layers (TOT) of the MMT and HTA.

Among clays type TOT, commonly named smectite, there is a great variety depending on the layer where the isomorphic replacement is carried out (tetrahedral, octahedral or both). If in the octahedral layer, 1/6 of the aluminum atoms are replaced for magnesium, and in the tetrahedral layer there are not substitutions, this clay is named montmorillonite (MMT).

One of the principal philosilicates 2:1 characteristics is its capacity to exchange its interlayer cations with other compounds; this process is named cationic exchange capacity (CEC) expressed in meq cation/100g clay. This property gives an endless of applications due to the possibility of including great variety of compounds between the philosilicates sheets that modifies their properties.

Clay's Modification

In general, many substances, such as oils, pollutants [13], amines, pesticides [14] and others [15,12] can be exchanged or inserted into the clay interlayer space modifying their characteristics. [16,17] The intercalations with alquil-amine compounds inside the clay increase the interlayer spacing and also modify its polarity transforming clay in a more hydrophobic compound, due to the alquil group. A scheme of this reaction is shown in the Figure 4, where it is observed how the protonated amine displaces cations through cationic exchange reaction present in the interlayer of clay. This is due, mainly, to the effect of added clorhydic acid, since the cations of interlayer space form very stable and soluble chlorides, allowing the cations to leave the interlayer, and the protonated amine penetrates on the interlayer replacing cations.

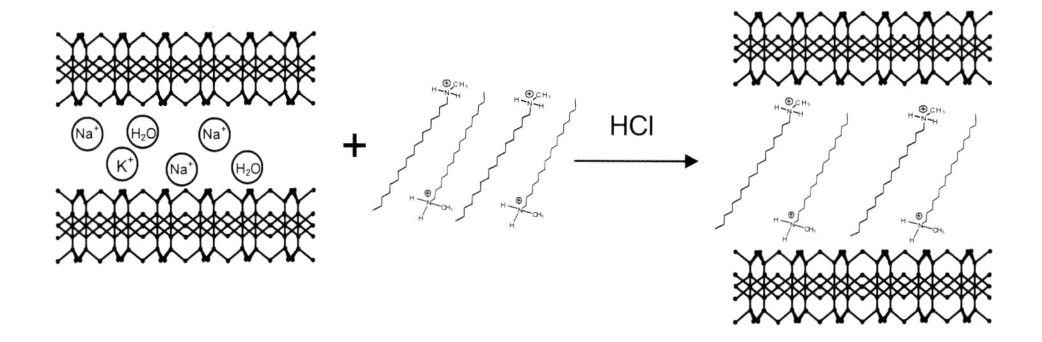

Figure 4. Schematic representation of the modification of clays with octadecylamine.

Studies, in this area, have shown that when the alquil-amine presents 18 atoms of carbon the biggest increases are gotten in the interlayer distance. [18,19] This is the vital importance to many applications and in particular for this research paper, since it wants to disperses sheets of clay in a nonpolar polymer like polypropylene. For this reason, in this chapter some variable of clay cationic exchange reactions with octadecylamine (ODA) are studied. Octadecylamine is a primary amine that has a C18 chain and accepts a proton, in an acid environment, remaining positively charged (ODA+). In this chapter, time and temperature reaction are variables taking into consideration to exchange cationic reaction.

Nanoparticulas Obtained by the Sol-gel Method

Among the principal difficulties in nanocomposites formation with clays, is its reduced interlayer space (1,0 nm), since, it makes difficult the polymer insertion between their sheets to generate inserted and exfoliated states. Another important characteristic is the lack of clays homogeneity and the presence of some ions like impurities that give coloration to nanocomposites, for example iron; or give toxicity like Cd and Cr. [20] For this reason, techniques to obtain synthetic clays or inorganic compounds with laminate structures [21,22] have been looked for, techniques that can modify the polymer properties when they are dispersed on the polymer, but without influencing their properties negatively. Here, the method of synthesis Sol-gel becomes one of the most important tools to obtain laminate structures of high purity. Certainly, the grade of structure organizations and their properties depend on the nature of their components, such as: organic, inorganic or hybrid systems in order to generating synergic interactions. [23,24,25] This synthesis method, also, allows modifying the morphology of particles that can be obtained such as: sheets, spheres, threads, etc. when template is used. [24]

Brinker and Sherer defined the Sol-gel method as the obtaining of materials through the preparation of a sol and the obtaining f gel by the solvent elimination. But, Since and Kolbe were who previously discovered the formation of silica mono-dispersed particles through hydrolysis and condensation of tetraethyl ortosilicate (TEOS) in ethanol with ammonia as catalytic compound. Since that moment numerous modifications have been developed and new variables have been introduced trying to have control on the size and the morphology of particles.

Breck and their collaborators in Mobil Oil Corporation were the pioneers in using surfactants as structural patternmakers (templates) for silicates synthesis and ordered aluminosilicates structurally. These materials were synthesized in a basic solution and were used surfactants of long alquilic chain, which were removed, later on, to obtain mesoporic silica. There are many "templates" used. In general, they are organic molecules and can be cationic, anionic or neuters. Also, copolymer with functionalities of different polarity can be used. When template is changed, the surfactant interaction (S) with the inorganic species (I) are modified (S^+I, $S^+X^-I^+$, S^0I^0, S^-I^+) affecting the properties of resulting material such us; pore wall thickness, superficial concentration of silanol groups, thermal stability and also the size of the pore. [24]

Many works (research papers) present the nanoparticles obtaining by the Sol-gel method and its later use in the nanocomposites formation [26,27,28,29,30], in the Moncada et, al, work, [31], a comparison between the works mentioned previously is made.

The possibility to obtain nanoparticles with different morphologies by means of the Sol-gel method makes it so much versatile for many applications, as it was mentioned previously. In figure 5, different morphologies are shown for the nanoparticles obtained by means of the Sol-gel method, the authors present the spherical nanoparticles formation (SNP), hybrid spherical nanoparticles (HSNP), and hybrid layer nanoparticles (HLNP), by means of variation in the synthesis conditions of the template agent like the polar and no-polar compounds are in general.

In order to generate silica particles with spherical morphology (SNP or HSNP), a modified version of the classical method of Stober [32] was applied. Instead of performing the sol–gel process of TEOS by adding it to an ethanol/NH4OH solution, prior to this step TEOS was slightly hydrolyzed in an ethanol/water azeotrope (4.4%H2O), forming short siloxane oligomers. Furthermore, template was added in order to obtain hybrid silica nanoparticles (HSNP) and to control the size and aggregation of the nanoparticles state. Figure 5 shows TEM images of particles obtained by the sol–gel method. For SNP, a spherical morphology with an approximate size of 10 nm can be seen with philiform aggregate formation. This unusual form could be made due to slightly prehydrolyzed TEOS before the addition to ammonia. The addition of a small amount of Template in the synthesis of HSNP allowed dispersed nanoparticles (10 nm) with a low degree of agglomeration to be obtained. Although also some discreet philiform aggregates (2–10 nanoparticles) are seen, they are dispersed without forming agglomerates. The probable way in which the template acts is by joining to the surface of the particles through R–NH2 ↔ HO–Si → R–NH+3 O–Si hydrogen bonds.

In this way the particles are capped with a layer allowing the particles to remain largely separated and dispersed. The template molecules attached to the surface of the silica nanospheres, surrounding them, give them the hybrid particles character. When greater amounts of template were added, the formation of a laminar micelle in the solution allowed the synthesis of laminar silica structures like HLNP. Figure 5 shows TEM images of these laminar particles in which the approximate size of a 300 nm × 300 nm sheets can be determined, but whose thickness cannot be measured exactly in this image.

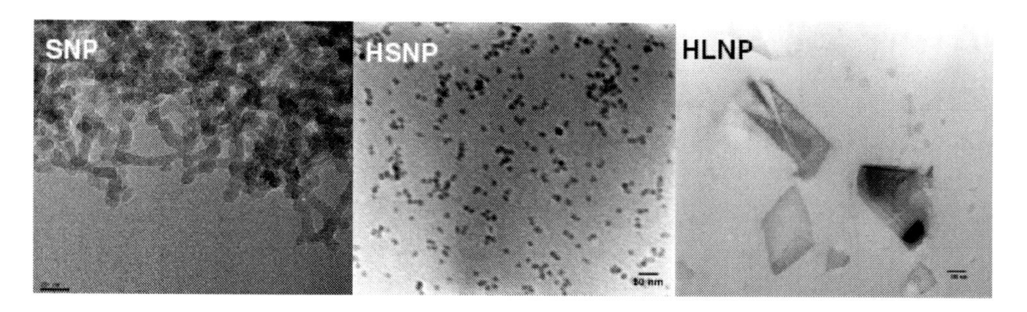

Figure 5. TEM images of nanoparticles obtained by Sol-Gel Method. SNP= Spherical nanoparticles, HSNP= hybrid spherical nanoparticles, HLNP= Hybrid layer nanoparticles.

Besides, nanoparticles previously named can have hybrid characteristic (organic-inorganic) such as HSNP or HLNP, and not hybrid characteristic such as SNP. This is presented in the figure 6. Where in the FT-IR spectrum can be observed the characteristic bands such inorganic elements (synthesis Sol-gel) as organic elements (template).

In the Table 1 are presented the assignments of the bands frequency.

The spectra of SNP and HSNP are characteristics of silica samples synthesized by sol–gel [33,34,35], but in the case of HSNP the spectrum also shows a shoulder at 2920 cm−1 corresponding to the C–H absorption from the template present in the sample. The broad band at 3000–3700 cm−1 is due to water and silanol groups engaged in hydrogen bonds. On the other hand, the spectrum of HLNP corresponds to a typical aluminosilicates where strong peaks at 2856 and 2920 cm−1 are attributed to template, which is present in considerable amounts in the particles. For more details of these particles see [31].

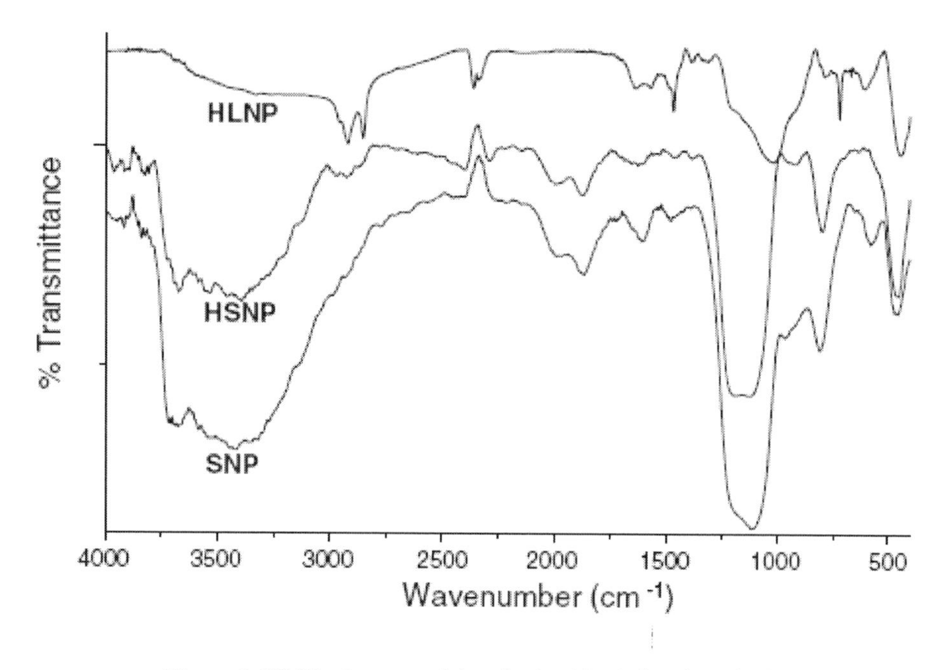

Figure 6. FT-IR of nanoparticles obtained by Sol-Gel method.

Table 1. Characteristic band of hybrid and unhybrid material obtained by Sol-Gel synthesis

Position of the band (cm^{-1})	Assignment			Reference
	SNP	HSNP	HLNP	
450	Bending Si-O-Si ____			[36,37,38]
720	------	------	"Rocking" -(CH_2)n-	[35,40]
800	____Flexion O-Si-O ____			[39,39]
960	Flexion Si-OH		-----	[39,41,44]
1000-1200	____Tension Si-O-Si ____			[39,40,41,44]
1210	----	----	Tension Al-O	[39,45]
1465	----	----	Flexion -C-H	[9]
1640			$-NH_3+$	[39]
1640	Flexion OH of the water			[39,44]
2918	----	$-CH_2-$____		[39,40,42]
3000-3700	OH of water and silica			-

Nanoparticles Obtain by Sol-gel Compared with Nanoparticles Obtained by Treatment of Clays

Nanoparticles obtained by Sol-gel method, like it was mentioned previously, show advantages in comparison with nanoparticles obtained by means of clay treatment, these are some of them:

- High purity and uniformity of chemical structure, because they are obtained by synthesis methods.
- Approval by legislation for direct contact with alive creatures (non toxicity), presenting it applications in such exigent industries with the legislation, for example: health areas, medicine, foods and drug.
- Obtaining of hybrid nanoparticles (organic-inorganic), it increases more the applications range.

Nanocomposites

Nanocomposite is the dispersion of nanometric size particles (nanoparticles) in a matrix, one of the first research to obtain nanocomposites was made by the TOYOTA company between the end of 80`s and the beginning of 90`s [40,41,42], with a clays exfoliation in a naylon-6 matrix. These results showed significative increments in an extensive range of properties to polymers reinforcement [43,44].

Later on to this discovery, a great explosion about nanocomposite researches has been developed, and researches were made in order to obtaining them with a great majority of polymers, such as: polypropylene [45,46,47], polyethylene [48], polystyrene [49], polyvinylclorure [50],acrilonitrile butadiene styrene copolymer (ABS) [51], polymetilmetacrylate [52] polyethylenterefthalate (PET) [53],ethylene-vinyl acetate copolymer (EVA) [54],polyacrilonitrile [55], polycarbonate [56], polyethylen oxide (PEO) [57],epoxy resin [58], polyamides [59], polycaprolactone [60], phenolic resin [61], rubers [62], polyurethanes [63], polyvinyl piridine [64]

The developed methods for the production of these nanocomposites are mainly three: mixed in solution, polymerization in-situ and mixed in fused.

Mixed in solution.[65] It consists in dissolving clay and polymer in an appropriate solvent and agitate the system, so that the polymer can introduce itself between the clay sheets. Then, the solvent is evaporated until obtaining a solid sample, in which is hoped to obtain the nanocomposite.

Polymerization In-Situ. [66] It consists in inserting the necessary catalytic system to the polymerization reaction in the inter-laminar region of clays, introducing the new system to the polymerization reactor and making the reaction. As the polymer chain grows, it will go separating the sheets until dispersing them.

Mechanic mixed in melt state. [67] It consists in using a mechanical mixer, which heats the polymer until the fused point. The mixer has molars that apply a force. When the clay is added the polymer is introduced between the clay's sheets.

There are three terms to describe the dispersion grade of clay in nanocomposites, as follows:

Nanocomposites Obtained Using Clays Modified

Where clays are used for the generation of nanoparticles, three possible morphological states can be found at the nanocompositos.

Tactoide state: When the polymer surrounds the clay structure.

Intercalated sate: when the polymer has the possibility to introduce itself between the clay's sheets given a highly ordered state that contain the clay's sheets and the polymer.

Exfoliated state: In contrast, this state is the dispersion of the clay's sheets in the polymeric matrix.

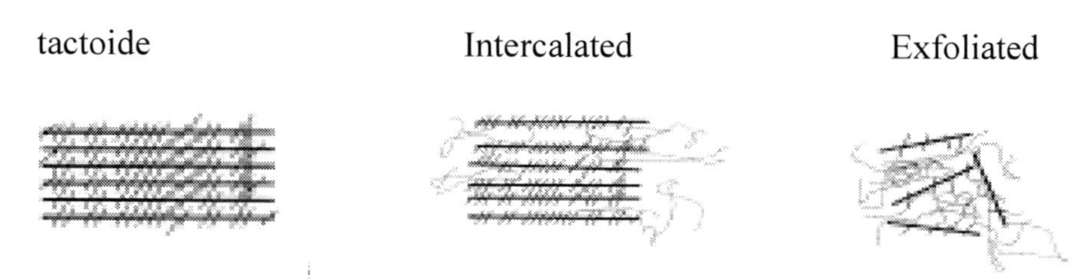

Figure 7. Different states of nanoparticles dispersion in a nanocomposites.

A graphic of these states is shown in the Figure 7, where its characteristics can be appreciated.

Depending on the characteristics of this mixture, it is necessary to reduce the repulsion in interface if the polymers are polar (nylon) or non-polar (PE, PP, etc) in order to minimize the repulsion force. For this, a third component named compatibilizer agent is used in the mixture in order to help with the compatibility between the two other materials. The compatibilizer agent more used has been maleic anhydride implanted in polypropylene and, in general, the most efficient mass relationship between this and the quantity of load has been 3/1. [68,69,70,71] This material has great importance because the modification in the properties not only depends of the particle size but also of an effective interaction between matrix and nanoparticles.

Compatibilizer Agent

Surface modification of hydrophilic clays through ion exchange reaction, normally with long chain alkyl ammonium ions increases the hydrophobic character as well as the interlayer spacing of these materials. [72,73,74] Therefore, the organically modified clays (O-clays) would be more compatible with non-polar polymers such as PP. However, in the case of polyolefin (PE, PP) the main difficulty encountered in preparing nanocomposites with O-clays is the lack of sufficient interaction between the hydrophobic polyolefin matrix and still relatively hydrophilic O-clays. [75,76,77] However, depending on the hydrophilic/hydrophobic balance of a polymer, the polymer chains could enter into the interlayer space of the clay and form either tactoid, intercalated or exfoliated nanocomposites. In the tactoid hybrid structure, the polymer chains mainly surround the stacks of the clay resulting in a material with properties similar to those of micro composites. In the intercalated hybrid structure, extended polymer chains are introduced between the silicate sheets, resulting in a multilayered structure consisting of alternating polymer and inorganic clay layers. In the exfoliated hybrid structure, the clay nanolayers are dispersed as single layers in the polymer matrix. However, the challenge is to obtain exfoliated nanocomposites since these materials show superior properties than those of intercalated or tactoid hybrid materials. Owing to the strong hydrophilic nature of the clays, low adhesion between clays and non-polar polymers causes a considerable decrease in the mechanical properties of the nanocomposites, as the interface becomes a weak point in the hybrid material. To overcome this limitation and in order to adequately disperse the O-clays in PP matrix, functionalized PP materials such as PP grafted with maleic anhydride has been used as compatibilizer agent.[78] Recent study carried out by Lertwimolnun et.al. [79] On the degree of dispersion of montmorillonite in PP nanocomposites, prepared by melt intercalation, showed that the dispersion of the clay in PP matrix was improved by addition of a modified PP with maleic anhydride as compatibilizer. However, this improvement was obtained for concentrations of the compatibilizer higher than 10 weight percent and up to 25 weight percent. In a more detailed study on this subject, the effect of the molecular weight and content of the maleic anhydride of PP grafted with this monomer have been studied by Wang et. al.[80] The results showed that not only the extent of grafting and the molecular weight of the compatibilizer but also the mixing temperature and processing conditions affected the structure of PP nanocomposites obtained with octadecylamine-modified montmorillonite. It was found that the exfoliation depended mainly on the intercalation capability of the compatibilizer as well as its composition in the nanocomposites. Other compatibilizer agents are study, as acrylic acid, [81] itaconic acid and its derivatives. [82,83] Itaconic acid is non-oil based dicarboxylic acid monomer. It is obtained through large-scale fermentation of agricultural wastes such as molasses, a sub-product of sugar industry. Due to its double functionality, itaconic acid and its derivatives offer interesting possibilities as polar functional monomers for the modification of polyolefin. The scheme of grafted reaction of itacónico acid in polypropylene is shown in the figure 8.

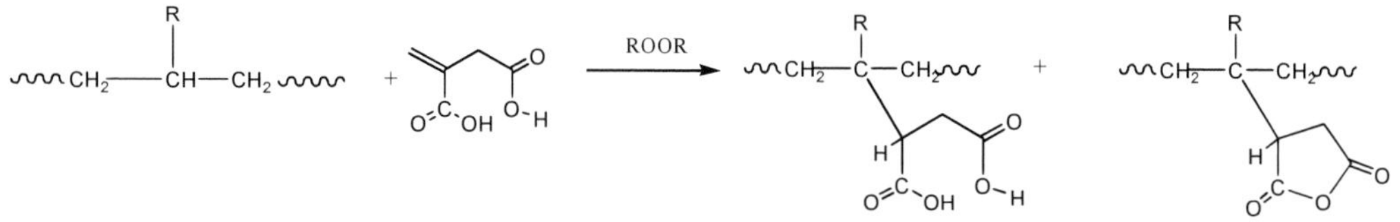

Figure 8. Schematic representation of the grafted reaction of itaconic acid in polypropylene. [46]

Nanocomposites Obtained Using Sol-gel Nanoparticles

Nanocomposites obtained using polypropylene with two different clays Montmorillonite (MMT) and Hectorite (HTA) and two different compatibilizer agents like compatibilizer commercial maleic anhydride (PP-g-MA) and the patented compatibilizer by this group: itaconic acid (PP-g-AI). They presented remarkable increments in mechanical nanocomposites properties in comparison with the polymer alone. [11]

In the figure 9, nanocomposites transmission microphotographs of polypropylene are shown using clays type Montmorillonite and Hectorite organically modified. TEM images show an effective nanocomposites formation and an effect of itaconic acid incorporation percentage (PP-g-$AI_{1,8}$ = 0.7% of grafting). They also show a better dispersion of clays nano sheets in the polymeric matrix when the clay type Hectorite was used, in comparison with the clay type Montmorillonite. This effect is more important in the mechanical properties, which will be shown below.

Table 2 shows mechanical properties of different nanocomposites obtained with polypropylene, two different clays (MMT and HTA) and itaconic acid implanted in polypropylene with different implant percentages. It is proved the nanocomposites formation by increasing the mechanical properties in comparison with the polymer without load. It is, also, observed an effect of itaconic acid implant percentage in the values of mechanical properties and consequently in the nanocomposites formation. [46]

When clay type MMT was used, it was observed a decrease of the mechanical property with the decrease of itaconic acid implant percentage. When clay type HTA was used the behavior was on the contrary, when the itaconic acid implant percentage was decreased, it was observed an increase in mechanical properties. The authors explain this tendency through different cationic exchange capacities of clays and their chemical interaction with the compatibilizer, which are shown in the Table 2.

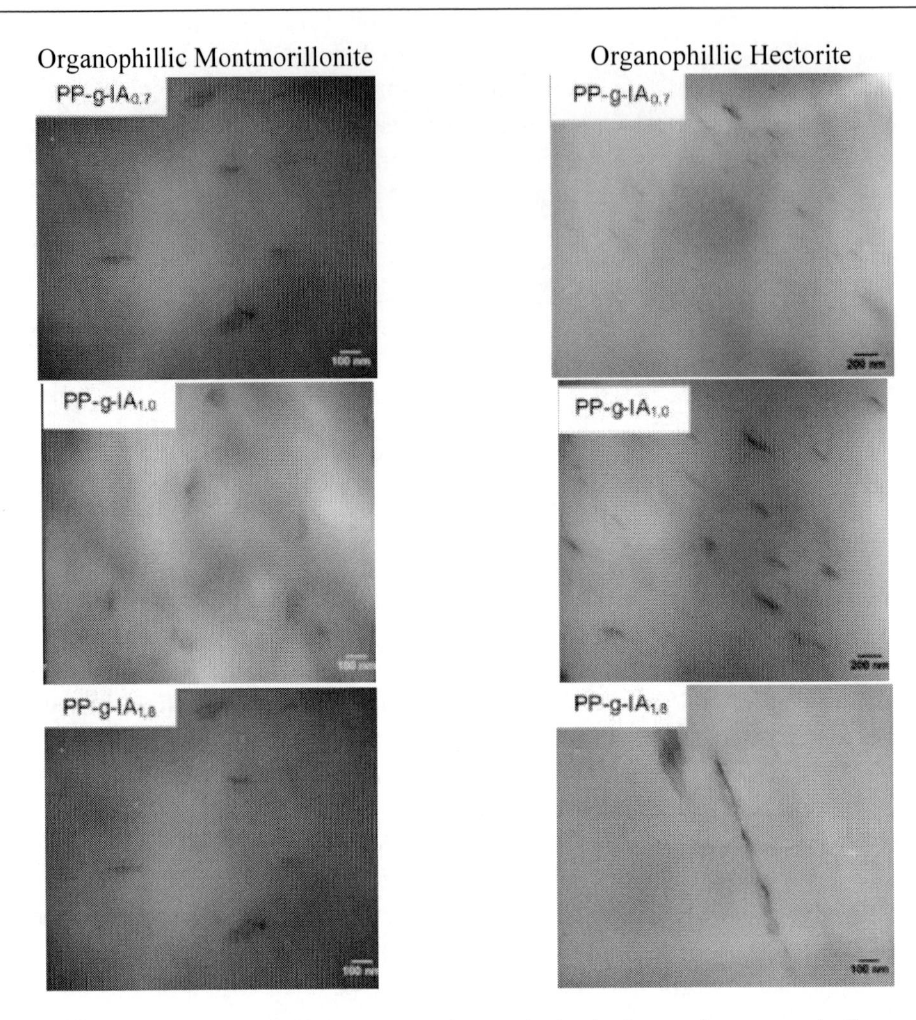

Figure 9. TEM images of nanocomposites obtained by using organically modified Montmorillonite/Hectorite and polypropylene grafting itaconic acid with different percentages of grafting.

Table 2. Mechanical properties of nanocomposites prepared by used PP, 1 wt -% of different chalice and 3% of functionalized PP grafted with different percentages of itaconic acid. (And, Modulus, (and, tensile strength; (, elongation at break)

PP-g-AI Wt.-%	Montmorillonite			Hectorite		
	E (MPa)	σ_y (MPa)	ε (%)	E (MPa)	σ_y (MPa)	ε (%)
1,8	2117 (±35)	40 (±1)	17	1791 ((±40)	49 (±2)	9
1,0	1941 (±33)	38 (±2)	8	2075 (±35)	48 (±1)	8
0,7	1900 (±42)	38 (±2)	8	2137 (±28)	51 (±2)	9

Unmodified PP: E= 1092 (±45) MPa, σ_y=30 (±2) MPa, ε= 20%.

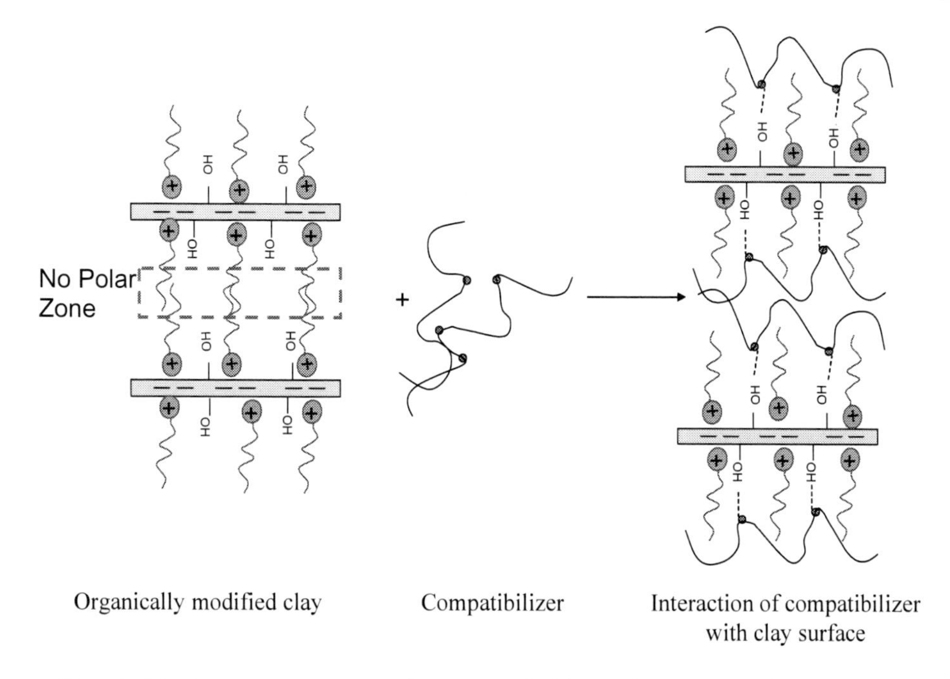

Figure 10. Model showing interaction of the organically modified clay with grafted PP used as compatibilizer.

Interaction of PP-g-IA with Nanoparticles

Grafted PP used as compatibilizer interacts with nanoparticles, for example clays, the interaction are through hydrogen bonding with OH groups on the surface of the clays as shown in Figure 10. As seen from this scheme, the polar groups of grafted PP interact with clay surface and therefore the interlayer space of the clay increases as the result of this interaction, generating a less polar region. This facilitates the entry of non polar PP chains into the clay gallery. [46]

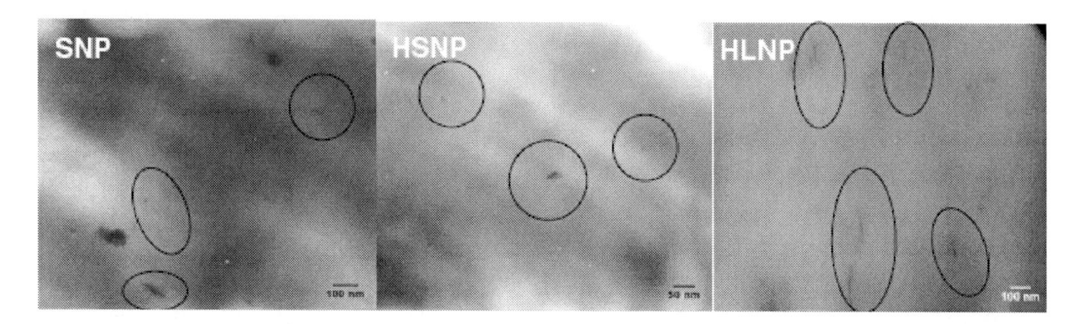

Figure 11. TEM images de nanocomposites obtained with 1% sol-gel nanoparticles, 3% de PP-g-MA and Polypropylene.

Nanocomposites Obtained Using Sol-Gel Nanoparticles

Recently, Chastek et al prepared silicate sheets with covalently attached C16-alkyl chains [84] and made nanocomposites with polystyrene samples of various molecular weights. A dramatic increase in Young's modulus of the nanocomposites compared with neat polystyrene was found, especially in the case of the polymers with the higher molecular weight. On the other hand, polypropylene nanocomposites with superficially treated silica nanoparticles were reported by Wu et al [85], showing remarkable improvements in the mechanical properties of the nanocomposites, pointing to the importance of the nanoparticles/matrix interface. Jain et al [86] worked on the modification of the surface of silica nanoparticles containing vinyl triethoxysilane groups that were dispersed in polypropylene, but the mechanical properties of the resulting materials are not mentioned in the paper. Many matrices have been used for the dispersion and evaluation of the modification of the properties of silica nanoparticles, such as polypropylene (PP) [87], high-density polyethylene (HDPE) [88], acrylic latex [89], polyethylene terephthalate (PET) [90] and polyethersulfone (PES) [91].

Nanocomposites generated with nanoparticles obtained previously (SNP, HSNP and HLNP) are going to be shown as follows and then it will be done a comparison with nanocomposites obtained with clays showed in the first section.

The figure 11 shows de TEM images of nanocomposites obtained with: 1% of Sol-Gel nanoparticles and 3% of compatibilizer agent (PP-g-MA) in a polypropylene matrix.

It is possible to see in the TEM images that spherical particles without hybrid composition (SNP) show small amassed of particles until 100 nm; these particles show sizes approximately of 10 nm (as it was shown previously). The nanocomposites obtained with the hybrid spherical particles (HSNP) show a better dispersion that the non-hybrid ones (SNP) with sizes approximately of 10 nm. This is the original size when the particles are alone. It shows the efficiency of the particles dispersion in the resin and so the contribution of the zone hybrid particle in the dispersion.

For the hybrid laminate particles (HLNP), an effective dispersion of laminates nanoparticles is shown, with dimensions of 10 nm and until 200nm (thickness and largess). The above show the efficiency of the nanoparticles obtained by Sol-gel. These nanoparticles without hybrid composition show good nanocomposites formation. When the hybrid function is established in nanoparticles, the dispersion in the resin is significantly improved; it changes of having nano-aggregates of nanoparticles of 100 to nanoparticles totally dispersed of 10 nm.

Mechanical Properties

The Table 3 shows the mechanical properties of nanocomposites obtained by Sol-Gel nanoparticles, grafted polypropylene with maleic anhydride and polypropylene as matrix.

It is evident the nanocomposites formation with all Sol-gel nanoparticles with increases in the module as the follows: since 40% for non-hybrid spherical nanoparticles (SNP), 48% for hybrid spherical nanoparticles (HSNP), until 62% for hybrid laminate nanoparticles (HLNP). Tensile strength (σ_y) has showed bigger increases: 57% for SNP, 73% for HSNP and 90% for HLNP.

Table 3. Mechanical properties of nanocomposites using, 1 wt.-% of nanoparticles obtained by Sol-Gel and 3% of PP-g-MA and different PPs. (Note: E, Modulus; σ_y, tensile strength; ε, elongation at break)

MFI PP	SNP			HSNP			HLNP		
	E (MPa)	σ_y (MPa)	ε (%)	E (MPa)	σ_y (MPa)	ε (%)	E (MPa)	σ_y (MPa)	ε (%)
13	1523(±22)	47(±2)	9	1620 (±22)	52(±1)	90	1770(±28)	57(±1)	10

Unmodified PP: E= 1092 (±45) MPa, σ_y=30 (±2) MPa, ε= 20%.

These values not only show the big efficiency in the nanocomposites formation generated by nanoparticles obtained by the Son-Gel method and the influence of the formation of hybrid systems, but also the big influence that presents the morphology of the nanoparticles in the mechanical properties of nanocomposites. It show that for laminate morphologies are more effective the increase of the mechanical properties.

Wu et al [13] in your publication are showing different concentrations of functionalized spherical silica nanoparticles were added to polypropylene and we compared them with hybrid nanoparticles of similar morphology (HSNP), presented by Moncada et, al. [REF] in the formation of nanocomposites in isotactic PP of MFI = 3. The higher changes in percentage of the mechanical properties for a 1% of load obtained in that work were approximately 17% for Young's modulus (E), where for pristine polymer E = 1390 MPa and polymer nanocomposites E = 1620 MPa and 5% for yield stress (σy), where for pristine polymer σy = 37 MPa and for polymer nanocomposites σy = 39 MPa. The changes percentage of the mechanical properties obtained in our work were 46% for Young's modulus (E), where for pristine polymer E = 1090 MPa and polymer nanocomposites E = 1590 MPa and 83% for yield stress (σy) where for pristine polymer σy = 30 MPa and for polymer nanocomposites σy = 55 MPa, clearly greater increases of these properties with silica nanoparticles obtained in our work was obtained. As to elongation at break (ε), in the work of Wu et al there is approximately a 300% increase where for pristine polymer ε = 90% and polymer nanocomposites ε = 320%, while in our work we found a 36% decrease of this property, where for pristine polymer ε = 250% and polymer nanocomposites ε = 160%. This shows the importance of the superficial modification of the nanoparticles on the mechanical properties of the nanocomposites, because it has a direct influence on the interfacial zone of the nanocomposites and on the energy transfer mechanism between the matrix and the nanoparticle.

The thermal stability of these nanocomposites also shows big increases for the nanocomposites compared with the polymer without load, in the polypropylene the temperature of decomposition is 319°C measured by DSC, and authors [REF] have reported values in nanocomposites of 350°C. This, amplifiers the applications range of this polymer so used at the moment and with a great expansion.

Table 4. Comparative mechanical properties of nanocomposites using 1% w/w of HLNP and Montmorillonite clays, 3% w/w PP-g-MA, and PP

MFI PP	HLNP			MMT[(a)]		
	E (MPa)	σ_y (MPa)	ε (%)	E (MPa)	σ_y (MPa)	ε (%)
3[(b)]	1607 (±23)	50 (±2)	42	1310 (±31)	34 (±1)	8
13[(b)]	1770 (±28)	57 (±1)	10	1415 (±23)	48 (±3)	5
3[(c)]	1711 (±35)	52 (±2)	50	1452 (±41)	43 (±1)	7

Note: E = Young's modulus, σy = elastic limit, ε = elongation at break.
Unmodified PP: E= 1092 (±45) MPa, σ_y=30 (±2) MPa, ε= 20%Values.
[(a)] taken from [45].
[(b)] Ziegler-Natta Polypropylene.
[(c)] Metallocene Polypropylene.

Comparison between Nanocomposites Obtained with Clays and Sol-gel Particles

Like it has been presented through this chapter, nanoparticles obtained by Sol-gel methods show advantages in the structure, composition and purity compared with nanoparticles obtained by clays.

In the Table 4, it is show a comparison between the properties of nanocomposites obtained by Sol-gel and clay nanoparticles, doing emphasis in mechanical properties.

For this, is done a comparison between laminates particles obtained by Sol-gel (HLNP) and the ones obtained by clays, principally with the Montmorillonite clay (MMT), because it is the most used in the studies. The obtainment conditions of these two nanocomposites were the same and were done by the same research group. In general, bigger increases are observed in the values of the mechanical properties in comparison with polymer without load, for nanocomposites generated with nanoparticles obtained by the Sol-gel method.

Comparison between PP-g-IA and PP-g-MA as Compatibilizer Agents in Nanocompositos Formation

It has been mentioned the big modifications in the mechanical properties of obtaining nanocomposites when the compatibilizer agent itaconic acid grafted in polypropylene (PP-g-IA) was used, now is shown a comparison of this one with the compatibilizer agent maleic anhydride grafted in polypropylene (PP-g-MA), that is the most used, conserving a similar percentage of grafted for the two products.

The table 5, shows the different values in the mechanical properties for obtaining nanocomposites, using MFI=26 Polypropylene as matrix, organically modified Montmorillonite and Hectorite as nanocomposites source, and the two different compatibilizer agents with similar percentage of grafted, itaconic acid grafted in polypropylene (0,7 % of graft) and maleic anhydride grafted in polypropylene (0,6% of graft).

Table 5. Mechanical properties of nanocomposites prepared by used PP of MFI=26, 1 wt.-% of different clays and 3% of PP grafted with 0,7% of itaconic acid (IA) or PP grafted with 0,6% of maleic anhydride (MA). (E, Modulus; σy, tensile strength; ε, elongation at break)

PP-g-X	Montmorillonite (OMMT)			Hectorite (OHTA)		
	E (MPa)	σ_y (MPa)	ε (%)	E (MPa)	σ_y (MPa)	ε (%)
[a]PP-g-$IA_{0,7}$	1900 (±42)	38 (±2)	8	2137 (±28)	51 (±2)	9
[b]PP-g-$MA_{0,6}$	1415 (±27)	38 (±3)	10	1607 (±32)	43 (±2)	10

Unmodified PP: E= 1092 (±45) MPa, σ_y=30 (±2) MPa, ε= 20%.
[a] Reference [31].
[b] Reference [46].

Considering the values for OM, and PP-g-$AI_{0.7}$ of E=1900 MPa, σy=38 MPa, and compared with the values of E=1415 MPa, σy=38 MPa when PP-g-$AM_{0,6}$ was used with the same clay. It is possible to appreciate the big difference in the obtained values. When grafted polypropylene with itaconic acid as compatibilizer agent was used, it presented the biggest increases. When OMMT clay was changed for OHTA and PP-g-$AI_{0.7}$ was used, values were: E=2137 MPa and σy=51 MPa, it compared with E=1607 MPa and σy=43 MPa for PP-g-$MA_{0.6}$. Again, the biggest increases were with PP-g-AI. These values show a bigger efficiency in the increment of the mechanical properties when grafted with itaconic acid polypropylene was used. It shows that in nanocomposites formation the itaconic acid grafted in polypropylene (PP-g-IA) is more effective that the maleic anhydride grafted in polypropylene (PP-g-MA).

This effect with better properties as compatibilizer agent of PP-g-AI in comparison with PP-g-AM can be explained with the chemical structures of the two compatibilizer agents studied yet. PP-g-AI show two groups free carbonic, with which can be done linkages bridges type of hydrogen with clay sheets. The oxygen of the compatibilizer with the hydrogen of the clay and vice versa (the oxygen of the clay with the hydrogen of the compatibilizer) are showing in the figure 12 generating six possible linkages points, in the other hand, PP-g-AM has an anhydride group that allows it to generate three possible linkage points by bridge of hydrogen between the compatibilizer oxygen and clay.

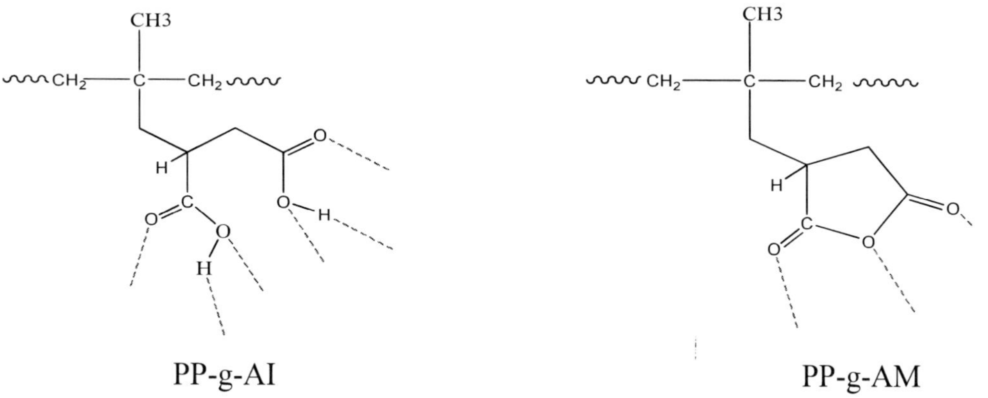

PP-g-AI PP-g-AM

Figure 12. Interactions bridge type of hydrogen generated by the different compatibilizer with the clay surface.

Therefore, when the number of possibilities of generated hydrogen bridges is bigger, the interaction of the compatibilizer agent with the clay is also bigger and better interface among the different materials (clay and PP) and also better compatibility, this is going to help, like it has been mentioned previously, to a bigger energy and stress transfer between the nanocomposites matrix and its load. All the above-mentioned is increased in great measure due to the great superficial area that nanoparticles have.

Nanoparticles Used as Sensor

The nanoparticles obtained in this chapter, can be functionalized to increase its applications and to meet the needs of the industry. The functionalization is possible using the Sol-gel technique, enabling the generation of hybrid materials and from this, to obtained materials with characteristic chemical that depending of functional group added to the structure of the nanoparticles can be generate new materials. For example materials with more bio-compatibility, smart and intelligent particles that can do a molecular identification of compound (toxical compound and viruses etc). These compounds are called chemical sensor, other sensors can be obtained by polymer [92,93] and oxides [94,95].

Sol-gel modification usually, the firs step un functionalizing nanoparticles involves building a self-ensambly monolayer (SAM) on the particles surface. In this approach, bifunctional molecules containing a hydrophilic head group and hydrophobic tail group absorb onto a substrate or an interface as closely packed monolayer, mostly thought the attraction between hydrophobic chains. The tail group and the head group can be chemical modified to contain functional group. SAMs are widely explored for engineering the surface and interfacial properties of materials such to wetting adhesion and friction. [96,97] These monolayer are also used to mediate the molecular recognition processes and to direct oriented crystal growth. [98] Chemical and physical modification of solid surface is extensively used in chromatography, chemistry analysis, catalysis, electrochemistry, electronic industry and the more actually used are for sensing and electronic devise.

Depending of substrate surface, two kinds of SAMs are commonly encountered. On metal surface such a gold and silver thiolated alkyl chains are often used, and the oxides surface such as silica, alkyl silanes are usually used. The silane and group hydrolyze first, and then become covalently bound to the substrate through a condensation reaction. The SAMs molecules may also be covalently linked with one another through siloxane group as a result of the hydrolyses of the silane group and their subsequent condensation. Because of the disordered nature of the substrate surface, SAMs on the oxide surface are usually disordered [99].

Use of Nano Sensor in the Industry

These sensors generated with nanometer size, can be to present applications in the broadest sectors of the industry, as medicine, agriculture, environmental science, construction, aerospace, even packing. Below will be presented few details of its application in the food and drug packaging industry.

Food and Drug Packaging

More than 200 companies around the world are developing nanomaterials in order to be used in food and pharmaceutical products packaging. United States is the main country in research, next Japan and China. In 2006 these investments increased in dollars, of $2.6 bn to $7.0 bn. It is expected that in 2010 this value can increase to $20.4 bn (49) and for the pharmacists industry in 2011 can be of $11 bn. It is predicted that in 2010 Asia will be the great market for these products, with the 50% of the world population. [100]

On the other hand, parameters like increment in the demand of surer feeding products (bio-terrorism), better relationship cost-efficiency, bigger control (producer) and opportune and effective information to the consumer, are taking to the development of packaging materials for foods and more demanding pharmaceutical products. Investigations in packaging are being carried out that allow the consumer to detect the possibility of microbiology, chemistry or physic contamination; and allow the producer better controls in their systems of production, transport and storage. This is possible to get with the called intelligent packaging, which, through sensors detect the parameters before mentioned by means of packaging and immediately, without analysis in laboratories. These sensors can be obtained using diverse technologies [101,102,103,104], among them, nanotechnology is predicted like more efficient and with more easiness to disperse it in the polymeric resins. But, this concept of intelligent packaging (IP) is not alone; it is ligature directly to other concepts like active packaging (AP) and smart packaging (SP) that will be enunciated next.

Generally, in literature it is possible to find some divergences among the packaging definitions above mentioned, (smart packaging SP, active packaging AP, and intelligent packaging IP). As follows: For Clarke (2001) IP is packaging that presents logical capacity, SP is one that communicates and AP is one that controls. For Brody and others (2001) IP is packaging systems that indicator and communicate and SP is the packaging that has the capacity of both AP and IP. For Rijk (2002) IP is one that monitories the conditions of packed food and gives information about the food quality during the transport and storage. The ambiguity showed in these definitions can cause difficulties in their interpretation and it is for this reason that the definitions of Kil L.Yan (Investigator of the Department of Foods Science of the University of Rutgers [105] (USA)) is considered the most appropriate, where SP are the packaging that gather AP and IP characteristics. IP are the packaging systems able to carry out intelligent functions (to detect, to indicator, to record, to communicate, with a scientific logic) in order to facilitate the decision of extending useful life, increasing security, improving quality, providing information and noticing about possible food problems. For AP, increasing food protection is a function.

This is, in a total packaging system, IP is responsible component to indicator the food environment and to give the information. AP is responsible component to make some action to protect the food (like an antimicrobial compound). It is important to say that IP and AP are not mutually excluding, some packaging systems can be classified as IP and others as AP or maybe both as SP. In an appropriate situation IP, AP and the traditional function of packaging are going to work in association to provide a total packaging solution. They always should depend on necessities and characteristic of packed product. The table 6 shows the main applications of each one of mentioned packaging.

The previous information's can be collected in the bars code, which allows to the producer knows always the quality of his product and to the consumer has more information about the product that wants to buy.

With the above-mentioned, it is possible to say that the traditional function of packaging is changing. It has been simply the material that contained the food and protects it of the external atmosphere. And it is becoming in a functional material that allows that as producer as consumer can verify the product they are selling or buying and with a nutritious quality and appropriate microbiological.

Intelligent Packaging by Supply Chain

The packaging technologies using systems RFID (intelligent packaging), are being presented as the most suitable for the identification and communication of the characteristics of the food, allowing know through the supplement chain of food, the state of preservation, as well as the location both on the shelf and in transport. The communication could be done at any time allowing for continuous monitoring of the food through network connection. In the future, the supplement chain will be principally of intelligent and active packaging integrates with communication system (RFID).

Table 6. Main applications of the active, intelligent and smart packaging [106]

Active Packaging AP	Intelligent Packaging IP	Smart parking SP
Antimicrobial	Microbial growth Indicator	Antimicrobial with indicator integrated
Oxygen absorbers	Time-temperature indicator	Oxygen absorbers with indicator integrated
Ethylene absorbers	Nutritional attributes Indicator	Ethylene absorbers with indicator integrated
Humidity absorbers	Indicator of gas concentration in packaging of modified atmosphere	Barrier plus with indicator integrated
Absorbers of off flavors, amines and aldehydes	Indicator of impact occurrence	Self - warming / cooling with indicator integrated
UV-light absorbers	Radio Frequency Indicator RFID	

The packages will protect the food without additives, inform about the product quality and history in every stage of the logistic chain, guide the journey of the package, reduce product loss, and will give real-time information to the consumer about the properties /quality/use of the product.

Future Trends in the World of Nanoparticles

The nanoparticles regardless of their provenance (clay, Sol-Gel, etc.) should see each needs more momentous and more demanding both chemical and selectivity of compound identification of organic substances such as viruses and bacteria live, or recognition of specific molecules as cancer cells, DNA, etc.. They must meet the requirements of interaction with the human body through functionalizations organic-inorganic enabling good communication and acceptance by living cells. Another application the need is of new physical functions as a high thermal and mechanical resistance, good compatibility with anywhere materials etc.

References

[1] Medeiros ES, Mattoso LH, Nanotecnologia. In: Duran N, Mattoso LHC, Morais PC. (Org.). Nanotecnologia: Introdução, preparação e caracterização de nanomateriais e exemplos de aplicação. São Paulo: Artliber, 2006;1: 1.
[2] Moriarty P, Nanotechnology: Radical new science or plus ça change?. *Nanotechnology Perceptions*. 2005;1(3):115-118.
[3] Scott F, Nanotechnology: Radical new science or plus ça change?- the debate. *Nanotechnology Perceptions*. 2005;1(3):119-146.
[4] Drexler KE. Molecular Engineering: An approach to the development of general capabilities for molecular manipulation. *Proc. Nad. Acad. Sci.* 1981; 78(9): 5275-5278.
[5] Merkle RC. Convergent assembly. *Nanotehcnology* 1997; 8(1): 18-22.
[6] Merkle RC, Parc X, Casing and assembler. Nanotechnology 1999;10:315-322.
[7] Merkle RC, Robert A, Freitas Jr. Theoretical analysis of a carbon-carbon dimer placement tool for diamond mechanosynthesis. *J. Nanosci. Nanotechnol.* 2003; 3: 319- 324.
[8] Hall JS, Architectural considerations for self-replicating manufacturing systems. *Nanotechnology* 1999; 10(3): 323-330.
[9] Herrmann PSP, Mattoso LH, A Nanotecnología no agronegócio, uma realidade na embrapa. XXXVI Congreso Brasilero de Engenharia agrícola, 2007.
[10] Kellehr K. http://www.thestreet.com/tech/internet/10279136.html, visited in December 2007.
[11] Moncada E, PhD thesis, Universidad de Chile, 2006.
[12] http://www.uclm.es/users/higueras/yymm/Arcillas.htm, Visited in Julio 2004.
[13] Ogawa M, *Chem Mater*. 1998; 10: 1382.
[14] Wang L.Q, Liu J, Exarhos G.J, Flanigam K.Y and Bordia R, *J. Phys. Chem. B.* 2000; 104: 2810.
[15] Kunyima B, Viaene K, Hassan Khalil M, Schoonheydt R.A, *Langmiur,* 1990; 6: 482.
[16] Ray S.S, Okamoto M, *Prog. Polym. Sci*, 2003; 28: 1539.
[17] Zanneti M, Camino G, Reichert P, Miilhaupt R, *Macromol. Rapid Comun*, 2001; 22: 176.
[18] http//www. nanomat.de Visited April 2007
[19] Alexandre M, Dubois P, *Materials Science and Engineering,* 2000; 28: 1.
[20] Carrado K.A, *Applied Clay Science*, 2000;17:1.
[21] Kim D.W, Blumstein A, Kumar J, Tripathy S.K, *Chem. Mater.*, 2001;13: 243.
[22] Shimojima A, Umeda N, Kuroda K, *Chem. Mater.*, 2001;13: 3610.
[23] Hench Larry L, West Jon K, *Chem. Rev.*, 1990; 90; 33.
[24] Galo J. de A. Llia-Soller G.J, Sanchez C, Lebeau B, *Chem. Rev.*, 2002; 102: 4093.

[25] Hong-Ping L, Chung-Yuan M, *Acc. Chem. Res.*, 2002; 35: 927.
[26] Soler-Illia G. J, Sanchez C, Lebeau B and Patarin J, *Chem. Mater.* 2002; 102: 4093.
[27] Ogawa M, *Langmuir*, 1997; 13: 1853.
[28] Yun S.K and Maier J, *Inorg. Chem.*, 1999; 38: 545.
[29] Chastek T.T, Stein A and Macosko C, *Polymer,* 2005; 46: 4431.
[30] Wu C.L, Zhang M.Q, Rong M.Z and Friedrich K, *Compos. Sci. Technol.* 2005; 65: 635.
[31] Moncada E, Quijada R, Retuert J. Nanoparticles prepared by the Sol-Gel method and their use in the formation of nanocomposites with polypropylene, *Nanotechnology* 2007; 18: 335606.
[32] Stober W, Fink A and Bohn E, *J. Colloid Interface Sci.*, 1968; 26: 62.
[33] Ellis T.S and D'Angelo J.S, *J. Appl. Polym. Sci.*, 2003; 90: 1639.
[34] Hernandez C and Pierre A.C, *Langmuir*, 2000; 16: 530.
[35] Gao Y, Choudhury N.R, Dutta N, Matisons J, Reading M and Delmotte L, *Chem. Mater.*, 2001; 13: 3644.
[36] Hernandez C, and Pierre A.C, *Langmuir*, 2000; 16: 530.
[37] Transfetetti B.C and Davanzo C.U, *Macromolecules,* 2004; 37: 459.
[38] Nakanishi K, Infrared Absortion Spectroscopy, Ed. Holden-Day, San Francisco, 1967. 39.
[39] Lopez T, Asomoza M, Gómez R, *Journal of Non-Crystalline Solids*, 1992; 147&148: 769.
[40] Okada A, Fukushima Y, Kawasumi M, Inagaki S, Usuki A, Sugiyama S, et al. Inventors 1986. Composite material and process for manufacturing same. United Estate Patent 4,739,007 19 Abril 1988.
[41] Kawasumi M, Kohzaki M, Kojima Y, Okada A, Kamigaito O. Inventors 1988 Process for producing composite material. United Estate Patent 4,810,734. 7 March 1989.
[42] Usuki A, Kawasumi M, Okada A. Synthesis of nylon 6-clay hybrid, *J. Mater. Res.* 1993; 8(5): 1179-1184.
[43] Yang F, Yingar R, Nelson G.L, Flammability of polymer-clay and polymer-silica nanocompositos. *Journal of fire Sci,* 2005; 23: 209-226.
[44] Kojima A, Yamada Ki, Fujii T, Hirata M. Improvement in adhesive-free adhesion by the use of electrostatic interactions between polymer chains grafted onto polyethylene plates. J. *Appl. Polym. Sci.* 2006;101(4): 2632-2638.
[45] Moncada E, Quijada R, Retuert J. Comparative effect of metallocene and Ziegler-Natta polypropylene on the exfoliation of Montmorillonite and Hectorite clays to obtain nanocompositos. *J. Appl. Polym. Sci.* 2007; 103; 698 -706.
[46] Moncada E, Quijada R, Yazdani-Pedram M. Usse of PP grafted with itaconic acid as new compatibilizer for PP/Clays nanocomposites. *Macromol. Chem. Phys.* 2006; 207: 1376-1386.
[47] Moncada E, Quijada R, Yazdani-Pedram M, Inventors 2006. Patent PCT/IB2007/003008
[48] Zhao C, Qin H, Gong F, Feng M, Zhang S and Yang M, Mechanical, thermal and flammability properties of polyethylene/Clay nanocompositos. *Polym. Degrad. Stabil* 2005; 87(1): 183-189.

[49] Mohanty S, Nayak S.K. Melt blended polystyrene/layered silicate nanocomposites: effect of Clay modification on the mechanical, thermal, morphological and viscoelastic behavior. *J. Thermoplast. Compo. Mater,* 2007; 20(2): 175-193.

[50] Xu W, Zhou Z, Ge M, Pan W. Polyvinyl chloride/Montmorillonite Nanocomposites: glass transition Temperature and mechanical properties. *Journal of Thermal Analysis and Calorimetry,* 2004;78 (1): 91-99.

[51] Alma P, Patiño S, Ramos LF. Morphological and thermical properties of ABS/Montmorillonite nanocomposites using ABS with different AN content, *Macromol. Mater. Eng,* 2007; 299(3): 302-309.

[52] Guo-An Wang, Cheng-Chien Wang, Chuh-Yung Chen. Preparation and characterization of cayered double hydroxides – PMMA nanocomposites by solution polymerization, *Journal of Inorganic and Organometallic Polymers and Materials,* 2005; 15(2): 239-251.

[53] Kráčalík M, Mikešová J, Puffr R, Baldrian J, Thomann R, Friedrich C. Effect of 3D structures on recycled PET/organoclay nanocomposites. *Polymer Bulletin,* 2007; 58(1): 313-319.

[54] Pasanovic-Zujo V, Gupta R.K, Bhattacharya S.N. Effect of vinyl acetate content and silicate loading on EVA nanocomposites under shear and extensional flow. *Rheologica Acta,* 2004; 43(2): 99-108.

[55] Tianshi Yu, Jiaping Lin, Jiafu Xu, Tao Chen, Shaoliang Lin. Novel polyacrylonitrile nanocomposites containing Na-Montmorillonite and nano SiO2 particle. *Polymer,* 2005; 46(15): 5695-5697.

[56] Yong Gao, Peng He, Jie Lian, Lumin Wang, Dong Qian, Jian Zhao, et al. Improving the mechanical properties of polycarbonate nanocomposites with plasma-modified carbon nanofibers. *Journal of Macromolecular Sciences, Part B: Physics,* 2006; 45: 671–679.

[57] Marc X, Reinholdt, R, Kirkpatrick J, Pinnavaia T. Montmorillonite-poly(ethylene oxide) nanocomposites: interlayer mlkali Metal behavior. *Phys. Chem. B,* 2005; 109 (34): 16296 -16303.

[58] Lijia P, Pingsheng H, Gang Z, Dazhu C. PbS/epoxy resin nanocomposite prepared by a novel method. *Materials Letters,* 2004; 58(1-2):176-178.

[59] Contreras V, Cafiero M, Da Silva S, Rosales C, Perera R, Matos M. Characterization and tensile properties of ternary blends with PA-6 nanocomposites. *Polymer Engineering & Science,* 2006; 46(8): 1111-1120.

[60] Khalid Saeed, Soo-Young Park. Preparation and properties of multiwalled carbon nanotube/polycaprolactone nanocompositos, *Journal of Applied Polymer Science,* 2007; 104(3): 1957-1963.

[61] Zhang Y, Lee SH, Mitra Y, Toghiani H, Pittman C. Phenolic resin/octa(aminophenyl)-T8-polyhedral oligomeric silsesquioxane (POSS) hybrid nanocomposites: synthesis, morphology, thermal and mechanical properties. *Journal of Inorganic and Organometallic Polymers and Materials,* 2007; 17(1): 159-171.

[62] Valentini L, Biagiotti J, Kenny J.M , López Manchado M.A. Physical and mechanical behavior of single-walled carbon nanotube/polypropylene/ethylene-propylene-diene rubber nanocompositos, *Journal of Applied Polymer Science,* 2003; 89(10): 2657-2663.

[63] Pattanayak A, Jana S. Properties of bulk-polymerized thermoplastic polyurethane nanocompositos, *Polymer,* 2005; 45(10): 3394-3406.
[64] Reiko Saito, Tadakuni Tobe. Synthesis of poly(vinyl pyridine)-silica nanocomposites using perhydropolysilazane, *J. Appl. Polym. Sci.,* 2004; 93(2): 740-757.
[65] K. Yano, *J. Polym. Sci., Part A: Polym. Chem.,*1993; 31(10): 2493.
[66] Heinemann J, Reichert P, Thomas R, Mulhaupt R, *Macromol. Rapid. Commun.*, 1999; 20: 423.
[67] Vaia R, Giannelis A, *Chem. Mater.*, 1993; 5(12): 1694.
[68] Hasegawa N, Kawasumi M, Kato M, Usuki A, Okada A, *J. Appl. Polym. Sci.*, 1998; 67: 87.
[69] Hasegawa N, Kawasumi M, Kato M, Usuki A, Okada A, *Macromolecules*, 1997; 30: 6333.
[70] Ye W, Feng-B C, Yann-C L, Kai-C W, *Composites Part B: engineering,* 2004; 35: 111.
[71] Ye W, Feng-B C, Kai-C W, *Journal of Applied Polymer Science,* 2005; 97: 1667.
[72] Suprakas R, Okamoto M, *Prog. Polym. Sci.,*2003; 28: 1539.
[73] Silva C.R, Fonseca M.G, Barone J.S, and Airoldi C, *Chem. Mater.*, 2002; 14: 175.
[74] Ishida H, Blackwell S.C and Blackwell J, *Chem Mater.*, 2000; 12: 1260.
[75] Zhu J, Uhl F.M, Morgan A.B and Wilkie C.A, *Chem. Mater.,* 2001; 13: 4649.
[76] Koo C.M, Ham H.T, Kim S.O, Wang K.H, *Macromolecules,* 2002; 35: 5116–5122.
[77] Zanneti M, Camino G, Reichert P, Miilhaupt R, *Macromol. Rapid Comun.,* 2001; 22: 176.
[78] Sinha S, Okamoto M, *Prog. Polym. Sci.*, 2003; 28: 1539.
[79] Lertwimolnun W, Vargas B, *Polymer*, 2005; 46: 3462.
[80] Wang Y, Chen F.B, Wu K.C, *J. Appl. Polym. Sci.*, 2005; 97: 1667.
[81] Moad G, *Prog. Polym. Sci.*, 1999; 24: 81.
[82] Yazdani-Pedram M, Vega H, Quijada R, *Macromol. Chem.Phys*., 1998; 199: 2495.
[83] Yazdani-Pedram M, Vega H, Quijada R, *Polymer*, 2001; 42(10): 4751.
[84] Chastek T T, Stein A and Macosko C, *Polymer,* 2005 46 4431.
[85] Wu C.L, Zhang M.Q, Rong M.Z and Friedrich K, *Compos.Sci. Technol.,* 2005; 65: 635.
[86] Jain H, Goossens F, Picchioni P, Magusin B, Mezari M and Van D, *Polymer,* 2000; 46: 6666.
[87] Wu C.L, Zhang M.Q, Rong M.Z and Friedrich K, *Compos.Sci. Technol.,* 2002; 62: 1327.
[88] Zhang M.Q, Rong M.Z, Zhang H.B and Friederich K, *Polym. Eng. Sci.,* 2003; 43: 490
[89] Nicolas L and Narkis M *Polym. Eng. Sci.* 1971; 11: 194.
[90] Kr´aˇcal´ık M, Mikeˇscov´a J, Puffr R, Baldrian J, Thomann R and Friedrich C, *Polymer Bulletin,* 2007; 58: 313.
[91] Jana S.C and Jain S, *Polymer,* 2001; 42: 6897.
[92] Riul A, Malmegrim R.R, Fonseca F.J, Matosso L.H, Nano-Assembled Films for taste sensor application, *Artf. Organs*. 2003; 27:(5), 469-472.
[93] Dos Santos D, Riul A, Fonseca F.J, Mattoso L.H, A layer –by-layer film of chitosan in a taste sensor application, *Macromol. Biosci.*, 2003; 3: 591-595.

[94] Pacquit A, Tong Lau, Doamond D, Smart packaging for monitoring of fish freshness, *Proc. of SPIE,* 2005; 5826: 545-550.
[95] Seok-In B, Gravure-printed colour indictor for monitoring kimchi fermentation as a novel intelligent packaging, *Packag. Technol Sci.*, 2002; 15: 155-160.
[96] Whitesides G.M, Self-assembling materials, *Sci. Am.* 1995; 273: 146-149.
[97] Ulman A, Formation and structure of self-assembled monolayer, *Chem. Rev.*, 1996; 96: 1533-1554.
[98] Bunker B.C, Rieke P.C, Tarasevich B.J, Cambell A.A, Fryxell G.E, Ceramic thin films formation on functionalized interface through biomimetic processing, *Sicence,* 1994; 264: 48-55.
[99] Self-Assembled Nanostructues: Nanostruture Science and Technology, (David J. Lockwood) New York, United States of America. Jin Zhang, Zhong-lin Wang, Jun Liu, Shaowei Chem, and Gang-yu Liu. ISBN:0-306-47299-6.
[100] Moraru, C, Panchapakesan C, Takhistov K. Nanotechnology: new frontier in food science. *Food Technology.*, 2003; 57: 25-28.
[101] Das R, Selke S, Harte J. Development of electronic nose method for evaluation of HDPE flavour characteristics, correlated with organoleptic testing. *Packaging Technology and Science,* 2007; 20(2): 125-136.
[102] Seok-In Hong. Gravure-printed color indicators for monitoring Kimchi fermentation as a novel intelligent packaging. *Packag. Technol. Sci.*, 2002;15:155-160.
[103] O'Mahony F, O'Riordan T, Papkovskaia N, Ogurtsov V, Kerry J, Papkovsky D. Assessment of oxygen levels in convenience-style muscle-based sous vide products through optical means and impact on shelf-life stability. Packaging Technology and *Science,* 2004; 17(4): 225-234.
[104] Bodenhamer WT. Inventor 2002. Toxin Alert. United Estate Patent 6,376,204. 18 April 2004.
[105] Yam K.L. Takhistov P.T. and Miltz J. Intelligent packaging: concepts and applications. *Journal of Food Science.* 2005; 70(1): R1-10.
[106] Butler P. Smart packaging – Intelligent packaging for food, beverages, pharmaceuticals and household products: http://www.azom.com/details.asp?ArticleID=2152. visited April 2007.

In: Nanoparticles: New Research
Editor: Simone Luca Lombardi, pp. 335-362
ISBN: 978-1-60456-704-5

Chapter 11

THE APPLICATIONS OF NANOPARTICLES IN ELECTROCHEMISTRY

Ruhai Tian and Jinfang Zhi
Technical Institute of Physics and Chemistry, Chinese Academy of Sciences, No.2, Beiyitiao, Zhong-guan-cun, Haidian District, Beijing, 100080, P. R. China

Abstract

Combination of Nano-sciences & techniques and electrochemistry makes electrochemical techniques are promising in many aspects, such as electroanalysis, bio-analysis, power & energy, etc. At the same time, the introduction of Nano-sciences & techniques in electrochemistry also expands the unknown filed of the interdiscipline.

In this comment paper, we will discuss the applications of nanoparticles (NPs) in electrochemistry including the following threads 1) why scientists introduce the NPs into electrochemistry; in this part, we summarize the develop of Nano-sciences & techniques; mainly discuss some basic theories and concepts about NPs and electrochemistry and how scientists use NPs in electrochemistry; 2) progresses of NPs application in electrochemistry , in this part, we summarize the achieved marvelous achievements in this area; 3) challenge of the NPs application in electrochemistry, in this part, we summarize the challenges that electrochemists are encountering with and bring up our opinions and strategies.

1. Introduction

Electrochemical methods are widely adopted techniques because of their essential significance not only in scientific research but also in industry. [1] Actually, the field of electrochemistry is a very large domain, encompassing fundamental theories (e.g. electron transfer, electrolyte diffusion etc), phenomena (e.g., electrophoresis and corrosion), technologies and practical applications (e.g. electrochromics, electrolysis, sensors, batteries, fuel cells, supercapacitors, etc). Electrochemistry related publications such as journals, monographs, patents and conferences are largely published.

On the other hand, nano-sciences & techniques have been developed a lot since Dr Richard P. Feynman's famous prophetic lecture "There's Plenty of Room at the Bottom" in

1959. Along with the development and advancement of nano-related theories and techniques, the programmable and reproduceable fabrications of nano-materials and nano-structures are becoming more and more reliable. Nano-sciences & techniques are dramatically infiltrating to experimental sciences, such as chemistry, physics, biology, materials, electronics, etc. To this day, more and more organizations and scientists are paying more and more attentions on nano-sciences & techniques. Nano-sciences & techniques are opening new doors for not only sciences & techniques but also industries.

Nano-sciences & techniques is one of the developing areas at top speed due to a wide variety of potential applications in biomedical, optical, material, and electronic fields. Nano-related organizations, journals were established largely and quickly after 90s of last century and numerous of nano-related achievements have been published. "Nanoparticle" is a descriptive name of small particles with at least one dimension is less than 100 nm, which consists of nanospheres, nanoreefs, nanoboxes, nanocrystals, nanosheets, etc. The unconscious application of NPs started from 9th century in pottery and art so as to get glittering effect or beautiful colors (this is obviously based on the optical properties of NPs). As early as 1857, Faraday consciously synthesized very stable gold nanoparticle colloidal solution through solution process but people didn't know the details of "nano" at that time. After the development of advanced instruments (Scanning electronic microscope/SEM, Transmission electron microscopy/TEM, X-ray Photoelectron Spectroscopy/XPS, atomic force microscopy/AFM, dynamic light scattering/DLS, powder x-ray diffractometry/XRD, etc.) in the early of 20th century, scientists got the tools to develop nano-related sciences and technologies.

Today, we know that the small size of NPs results in some fancy properties (electronic band structure change, crystal structure change etc) and size-effects (quantum confinement effect in semiconductor particles, surface plasmon resonance in some metal particles especially on the surface and superparamagnetism in magnetic materials, etc). No other reasons than these fancy properties attracted more and more scientists to focus on nano-related researches.

In this chapter, we mainly discuss the applications of NPs in electrochemistry. It is not only impossible to cover all the NPs-related electrochemistry in this small chapter, but also unnecessary because a lot of reviews about NPs related electrochemistry have been published.

Reference [2,3] summarized the applications of NPs in electroanalysis.

Reference [4] summarized the applications of Carbon nanotubes (CNTs) in the DNA sensor.

Reference [5] reviewed the applications of CNTs in electrochemistry.

Reference [6] reviewed nanowire-based electrochemical biosensors.

Reference [7] reviewed the application of NPs in energy fields.

Moreover, some special topics were also published , e.g. prof. Erkang Wang reviewed the advances in synthesis and electrochemical applications of gold nanoparticles (AuNPs). [8]

We hope this chapter about the NPs applications in electrochemistry could be helpful to those interested researchers. At the same time, we are looking forward to see more achievements and publications in this area. In this chapter, we discuss two parts of contents: 1) the fabrication of NPs 2) applications of NPs in electrochemistry. The second part is the focal point we want to discuss. We will just discuss the applications of NPs in electrochemical analysis and energy.

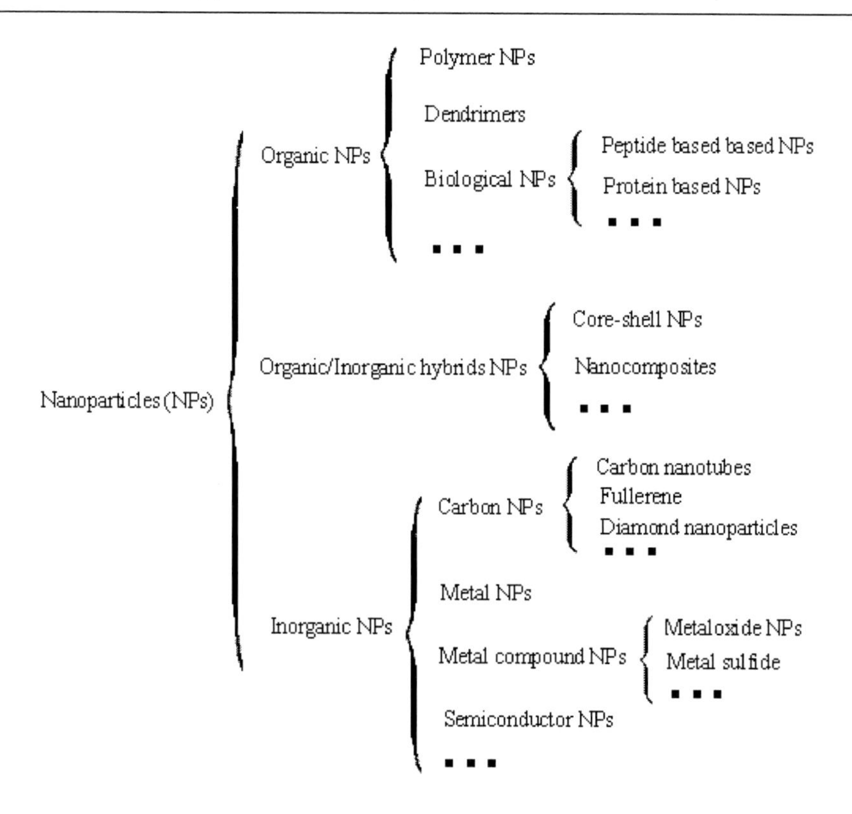

Chart 1. The classification of NPs.

2. The NPs Fabrication

Synthesis and applications of NPs have witnessed tremendous growth during the past decades. Here, we can just simply classify NPs as shown in chart 1. Inorganic NPs are most widely used nanoelements in electrochemistry because of their excellent conductivity, catalytic abilities and stabilities. In this chapter, we just discuss the applications of inorganic NPs in electrochemistry. The applications of inorganic NPs mainly focus on metal NPs; carbon NPs; semiconductor NPs; other nanostructures materials.

2.1. 0-Dimensional Inorganic Nanostructure Materials (Metal NPs)

The main approaches used to fabricate 0-dimensional metal NPs for the applications in electrodes were either the evaporation or direct electrochemical deposition of metal onto an electrode surface or bottom-up self-assembly of metal NPs by solution processes.

Solution Processes

Fabrication of NPs through solution process is most popular because of good reproducibility, easy control, high dispersion and narrow size distribution. Frens [9] method and phase-transfer [10,11] method are widely used in solution synthesis of metal NPs. The difference between these two methods is that metal NPs synthesized by Frens method are surface charged but NPs synthesized by phase-transfer are more surface-functional because of the capping organic layer. Synthesis of colloids AuNPs [9,10,12-19], cobalt NPs (CoNPs) [20] via chemical reduction methods has been reported. Multifarious solution process, such as microemulsions [21], reverse micelles and organometallics [22] were also developed to synthesize silver NPs. Selective structured platinum NPs were also synthesized in solution with different reagent as reduction agent [23,24]

Electrochemical Methods

Electrochemical method is another common method used in the fabrication of metal NPs, especially nanostructured transition metal clusters [25]. The advantage of this method is the particles size, shape, surface properties and product quantity could be easily controlled [26-28]. For electrochemistry research, metal NPs can be directly deposited onto electrode surface without more subsequent immobilization procedures. Manfred T. Reetz reported the synthesis of tetra-alkyl-ammonium-stabilized Pd and Ni clusters by a simple electrochemical process [26]. The size of particles could be controlled by variation of current density and the processes were high yield and absence of side products. Reference [27] reported the electrochemical synthesis of gold nanorods with different mean aspect ratio. L. Rodrı´guez-Sa´nchez reported the synthesis of 2-7nm silver NPs with silver as sacrificial anode, and tetrabutylammonium bromide as stabilizing salt [28]. Direct deposition of silver NPs (AgNPs) [29,30], platinum NPs (PtNPs) [31-35], nickel NPs (NiNPs) [36], ruthenium composites (RuNPs) [37], chromium NPs (CrNPs) [38], palladium NPs (PdNPs) [26,39], CdS NPs [40] on electrode surface was also studied.

Other Methods

Many other methods are also used for the fabrication of metal NPs such as radio-frequency sputtering [41-45], electron beam lithography [46], sonochemical synthesis [47], chemical vapor deposition [48], laser evaporation [49,50] Sonoelectrochemical method [51], etc.

2.2. 1-Dimensional Inorganic Nanostructure Materials

The most widely used 1-D nanostructure materials in electrochemistry are CNTs, carbon wires, silicon nanowire (SiNWs), conducting polymer wires (CP-NWs), etc. The basic construction methods of 1-D nanostructure materials including:

1. Morphology template auxiliary direct formation (porous Al_2O_3, DNA etc)
2. Symmetric growth of seed at liquid/solid, vapor/solid, vapor/liquid/solid interface
3. Selective etching according to the intrinsically properties of different crystalline face
4. Capping agents kinetically controlled growth of seed
5. Dip-pen nanolithography (AFM, STM tips)
6. Other methods

Reference [6] reviewed the fabrication, assembly/alignment and surface functionalization of 1-D nanostructures (CNTs, SiNWs, CP-NWs). As one of the 1-D nanostructures, because of their remarkable mechanical, chemical, thermal and electrical properties [52-54], CNTs have been developed a lot since its discovery in 1991 [55,56]. CNT structure can be thought as rolled-up rectangular sheet of graphite and can be divided into single-walled (SWNTs) and multi-walled (MWNTs). MWNTs are coaxial CNTs with 0.34nm of interlayer space. The diameters of CNTs range from 0.4nm to 3nm for SWNTs and from 1.4nm to 100 nm for MWNTs. The excellent properties of CNTs results from their tube-shape rolled graphite structure and their symmetric and electronic structure. CNTs electrodes can be fabricated through 1) polymer confining 2) paste conglutination 3) layer by layer assembly 4) matrix growth and sol-gel matrix assembly [5].

3. The Application Fields of NPs in Electrochemistry

3.1. Electrochemical Analysis

Electrochemical analysis is one of the most sensitive analytical methods. [1] The applications of NPs in electrochemical analysis mainly focus on the utilization of the nano-effects of NPs: catalysis abilities, large specific surface area (surface area to volume), etc. In Reference [3], Christine M. Welch and Richard G. Compton reviewed the applications of metal NPs (gold, silver, platinum, nickel, copper, iron, etc) in electroanalysis. Here, we discuss the applications of NPs in electrochemical analysis in two aspects:

3.1.1. As Nanoelectrode or Arrays

Nanoelectroces (NEs) and nanoelectrodes arrays (NEAs) which possess 1-100nm dimension range show great advantages with respect to traditional macro electrodes. The large mass transport mode of electrolyte on NEs&NEAs results in the low Faradic capacitance, fast electrochemical response and high signal-to-noise ratio, which makes them very suitable in the trace analysis and ultra fast electron exchange kinetic research [57,58]. Moreover, the small size of NEs&NEAs makes them suitable in the analysis of small volume sample and easy to assemble them as chips. Based on these fancy and unique properties, NEs&NEAs can be used as electrochemical detector in high-performance liquid chromatography (HPLC) and capillary electrophoresis (CE) [59,60].

NPs are widely used as elements for the fabrication of ultramicro-electrode and arrays even the nanoelectrode and arrays. Comparing with macroelectrode, ultramicroelectrode, ultramicroelectrode arrays, NEs&NEAs possess great advantages, such as lower ohmic drop of potential, shorter time to get the steady-state signal, enhancement of signal-to-noise, etc. [58] NEs&NEAs based on the NPs immobilization and assembly have been proved to be ultra-sensitive and catalytic. [61] Here, the definition of nanoelectrode is electrode with a critical dimension in the range of nanometers, where the critical dimension is not an exact number. According to the agreed consensus in nanoscience and technologies, we define the critical dimension of nanoelectrode is less than 100nm.

3.1.1.1. Gold Nanoparticles (AuNPs) as NEs&NEAs Element

In metal NPs, AuNPs were fabricated very early [62] , in 1857, (benzene was discovered in 1825 and its structure was given in 1865) and studied almost most clearly [16]. Allen J. Bard etc reported the fabrication and characterization of self-assembled spherical gold ultramicroelectrodes in the size of 1-30μm with 8nm AuNPs as basic elements through the bottom-up approach. [63] Figure 1A shows the fabrication procedures of gold ultramicroelectrodes based on the self-assembly of individual AuNPs. Figure 2B is the SEM images of different sizes of gold ultramicroelectrodes constructed. They found the ultramicroelectrode geometries and sizes were perfectly controllable and reproducible. The ultramicroelectrode they made had the electrochemical properties of metallic gold and showed ideal microelectrode behavior both in aqueous and acetonitrile electrolyte solution.

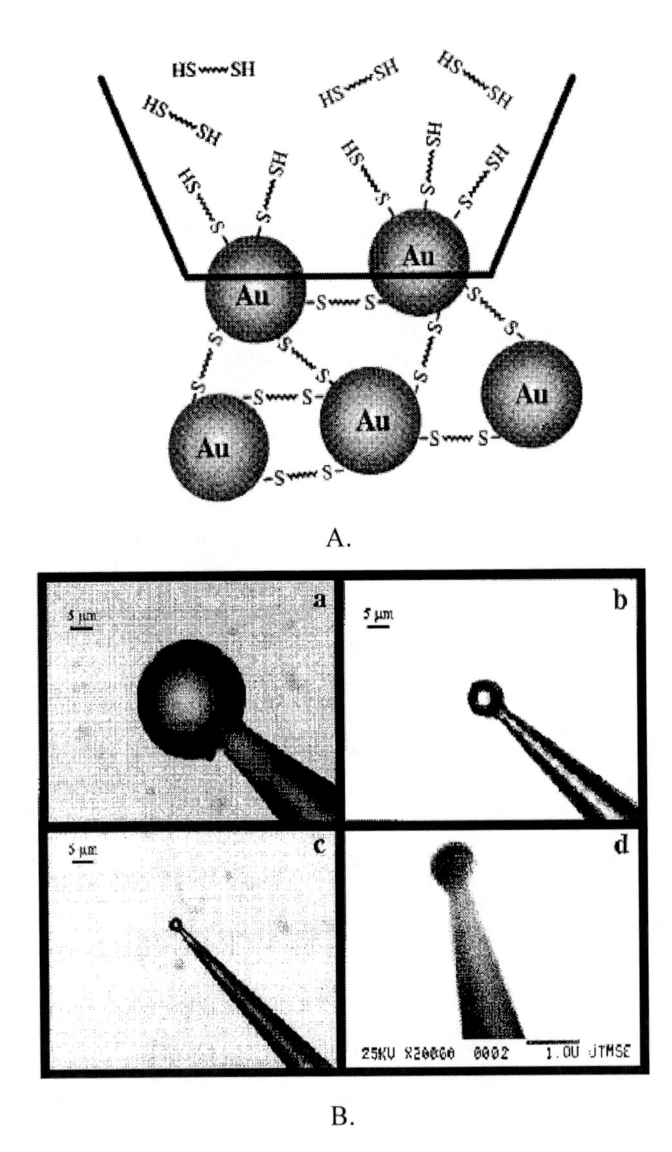

Figure 1. A) Schematic picture of the self-assembly process leading to the formation of microspheres at the end of the micropipet tip and B) optical (a-c) and SEM (d) micrographs of self-assembled UMEs. Electrode diameter, (a) 26, (b) 8, (c) 3, and (d) 0.9 μm. (Reprinted from Ref [63] Copyright 1997 American Chemical Society).

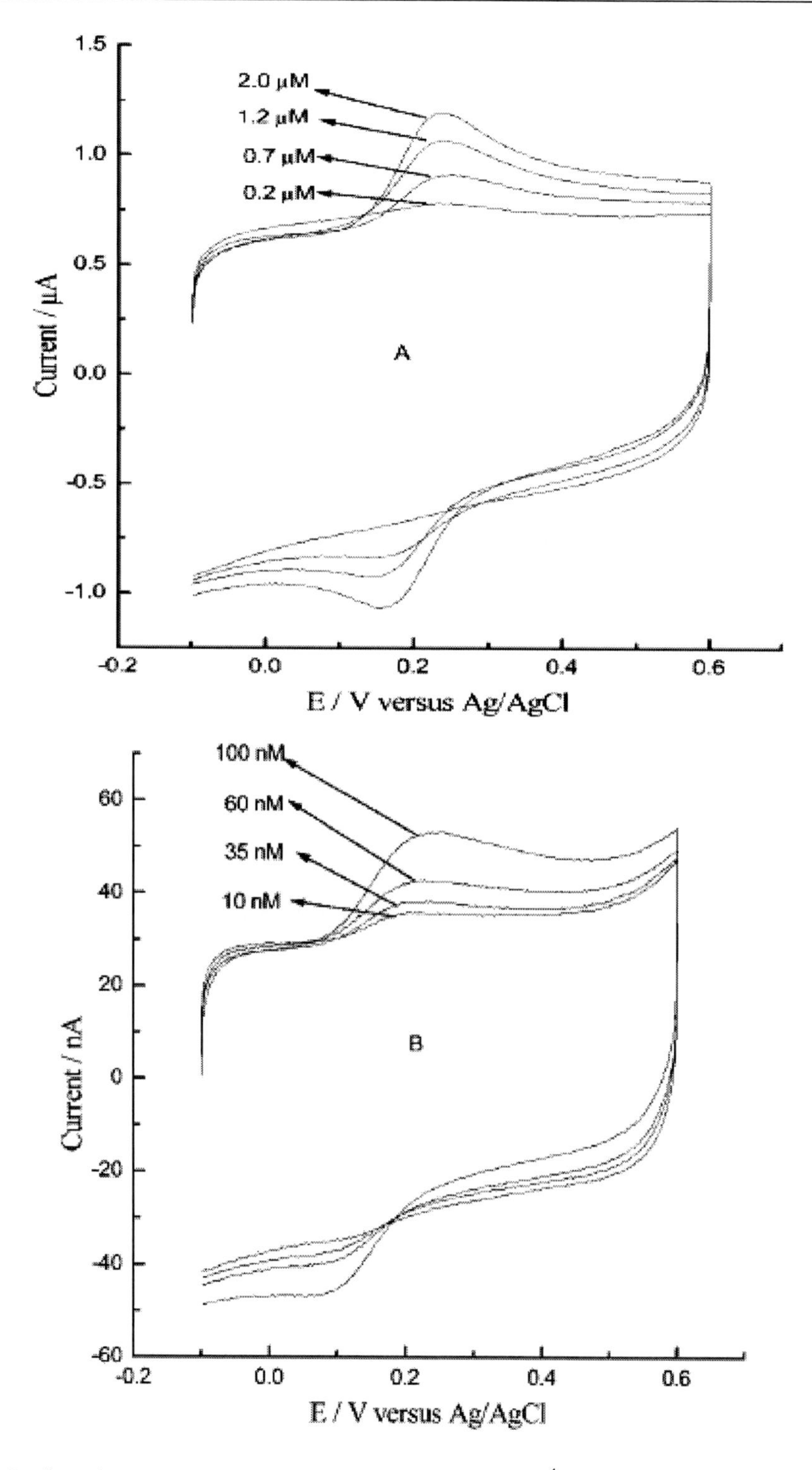

Figure 2. Cyclic voltammograms at 50 mV/s in aqueous $Fe(CN)_6^{4-}$ at (A) a gold macroelectrode in 50 mM KNO_3 and (B) the NEEs obtained by self-assembly of AuNPs for 9minutes in 50 mM KNO_3. $Fe(CN)_6^{4-}$ concentrations are as indicated. (Reprinted from Ref [64] Copyright 1997 American Chemical Society).

Erkang Wang et al reported the fabrication of gold nanoelectrode arrays through the assembly of colloidal AuNPs on (3-Mercaptopropyl)-trimethoxysilane (MPTMS) functionalized ITO glass surface. [64] They found that the assembly processes were high controllable and the size, shape and interspacing of as-prepared NEAs are highly flexible. Their electrochemical analysis experiments indicated that the detection limit of $Fe(CN)_6^{4-}$ on NEAs obtained by AuNPs assembly for 9 minutes were 20 times lower than at gold macroelectrode (Figure 2). Jinfang Zhi and Ruhai Tian reported the modification of AuNPs on functional boron doped diamond (BDD) electrode surface and investigated its basic electrochemical properties [65]. They found the AuNPs surface charge strongly affect the electron exchange between electrode and electrolyte and the electrode characterizations could be controlled by the AuNPs size and modification density [65,66].

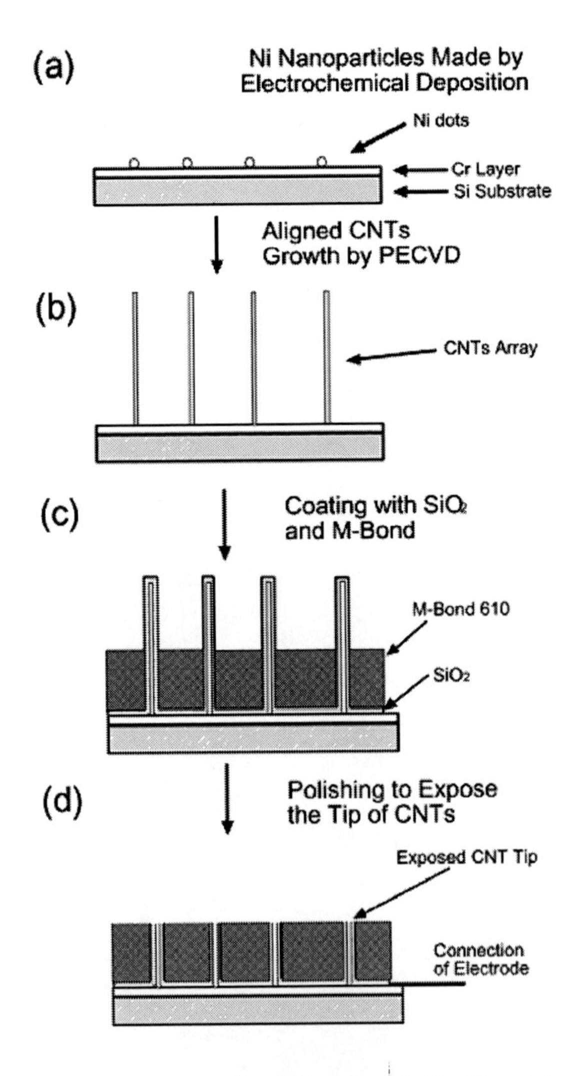

Figure 3. Fabrication procedures of CNTs electrode arrays a) Ni NPs catalysis electrodeposition; b) aligned CNTs growth; c) coating of SiO_2 on aligned CNTs array d) polishing the surface to expose CNTs (Reprinted from Ref [69] Copyright 2000 American Chemical Society).

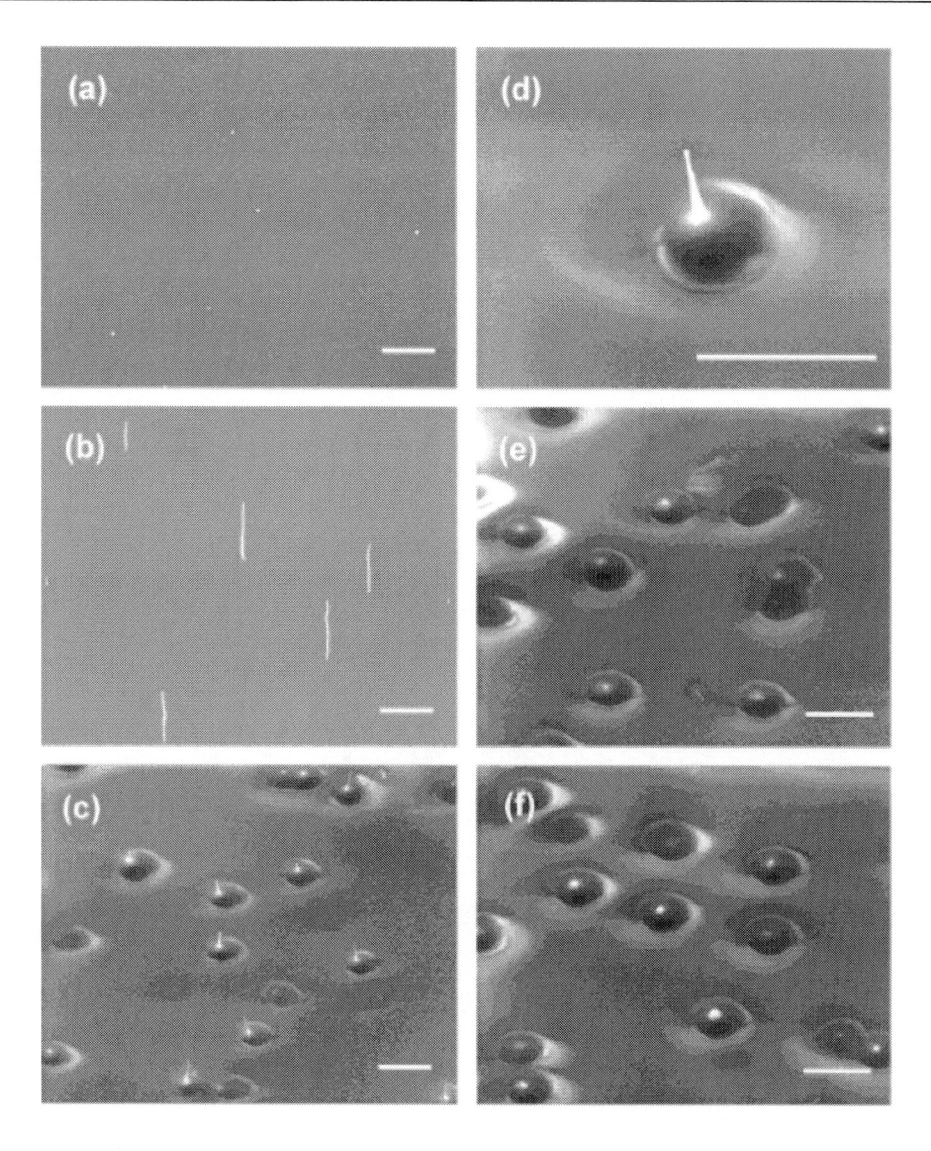

Figure 4. Scanning electron microscope images of a) Electrodeposited Ni nanoparticles; b) low site density aligned CNTs array; c) CNTs array coated with SiO_2 and an epoxy layer, d) close-up look at a single half-embedded CNT; e) CNTs afterpolishing; and f) second electrodeposition of Ni nanoparticles onthe broken CNTs only. (Reprinted from Ref [69] Copyright 2000 American Chemical Society).

3.1.1.2. Carbon Naontubes (CNTs) as NEs and NEAs Element

CNT is another kind of important element used in the NEs&NEAs construction [67,68]. Yuehe Lin and coworkers reported the fabrication of CNTs nanoelectrode array with vertically aligned single wall CNT as assembly element [69,70]. They fabricated highly dispersed and well aligned CNTs arrays as NEAs. Figure 3 shows the fabrication procedures of CNTs-NEAs on Chromium (Cr) coated silicon substrate. First, Cr layer was deposited onto the silicon surface as contact metal; then Ni-NPs were electrochemically deposited onto contact metal layer surface as CNTs growth catalyst. After the growth of aligned CNTs, the block material, SiOx or epoxy was coated onto the substrate surface to stuff the CNTs free areas. Finally, the substrate surface was polished to make the CNTs tips exposed. Figure 4 shows the SEM images of NEAs fabrication steps corresponding to the scheme shown figure 3. Moreover, they also studied the immobilization of glucose oxidase on the exposed CNT tips through the covalent amide linkages [71] (Figure 5). The glucose sensor they fabricated performed a selective electrochemical analysis of glucose in the presence of common interferents (e.g. ascorbic acids, uric, acetaminophen, etc). CNTs NEAs are also promising in the ultra-trace analysis of NADH, hydrogen peroxide, thiol-containing homocysteine, organophosphorus compound, alcohol, etc [72].

Jun Li and coworkers also reported the fabrication of integrated CNTs arrays on contact metals coated silicon surface with metal as CNTs growth catalyst [59]. By this way, they fabricated well-regulated CNTs bundle microelectrode arrays (Figure 6A, Figure 6B) and the electrode arrays typically performed as microelectrode arrays (Figure 6C). The detection limit of electroactive species could reach as low as a few nM on these electrodes. Otherwise, if the CNTs tips are functionalized with oligonucleotide probes or enzymes, this platform can be practicably applicable in DNA/RNA and bio-molecules analysis [73].

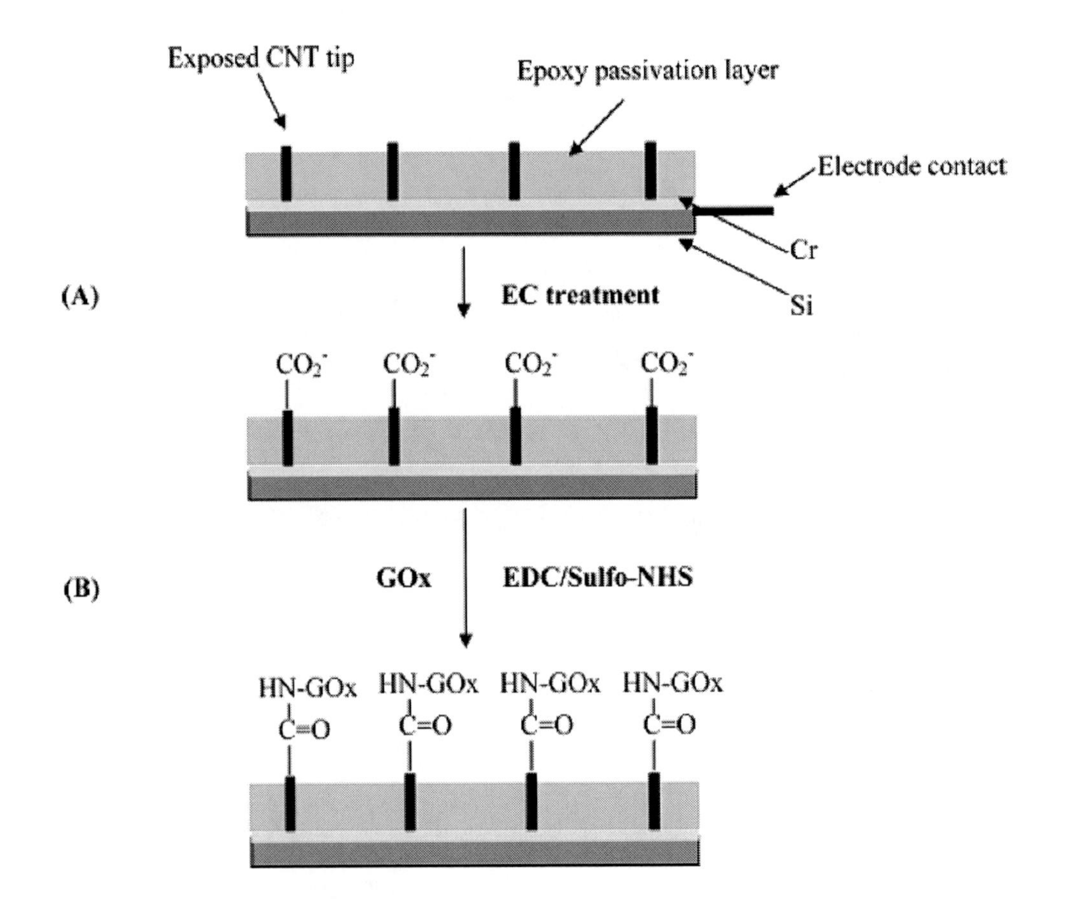

Figure 5. Fabrication of a glucose biosensor based on CNTs-NEAs (Reprinted from Ref [71] Copyright 2000 American Chemical Society).

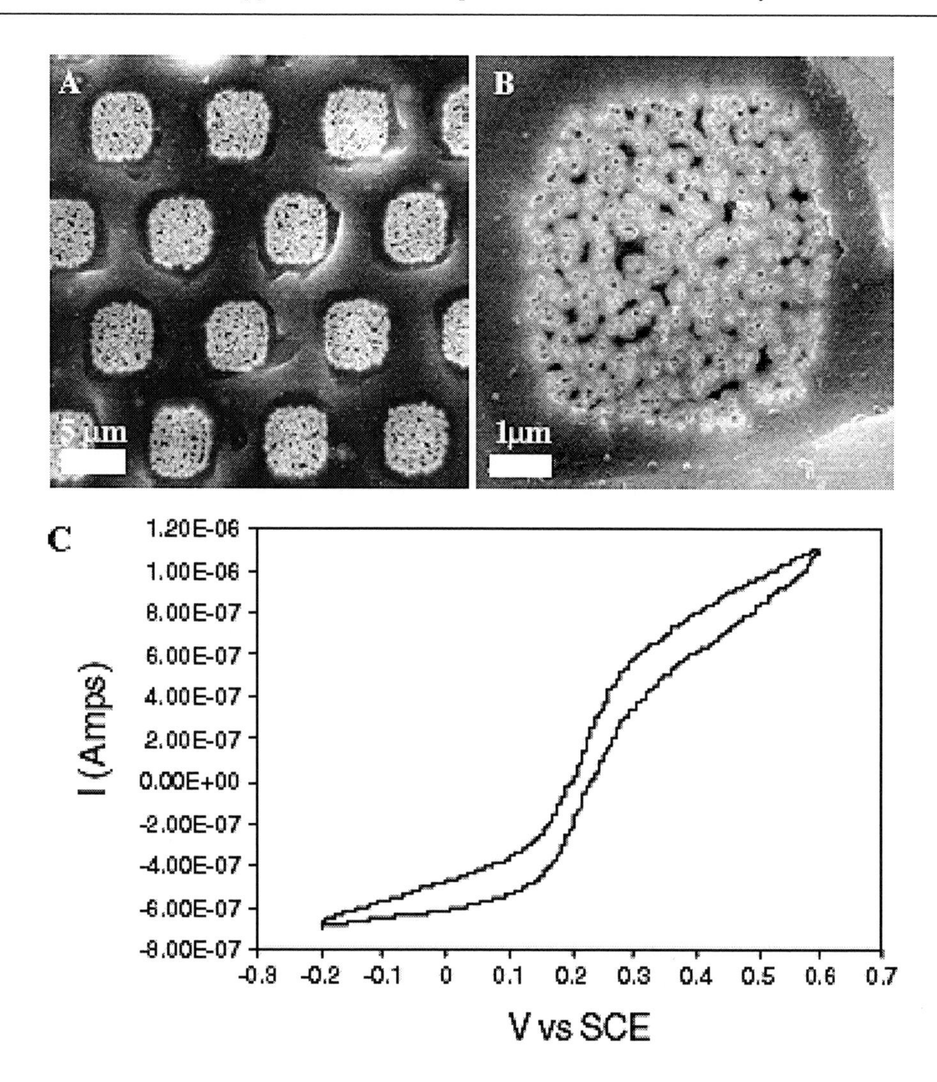

Figure 6. The top view of a CNTs bundle array (A) on silicon surface with 6 μm single-bundle-electrode diameter (B), the electrode array performs as microelectrode array, (C) CV scan of the CNTs bunch array in 1 mM $K_4Fe(CN)_6$/1 M KCl. Scan rate −10 mV s^{-1}. (Reprinted from Ref [59] Copyright 2004 Nanotechnology).

3.1.2. As Electrochemical Catalyst and Bio-label

One of the most significant applications of NPs in electroanalysis field is adopting them as electrochemical catalyst and bio-label. Applications of nanomaterials to biology or medicine include fluorescent biological labels; drug and gene delivery; bio detection of pathogens; detection of proteins; probing of DNA structure; tissue engineering; tumour destruction via heating (hyperthermia); separation and purification of biological molecules and Cells; MRI contrast enhancement; phagokinetic studies, etc [74]. The applications of NPS in bio and the combination of bio-techniques with electrochemistry techniques lead to the generation of some more precise and rapid analysis methods. This is one of the applications of NPs in electrochemistry.

3.1.2.1. Gold Nanoparticles (AuNPs)

AuNPs can be employed as a catalytic element in electrochemical analysis. Table 1 summarizes the applications of AuNPs in electroanalysis. AuNPs showed the excellent catalytic effect to the oxidation of dopamine. [75] Another research indicated that the electrochemical sensitivity of AuNPs modified carbon electrode to dopamine could be increased when some albumin was added during the preparation of AuNPs [19]. The selective detection of dopamine with AuNPs modified electrodes was also reported by many other groups [76-79].

The indium tin oxide (ITO) electrode modified by AuNPs (stabilized by dimethylamino pyridine) enables detection of nitric oxide. [80] Liu S et.al. reported the detection of phenol [81], glucose [82], hydrogen peroxide and nitrite [12,83,84] by the using of AuNPs modified electrodes. They found AuNPs could tune the electron transfer between electrode and myoglobin or hemoglobin which was immobilized onto the AuNPs surface.

S. Bharathi [85] fabricated the glucose oxidase enzyme/AuNPs/silicate network on the ITO glass surface by the co-deposition of these three elements. The glucose oxidase enzyme can catalyze the oxidation of glucose and AuNPs in the network can catalyze the oxidation and reduction of H_2O_2, the by-product of the enzymatic reaction. They also found that this kind of glucose sensor can eliminate the interference from acetaminlphen, ascorbic acid, dopamine etc because of the low operating potential. The electrochemical catalysis and detection of uric acid with AuNPs and gold nanorods modified ITO electrode were also studied [86,87].

J.M. Pingarrón constructed colloidal-gold/cysteamine modified carbon paste electrode (AuNPs-Cyst-CPE) and their studies showed that to the voltammetric determination of thio-compounds such as methionine, as well as of methionine-based peptides, AuNPs-Cyst-CPE exhibited improved electroanalytical characteristics when compared with colloidal-gold cysteamine-modified Au disk electrodes (AuNPs-Cyst-AuE) [88].

Table 1. Applications of AuNPs in electrochemical analysis

Compound	**Nitric oxide**	**Hydrogen peroxide**	**Dopamine**	**phenol**	**Glucose**
Detection limit	0.01mM [80] 0.7 μM [12]	4.6μM12 9.1×10^{-7} M. [83]	2×10^{-7} M [19,79] 0.13 μM [75]	6.1 nM [81]	0.01mM [82] 0.04ng [89]
Compound	Fructose	Lactose	Sucrose	Sorbitol	Arabinose
Detection limit	10ng [89]	9ng [89]	12ng [89]	9ng [89]	6ng [89]
Compound	Xylitol	Ascorbic acid	Serotonin (5-HT)	SO_2	Cholesterol
Detection limit	6ng [89]	[87]	8×10^{-7} M [79]	[90]	[91]
Compound	Norepinephrine	Uric acid	Arsenic-(III)	Methionine	
Detection limit	[87]	[86,87]	1 nM [92]	5.9×10^{-7}M [88]	

* In Flow injection and Liquid Chromatography.

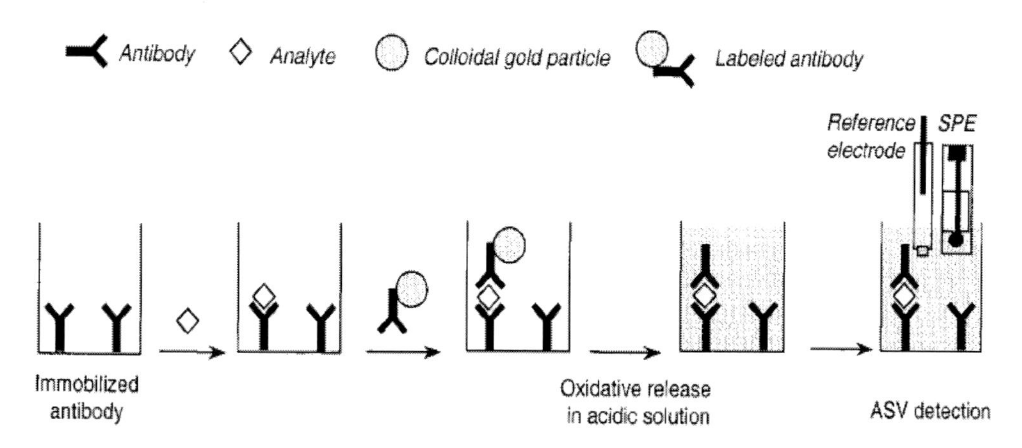

Figure 7. Schematic representation of noncompetitive heterogeneous electrochemical immunoassay based on a colloidal gold label. (Reprinted from Ref [93] Copyright 2000 American Chemical Society).

AuNPs were also used in bio-label and bio-electrochemical-analysis. Benoit Limoges group reported the sensitive electrochemical immunoassay which is based on the anodic stripping voltammetric detection of dissolved gold-label [93]. Figure 7 shows the principle of the procedures: arm-antibodies were absorbed onto the polystyrene micro-wells and goat IgG analyte was then captured by the antibody. The gold labeled antibody was employed to sandwich the goat IgG.. Finally, the gold was dissolved into acid solution for the subsequent electrochemical detection. By this method, 3 pM of immunoglobulin was determined. The exact idea was also used in the electrochemical detection of DNA [93].

Yuzhi Fang et.al. reported the colloidal AuNPs enhanced detection of ssDNA [94]. Reference [95,96] reported detection of DNA hybridization, based on the deposition of silver on AuNPs tags and a subsequent electrochemical stripping detection of dissolved silver. This method avoids the need of enzymes or fluorescence detection. The similar ideas were also used into the gene-analysis [97,98]

Reference [99] reported the reversible electrochemistry of horse heart cytochrome c at SnO_2 electrode modified with colloidal AuNPs. The electrochemical behavior of gold colloid and colloidal gold labeled anti-human IgG (immunogold) on carbon paste electrode was also studied [100]. Based on the electrochemical response of colloidal gold, low detection limits were achieved. And this methodology has been used in pharmacology research [101].

3.1.2.2. Carbon Nanotubes (CNTs)

CNT is a kind of widely used nanomaterial in electrochemistry because of its excellent and fancy electronic structure properties. Modification of CNTs onto traditional electrode surface has been widely reported. CNT plays very important role in the electron transfer between electrode and electrolyte. In most cases, CNT performs as electron tunnel and electrocatalyst. [5] High electrocatalysis of CNTs result from the surface defect sites and ends of nanotube [102,103]. CNTs modified electrodes and the nanocomposites consists of CNTs were employed in electrochemical applications such CNT/metal (oxide), CNT/organic layer, CNT/biomaterials [5].

Table 2 summarizes the application of CNTs in the electrochemical analysis of some inorganic and organic molecules.

Table 2. Applications of CNTs in electrochemical analysis

Compound	NPs type	Media	Detection limit
Nitric oxide	MWNT [104,105]	PBS (pH 7.0)	2×10^{-8}M [104] 8×10^{-8}M [105]
Hydrogen peroxide	MWNT [106-108]	PBS (pH 7.4)	50 μM [108]
Dopamine	SW/MWNT [108-112]	PBS (pH 7.4) [109] PBS (pH 7.1) [112] PBS (pH 5.0) [113]	0.5 μM [108] 1×10^{-7}M [110] 3×10^{-7}M [112] 2.5×10^{-10}M [113]
Serotonin	MWNT [110]	PBS (pH 7.0)	1×10^{-7}M [110]
Glucose	MWNT [109,114]	PBS (pH 7.4) [109] 0.1M NaOH [114]	0.6 mM [109] 1 μM [114]
Uric acid	SWNT [115]	PBS (pH 7.0)	
Cysteine	MWNT [116,117]	PBS (pH 7.0) [116] PBS (pH 7.8) [117]	2μM [116] 0.8μM [117]
N-acetylcysteine	MWNT [117]	PBS (pH 7.8) [117]	3.3μM [117]
Homocysteine	MWNT [117,118]	PBS (pH 7.8) [117] PBS (pH 7.0) [118]	0.75μM [117]
Epinephrine	SWNT [112]	PBS (pH 7.1) [112]	
Ascorbic acid	MWNT [119]	PBS (pH 7.0) [119]	
Glutathione	MWNT [117-119]	PBS (pH 7.8) [117] PBS (pH 7.0) [118]	2.9μM [117] 0.06μM [118]
NADH	MWNT [107,108]	PBS (pH 7.4)	2 μM [108]
Oxygen	MWNT [120]	0.1M NaOH	
Guanine	MWNT [121,122]	Acetate buffer (pH 5.0)	
Insulin	MWNT [122]	PBS (pH 7.4)	14 nM [122]
3,4-dihydroxyphenylaceticacid (DOPAC)	MWNT [113]	PBS (pH 5.0)	3×10^{-10}M [113]
Norepinephrine	MWNT [113]	PBS (pH 5.0)	2.5×10^{-10}M [113]
3-methoxy-4-hydroxyphenylglycol (MHPG)	MWNT [113]	PBS (pH 5.0)	5×10^{-10}M [113]
5-hydroxytryptamine (5-HT)	MWNT [113]	PBS (pH 5.0)	3.5×10^{-10}M [113]
homovanillic acid (HVA)	MWNT [113]	PBS (pH 5.0)	1.25×10^{-9}M [113]
5-hydroxyindoleacetic acid (5-HIAA)	MWNT [113]	PBS (pH 5.0)	6×10^{-10}M [113]
Catechol	MWNT [123,124]	PBS (pH 7.0)	

Because CNTs have large specific surface area, wide electrochemical window, flexible surface chemistry, conductivity, good stability, biocompatibility, they are also promising in the electrochemical DNA-sensors based on the electrochemistry. [4] Especially, the flexible surface chemistry make CNTs are attractive because their surface could be activated easily and versatilely, which facilitates the immobilization of bio-molecules on carbon surface and can also increase the immobilization yield. [125-127] Moreover, CNTs are very suitable in the applications of bioelectrochemistry and biosensors because the electron transfer can happen smoothly between CNTs and enzymes/proteins (glucose oxidase/GOx, hemeproteins, GRP, hemoglobin/Hb, myoglobin/Mb). The electron transfer between CNTs and biomolecules are mainly through the CNTs surface defect [128] and π -π communication [129]. The electrochemical detection limit of DNA with CNTs modified electrode can reach 1fg/ml [130] and 2.3×10^{-14} mol L^{-1131}. Table 3 summarizes the possible application of CNTs in bio analysis.

3.1.2.3. Nickel Nanoparticles (Ni NPs)\

Nickel electrode has been widely utilized in carbohydrate electrochemical analysis with alkaline solution as electrolyte. [132] Highly dispersed nickel-NPs were also modified onto graphite-like carbon film electrode surface for electroanalysis. This nickel NPs modified electrode had excellent electrocatalytic ability and good stability compared with a Ni-bulk electrode with regard to the electrooxidation of sugars. [43] The improved detection limits for the investigated sugars, namely, 20, 25, 50, and 37 nM for glucose, fructose, sucrose, and lactose, respectively (table 4).

Table 3. Electron transfer properties of proteins on CNTs (Reprinted from Ref [5] Copyright 2005 The Japan Society for Analytical Chemistry)

Protein	CNT electrode	$E^{0\prime}$/V	ET rate/s^{-1}	Analyte
Cyt. *c*	CNT paste	*ca.* 0.06	NA	NA
HRP	Poly-L-lysine/SWNT	NA	NA	H_2O_2
Hb	CNTPME	−0.278 (SCE)	0.062	H_2O_2
Hb	CNT film	−0.343 (SCE)	1.25 ± 0.25	H_2O_2
Mb	MWNT film	NA	NA	H_2O_2
Mb	MWNT film	−0.248 (SCE)	5.4	NO
MB	Oriented SWNT	−0.21 (SCE)	NA	H_2O_2
HRP	Oriented SWNT	−0.25 (SCE)	NA	H_2O_2
HRP	CNTPME	−0.643 (Ag/AgCl)	2.48 ± 0.03	H_2O_2
Cyt. *c*	SWNT film	NA	NA	Adenine
Catalase	SWNT film	−0.414 (SCE)	NA	H_2O_2
MP-11	MWNT film	−0.26 (Ag/AgCl)	38	H_2O_2,O_2
MP-11	Aligned SWNT/Au	−0.390 (Ag/AgCl)	2.8 - 3.9	NA
GOx	CTAB-CNT	−0.466 (SCE)	1.53 ± 0.45	NA
GOx	Aligned SWNTs	−0.422 (Ag/AgCl)	0.3	NA
GOx	Aligned SWNTs	−0.45 (SCE)	NA	Glucose
GOx	CNTPME	−0.659 (Ag/AgCl)	1.61 ± 0.3	NA
GOx	SWNT film	−0.441 (Ag/AgCl)	1.7	Glucose

$E^{0\prime}$, formal potential; NA, not available, MP-11, microperoxidase 11.

Table 4. Applications of Ni NPs in electrochemical analysis

Compound	Glucose	Fructose	Sucrose	Lactose
Detection limit	20nM [43]	25nM [43]	50nM [43]	37nM [43]

Table 5. Applications of copper NPs in electrochemical analysis

Compound	Hydrogen peroxide	o-diphenols (catechol, dopamine, pyrogallo)	Glucose	Oxygen	Nitric acid
Detection limit	0.97 μM [134]	Catechol, 3μM [135] Dopamine, 5μM [135]	40nM 0.8pmol [136]*	1ppm [133]	1.5 μM [137]

*in flow injection system

3.1.2.4. Copper Nanoparticles (Cu NPs)

Copper NP is another commonly used NPs in electrochemistry. The priority of CuNPs is its versatile surface state and surface reaction. The Cu_2O/CuO component generated on copper surface always participates in the electrochemical reaction through the hydrogen bond or complex interaction. Otherwise, because Cu_2O is a good p-type semiconductor, sometimes, photoelectrochemical reaction happens between excited Cu_2O and electrolyte. [133] Table 5 lists some examples of the applications of CuNPs in electrochemical analysis.

Table 6. detection of α amino acids at 100nm CuNPs modified screen-printed carbon electrode (Reprinted from Ref [138] Copyright 2004 The Royal Society of Chemistry)

Amino acid	Linear range/μM	R	Sensitivity/ nA μM^{-1}	D_L (S/N = 3) /nM	D_L (S/N = 3) /ng
Ala	5—500	0.9990	7.29	24	0.04
Arg	5—300	0.9960	5.97	587	2.47
	5—500	0.9894	5.05	695	2.93
Asn	5—300	0.9998	11.68	94	0.28
	5—500	0.9901	9.78	112	0.34
Asp	5—500	0.9951	5.98	2693	7.17
Cys	10—300	0.9955	6.41	875	2.12
Glu	5—500	0.9901	4.90	624	1.84
Gln	5—500	0.9962	7.40	221	0.65
Gly	5—500	0.9996	2.83	161	0.21
His	5—500	0.9961	12.07	44	0.14
Iso-Leu	5—500	0.9976	4.40	297	0.78
Leu	5—300	0.9978	7.44	231	0.61
Lys	5—500	0.9994	4.30	1890	7.23
Met	5—500	0.9929	8.99	102	0.30
Phe	5—500	0.9991	7.62	472	1.56
Pro	5—300	0.9988	3.70	2545	5.86
Ser	5—500	0.9914	7.58	1393	2.93
Thr	5—500	0.9954	8.46	429	1.02
Trp	5—500	0.9993	6.66	389	1.59
Tyr	5—500	0.9999	7.17	548	1.98
Val	5—500	0.9992	4.21	574	1.34

Copper NPs modified electrodes are also used in amino-acid analysis. 100nm copper NPs modified screen-printed carbon electrode was used in the detection of a-amino acids based on two step complexation between amino acid and copper surface [138]. They found the $Cu^{II}O$-amino acid complex on the 100nm copper surface played a very important role in the surface complexation processes. The reversible adsorption (on $Cu^{II}O$)/desorption (on $Cu^{I}O$) make 100nm Cu NPs modified screen-printed carbon electrode more stable and versatile in the amino-acid analysis. All 20 underivatized amino acids could be sensitively determined in buffer solution (table 6).

3.1.2.5. Other Nanoparticles

Platinum NPs (PtNPs), silver NPs (AgNPs) modified electrodes were also used in electroanalysis of some organic and inorganic molecules. [139,140,141] Table 7 gives some examples of the applications of other NPs in electroanalysis.

Table 7. Applications of other NPs in electrochemical analysis

Compound	Nitric oxide	Hydrogen peroxide	Dopamine	Glucose	Oxygen
Detection limit	$3\times^{-7}$M [141]	0.01μM [41,42] 25nM [88]	3×10^{-8} M [142]	0.5 μM [88]	
NPs	Silver NPs [141]	Platnium NPs [41,42,88] Iron NPs [44]	Fe_3O_4 NPs [142]	Platnium NPs [88]	Iridium NPs [143]

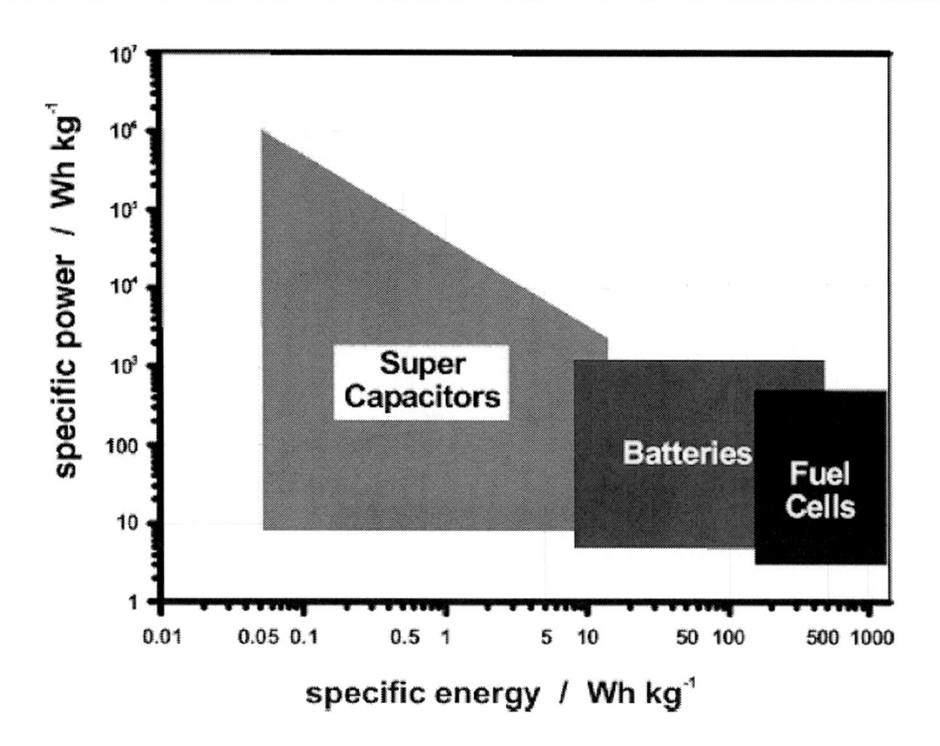

Figure 8. The different merits of different electrochemical power supplying system.

3.2. Power and Energy

Electrochemical energy storage and conversion include batteries, fuel cells, and electrochemical capacitors (ECs), etc [144]. Batteries and fuel cells are based on the chemical energy conversion to electrical energy and electrochemical capacitors are based on capacitors result from the generation and destruction of electric double layer or solid surface reaction. The different working mechanisms lead to the different characterizations of these three power supplying systems. Figure 8 shows the difference: batteries and fuel cells can provide long-time power but supercapacitors can provide high power energy. Single electrochemical power source can't supply both high energy and high power but the best system with high power and high energy can be achieved when the available electrochemical power systems are combined.

The application of NPs in power and energy is one of the most promising and active fields. Reference [7] reviewed the application of NPs in power and energy. The authors gave the fundamental principles and some very good example about the application of NPs in power and energy. The advantage of NPs in power and energy field is its catalysis and ultra-high specific surface area. All the researches of this field are all based on these two fundaments. The studies in this area focus on 1) trying to increase the activity per unit area 2) decreasing the necessary amount of expensive catalyst 3) exploiting the new process to get new materials 4) engineering design.

3.2.1. Fuel Cell

The applications of NPs in fuel are mostly based on their high catalytic abilities. Some catalytic reactions just happen on the specific crystal face, surface step sites or lattice defects. NPs possesses large specific surface area (surface area to the volume), numerous free valence positions to the cluster, which makes NPs are very suitable for catalyst. NPs-catalysis has electronic effects, geometric effects and support effects [7]. Electronic effects arise from the electronic structure compactness. For metal NPs, especially when the size of nanoparticle was smaller than 5nm, nanoparticle shows totally different electronic structure comparing with bulk metal. The small number of atoms involved in the electron bond formation leads to strong valence electrons localization. The heavy electrons localization can result in the smaller lattice constant with respect to bulk metal. The smaller lattice constant results in the high surface catalytic ability [145,146]. The more the nanoparticle is small, the more the electronic structures are different. The geometric effect results from the morphological properties of nanoparticle which strongly affect the catalytic activity during the reaction process. The supporting material always affects the catalytic activity of NPs because the interaction of catalytic NPs with supporting material may lead to the electronic properties change of those atoms close to the supporting surface [147].

3.2.1.1. PtNPs

The most widely used catalysts in fuel cell is platinum containing metal or alloys [148]. PtNPs were well-known catalytic to the reduction of oxygen in both acid and alkaline media [149-155] and platinum alloys were also employed and developed as highly-active catalyst.

Pt_xCr_y, Pt_xCo_y, Pt_xNi_y alloy and PtV/C composite were proved highly active to the reduction of oxygen [37,156-158]

Platinum metal is also used as catalyst in the oxidation of methanol. The deadly drawback of platinum material as methanol oxidizing electrode is the irreversible adsorption of monoxide carbon which results in the electrode poisoning. A way to resolve this problem is to synthesize the alloy NPs [159] . To solve this problem, gold-platinum NPs were studied as catalyst for the oxidation of methanol [160]. Other platinum containing alloy NPs (Pt-Sn, Pt-Ru) were also studied in the oxidation of methanol. [37] Ruthenium, tin, molybdenum were co-alloyed with platinum to make CO tolerate catalyst [161,162]. Research indicated the metal oxide (RuO_x) played very important roles for the tolerance of CO) Moreover, platinum-alloy NPs were also employed as anode catalysts in the hydrogen oxidation in the presence of CO.

3.2.1.2. AuNPs

AuNP is another kind of promising catalytic NP in fuel cell. AuNPs were proved to be catalytic for the reduction of oxygen/hydrogen peroxide to water directly in acid [163-168] or base solution [169,170] and in the oxidation of carbon monoxide or methanol. Reference [171] reported the catalytic activation of core-shell assembled AuNPs as catalyst for methanol electro-oxidation; they found the oxidized layer plays a very important role in this process.

3.2.1.3. Carbon NPs

Because of economic advantages and excellent properties of carbon materials [172] such as accessibility, low price, good electric conductivity, good thermal conductivity, low density, corrosion resistance, oxidation resistance and reduction resistance, carbon electrolytes were widely used as catalyst supporting materials [148]. The other forms of carbon (nano-horns, nano-tubes, nano-fibers, nano-sheets, powers, clothes or papers) nanostructures were also adopted as catalyst supporting materials [173,174] Moreover, carbon nanotubes are also used in the hydrogen storage [175-177].

3.2.2. Supercapacitor

A supercapacitor is an electrochemical capacitor that offers very high capacitance in a small package [178]. They are able to store a large amount of charge (energy) that can be released

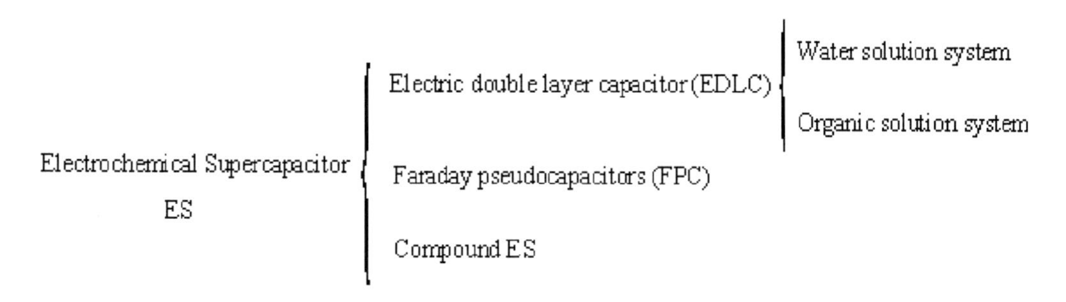

Chart 2. The classification of electrochemical supercapacitors.

very quickly. This means they are superior in short term, high-energy applications, such as when an appliance is switched on or an electric car accelerates. The electrochemical supercapacitors (ES) are also called Electrochemical Capacitors, (EC), Supercapacitors or Ultracapacitors. ES can be classified as chart 2 shows.

The merits of ES including:

1. Quick recharge and discharge (seconds);
2. Super capacitance
   a. Gravimetric specific energy >2.5Wh/kg
   b. Gravimetric specific power>500W/kg;
3. Superior recyclability (over 500 000 recharge cycles);
4. No memory effect;
5. Free size and shape;
6. Low-toxicity

Applications of ES including:

1. Hybrid-electric vehicles
2. In notebook computers
3. Industrial actuators and controls
4. Uninterruptible power supplies and power quality devices
5. Automotive auxiliary systems and engine starting
6. Two-way pagers, cellular phones and other wireless communications devices
7. Digital cameras, power tools, etc
8. Motive and standby power.
9. Military affairs, to apply very high power instantly

High specific surface area materials such as CNTs, carbon powder, carbon black are commonly used electrode materials. FPC is based on the rapid reaction on solid surface. Metal oxides NPs are commonly used as electrode materials. This kind of ES system possesses good conductivity, high electric capacitor to weight, long life, but high price and toxicity. Compound ES is combining EDLC and FPC together through proper design for given purpose. Because the construction of compound ES is most an engineering issue, we don't discuss compound ES here.

The capacitance of double layers is typically dozens of $\mu F\ cm^{-2}$ grade. If high surface area electrodes are employed for capacitor devices, the capacitance could reach hundreds of Faradays. Electrochemical supercapacitor has revolutionized enormous energy storage if high specific-surface area materials were used as electrode materials [178]. High specific surface area of carbon powders [179] and hybrid materials are utilized in commercial electrochemical capacitors.

3.2.2.1.CNTs

CNTs were firstly utilized as electrode for electrochemical supercapacitors in 1997 [180]. They prepared multi-walled CNTs with specific surface of 430 $m^2\ g^{-1}$ and the maximum specific capacitance and energy density reached 113 F/g and 0.56Wh/kg correspondingly. Other researches also demonstrated the CNTs are very suitable for the supercapacitor devices. [181,182]

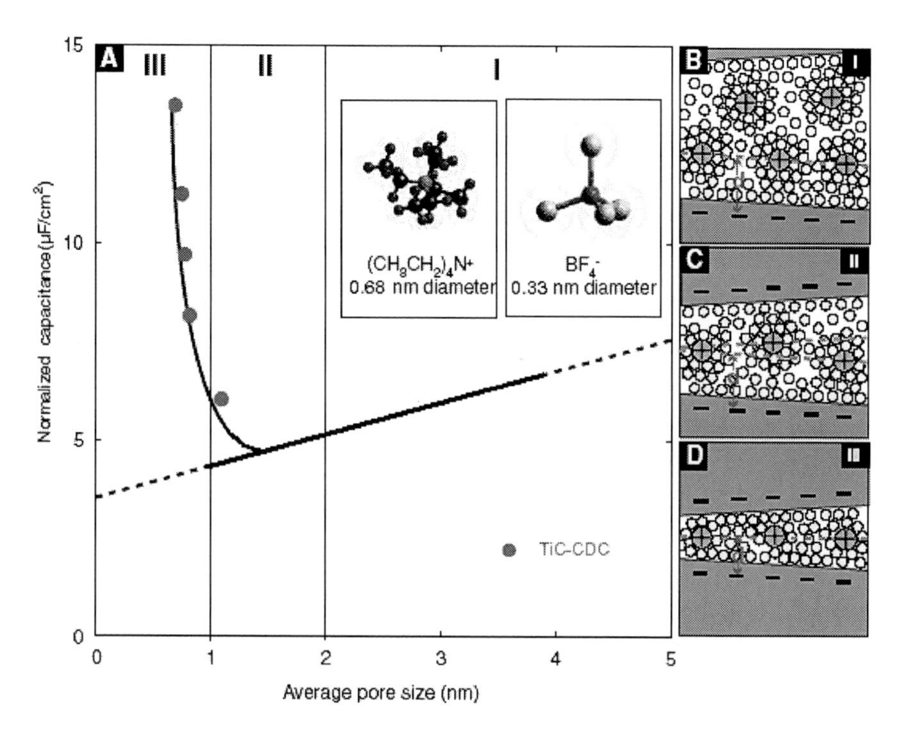

Figure 9. (A) Plot of specific capacitance normalized by BET SSA for nanoporous TiC-CDC with average diameter from ~0.6 nm to 1.25 nm (red dot) in contrast to traditional result (straight line) Drawings of solvated ions residing in pores with distance between adjacent pore walls (B) greater than 2 nm, (C) between 1 and 2 nm, and (D) less than 1 nm illustrate this behavior schematically. The capacitor can be calculated with equation $C=\varepsilon A/d$; ε is the electrolyte dielectric constant; A is the electrode area accessible to electrolyte ions, d is the electrode separation distance. (Reprinted from Ref [183] Copyright 2006 Science).

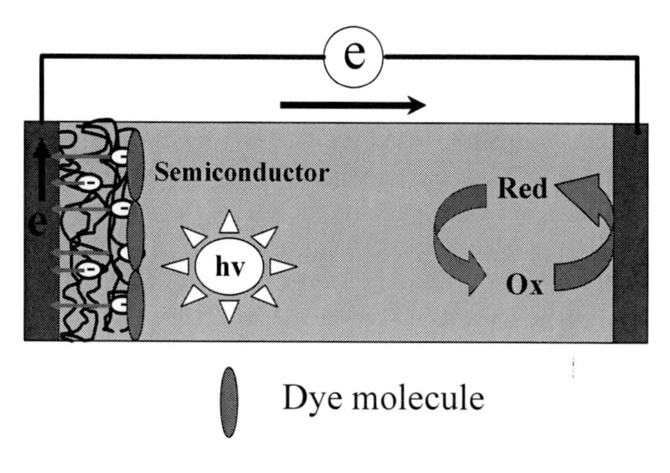

Figure 10. Photocurrent generation process in DSSCs.

3.2.2.2. Mesoporous Hybrid Nano-materials

C. Portet etc studied the supercapacitor with pore size smaller than 1nm porous material as electrode [183]. They synthesized the titanium carbide-derived carbon (TiC-CDC) with pore size range from 0.6nm to 2.25nm (with 0.05nm accuracy). They found when the nanopore size is smaller than the solvated ion, the specific capacitor increase dramatically because of the compact and squeezing structure of solvated ion in the pores (figure 9). The increase of capacitor arises from the dramatic decrease of virtual counter electrode distance (d). Their report firstly demonstrated 1) the charge storage in pores smaller than the size of solvated electrolyte ions could lead to specific capacitor increase 2) ions could squeeze into nanopores whose size is similar or even smaller than the ion size (this result beyond the exception). This amazing and unbelievable research makes super-capacitor become explicitly promising power device.

3.2.3. Solar Cell

The most commonly used NPs in nanocrystalline solar cells are semiconductive NPs, such as TiO_2, CdS, ZnS, CdSe, CNTs, C_{60}, etc [184]. Photosensitized semiconductor solar cells have different fundamental principles in compare to p-n junction solar cell. In reference to high-cost silicon solar, the nanocrystalline solar cells are more attractive because of low-cost and easy fabrication. Dye-sensitized nanocrystalline solar cells (DSSCs) system is one of the examples. The most widely studied semiconductive NPs in solar cell are titanium dioxide (TiO_2) NPs. Grätzel firstly reported the dye-sensitized TiO_2 NPs solar cell in 1991 and the efficiency can get about as high as 7.5%. [185]

Figure 10 shows the structure of typical DSSCs. DSSCs consist of anode, cathode and electro-active electrolyte. The irradiation light is absorbed by dye molecules and the excited electrons transfer to the conduction band of semiconductor nanocrystals. The positively charged dye molecules immediately react with electro-active electrolyte and get compensated. At the same time, the collected electrons on anode (semiconductor electrode) go through the out loop and generate the photocurrent (The electro-active electrolyte gets electrons from the cathode).

4. Conclution

Similarly to that the bulk electrodes used so far, NPs modified electrode (including gold, platinum, copper and nickel) become deactivated through fouling [76-78,140,186]. The NPs surface fouling is a deadly and serious problem for the application of NPs in electrochemistry. The possible approach to solve this problem is to design NPs structure and surface functionalization. Reference [187] described the cleaning of platinum NPs without loss of crystalline surface structure. The NPs could be electrochemically refreshed in sulfuric acid solution after cyclic voltammograms. NPs doped carbon paste electrode surface could be renewed just by removing the top surface [12,19,187].

The other problem of NPs in electrochemistry is the reproducibility. One of the drawbacks of electrochemical methods is the bad reproducibility which arises from the difficult reproduction of electrode surface. It seems that the introduction of NPs in electrochemistry aggravates this situation but not making it better. That's because the bad reproducibility of NPs modification or doping as electrode element. To get better reproducibility, the precise fabrication of nanostructure and functional surface is necessary. The related techniques are still under development.

Another topic of NPs in electrochemistry is the safety, which is a comprehensive topic in NPs applications. Because of small size, NPs can penetrate cell membranes and integrate themselves into larger molecules. The fatalness of the invasion of NPs to bio-system is still under investigation and attracting more and more attentions.

References

[1] Allen J. Bard, L. R. F. *John Wiley & Sons, Inc.* 2001.
[2] David Herna¬ndez-Santos, M. B. G. l.-G., Agustln Costa Garcla *Electroanalysis* 2002, 14, 1225-1235.
[3] Compton, C. M. W. R. G. *Anal Bioanal Chem* 2006, 384, 601-619.
[4] Pingang He, Y. X., and Yuzhi Fang *Microchim Acta* 2006, 152, 175.
[5] Kuanping Gong, Y. Y., Meining Zhang, Lei Su, ShaoXiong , Lanqun Mao *Anal Sci* 2005, 21, 1383-1393.
[6] Adam K. Wanekaya, W. C., Nosang V. Myung, Ashok Mulchandani *Electroanalysis* 2006, 18, 533 - 550.
[7] Fabio Raimondi, G. n. G. S., R diger K tz, and Alexander Wokaun *Angew. Chem. Int. Ed.* 2005, 44, 2190-2209.
[8] Shaojun Dong, E. W. *Anal. Chim. Acta* 2007 598, 181-192.
[9] Frens, G. *Nature Phys. Sci.* 1973, 241, 20-22.
[10] Mathias Brust, M. W., Donald Bethell, David J. Schiffrin, Robin Whyman *J. Chem. Soc., Chem. Commun.* 1994, 7, 801-802.
[11] Allen C. Templeton, W. P. W., and Royce W. Murray. *Acc.Chem.Res* 2000, 33, 27-36.
[12] Liu S, J. H. *Electroanalysis* 2003, 15, 1488.
[13] Song Y, J. V., McKinney C, Donkers R, Murray RW *Anal Chem* 2003, 75, 5088.
[14] Maye M M, L. J., Lin Y, Engelhard MH, Hepel M, Zhong C-J *Langmuir* 2003, 19, 125.
[15] Wilcoxon JP, M. J., Provencio P *Langmuir* 2000, 16, 9912.
[16] Daniel, M.-C. A., D. *Chem. Rev.* 2004, 104, 293.
[17] Sutherland, W. S. W., J. D. *J. Colloid. Intelface Sci.* 1992, 148, 129-141.
[18] Kenneth R. Brown, D. G. W., and Michael J. Natan *Chem. Mater.* 2000, 12, 306-313.
[19] Miscoria SA, B. G., Rivas GA *Electroanalysis* 2005, 17, 1578.
[20] Becker JA, S. R., Festag R, Ruland W, Wendorff JH, Pebler J, Quaiser SA, Helbig W, Reetz MT *J Chem Phys* 1995, 103, 2520.
[21] Khilar, R. P. B. a. K. C. *Langmuir* 2000, 16, 905-910.
[22] Abhijit Manna, T. I., Masayasu Iida, and Naoko Hisamatsu *Langmuir* 2001, 17, 6000-6004.
[23] Vidal-Iglesias FJ, S.-G. J., Rodriguez P, Herrero E, Montiel V, Feliu JM, Aldaz A *Electrochem Commun* 2004, 6, 1080.
[24] Harriman A, M. G., Neta P, Richoux MC, Thomas JM *J Phys Chem* 1988, 92:, 1286.
[25] Thomas Hirsch, M. Z. *Angew. Chem. Int. Ed.* 2005, 44, 6775 -6778.

[26] Helbig, M. T. R. a. W. *J. Am. Chem. Soc.* 1994, 116, 7401-7402.
[27] Yu-Ying Yu, S.-S. C., Chien-Liang Lee, and C. R. Chris Wang *J. Phys. Chem. B* 1997, 101, 6661-6664.
[28] L. Rodrı´guez-Sa´nchez, M. C. B., and M. A. Lo´pez-Quintela *J. Phys. Chem. B* 2000, 104, 9683-9688.
[29] Zoval JV, S. R., Biernacki PR, Penner RM *J Phys Chem* 1996, 100, 837.
[30] Ueda M, D. H., Anders A, Kneppe H, Meixner A, Plieth W *Electrochim. Acta* 2002, 48, 377.
[31] Adora S, S. J. P., Soldo-Olivier Y, Faure R, Chainet E, Durand R *ChemPhysChem* 2004, 5, 1178.
[32] Liu S, T. Z., Wang E, Dong S *Electrochem Commun* 2000, 2, 800.
[33] Lu G, Z. G. *J Phys Chem B* 2005, 109, 7998.
[34] He Z, C. J., Liu D, Tang H, Deng W, Kuang Y *Mater Chem Phys Lett* 2004, 85, 396.
[35] Maillard F, E. M., Cherstiouk OV, Schreier S, Savinova E, Stimming U *Faraday Discuss* 2004, 125, 357.
[36] Sarkar DK, Z. X., Tannous A, Louie M, Leung KT *Solid State Commun* 2003, 125, 365.
[37] Leger, J. M. *Electrochim. Acta* 2005, 50, 3123.
[38] Przenioslo R, S. I., Rousse G, Hempelmann R: *Phys Rev B* 2002, 66, 014404/1.
[39] Reetz MT, M. M. *Adv Mater* 1999, 11, 773.
[40] Y. J. Yang, L. Y. H., H. Xiang *Russian Journal of Electrochemistry* 2006, 42, 954-958.
[41] You T, N. O., Tomita M, Hirono S *Anal Chem* 2003, 75, 2080.
[42] You T, N. O., Horiuchi T, Tomita M, Iwasaki Y, Ueno Y, Hirono S *Chem Mater* 2002, 14, 4796.
[43] Tianyan You, O. N., Zilin Chen, Katsuyoshi Hayashi, Masato Tomita, and Shigeru Hirono *Anal Chem.* 2003, 75, 5191-5196.
[44] Guo L, H. Q., Li X-y, Yang S *Phys Chem Chem Phys* 2001, 3, 1661.
[45] Dubois E, C. J. *Langmuir* 2003, 19, 10892.
[46] Grunes J, Z. J., Anderson EA, Somorjai GA *J Phys Chem B* 2002, 106, 11463.
[47] Suslick KS, C. S., Cichowlas AA, Grinstaff MW *Nature* 1991, 353, 414.
[48] Serp P, F. R., Kihn Y, Kalck P, Faria JL, Figueiredo JL *J Mater Chem* 2001, 11, 1980.
[49] Heiz U, V. F., Sanchez A, Schneider WD *J. Am. Chem. Soc.* 1998, 120, 9668.
[50] U, H. *Appl Phys A* 1998, 67, 621.
[51] Liu Y-C, L. L.-H. *Electrochem Commun* 2004, 6, 1163.
[52] Dimitrios Tasis, N. T., Alberto Bianco, and Maurizio Prato *Chem. Rev.* 2006, 106, 1105-1136.
[53] P.M.Ajayan *Chem. Rev.* 1999, 99, 1787.
[54] J.C.Charlier *Acc.Chem.Res* 2002, 35, 1063.
[55] Iijima, S. *Nature* 1991, 354, 56.
[56] S.Iijima, T. I. *Nature* 1993, 363, 603.
[57] Rubinstein, E. S. a. I. *J. Phys. Chem.* 1987, 91, 6663.
[58] Stulik, K., Amatore, C., Holub, K., Marecek, V., Kutner, W *Pure & Appl.Chem.* 2000, 72, 1483.
[59] Alan M Cassell, Q., Brett A Cruden, Jun Li, Philippe C Sarrazin, Hou Tee Ng, JieHan, and M Meyyappan *Nanotechnology* 2004, 15, 9.
[60] Baranski., A. S. *J. Electroanal. Chem.* 1991, 300, 309

[61] W.M.Arrigan, D. *Analyst.* 2004, 129, 1157.
[62] Faraday, M. *Philos. Trans. R. Soc. London* 1857, 147, 145.
[63] Christophe Demaille, M. B., Michael Tsionsky, and Allen J. Bard *Anal Chem.* 1997, 69, 2323-2328.
[64] Wenlong Cheng, S. D., and Erkang Wang, *Anal Chem.* 2002, 74, 3599.
[65] Ru-hai Tian, T. N. R., Yasuaki Einaga, and Jin-fang Zhi *Chem. Mater.* 2006, 18, 939-945.
[66] Ru-hai Tian, J.-f. Z. *Electrochem Commun* 2007, 9, 1120-1126.
[67] Jun Li, J. E. K., Alan M. Cassell, Hua Chen, Hou Tee Ng, Qi Ye, Wendy Fan, Jie Han, M. Meyyappan *Electroanalysis* 2005, 17, 15-27.
[68] Wang, J. *Electroanalysis* 2005, 17, 7-14.
[69] Y. Tu, Y. L., Z. F. Ren *Nano Letters* 2003, 3, 107-109.
[70] Yi Tu, Y. L., Wassana Yantasee, Zhifeng Ren *Electroanalysis* 2005, 17, 79-84.
[71] Yuehe Lin, F. L., Yi Tu, and Zhifeng Ren *Nano Letters* 2004, 4, 191-195.
[72] Yuehe Lin, W. Y., and Joseph Wang *Frontiers in Bioscience* 2005, 10, 492-505.
[73] Jessica Koehne, J. L., Alan M. Cassell, Hua Chen, Qi Ye, Hou Tee Ng, Jie Han and M. Meyyappan *J. Mater. Chem* 2004, 14, 676-684.
[74] Salata, O. *Journal of Nanobiotechnology* 2004, 2, http://www.pubmedcentral.nih.gov/articlerender.fcgi?artid=419715.
[75] Raj CR, O. T., Ohsaka T *J. Electroanal. Chem* 2003, 543, 127.
[76] Pihel K, W. Q., Wightman RM *Anal Chem* 1996, 68, 2084.
[77] Malem F, M. D. *Anal Chem* 1993, 65, 37.
[78] Montenegro MCBSM, S. M. *J Pharm Sci* 2000, 89, 876.
[79] Liping Lu, S., Wang, Xiangqin Lin *Anal Sci* 2004, 20, 1131-1135.
[80] Yu A, L. Z., Cho J, Caruso F *Nano Letters* 2003, 3, 1203.
[81] Liu S, Y. J., Ju H *J Electroanal Chem* 2003, 540, 61.
[82] Liu S, J. H. *Biosens. Bioelect* 2003, 19, 177.
[83] Lei Zhang, X. J., Erkang Wang, Shaojun Dong *Biosens. Bioelect* 2005, 21, 337-345.
[84] Xiaojun Han, W. C., Zheling Zhang, Shaojun Dong, Erkang Wang *Biochimica et Biophysica Acta* 2002, 1556, 273- 277.
[85] S. Bharathi, M. N. *Analyst* 2001, 126, 1919-1922.
[86] Jingdong Zhang, M. K., Munetaka Oyama *Electrochem Commun* 2004, 6, 683-688.
[87] Liping Lu, X. L. *Anal Sci* 2004, 20, 527-530.
[88] L. Agü´ı, J. M., P. Yáñez-Sedeño, J.M. Pingarrón *Talanta* 2004, 64, 1041-1047.
[89] I. G. Casella, A. D., E. Desimoni *Analyst* 1996, 121, 249.
[90] C.-Y. Chiou, T.-C. C. *Electroanalysis* 1996, 8, 1179.
[91] A. L. Crumbliss, J. S., R. W. Henkens, J. Zhao, J. P. O'Daly *Biosens. Bioelect* 1993, 8, 331.
[92] Xuan Dai, O. N., Michael E. Hyde, and Richard G. Compton *Anal Chem.* 2004, 76, 5924-5929.
[93] M. Dequaire, C. D., B. Limoges *Anal Chem.* 2000, 72, 5521-5528.
[94] Hong Cai, C. X., Pingang He, Yuzhi Fang *J. Electroanal. Chem.* 2001, 510, 78-85.
[95] Joseph Wang, R. P., and Danke Xu *Langmuir* 2001 17, 5739-5741.
[96] Wang, J. *Chem. Eur. J.* 1999, 5, 1681.
[97] J. Wang, D. X., A.-N. Kawde, R. Polsky *Anal Chem.* 2001, 73, 5576.
[98] J. Wang, R. P., D. Xu *Langmuir* 2001, 17, 5739.

[99] K. R. Brown, A. P. F., M. J. Natan *J. Am. Chem. Soc.* 1996, 118, 1154.
[100] M. B. Gonzalez Garcìa, A. C.-G. *Bioelectrochem. Bioenerg* 1995, 38, 389.
[101] M. B. Gonzalez Garcìa, C. F. n.-S. n., A. Costa-Garcìa *Biosens. Bioelect.* 2000, 15, 315.
[102] Craig E. Banks, T. J. D., Gregory G. Wildgoose and Richard G. Compton *Chem Comm* 2005, 7, 829-841.
[103] Craig E. Banks, R. R. M., Trevor J. Davies and Richard G. Compton *Chem Comm* 2004 1804-1805.
[104] Yazhen Wang, Q. L., Shengshui Hu *Bioelectrochemistry* 2005, 65, 135.
[105] F.Wu, G. Z., X.Wei *Electrochem.Commun* 2002, 4 690.
[106] M.Musameh, N. S. L., J.Wang *Electrochem Commun* 2005 7 14.
[107] J. Wang, M. M. *Anal Chem.* 2003, 75 2075.
[108] F.Valentini, S. O., M.L.Terranova,A.Amine, G.Palleschi *Sens, Actuators, B* 2004, 100, 117.
[109] M.D.Rubianes, G. A. R. *Electrochem.Commun* 2003, 5, 689.
[110] K.Wu, J. F., S.Hu *Anal.Biochem* 2003, 318 100.
[111] W.C.Poh, K. P. L., W.De Zhang, S.Triparthy, J.Ye, F.Sheu *Langmuir* 2004, 20, 5484.
[112] H.Zhang, X. W., L.wang, Y.Liu, C.Bai *Electrochim. Acta* 2004, 49, 715.
[113] W.Zhang, Y. X., S.Ai, F.Wan, J.Wang, L.Jin, J.Jin *J.Chromatogr. B* 2003, 791, 217.
[114] J.Ye, Y. W., W.Zhang *Electrochem Commun* 2004, 6, 66-70.
[115] W.C.Poh, K. P. L., W.De Zhang, S.Triparthy, J.Ye, F.Shey *Langmuir* 2004, 20, 5484.
[116] Y.zhao, W. Z., H.Chen and Q.Luo *Sens, Actuators, B* 2003, 92, 279.
[117] G.Chen, L. Z., J.Wang *Talanta* 2004, 64, 1018.
[118] K.Gong, Y. D., S.Xiong, Y.Chen, L.Mao *Biosens. Bioelect* 2004 20 253.
[119] K.Gong, M. Z., Y.Yan, L.Su, L.Mao, S.Xiong, Y.Chen *Anal Chem.* 2004, 76 6500.
[120] M.Zhang, Y. Y., K.Gong, L.Mao, Z.Guo, Y.Chen *Langmuir* 2004 20 8781.
[121] M.L.Pedano, G. A. R. *Electrochem.Commun* 2004, 6, 10.
[122] J.Wang, M. M. *Anal.Chim.Acta* 2004, 511, 33.
[123] Z.Xu, X. C., X.Qu, S.Dong *Electroanalysis* 2004, 16, 684.
[124] D.Nematollahi, M. A., S.W.Husainc *Electroanalysis* 2004, 16, 1359.
[125] He P G, D. L. M. *Chem Comm* 2004, 3, 348.
[126] Taft B J, L. A. D., etc *J. Am. Chem. Soc.* 2004, 126, 12750.
[127] Xin Lu, Z. C. *Chem. Rev.* 2005, 105, 3643-3696.
[128] Y.zhao, W. Z., H.Chen, Q.Luo, S.F.Y.Li *Sens, Actuators, B* 2002, 87, 168.
[129] J.J.Davis, K. S. C., B.R.Azamian, C.B.Bagshqw, M.L.H.Green *Chem.Eur.J.* 2004, 9, 3732.
[130] Wang J, L. G. D., Jan M R *J. Am. Chem. Soc.* 2004, 126, 3010.
[131] Chen G F, Z. J., Tu Y H etc *Anal.Chim.Acta* 2005, 533, 11.
[132] Buchberger, W. W., K.; Breitwieser, C. Fresenius' Z *Anal Chem* 1983, 315, 518-520.
[133] Zen J-M, S. Y.-S., Chung H-H, Hsu C-T, Kumar AS *Anal Chem* 2002, 74, 6126.
[134] Zen J-M, C. H.-H., Kumar AS *Analyst* 2000, 125, 1633.
[135] Zen J-M, C. H.-H., Kumar AS *Anal. Chem.* 2002, 74, 1202.
[136] I. G. Casella, M. G., M. R. Guascito, T. R. I. Cataldi *Anal.Chim.Acta* 1997, 357, 63.
[137] Welch CM, H. M., Banks CE, Compton RG *Anal Sci* 2005, 21, 1421.
[138] Jyh-Myng Zen, C.-T. H., Annamalai Senthil Kumar, Hueih-Jing Lyuu and Ker-Yun Lin *Analyst* 2004, 129, 841-845.

[139] Elliot JM, B. P., Bartlett PN, Attard GS *Langmuir* 1999, 15, 7411.
[140] Hrapovic S, L. Y., Male KB, Luong JHT *Anal Chem* 2004, 76, 1083.
[141] Gan X, L. T., Zhu X, Li G *Anal Sci* 2004, 20, 1271.
[142] Fang B, W. G., Zhang W, Li M, Kan X *Electroanalysis* 2005, 17, 744.
[143] Goto T, O. T., Hirai T *Scr Mater* 2001, 44, 1187.
[144] Martin Winter, R. J. B. *Chem. Rev* 2004, 104, 4245-4269.
[145] B. Hammer, J. K. N. *Adv. Catal* 2000, 45, 71.
[146] M. Valden, X. L., D. W. Goodman *Science* 1998, 28, 1647-1650.
[147] M. Baumer, H.-J. F. *Prog. Surf. Sci.* 1999, 61, 127.
[148] D. A. Landsman, F. J. L. *Handbook of Fuel Cells, Wiley, New York* 2003, 4.
[149] JM, L. *Electrochim. Acta* 2005, 50, 3123.
[150] Shen Y, L. J., Wu A, Jiang J, Bi L, Liu B, Li Z, Dong S *Langmuir* 2003, 19, 5397.
[151] Zhang L, C. B., Samulski ET *Chem Phys Lett* 2004, 398, 505.
[152] Nagahara Y, H. M., Yoshimoto S, Inukai J, Yau S-L, Itaya K *J Phys Chem B* 2004, 108, 3224.
[153] L. Genie¡ s, R. F., R. Durand *Electrochim. Acta* 1998, 44, 1317.
[154] O. Antoine, Y. B., R. Durand *J. Electroanal. Chem* 2001, 499, 85.
[155] A. Gamez, D. R., P. Gallezot, F. Gloaguen, R. Faure, R. Durand *Electrochim. Acta* 1996, 41, 307.
[156] Stonehart, P. *Ber. Bunsenges. Phys. Chem* 1990, 94, 913.
[157] U. A. Paulus, A. W., G. G. Scherer, T. J. Schmidt, V.Stamenkovic, N. M. Markovic, P. N. Ross *Electrochim. Acta* 2002, 47, 3787.
[158] U. A. Paulus, G. G. S., A. Wokaun, T. J. Schmidt, V. Stamenkovic, V. Radmilovic, N. M. Markovic, P. N. Ross *J. Phys. Chem. B* 2002, 106, 4181.
[159] Arenz M, M. K., Stamenkovic V, Bilzanac BB, Tomoyuki T, Ross PN, Markovic NM *J. Am. Chem. Soc.* 2004, 127, 6819.
[160] Y. Lou, M. M. M., L. Han, J. Luo, C.-J. Zhong *Chem Comm* 2001, 473.
[161] F. Hajbolouri, B. A., G. G. Scherer, A. Wokaun *Fuel Cells* 2004, 4, 160.
[162] C. Roth, N. M., F. Hahn, J.-M. Leger, C. Lamy, H. Fuess *J.Electrochem. Soc.* 2002, 149, 433.
[163] El-Deab MS, O. T. *Electrochim. Acta* 2002, 47, 4255.
[164] El-Deab MS, O. T., Ohsaka T *J Electrochem Soc* 2003, 150, A851.
[165] Mohamed S. El- Deab 1, T. O. *Electrochem Commun* 2002, 4 288-292.
[166] Yanrong Zhang, S. A., Sachio Yoshihara, Takashi Shirakashi *Electrochim. Acta* 2003, 48, 741-747.
[167] Ave Sarapuu, K. T. e. *Electrochem Commun* 2001, 3 446-450.
[168] Ichizo Yagi, T. I., Kohei Uosaki *Electrochem Commun* 2004, 6, 773-779.
[169] Mohamed S. El-Deab, T. O. *J. Electroanal. Chem.* 2003, 553, 107-115.
[170] Y. Zhang, V. S., Nakazawa, S. Yoshihara* , T. Shirakashi *Electrochim. Acta* 2004, 49, 5235-5240.
[171] Jin Luoa, M. M. M., Yongbing Loua, Li Hana, Maria Hepel b, Chuan Jian Zhonga, *Catalysis Today* 2002, 77, 127-138.
[172] Kinoshita, K. *Wiley, New York* 1988.
[173] T. Vad, F. H., H. G. Haubold,G. G. Scherer, A.Wokaun *J. Phys. Chem. B* 2004, 108, 12442.

[174] T. Yoshitake, Y. S., S. Kuroshima, H. Kimura, T. Ichihashi, Y. Kubo, D. Kasuya, K. Takahashi, F. Kokai, M.Yudasaka, S. Iijima *Phys. B* 2002, 323, 124.

[175] Froudakis, G. E. *Pev. Aav. Mater. Sci* 2003, 5, 259-264.

[176] Dillon, A. C. a. M. J. H. *Applied Physics A: Mateials Science & Processing* 2001, 72, 133-142.

[177] Dillon, A. C., et al *Nature* 1997, 386, 377-379.

[178] Conway, B. E. *Kluwer Academic/Plenum Publishers, New York* 1999.

[179] R. Kotz, M. C. *Electrochim. Acta* 2000, 45, 2483.

[180] C. Niu, E. K. S., R. Hoch, D. Moy, H. Tennent *Appl. Phys.Lett.* 1997, 70, 1480.

[181] Chong-yang Liu, A. J. B., Fred Wudl, Iris Weitz and James R. Heath *Electrochemical and Solid-State Letters* 1999, 2, 557

[182] L. Diederich, E. B., P. Piseri, A. Podesta, P. Milani, A. Schneuwly and R. Gallay *Appl. Phys.Lett.* 1999, 75, 2662.

[183] J. Chmiola, G. Y., Y. Gogotsi, C. Portet, P. Simon, P. L. Taberna *Science* 2006 313, 1760.

[184] Sandeep Kumar, G. D. S. *Microchim Acta* 2007, in press.

[185] M.K. Nazeeruddin, A. K., I. Rodicio, R. Humphrybaker, E. Muller, P. Liska, N. Vlachopoulos, M. Gra¨ tzel *J. Am. Chem. Soc.* 1993, 115, 6382.

[186] Zen J-M, H. C.-T., Kumar AS, Lyuu H-J, Lin K-Y *Analyst* 2004, 129, 841.

[187] J. Solla-Gullon, V. M., A. Aldaz, J. Clavilier, J *Electroanal. Chem* 2000, *491*, 69.

In: Nanoparticles: New Research
Editor: Simone Luca Lombardi, pp. 363-371
ISBN: 978-1-60456-704-5

Chapter 12

FIELD ENHANCED THERMAL CONDUCTIVITY IN HEAT TRANSFER NANOFLUIDS WITH MAGNETIC COATING ON CARBON NANOTUBES

Brian Wright, Dustin Thomas, Haiping Hong
Department of Material and Metallurgical Engineering, South Dakota School of Mines and Technology, Rapid City, SD 57701, USA
Lori Groven, Jan Puszynski
Department of Chemical and Biological Engineering, South Dakota School of Mines and Technology, Rapid City, SD 57701, USA
Edward Duke
Department of Geology and Geological Engineering, South Dakota School of Mines and Technology, Rapid City, SD 57701, USA
Xiangrong Ye, Sungho Jin
Department of Mechanical & Aerospace Engineering, University of California, San Diego, La Jolla, CA 92093, USA
Pauline Smith, Walter Roy
Army Research Lab, Aberdeen Proving Ground, MD 21005, USA

Abstract

In this paper, we report that the thermal conductivity (TC) of heat transfer nanofluids containing Ni coated single wall carbon nanotube (SWNT) can be enhanced by applied magnetic field. A reasonable explanation for these interesting results is that Ni coated nanotubes form aligned chains under applied magnetic field, which improves thermal conductivity via increased contacts. On longer holding in magnetic field, the nanotubes gradually move and form large clumps of nanotubes, which eventually decreases the TC. When we reduce the magnetic field strength and maintain a smaller field right after TC reaches the maximum, the TC value can be kept longer compared to without magnetic field. We attribute gradual magnetic clumping to the gradual cause of the TC decrease in the magnetic field. We also found that the time to reach the maximum peak value of TC is increased as the applied magnetic field is reduced. Scanning Electron Microscopy (SEM) images show that the Ni coated nantubes are aligned well under the influence of a magnetic field. Transmission Electron Microscopy (TEM) images indicate that nickel remains attached onto the nanotubes after the magnetic field exposure.

Keywords: Thermal conductivity, Carbon nanotube, Ni coated, Magnetic field.

1. Introduction

The discovery of carbon nanotubes (CNT) has instigated tremendous research efforts in recent years, due to their promising thermal, electrical, mechanical, and functional properties. For example, it has been reported that single wall carbon nanotubes exhibit a thermal conductivity (TC) value as high as 2000 – 6000 W/m.K.[1] under an ideal circumstance. By contrast typical heat transfer fluids like water and oil have TC values of only 0.6W/m.K. and 0.2 W/m.K. respectively.

We prepared a composite fluid containing high-TC materials of CNTs and magnetic-field-responsive magnetic nanoparticles with an aim to increase the thermal conductivity of the fluid. Such fluids with enhanced thermal conductivity are useful for a variety of applications such as heat transfer coolants and lubricants. [2,3] The fluids containing carbon nanotubes (called "nanofluids" hereafter) should exhibit substantially improved TC values. [4-6] However, a simple composite structure turned out to be not effective in enhancing the TC of nanofluids. At low nanotube percentage loading, no significant improvements in TC were reported. At the loading of 1 vol% CNTs (~1.4wt %), there is about 10-20% TC increase reported by different groups.[7,8] However, at such a high concentration, the fluid already lost most of its fluidity and became mud-like, thus making the fluid much less useful for coolant & lubricant applications.

A possible explanation for this lack of sufficient TC increase through compositing with carbon nanotubes may be attributed to a lack of alignment and orientation in the fluids when the carbon nanotubes are irregularly positioned in the fluids with only a random and infrequent chance for them to contact each other. Therefore, only very high concentrations of CNTs produce noticeable TC improvements. Therefore, there is a need to introduce nanotube-nanotube physical contacts in the fluids in order to increase the thermal (and electrical) conductivity significantly, while maintaining the nanotube concentration relatively low to ensure a desirably low fluid viscosity.

We introduce a new concept of incorporating magnetically sensitive metal or metal oxide nanoparticles in a carbon-nanotube-containing fluid. Magnetic particles in a liquid medium can have a variety of configurations depending on the nature of magnetic particles and the field strength.[9-10] Under a relatively strong magnetic field, small magnetic particles form connected networks and also tend to get somewhat oriented toward the field direction, which also moves the carbon nanotubes nearby and induce more physical contacts. Thus improved thermal conductivity is anticipated. We incorporated carbon nanotubes and magnetic-field-sensitive nanoparticle Fe_2O_3 into water under a magnetic field and the results confirmed this assumption. [11]

Based on the above results, we estimated that the metal encapsulated carbon nanotubes [12] dispersed in water, water/ethylene glycol or oil with the help of appropriate chemical surfactant, could also significantly increase the TC value under a magnetic field if the coating metal is magnetic. The coated layer configuration of the magnetic material could be more beneficial than the presence of isolated magnetic nanoparticles as small magnetic particles tend to be superparamagnetic with a weak response to applied magnetic field.

In this paper, we investigated nanofluid systems containing 0.01 and 0.02 wt% of Ni coated, magnetic nanotubes in water under different magnetic field strengths. Understanding of thermal conductivity behavior of such a composite material system in magnetic field would be valuable in analysis and synthesis of nanofluids.

2. Experimental

Single wall carbon nanotubes (SWNT) were purchased from Helix Material Solutions Inc (Helix, Richardson, Texas). Chemical surfactant sodium dodecylbenzene sulfonate (SDBS) was purchased from Sigma Aldrich.

Ni coated SWNTs were prepared using the following methodology:

1. Oxidation: HNO_3: H_2SO_4 (3:1) reflux at 110 °C for 6 hrs.
2. Sensitization and Activation: Immersion in 0.1 M $SnCl_2$-0.1 M HCl for 30 min, immersion in 0.0014 M $PdCl_2$-0.25 M HCl for 30 min.
3. Electroless Plating: Introduction into electroless bath for 25 min with composition as listed in Table 1.

The Ni coated SWNTs were put into water together with the appropriate amount of chemical surfactant sodium dodecylbenzene sulfonate (SDBS). Sonication was performed using a Branson Digital Sonifier, model 450. Magnetic field was provided by a pair of spaced apart Ba-ferrite magnet plates (4 x 6 x 1 inch dimension) and placing the sample in the middle of the gap between the magnets.

Table 1. Electroless bath conditions

Chemical	Concentration (g/L)
$NiSO_4 \cdot 6H_2O$	25
$Na_3C_6H_5O_7 \cdot 2H_2O$	5
$NaH_2PO_2 \cdot H_2O$	15
NH_4Cl	60
$Pb(NO_3)_2$	2.5
Bath temperature	20°C
pH (adjusted w/ NH_4OH)	~8.0

The microstructure was taken using an Olympus VANOX-T microscope. The magnification was 500X. Scanning electron microscopy (SEM) images were acquired using the backscattered electron detector on a Zeiss Supra40VP variable pressure system. Transmission electron microscopy (TEM) images were acquired with a Hitachi H-7000 FA.

The pH values of the fluid containing the Ni coated SWNTs were measured using a Denver instrument UP-10 pH/mV meter. The thermal conductivity data was obtained by the Hot Disk™ thermal constants analyzer [13], using the following parameters: measurement depth 6 mm, room temperature, power 0.012W, measurement time 15s, sensor radius 3.189mm, TCR 0.0471/K, disk type kapton, temperature drift rec yes. The magnetic field intensity was recorded by F.W. Bell Gaussmeter Model 5060.

Dispersion stability of the Ni coated nanotubes in fluid was observed with visual inspection.. We put the nanofluids in a see-through glass beaker and observed if there was any precipitation at the edge and/or bottom of the glass beaker.

3. Results and Discussion

Figure 1 shows the optical microstructure of 0.01wt % of Ni-coated carbon nanotube in DI water. Because of the heavy Ni, 0.01 wt % of Ni-free nanotubes will contain several times larger number of nanotubes than 0.01 wt % of Ni-coated nanotubes. Relatively large dark spots which may be attributed to the agglomerated nanotubes or other particles can barely be observed. This indicates that the Ni coated nanotubes are dispersed quite well in the water with the help of the chemical surfactant used.

Figure 1. Optical microscope picture of 0.01wt % of Ni coated carbon nanotubes in DI water.

The thermal conductivity (TC) versus time plot for various concentrations of Ni coated nanotubes (0.01 wt% and 0.02 wt%) in magnetic fields (two different intensities) are shown in Figure 2. Without a magnetic field, the TC value of Ni coated SWNT loaded nanofluids is around 0.63 ~ 0.64 W/m.K. and stays mostly constant versus the time. Because of very few contacts between Ni coated carbon nanotubes, the TC value is essentially the value for the DI water itself. Figure 2a shows the TC versus time at a 51mm gap separation between the two magnets (0.62 kOe magnetic field) with two concentrations (0.01 wt% and 0.02 wt%), and Figure 2b shows the TC versus time at a 91mm gap (0.38 kOe) with the same concentrations. Under the magnetic field, thermal conductivity shows very interesting behavior. TC initially increases with time but eventually reaches a peak. On longer time holding in magnetic field, the Ni coated nanotubes gradually move and form bigger clumps of CNTs, thus a decrease in the TC. The maximum TC value is around of 1.10 W/m.K. in 51mm gap (0.62 kOe) at 0.02 wt% and 1.01 W/m.K. in 91mm gap (0.38 kOe) at 0.02wt%, about 75% and 60% higher than the value of DI water. For both the 51mm gap and the 91mm gap, the TC of 0.02 wt% nanofluids is obviously higher than that of 0.01 wt% sample, and the peak of maximum TC is slightly left-shifted versus time.

These results strongly support our hypothesis that a good alignment and orientation of conducting materials in a fluid are critical and essential to the enhancement of TC of the composite fluid. However, in the case of nano particle containing composite fluid, the increase ratio of TC is not as significant due to the thermal contact resistance in the nanofluids [14].

The next step was to determine what happens to the TC value of nanofluids containing Ni coated single wall carbon nanotube if the external magnetic field is reduced and maintained at a much smaller field once the TC reaches the maximum. Figure 3 is a plot of thermal conductivity versus time in a nanofluid containing 0.01wt% Ni coated SWNTs.. A) after placing the nanofluid in a 0.62 kOe magnetic field for 2-3 minutes to reach the maximum TC, the magnetic field was reduced to 0.18 kOe and maintained. B) after placing the nanofluid in a 0.62 kOe magnetic field for 2-3 minutes to reach the maximum TC, removed the magnetic field completely. C) placed the nanofluid in a 0.18 kOe magnetic field all the time. It is clearly seen from Figure 3 that the TC of nanofluids under somewhat reduced magnetic field after TC reaches the maximum is higher than that for removed magnetic field after TC reaches the maximum. The results indicate that the gradual magnetic clumping was the cause of the TC decrease. To better understand this, measurements of TC versus time in a constant weak magnetic field (0.18 kOe) were conducted as for the sample C in Figure. 3, the result of which was a relative low TC value. Further optimization or time-dependent programming of the applied magnetic field is likely to yield more desirable thermal conductivity as a function of time.

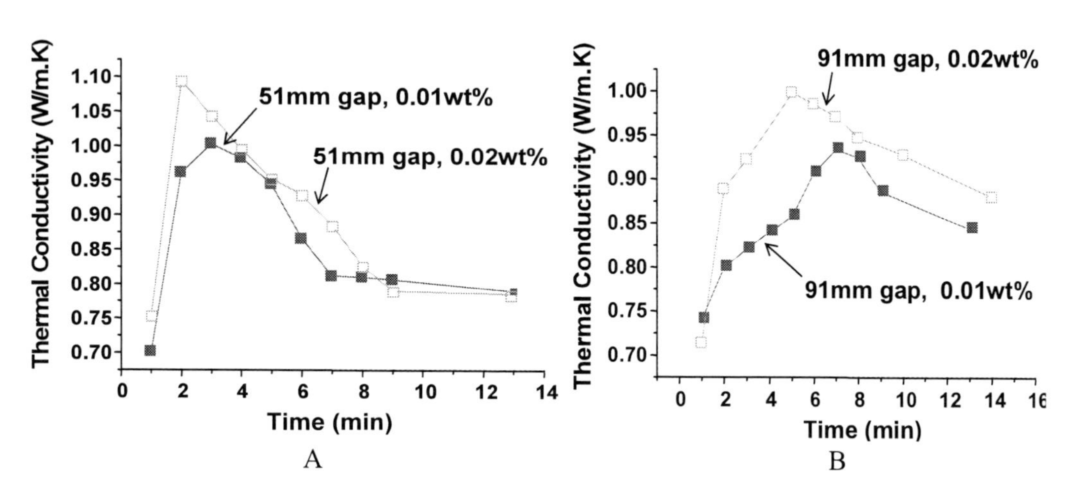

Figure 2. Thermal conductivity versus time in different concentrations at two different magnetic field intensities. A) 51mm gap (0.62 kOe) . B) 91mm gap (0.38 kOe).

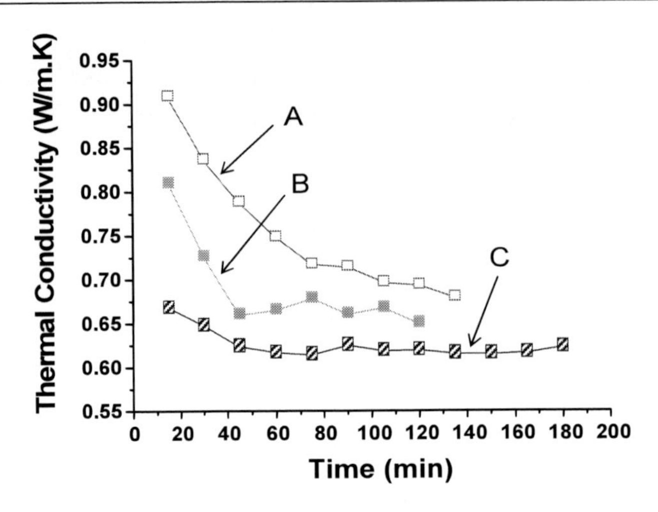

Figure 3. Thermal conductivity versus times in 0.01wt% Ni-coated SWNT loading nanofluids. A) after putting fluids in 0.62 kOe magnetic field for 2-3 mins to reach the maximum TC, reduce the magnetic field to 0.18 kOe and maintain. B) after putting fluids in 0.62 kOe magnetic field for 2-3 minutes to reach the maximum TC, withdraw the magnetic field. C) put the fluids in 0.18 kOe magnetic field all the time.

It is interesting to note that if one extrapolates the curve A and B to the zero time, two curves would come to the similar starting point, around 1.00 W/m.K. Actually this more or less coincides with the maximum TC value of the nanofluid under 0.62 kOe magnetic field.

Also, it is interesting to note that different magnetic field strength influences the maximum TC value, the time to reach the maximum, and the shape of the TC vs time curve. Figure 4 shows thermal conductivity versus time at different magnetic field strengths (as adjusted by changing the distance between the pair of magnets as shown in Table 2). , A) 51mm gap. B) 91mm gap. C) 141mm gap. It is seen that the shorter the gap distance (i.e., the higher the applied magnetic field is), the shorter the time to reach the maximum TC. At 35mm gap, it is impossible to observe the maximum TC value. This is as anticipated since the shorter distance means a stronger magnetic field, which enables to very quickly assemble and align the nanotubes at a rate quicker than we are able to measure the thermal conductivity with our current equipment.

The quantitative data correlating the magnetic field intensity, time to reach TC maximum and TC value at different gap distances between the two magnets are listed in Table 2.

Table 2. Magnetic field dependence of the time to reach peak TC and the value of TC maximum at the concentration of 0.02 wt% Ni coated SWNT

Distance between two magnets (mm)	Magnetic field (kOe)	Time to reach TC maximum (min)	TC maximum value (W/m.K)
35	0.86	<2	N/A
51	0.62	2	1.10
91	0.38	5	1.01
141	0.18	7	0.95

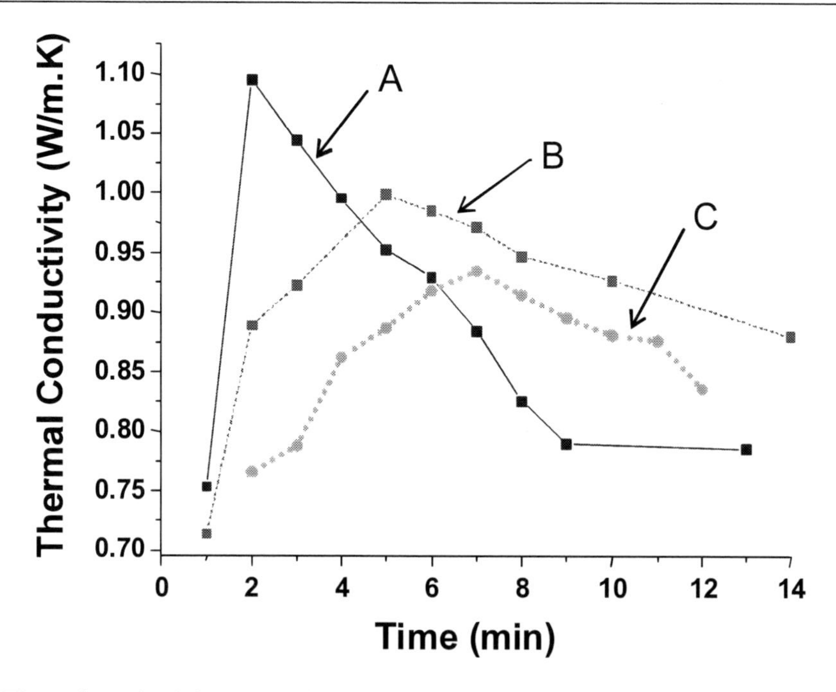

Figure 4. Thermal conductivity versus time at different gap distances between two magnets (different magnetic field intensity) at 0.02 wt% concentration. A) 51mm gap. B) 91mm gap. C) 141mm gap.

From Figure 5, it is clearly seen that under the magnetic field, the nanotubes align well in the magnetic field direction due to the magnetic moment of the Ni coating. The red arrow represents the magnetic field direction. The black material represents the carbon nanotube. Small bright particles are images from some portion of the coated Ni metal. Since Ni is much heavier than carbon, Ni appears brighter in the backscattered electron image. Energy-dispersive X-ray analysis showed that the bright particles contain Ni element.

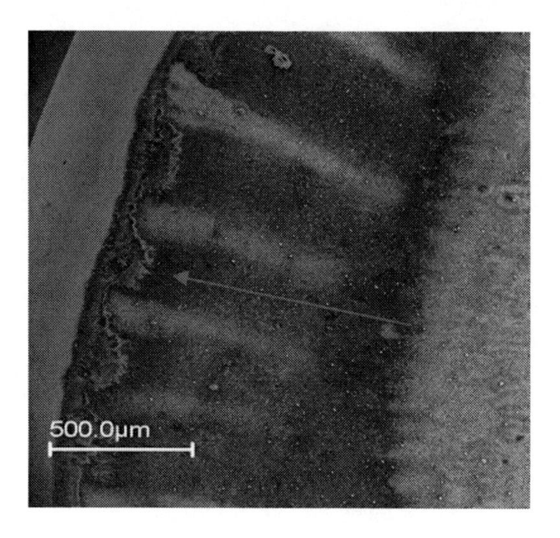

Figure 5. Backscattered electron SEM image of 0.01wt% Ni coated single wall carbon nanotubes aligned with magnetic field. Scale bar is 500μm. Red arrow is magnetic field direction.

To prepare the SEM sample, drops of nanofluids containing Ni coated nanotubes were placed on the SEM sample holder (usually made by Al) and left to dry under the magnetic field, see Figure. 5. Since long time exposure to a magnetic field will reduce the TC value, the samples were only placed under a magnetic field for a short time (2-4 mins). During this time, the samples could reach substantially improved TC values, indicating that nanotubes and nano particles align very well, this is evident from Figure 5. No significant change was observed in the thermal conductivity of the fluids after the magnetic field was removed, i.e., after ~10 minutes.

In order to make sure that all these interesting results are solely due to the Ni-coated SWNT, and not Ni particles detached from the nanotubes, several Transmission Electron Microscopy (TEM) pictures were taken. Figure 6 shows TEM images of carbon nanotube. A) As received, uncoated Ni. B) Ni coated before the experiment. C) Ni coated after the experiment. It is clearly seen that Ni is still attached to the nanotube after the experiment. All the results reflect the magnetic nature of Ni coated SWNT.

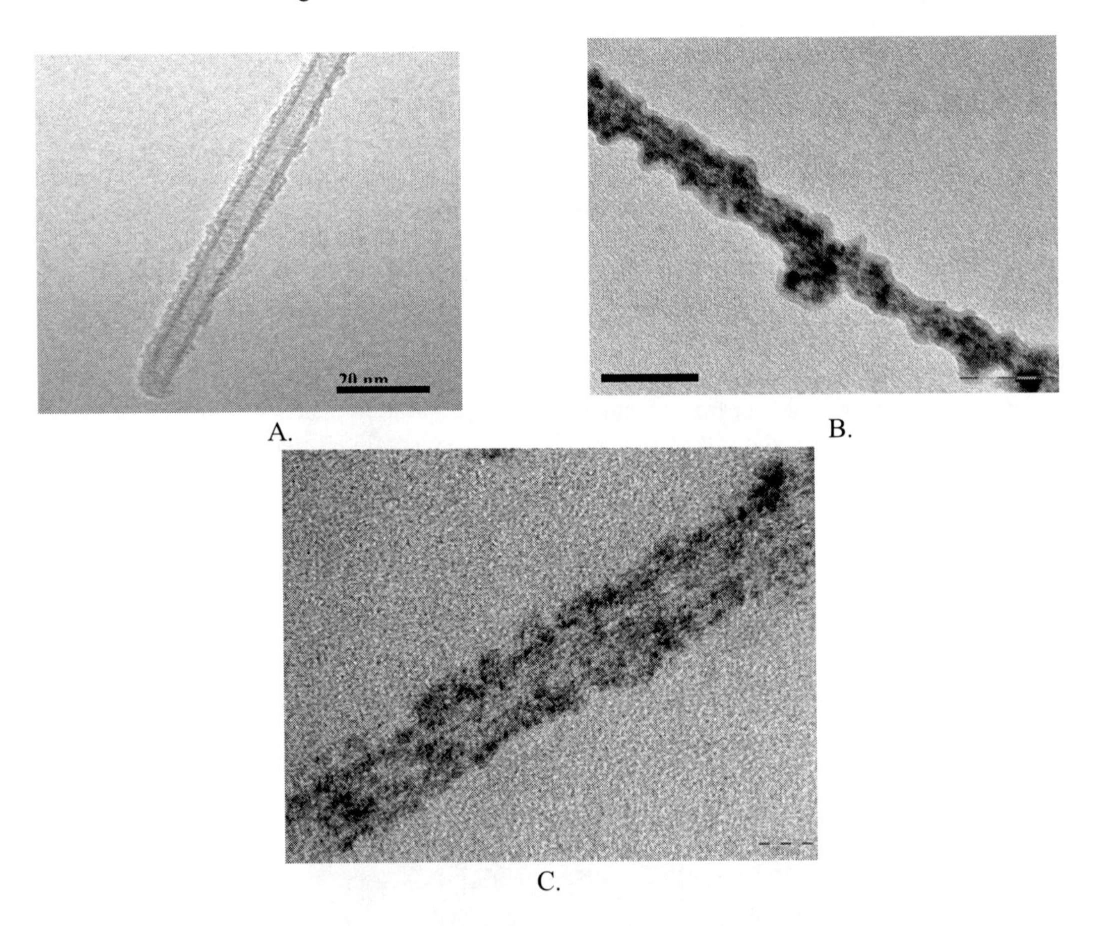

Figure. 6. TEM images of carbon nanotube. A) As received, uncoated Ni. B) Ni coated before the experiment. C) Ni coated after the experiment. Scale bar is 20nm.

4. Conclusion

In summary, we have successfully demonstrated that the thermal conductivity (TC) of the heat transfer nanofluids containing Ni-coated single wall carbon nanotubes (SWNTs) can be significantly enhanced in the presence of applied magnetic field. The explanation for these interesting results is that Ni coated nanotubes form aligned chains under an applied magnetic field, resulting in improved contacts and hence improved thermal conductivity. On longer holding in magnetic field, the nanotubes gradually move and form large clumps of nanotubes, thus decreasing the TC. Reducing the magnetic field strength and maintaining a much smaller field right after TC reaches the maximum value enables to keep the TC value longer compared to the case of complete magnetic field removal.

We also found that the time to reach the maximum peak value of TC is increased as the applied magnetic field is reduced by increasing the gap between two magnets. Scanning Electron Microscopy (SEM) pictures show that the Ni coated nantubes are aligned well under the influence of a magnetic field.

Acknowledgments

H. Hong would like to thank Army Research Lab (Cooperative agreement DAAD19-02-2-0011) for financial support. Stimulating discussion with Prof. G.P. Peterson of University of Colorado, Boulder is acknowledged.

References

[1] S. Berber, Y. K. Kwon, and D. Tománek, *Phys. Rev. Lett.* **84**, 4613 (2000).
[2] H. Hong, J. Wensel, F. Liang, W. E. Billups, W. Roy, J. Thermophys. *Heat Trans.* **21**, 234 (2007).
[3] P. Keblinski, J. A. Eastman, D. G. Cahill, *Materials Today*, **8**(6), 36 (2005).
[4] X. Wang, X. Xu, S. Choi, J. Thermophys. *Heat Trans.* **13**(4), 474 (1999).
[5] S. Choi, Z. Zhang, W. Yu, F. E. Lockwood, E. A. Grulke, *Applied Phys Letters,* **79**(14), 2252 (2001).
[6] BH. Kim, GP. Peterson, *J. Thermophys. Heat Trans.* **21**, 451 (2007).
[7] H. Xie, H. Lee, W. Youn, M. Choi, *Journal of Applied Physics,* **94** (8), 4967 (2003).
[8] H. Hong, J. Wensel, S. Peterson, W. Roy, *Polymeric Materials: Science & Engineering,* **95**, 1076 (2006).
[9] S. Jin and M. McCormack. *J. Electronic Materials* **23**, 8 (1994).
[10] X. R. Ye, C. Caraio, C. Wang, and J. B. Talbot, S. Jin, *J. Nanosci. and Nanotechnol* **6**, 852 (2006).
[11] H. Hong, B. Wright, J. Wensel, S. Jin, X. Ye, W. Roy, *Synthetic Metal,* **157**, 437 (2007).
[12] Y. Zhang, N. W. Franklin, R. J. Chen, and H. Dai, *Chem. Phys. Lett.* **331**, 35 (2000).
[13] Detail information see www.hotdisk.se.
[14] B. Wright , D. Thomas, H. Hong , L. Groven, J. Puszynski, E. Duke, X. Ye , S. Jin, *Applied Physics Letter,* **91**, 173116 (2007).

In: Nanoparticles: New Research
Editor: Simone Luca Lombardi, pp. 373-386
ISBN: 978-1-60456-704-5

Chapter 13

POLYMERIC NANOPARTICLES FOR ORAL DELIVERY OF PROTEIN DRUGS

Ji-Shan Quan, Hu- Lin Jiang, Jia-Hui Yu, Ding-Ding Guo, Rohidas Arote, Yun-Jai Choi and Chong-Su Cho [*]
Department of Agricultural Biotechnology, Seoul National University,
Seoul 151-921, South Korea

Abstract

Proteins and peptides as therapeutic agents have been used for clinical applications. While due to their short *in vivo* half life, easily undergoing hydrolysis, low permeability to biological membrane barriers and poor stability in gastrointestinal tract, parenteral administration such as subcutaneous, intravenous, or intramuscular injections is the common way for these drugs delivery. Alternative injection, the oral route presents many advantages such as avoidance injection pain and discomfort, elimination of possible infections caused by the use of needles. Therefore, many strategies to improve their oral delivery efficiency have been studied. One of the most potential strategies is usage of micro- and nanoparticles of functional polymers as protein drug carriers. This chapter summarizes the nanoparticles based on different polymers as an oral administration system for protein and peptide drugs delivery. It primarily describes the pH-sensitive nanoparticles. Depending on the fact that pH of the human gastrointestinal tract increases progressively from the stomach to the intestine, nanoparticles remain insoluble in the stomach and disintegrate at the higher pH. Secondly, mucoadhesive polymer-coated nanoparticles are described, which can intensify the contact between the polymer and the mucus surface either by non covalent bonds or covalent disulfide bonds. Thirdly, bacteria degradable polymer-coated nanoparticles for colon delivery are briefly reported. Fourthly, a pH-sensitive and mucoadhesive polymer based on nano- and microparticles is included. Finally, some success of oral protein drug *in vivo* deliveries are enumerates. All of these nanoparticles can be expected to improve the proteins absorption.

Keywords: Nanoparticles; Protein drugs; Oral delivery; Mucoadhesive; pH-sensitive

[*] E-mail address: chocs@plaza.snu.ac.kr. Tel: +82 2 880 4636; Fax: +82 2 875 2494. (To whom should be addressed.)

1. Introduction

With the advances in biotechnology and genetic engineering, peptides and proteins are attracting increasing interest for applying to the therapeutic agents for treatment of numerous diseases. These proteins have great potential to cure diseases with negligible side effects and merely symptomatic treatment [1]. Generally these proteins possesses shorter half lives *in vivo*, easily undergoes hydrolysis, low permeability to biological membrane barriers and also poor stability in gastrointestinal tract. Owing to above fact, these protein drugs are administered parenterally such as subcutaneous, intravenous, or intramuscular injections. These administration routes also reveal some disadvantages such as discomfort due to repeated and prolonged dosage regimes and difficulty to self-administration, and may cause infections by injectable dosage forms which could not be ignored [2]. Oral route of administration is so far easiest route for these protein drugs which avoids injection pain and elimination of the possible infections caused by improper sterilization techniques. However, proteins and peptides show poor bioavailability after oral administration. To improve their bioavailability various strategies have been employed such as:

a. Chemical modification of these peptides and proteins for physicochemical properties: structure modifications of these macromolecules can provide several advantages such as improvement in the intestinal absorption [3], enhancement of the stability [4] and protection from degradation by proteases [5].
b. Addition of novel functionality to the proteins and peptides: introduction of novel functionality to peptides and proteins using transport-carrier molecules that are recognized by endogenous cellular-transport systems in the gastrointestinal (GI) tract to increase the intestinal absorption [1, 7].
c. Administration of the protein drugs with the penetration enhancers or protease inhibitors: penetration enhancers can modify the barrier properties of the mucosa; as a result the drug absorption is promoted [8]. The protease inhibitors suppress the enzyme activity, thereby avoiding the degradation in GI by enzymes [9].
d. Utilization of the delivery carriers: liposome, nano-/ microparticles, micelles, and etc. protection of the proteins from degradation by stomach acid or enzymes. And they also can improve the absorption of proteins and peptides [10].

These approaches are successfully used to increase the bioavailability of protein drugs in the laboratory although it still shows some safety concerns when used in clinics. For example, the use of enzyme inhibitors in long-term therapy remains questionable because of possible absorption of unwanted proteins, disturbance of the digestion of nutritive proteins and stimulation of protease secretion as a result of feedback regulation [1] and some surfactants which enhance the absorption of poor absorptive drugs can also cause the mucosal damage in the gastrointestinal tract [11].

Carriers for oral protein and peptide delivery were found attractive owing to their property to protect proteins from degradation by the stomach acid or by the proteolytic enzymes in the GI tract, and especially some nanoparticles can deliver antigens to the payers patch for oral immunization [12, 13, and 14]. Polymeric nanoparticles have been studied as carriers for oral drug delivery since the latter half of the 1980s [15]. They include solid particles ranging in size from 10 nm to 1000 nm (1 μm) and consist of macromolecular materials in which the active principle (drug or biologically active material) is dissolved, entrapped, or encapsulated, and/or to which the active material is adsorbed or attached [16]. Of the delivery systems, polymeric nanoparticles have received more attention because of the subcellular size which makes them to cross the epithelium than do microparticles, and have the stability in biologic fluids as well as during storage than the liposomes have [17]. Polymeric nanoparticles have been used as oral drug carriers due to several reasons [18]:

1. Improvement of the bioavailability of drugs with poor absorption characteristics
2. Prolongation of the residence time of drugs in the intestine
3. High dispersion at the molecular level and consequently increase of absorption
4. Delivery of vaccine antigens to gut-associated lymphoid tissue
5. Control of the release of the drugs
6. Targeting of therapeutic agents to a particular organ and thus reducing toxicity
7. Reduction of the GI mucosal irritation caused by drugs
8. Assurance of the stability of drugs in the GIT

In this chapter, we will focus on the nanoparticles based on different functional polymers as an oral administration system for protein and peptide drugs delivery.

2. pH-Sensitive Polymeric Nanoparticles for Protein and Peptide Drugs Delivery

The pH-dependent systems are based on the fact that pH of the human gastrointestinal tract increases progressively from the stomach (pH 1-2), small intestine (pH 5-7) at the site of digestion and it increases to 7-8 in the distal ileum. pH-dependent polymers, containing ionizable groups in their backbone that can accept or donate protons in response to the environmental change in pH. The degree of ionization in a polymer bearing weakly ionizable groups is dramatically altered at a specific pH. This rapid change in net charge of pendant groups causes an alternation of the hydrodynamic volume of the polymer chains. The transition from collapsed state to expanded state is explained by the osmotic pressure exerted by mobile counter ions neutralizing the network charges. Thus these polymer- coated particles swell at high pH, further release the active drug more.

Hydrogels have been extensively exploited for biomedical applications due to their high water content and excellent biocompatibility The pH-sensitive hydrogels containing pendant acidic or basic groups such as carboxylic acids, sulphonic acids, primary amines, or ammonium salts which change ionization in response to change in the pH have become the subject matter of major interest for use as carriers in oral drug delivery research [19].

Weak polyacids (or polybases), which undergo an ionization/deionization transition from pH 4-8, are utilized as pH-responsive polymers such as poly(acrylic acid) (PAAc) and poly(methacrylic acid) (PMAAc). Their carboxylic pendant groups accept protons at low pH, while releasing them at high pH. This gives a momentum along with the hydrophobic interaction to govern precipitation or solubilization of molecular chains, deswelling/swelling of hydrogels, or hydrophobic/hydrophilic characteristics of surfaces. Therefore, at low pH (1–

2) these materials shrink and the encapsulated drug can be protected from the acidic environment of stomach due to its limited release. As they enter the small intestine, the pH changes from acidic to neutral (6 to 7.4) lead the matrix to swell and release the encapsulated drug [20].

The pH-sensitive polymers containing sulfonamide groups (derivatives of *p*-aminobenzene sulfonamide) are the weak polyacids that shows various pKa values ranging from pH 3 to 11, resulting from different pendant substituents at the sulfoneamide group acting as electron withdrawing or donating groups [21, 22]. The hydrogen atom of the amide nitrogen can be readily ionized to form a weak polyacid.

When these polymers including poly (methacrylic acid-co-methylacrylate) copolymers, such as Eudragit® E100, Eudragit®, L100-55, Eudragit®, L100, Eudragit®, S100 are used as particulate carriers with incorporation of CyA using an adaptation of the quasi-emulsion solvent diffusion technique. These polymers can be dissolved rapidly upon deprotonation of carboxylic acid groups at specific pH values. Thereby the release profiles of these nanoparticles exhibit significant pH-sensitivity, making CyA mainly released at its specific absorption part of the gastrointestinal tract, decreasing the degradation by gastric acid and the first-pass metabolism by gastrointestinal enzymes and increase the oral bioavailability of CyA [23].

Insulin and bovine serum albumin (BSA) are poorly suited for the purpose of the oral delivery due to their susceptibility towards enzymatic degradation and extremely poor diffusion across intestinal mucosa. But when these two kinds of drugs were incorporated into the polymeric nanoparticles based on the pH-sensitive polymethacrylic acid–chitosan–polyethylene glycol (PCP), release characteristics of model proteins were largely dependent on pH of the medium: less than 15 % of loaded-BSA and around 15 % of loaded-insulin was released at acidic pH in the first 2 h of study. On the other hand, at pH 7.4, more than 90% of loaded-BSA was released in 5 h and the insulin from PCP nanoparticles was completely released in 3 h. Thus these pH-sensitive polymeric nanoparticles can release drug at a specific pH within the GIT, as close as possible to the absorption window of the drug. It shows that these polymeric nanoparticles can serve as good candidate for oral peptide delivery [24].

Biocompatible nanoparticles from vinylpyrrolidone and acrylic acid monomers crosslinked with N, N'-methylene bisacrylamide were prepared [19] and FITC-dextran was used as a marker compound which was entrapped in these nanoparticles. It was found that the nanoparticles exhibited an initial burst of about 60 and 40% release for the first day at pH 10 and pH 7.0 solutions, respectively. In contrast, the solution of pH 1.0 showed release of only 2–3% of the dextran for the first day. The swelling behaviors of the solid copolymer gels were studied by measuring the increase in weight of the polymer. Swelling was increased quickly with the time of treatment, which was faster in the initial steps, reaching a constant value after a time depending on the pH of the medium and the time to reach equilibrium in acidic solution was much longer than that required in neutral and alkaline solutions [19].

3. Mucoadhesive Polymeric Nanoparticles for Protein Drugs Delivery

Over the past few decades, there has been considerable interest in developing new routes, alternative to injection, for delivering the macromolecules such as proteins and peptides [25]. However, peptide and protein drugs get degraded before they reach the blood stream which in turn unable to cross the mucosal barriers [26]. The mucoadhesive polymer-coated nanoparticles as colloidal carriers are a promising approach to solve these problems. Especially the mucoadhesive nanaoparticulate system is one of the most attractive applications for the oral administration of peptide or protein drugs because peptide or protein drugs are intrinsically poorly absorbable owing to their high molecular weight and hydrophilicity and they are also susceptible to enzymatic degradation in the gastrointestinal tract whereas the entrapped drug can be protected from enzymatic degradation using muchoadhesive nanoparticles [27]. In this context, we will focus the basic mechanisms by which mucoadhesive can adhere to a mucous membrane and peptide or protein drug delivery with mucoadhesive nanoparticulate systems.

3.1. Mechanisms of Mucoadhesion

3.1.1. Mucous Membranes

Mucous membranes are the moist surfaces lining the walls of various body cavities such as the gastrointestinal and respiratory tracts. They consist of a connective tissue layer above which is an epithelial layer, the surface of which is made moist usually by the presence of a mucus layer [28]. The epithelia may be either single layered or multilayered/stratified. The former contains goblet cells which secrete mucus directly onto the epithelial surfaces, the latter contains, or is adjacent to tissues containing specialized glands such as salivary glands that secret mucus onto the epithelial surface. Mucus is present as either a gel layer adherent to the mucosal surface of as a luminal soluble or suspended form. The major components of all mucus gels are mucin glycoproteins, lipids, inorganic salts and water, the latter accounting for more than 95% of its weight, making it a highly hydrated system [29]. The mucin glycoproteins are the most important structure-forming component of the mucus gel, resulting in its characteristic gel-like, cohesive and adhesive properties. The thickness of this mucus layer varies on different mucosal surface, from 50 to 450 μm in the stomach [30, 31], to less than 1 μm in the oral cavity [32]. The major functions of mucus are that of protection and lubrication.

3.1.2. The Mucoadhesive/Mucosa Interaction

For adhesion to occur, molecules must bond across the interface. These bonds arises in the following ways [28].

(1) Ionic bonds: where two oppositely charged ions attract each other via electrostatic interactions to form a strong bond.

(2) Covalent bonds: where electrons are shared, in pairs, between the bonded atoms in order to "fill" the orbitals in both. These are also strong bonds.
(3) Hydrogen bonds: here a hydrogen atom, when covalently bonded to electronegative atoms such as oxygen, fluorine or nitrogen, carries a slight positively charge and is therefore attracted to other electronegative atoms. The hydrogen can therefore be thought of as being shared, and the formed bond is generally weaker than ionic or covalent bonds.
(4) Van-der-Waals bonds: these are some of the weakest forms of interaction that arise from dipole-dipole and dipole-induced dipole attractions in polar molecules, and dispersion forces with non-polar substances.
(5) Hydrophobic bonds: more accurately described as the hydrophobic effect, these are indirect bonds that occur when non-polar groups are present in an aqueous solution. Water molecules adjacent to non-polar groups form hydrogen bonded structures, which lowers the system entropy. There is therefore an increase in the tendency of non-polar groups to associate with each other to minimize this effect.

3.2. The Mucoadhesive Materials

The most widely investigated group of mucoadhesives is hydrophilic macromolecules containing numerous hydrogen bond forming groups [33, 34, and35]. Their initial use as mucoadhesives was in denture fixative powders or pastes. The presence of hydroxyl, carboxyl or amine groups on the molecules favors adhesion. They are called "wet" adhesives in that they are activated by moistening and will adhere non-specifically to many surfaces [36]. Once activated, they will show stronger adhesion to dry inert surfaces than those covered with mucus. Unless water uptake is restricted, they may overhydrate to form slippery mucilage. Like typical hydrocolloid glues, once formed, the adhesive joint is allowed to dry which in turn forms very strong adhesive bond. Typical examples are carbomers [37], chitosan [38], sodium alginate [39] and the cellulose [40] derivatives.

3.3. Mucoadhesive Polymeric Nanoparticles for Peptide or Protein Drugs Delivery

3.3.1. Mucoadhesive Polymeric Nanoparticles

Nanoparticles of biodegradable and biocompatible polymers are good candidates for particulate carriers to deliver peptide or protein drugs. Such particles are expected to be adsorbed in an intact form in the gastrointestinal tract after oral administration [41].

Chitosan is as a widely available mucoadhesive polymer able to increase cellular permeability and improve the bioavailability of orally administered peptide or protein drugs [42]. The mucoadhesive properties of chitosan have been illustrated by its ability to adhere to porcine gastric mucosa *in vitro* [43]. In the interactions between chitosan and mucus, the primary mechanism of action at the molecular level was found to be electrostatic [44]. This mucoadhesive property of chitosan was even significantly further improved by the immobilization of thiol groups on the polymer. The immobilization of thiol bearing moieties on the chitosan backbone yields the thiolated chitosan. The derivatization of the primary amino groups, mainly at the 2-position of the glucosamine subunits, with coupling reagents bearing thiol groups leads to the formation of thiolated chitosans [45]. Mainly four type of thiolated chitosans have been synthesized so far. They are chitosan-thioglycolic acid conjugates [46, 47, and 48], chitosan-cysteine conjugates [49], chitosan-4-thio butyl-amidine (chitosan-TBA) conjugates [50] and chitosan-thioethylamidine (TEA) derivative [51]. The mucoadhesive and permeation enhancing properties of chitosan are improved by thiolation [47, 49, 50, and 52]. The improved mucoadhesive properties of thiolated chitosans are mainly due to the formation of covalent bonds between thiol groups of the polymer and cysteine-rich subdomains of glycoproteins in the mucus layer [53]. These covalent bonds are supposedly stronger than non-covalent bonds, such as ionic interactions of chitosan with anionic substructures of the mucus layer.

3.3.2. Surface Modification of Nanoparticles with Mucoadhesive Polymers

One of the most promising strategies for designing mucoadhesive particulate systems is surface modification, or coating, of the drug carrier particles with mucoadhesive polymers. In reviewing the surface modification of colloidal particles, a number of investigations of the polymer-coating of liposomes have been published. Sunamoto et al. demonstrated an improvement in the chemical and physical stability of polymer-coated liposomes prepared with polysaccharide derivatives such as mannan or amylopectin [54, 55]. So far several compounds such as poloxamer [56], polysorbate 80 [57], carboxymethyl chitin [58], carboxymethyl chitosan [59] and dextran derivatives [60] have been reported to be suitable for the surface-coatings of the liposomal system. Kawashima et al. also reported that an increase in residence time and intimate contact of the chitosan-coated nanoparticles with the wall of the small intestine significantly improved the degree of drug absorption although they reported an enhancement of nanoparticles taken up by gut tissue [61].

The mucoadhesive nanoparticulate system was found to be useful for peptide or protein drugs to improve oral mucosal delivery due to their prolonged retention in the gastro-intestinal tract and excellent penetration into the mucus layer.

4. Bacteria Degradable Polymeric Particles for Protien and Peptide Delivery

Colon as a target site for protein and peptide drugs received an increasing interest because it offers distinct advantages. The colon has relatively low activities of proteolytic enzymes. Thus, the peptides and proteins can be more effectively absorbed in the colon. And there is around 10^{11}-10^{12}CFU ml^{-1} of microflora in the colon, which is much more than the other region of GI tract [62]. These bacteria utilize the foods that have not been digested in the small intestine, for instance di, tri-polysaccharides, and mucopolysaccharides [63], as a source of carbon produce a wide range of reductive and hydrolytic enzymes. These include a-arabinosidase, azoreductase, β- glucuronidase, β-galactosidase, β-xylosidase, nitroreductase, deaminase, urea hydroxylase etc [64]. Therefore, proteins and peptides can be encapsulated in the polymers which are refractory in the stomach and small intestine yet specifically degraded by those enzymes in the colon [65].

4.1. Azo Polymer-Coating Particles

Varied azo polymers [65, 66] have been synthesized and evaluated as coating materials for colon targeting. When the polymers, crosslinked by the azo aromatic group R-C_6H_4-N=N-C_6H_4-R, are used for coating, particles can protect the drugs from digestion in the stomach and small intestine. Once they reached at the colon, the microflora reduced the azo bonds, and broke the crosslinks to form a pair of aromatic amines, R-C_6H_4-NH_2 + H_2N--C_6H_4-R. The polymer is degraded and the drug will be released [67]. However, the use of azo polymers and other synthetic polymers need to concern about the safety of these polymers. And a detailed toxicological study should be performed before being used commercially [62].

4.2. Polysaccharides-Coating Particles

Polysaccharide-coated particles can also be used as colon targeting carriers for protein and peptide drugs. As we known, the polysaccharides can be degraded by the polysaccharidase in the colon, which is the main reason for these natural compounds used as colon specific carriers. They also have some other advantages including low prices and wide availability, as well as they can be modified chemically and biochemically, highly stable, safe and etc. Polysaccharides such as pectin, chitosan, inulin, dextran etc have been studied for this purpose. Pectins are nonstarch, linear polysaccharides that consist of α-1,4-D-galacturonic acid and 1,2 D-rhamnose with D-galactose and D-arabinose side chains having average molecular weights between 50,000 to 150,000 [68]. One major problem with the pectin and also with other polysaccharides is the high water solubility that would lead to drug early release in the tracts of the upper GI tract [69]. Therefore, they must be made water-insoluble by crosslinking or hydrophobic dramatization [68]. Atyabi and coworkers [70] prepared bovine serum albumin-loaded pectinate beads by adding the mixture of polymer and drug to the calcium chloride solution via a nozzle. The spheres were formed by ionotropic reaction between the divalent calcium ions and the negatively charged carboxyl groups of the pectin molecules. It was observed when the enzyme existed in the dissolution medium, BSA content released from the calcium pectinate beads was much faster.

5. pH-Sensitive and Mucoadhesive Polymereric Particles for Protein and Peptide Delivery

pH-sensitive polymer and mucoadhesive polymer-coating particles have been described at the above. While, Eudragit-cysteine conjugate which have both pH-dependent and mucoadhesive character was synthesized by Quan et al. [71]. The amount of cysteine in the conjugate calculated by ^{1}H nuclear magnetic resonance (^{1}H NMR) was about 8 mol-percent. Furthermore, the polymer was used to coat chitosan microspheres. The release rate of BSA from BSA-loaded thiolated Eudragit-coated chitosan microspheres (TECMs) at pH 7.4 PBS solution was significantly higher than at pH 2.0 PBS solution. Gamma camera imaging of Tc-99m labeled microsphere distribution in rats after oral administration suggested that TECMs had comparatively stronger mucoadhesive characters than either Eudragit microspheres or chitosan microspheres. Therefore, thiolated Eudragit-coated chitosan microspheres will be a good carrier for oral protein drug delivery.

6. In vivo Results of Functional Polymeric Nanoparticles

For oral delivery of the protein and peptide drugs, the functional polymeric nanoparticles have been demonstrated to possess excellent property *in vitro*. However, it should be testified *in vivo* to see the promising relationship between two of them, as the *in vivo* condition has lots of difference from the *in vitro*. Many papers reported that functional polymeric nanoparticles as an oral dosage form can significantly improve bioavailability of some protein or peptide drugs. Jaeghere et al. reported that pH-sensitive nanoparticles prepared by the poly (methacrylic acid-co-methacrylate) copolymer Eudragit® . L100-55 improved the oral bioavailability of HIV-1 protease inhibitor-CGP 70726 in dogs [72]. Similar result was obtained by Dai et al., when CyA-loaded pH-sensitive nanoparticles made by Eudragit copolymer was given to SD rats, in spite of some difference of relative bioavailability for each polymer [23].

Mucoadhesive nanoparticles for oral delivery of protein and peptide drugs also have been evaluated by *in vivo*. Pan et al. [73] and Ma et al. [74] used chitosan nanoparticles (CNS) to incorporate insulin. These formulations were orally administered to streptozotocin-induced diabetic rats. After orally administration, the decrease in plasma glucose levels was significantly different from that induced by the control insulin solution, and they sustained the serum glucose at pre-diabetic levels for 15h and 11h, respectively. The average pharmacological bioavailability relative to SC injection of insulin solution was up to 14.9% [73]. Confocal micrographs showed strong interaction between rat intestinal epithelium and chitosan nanoparticles after 3 h post-oral administration [74]. Sarmento et al. [75] also successfully delivered the insulin by alginate/chitosan nanoparticles by oral administration. The basal serum glucose levels lowered were more than 40% with 50 and 100 IU/kg doses and hypoglycemia was sustained for over 18 h.

7. Conclusion

The development of a dosage form that improves the oral absorption of peptide and protein drugs with an extremely low bioavailability because of instability in the gastro-intestinal tract and low permeability through the intestinal membrane is one of the greatest challenges in the pharmaceutical field. From the discussion mentioned above, functional polymer-coated nanoparticulate systems were found to be a promising strategy for oral peptide or protein drug carriers as these materials possesses excellent absorption enhancement/proteolytic enzyme inhibition capabilities, which avoid the proper drugs' potential risks of degradation by gastric acid and the first-pass metabolism by gastrointestinal enzymes. The same bioavailability improving effects can be obtained using mucoadhesive material, pH-dependent material or bacteria degradable polymers/copolymer as carriers due to their unique characterizations. Overall the polymeric nanopaticles have high potential in protein drug delivery. However, the success of these special vehicles is greatly dependent upon the availability of superior nontoxic materials with high biocompatibility. Also, methods for preparing these nanoparticles need to be developed persistently. Moreover, additional *in vivo* experiments must be performed to substantiate the success of polymeric drug delivery systems in the real life applications.

Acknowledgements

We would like to acknowledge Biogreen 21 (20050401034696) for providing funding. Ji-Shan Quan and Hu-Lin Jiang were supported by BK21 program.

References

[1] Morishita, M. and Peppas, N.A., (2006). Is the oral route possible for peptide and protein drug delivery? *Drug Discovery Today,* **11**, 905-910.

[2] Hamman, J.H., Enslin, G.M. and Kotzé, A.F., (2005).Oral delivery of peptide drugs: barriers and developments. *Bio Drugs,* **19**, 165-177.

[3] Fujita, T., Fujita, T., Morikawa, K., Tanaka, H., Iemura, O., Yamamoto, A. and Muranishi, S., (1996). Improvement of intestinal absorption of human calcitonin by chemical modification with fatty acids: Synergistic effects of acylation and absorption enhancers. *Int. J. Pharm.*, **134**, 47-57.

[4] Calceti, P., Salmaso, S., Walker, G. and Bernkop-Schnurch, A., (2004). Development and invivo evaluation of an oral insulin-PEG delivery system. *Eur. J. Pharm. Sci.,* **22** , 315-323.

[5] Bundgaard, H. and Møss, J., (1990). Prodrugs of peptides. 6. Bioreversible derivatives of thyrotropin-releasing hormone (TRH) with increased lipophilicity and resistance to cleavage by the TRH-specific serum enzyme. *Pharm Res.,* **7**, 885-892.

[6] Lee, C.H., Woo, J.H., Cho, K.K., Kang, S.H., Kang,S.K. and Choi, Y.J., (2007). Expression and characterization of human growth hormone-Fc fusion proteins for transcytosis induction. *Biotechnol. Appl. Biochem.,* **46**, 211-217.

[7] Eaimtrakarn, S., Rama Prasad, Y.V., Ohno, T., Konishi, T., Yoshikawa, Y., Shibata, N. and Takada, K., (2002). Absorption enhancing effect of labrasol on the intestinal absorption of insulin in rats. *J. Drug Target,* **10**, 255-260.

[8] Fix, J.A., (1987). Absorption enhancing agents for the GI system. *J. Control. Release,* **6**, 151-156.

[9] Lee, V.H.L., (1990). Protease ihibitors and penetration enhancers as approaches to modify peptide absorption. *J. Control. Release*, **13**, 213-223.

[10] Sakuma, S., Hayashi, M. and Akashi, M., (2001). Design of nanoparticles composed of graft copolymers for oral peptide delivery. *Adv Drug Deliv Rev,* **47**, 21-37.

[11] Oberle, R.L., Moore, T.J. and Krummel, D.A.P., (1995). Evaluation of mucosal damage of surfactants in rat jejunum and colon. *J. Pharmacol. Toxicol. Methods,* **33**, 75-81.

[12] Jani, P.U., Florence, A.T., and McCarthy, D.E., (1992). Further histological evidence of the gastro-intestinal absorption of polystyrene nanospheres in the rat. *Int. J. Pharm.,* **84**, 245-252.

[13] Florence, A.T., Hillery, A.M., Hussain, N. and Jani, P.U., (1995). Nanoparticles as carriers for oral peptide absorption: studies on particle uptake and fate. *J. Control. Release*, **36**, 39-46.
[14] des Rieux A, Fievez, V., Garinot, M., Schneider, Y.J. and Préat, V., (2006). Nanoparticles as potential oral delivery systems of proteins and vaccines: A mechanistic approach. *J. Control. Release*, **116**, 1-27.
[15] Sakuma, S., Hayashi, M. and Akashi, M., (2001). Design of nanoparticles composed of graft copolymers for oral peptide delivery. *Adv Drug Deliv Rev,* **47**, 21-37.
[16] Kreuter, J., (1996). Nanoparticles and microparticles for drug and vaccine delivery. *J. Anat.,* **189**, 503-505.
[17] Pinto, Reis C., Neufeld, R.J., Ribeiro, A.J. and Veiga, F., (2006). Nanoencapsulation II. Biomedical applications and current status of peptide and protein nanoparticulate delivery systems. *Nanomedicine: Nanotechnology, Biology, and Medicine,* **2**, 53-65.
[18] Shah, R.B., Ahsan, F. and Khan, M.A., (2002). Oral delivery of proteins progress and prognostication. *Drit. Rev. ther. Drug carrier syst.,* **19**, 135-169.
[19] Sahoo, S.K., De, T.K., Ghosh, P.K. and Maitra, A.J., (1998). pH- and thermo-sensitive hydrogel nanoparticles. *Colloid Interface Sci.,* **206**, 361-368.
[20] Gil, Eun Seok. and Hudson, Samuel M., (2004). Stimuli-reponsive polymers and their bioconjugates. *Prog. Polym. Sci.,* **29**, 1173–1222.
[21] Park, S.Y. and Bae, Y.H., (1999). Novel pH-sensitive polymers containing sulfonamide groups. *Macromol Rapid Commun,* **20**, 269–273.
[22] Kang, S.I. and Bae, Y.H., (2003). A sulfonamide based glucose-responsive hydrogel with covalently immobilized glucose oxidase and catalase. *J. Control. Release,* **86**, 115–121.
[23] Dai, J., Nagai, T., Wang, X., Zhang, T., Meng, M. and Zhang, Q., (2004). pH-sensitive nanoparticles for improving the oral bioavailability of cyclosporine A. *Int. J. Pharm.,* **280**, 229-240.
[24] Sajeesh, S. and Sharma, C.P., (2006). Novel pH responsive polymethacrylic acid-chitosan-polyethylene glycol nanoparticles for oral peptide delivery. *J Biomed Mater Res B Appl Biomater,* **76,** 298-305.
[25] Cui, F.Y., Qian, F. and Yin, C.H., (2006). Preparation and characterization of mucoadhesive polymer-coated nanoparticles. *Int. J. Pharm.,* **316**, 154-161.
[26] Prego, C., Garci a, M., Torres, D., and Alonso, M.J., (2005). Transmucosal macromolecular drug delivery. *J. Control. Release*, **101**, 151-162.
[27] Hirofumi, T., Hiromitsu, Y. and Yoshiaki, K., (2001). Mucoadhesive nanaoparticulate systems for peptide drug delivery. *Adv Drug Deliv Rev,* **47**, 39-54.
[28] John, D. S., (2005). The basics and underlying mechanisms of mucoadhesion. *Adv Drug Deliv Rev,* **57**, 1556-1568.
[29] Marriott, C. and Gregory, N.P., (1990). Mucus physiology and pathology, in: V. Lanaerts, R. Gurny (Eds.). *Bioadhesive Drug Delivery Systems*, CRC Press, Florida, 1-24.
[30] Allen, A., Cunliffe, W.J., Pearson, J.P., and Venables, C.W., (1990). The adherent gastric mucus gel barrier in man and changes n peptic ulceration. *J. Intern. Med,.* **228**, 83-90.
[31] Kerss, S., Allen, A., and Garner, A., (1982). A simple method for measuring the thickness of the mucus gel layer adherent to rat, frog, and human gastric mucosa:

influence of feeding, prostaglandin, N-acetylcysteine and other agents. *Clin. Sci.,* **63**, 187-195.

[32] Sonju, T., Cristensen, T.B., Kornstad, L., and Rolla, G., (1974). Electron microscopy, carbohydrate analysis and biological activities of the proteins adsorbed in two hours to tooth surfaces in vivo. *Caries Res.,* **8**, 113-122.

[33] Smart, J.D., Kellaway, I.W. and Worthington, H.E.C., (1984). An in vitro investigation of mucosa-adhesive materials for use in controlled drug delivery. *J. Pharm. Pharmacol.,* **36**, 295-299.

[34] Smart, J.D., (1993). Drug delivery using buccal adhesive systems. *Adv Drug Deliv Rev,* **23**, 253-270.

[35] Lee, J.W., Park, J.H. and Robinson, J.R., (2000). Bioadhesive-based dosage forms: the next generation. *J. Pharm. Sci.,* **89**, 850-866.

[36] Ben Zion, O. and Nussinovitch, A., (1997). Physical properties of hydrocolloid wet glues. *Food Hydrocoll.,* **11**, 429-442.

[37] Llabot, J.M., Manzo, R.H. and Allemandi, D.A., (2004). Drug release from carbomer:carbomer sodium salt matrices with potential use as mucoadhesive drug delivery system. *Int J Pharm.,* **276**, 59-66.

[38] George, M. and Abraham, T.E., (2006). Polyionic hydrocolloids for the intestinal delivery of protein drugs: alginate and chitosan--a review. *J. Control. Release,* **114**, 1-14.

[39] Ryu, J.M., Chung, S.J., Lee, M.H., Kim, C.K. and Shim, C.K., (1999). Increased bioavailability of propranolol in rats by retaining thermally gelling liquid suppositories in the rectum. *J. Control. Release,* **59**, 163-172.

[40] Miyazaki, Y., Ogihara, K., Yakou, S., Nagai, T. and Takayama, K., (2003). Bioavailability of theophylline and thiamine disulfide incorporated into mucoadhesive microspheres consisting of dextran derivatives and cellulose acetate butyrate. *Biol Pharm Bull.* **26**, 1744-1747.

[41] Florence, A.T., Hillery, A.M., Hussain, N. and Jani, P.Y., (1995). Nanoparticles as carriers for oral peptide absorption: studies on particle uptake and fate. *J. Control. Release,* **36**, 39-46.

[42] Bowman, K. and Leong, K.W., (2006). Chitosan nanoparticles for oral drug and gene delivery. *Int. J. Nanomedicine,* **1**, 117-128.

[43] Gåserød, O., Jolliffe, I.G., Hampson, F.C., Dettmar, P.W. and Skjåk-Bræk, G., (1998). The enhancement of the bioadhesive properties of calcium alginate gel beads by coating with chitosan. *Int. J. Pharm.,* **175**, 237-246.

[44] Deacon, M.P., McGurk, S., Roberts, C.J., Williams, P.M., Tendler, S.J., Davies, M.C., Davis, S.S. and Harding, S.E., (2000). Atomic force microscopy of gastric mucin and chitosan mucoadhesive systems. *Biochem. J.,* **348**, 557-563.

[45] Schnürch, A.B., Hornof, M. and Guggi, D., (2004). Thiolated chitosans. *Eur. J. Pharm. Biopharm.,* **57**, 9-17.

[46] Schnürch, A.B. and Hopf, T.E., (2001). Synthesis and in vitro evaluation of chitosan–thioglycolic acid conjugates. *Sci. Pharm.,* **69**, 109-118.

[47] Kast, C.E. and Schnürch, A.B., (2001). Thiolated polymers-thiomers: development and in vitro evaluation of chitosan–thioglycolic acid conjugates. *Biomaterials,* **22**, 2345-2352.

[48] Hornof, M.D., Kast, C.E. and Schnürch, A.B., (2003). In vitro evaluation of the viscoelastic behavior of chitosan-thioglycolic acid conjugates. *Eur. J. Pharm. Biopharm.,* **55**, 185-190.

[49] Schnürch, A.B., Brandt, U.M. and Clausen, A.E., (1999). Synthesis and in vitro evaluation of chitosancysteine conjugates. *Sci. Pharm.,* **67**, 196-208.

[50] Schnürch, A.B., Hornof, M. and Zoidl, T., (2003). Thiolated polymers-thiomers: modification of chitosanwith 2-iminothiolane. *Int. J. Pharm.,* **260**, 229-237.

[51] Kafedjiiski, K., Krauland, A.H., Hoffer, M.H. and Schnurch, A.B., (2005). Synthesis and in vitro evaluation of a novel thiolated chitosan. *Biomaterials,* **26**, 819-826.

[52] Langoth, N., Guggi, D., Pinter, Y. and Schnürch, A.B., (2003).Thiolated chitosan: in vitro evaluation of its permeation enhancing. *Properties, 30th Annual Meeting and Exposition of the Controlled Release Society, Glasgow, UK,* 034.

[53] Leitner, V.M., Walker, G.F. and Schnürch, A.B., (2003). Thiolated polymers: evidence for the formation of disulphide bonds with mucus glycoproteins. *Eur. J. Pharm. Biopharm.,* **56**, 207-214.

[54] Sunamoto, J., Iwamoto, K., Takada, M., Yuzuriha, T. and Katayama, K., (1983). Improved drug delivery to target specific organs using liposomes as coated with polysacchrides. *Polymer Sci. Technol.,* **23**, 157-168.

[55] Sunamoto, J., Iwamoto, K., Takada, M., Yuzuriha, T. and Katayama, K., (1984). Polymer coated liposomes for drug delivery to target specific organs. In: J.M. Anderson and S.W. Kim, Editors, *Recent Advances in Drug Delivery Systems,* 153-162.

[56] Jamshaid, M., Farr, S.J., Kearney, P. and Kellaway, I.W., (1988). Poloxamer sorption on liposomes: comparison with polystyrene latex and influence on solute efflux. *Int. J. Pharm.,* **48**, 125-131.

[57] Kronberg, B., Dahlman, A., Carlfors, J., Karlsson, J., and Artursson, P., (1990). Preparation and evaluation of sterically stabilized liposomes: colloidal stability, serum stability, macrophage uptake, and toxicity. *J. Pharm. Sci.,* **79**, 667-671.

[58] Dong, C. and Rogers, J.A., (1991). Polymer-coated liposomes: stability and release of ASA from carboxymethyl chitin-coated liposomes. *J. Control. Release,* **17**, 217-224.

[59] Alamelu, S. and Rao, K.P., (1991). Studies on the carboxymethyl chitosan-containing liposomes for their stability and controlled release of dapson. *J. Microencapsul.,* 8, 505-515.

[60] Elferink, W.G.L., de Wit, J.G., In't Veld, G., Reichert, A., Driessen, A.J.M., Ringsdorf, H. and Konings, W.N., (1992). The stability and functional properties of proteoliposomes mixed with dextran derivatives bearing hydrophobic anchor groups. *Biochim. Biophys. Acta.,* **1106**, 23-30.

[61] Kawashima, Y., Yamamoto, H., Takeuchi, H. and Kuno, Y., (2000). Mucoadhesive DL-lactide/glycolide copolymer nanospheres coated with chitosan to improve oral delivery of elcatonin. *Pharm Dev Technol.,* **5**, 77-85.

[62] Sinha, V.R. and Kumaria, R., (2003). Microbially triggered drug delivery to the colon. *Eur J Pharm Sci.,* **18**, 3-18.

[63] Rubinstein, A., (1990). Microbially controlled drug delivery to the colon. *Biopharm. Drug Dispos.,* **11**, 465-475.

[64] Kinget, R., Kalala, W., Vervoort, L. and Mooter, G.U., (1998). Colonic drug targeting. *J. Drug Target,* **6**, 129-149.

[65] Tozaki, H., Nishioka, J., Komoike, J., Okada, N., Fujita, T., Muranishi, S., Kim, S.H., Terashima, H. and Yamamoto, A., (2001). Enhanced absorption of inslin and ($Asu^{1,7}$) Eel-calcitonin using novel azopolymer-coated pellets for colon-specific drug delivery. *J Pharm Sci.,* **90**, 89-97.

[66] Kimura, Y., Makita, Y., Kumagai, T., Yamane, H. and Kitao, T., (1992). Degradation of azo-containing polyurethane by the action of intestinal flora: Its mechanism and application as a drug delivery system. *Polymer,* **33**, 5294-5299.

[67] Saffran, M., Kumar, G.S., Savariar, C., Burnham, J.C., Williams, F. and Neckers, D.C., (1986). A new approach to the oral administration of insulin and other peptide drugs. *Science,* **233**, 1081-1084.

[68] Chourasia, M.K. and Jain, S.K., (2004). Polysaccharides for colon targeted drug delivery. *Drug Delivery,* 129-148.

[69] Sinha, V.R. and Kumria, R., (2003). Microbially triggered drug delivery to the colon. *Eur J Pharm Sci.,* **18**, 3-18.

[70] Atyabi, F., Inanloo, K. and Dinarvand, R., (2005). Bovine serum albumin-loaded pecinate beads as colonic peptide delivery system: preparation and in vitro characterization. *Drug Delivery,* **12**, 367-375.

[71] Quan, J.S., Jiang, H.L., Choi, Y.J., Yoo, M.K. and Cho, C.S., (2007). Thiolated Eudragit-coated chitosan microspheres as an oral drug delivery system. *Key Engineering Materials,* 342-343, 445-448.

[72] De Jaeghere, F., Allémann, E., Kubel, F., Galli, B., Cozens, R., Doelker, E. and Gurny, R., (2000). Oral bioavailability of a poorly water soluble HIV-1 protease inhibitor incorporated into pH-sensitive particles: effect of the particle size and nutritional state. *J. Control. Release,* **68**, 291-298.

[73] Pan, Y., Li,Y.J., Zhao, H.Y., Zheng ,J.M., Xu, H., Wei, G., Hao, J.S. and Cui, F.D., (2002). Bioadhesive polysaccharide in protein delivery system: chitosan nanoparticles improve the intestinal absorption of insulin in vivo. *Int J Pharm.,* **249**, 139-147.

[74] Ma, Z.S., Lim, T. M. and Lima, L.Y., (2005). Pharmacological activity of peroral chitosan–insulin nanoparticles in diabetic rats. *Int J Pharm.,* **293**, 271–280.

[75] Sarmento, B., Ribeiro, A., Veiga, F., Sampaio, P., Neufeld, R. and Ferreira, D., (2007). Alginate/chitosan nanoparticles are effective for oral insulin delivery. *Pharm Res. Dec.,* **24**, 2198-2206.

INDEX

B

C

D

E

F

G

H

I

J

K

L

M

N

O

P

Q

R

S

T

U

V

W

X

Y

Z